J-Aggregates

Volume 2

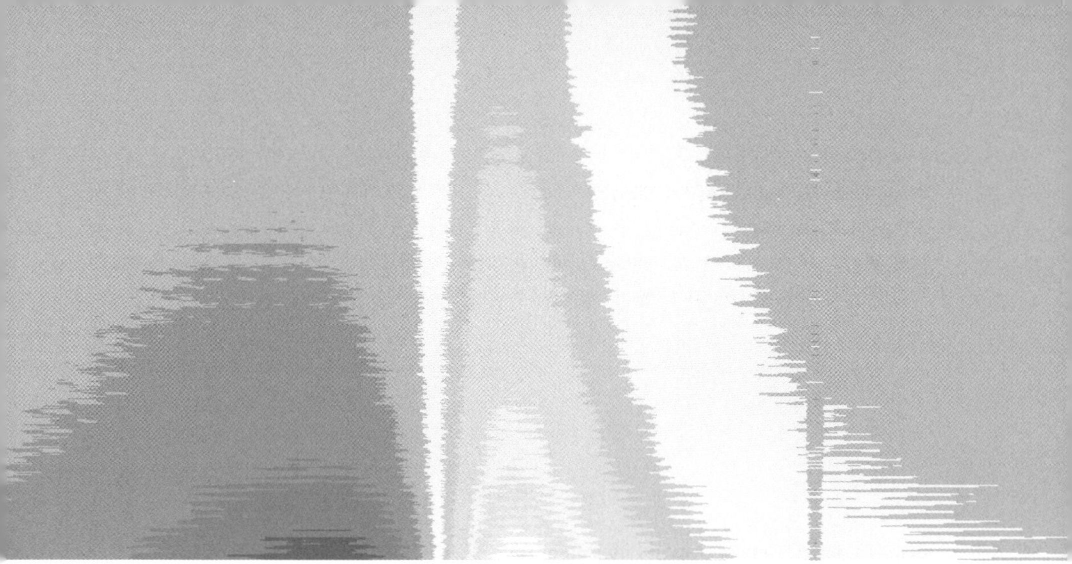

J-Aggregates

Volume 2

Editor

Takayoshi Kobayashi

University of Electro-Communications, Japan

World Scientific

NEW JERSEY • LONDON • SINGAPORE • BEIJING • SHANGHAI • HONG KONG • TAIPEI • CHENNAI

Published by

World Scientific Publishing Co. Pte. Ltd.
5 Toh Tuck Link, Singapore 596224
USA office: 27 Warren Street, Suite 401-402, Hackensack, NJ 07601
UK office: 57 Shelton Street, Covent Garden, London WC2H 9HE

British Library Cataloguing-in-Publication Data
A catalogue record for this book is available from the British Library.

J-AGGREGATES
Volume 2

ISBN-13 978-981-4365-74-1
ISBN-10 981-4365-74-2

Printed in Singapore by B & Jo Enterprise Pte Ltd

Preface

In 1996, "J-aggregates" was published by World Scientific under my editorship. Since then, research on molecular aggregates has been particularly active, as will be seen by the contents of this book. Nanostructured materials, such as quantum-wells and -dots [1], Langmuir-Blodgett films and polymers [2], became the focus of condensed matter physics. The unique optical and dynamic properties due to the low dimensionality of these systems and discreetness of their electronic structures induced by the quantum confinement of these materials, received much attention. At the time, it seemed appropriate to edit a book on the aggregates of dye molecules, in order to compare the similarities and differences among the low-dimensional systems that were being studied so extensively. 15 years later, new low dimensional systems appear: they are carbon systems, including fullerenes, carbon nanotubes and graphemes, which are zero-, one- and two-dimensional systems [3-5].

And so, it has become important to summarize the current status of the research on organic molecule aggregates. I hope that this book will prove useful to those working on semiconductor quantum dots, quantum wires and quantum wells, as well as carbon fullerenes, carbon nanotubes and graphenes. I also hope that this will be a platform to share information and technology, the concepts and the phenomena, including application possibilities.

[1] Physics Today, 46 (1993).
[2] Special issue, J. Phys. Chem. 96, Conjugated Conducting Polyemrs, ed. H. Kiess (Springer, Berlin, 1982).
[3] S. Iijima, J. Crystal Growth, 50: 675 (1980).
[4] J.W. Mintmire, B. I. Dunlap, C. T. White, Phys. Rev. Lett. 68, 631(1992).
[5] Dekker, Cees (1999). *Physics Today* **52,** 22 (1999)

Takayoshi Kobayashi
At The University of Electro Communications, Chofu, Tokyo, Japan
28/07/2011

Contents

CHAPTER 1

ULTRAFAST REAL-TIME VIBRATIONAL DYNAMICS IN J-AGGREGATES

Takayoshi Kobayashi[1-4,*], Juan Du[1,2], and Yuichiro Kida[1,2]

[1]*Advanced Ultrafast Laser Research*
The University of Electro-Communications, 1-5-1 Chofugaoka, Chofu, Tokyo 182-8585, Japan

[2]*JST, CREST, K's Gobancho, 7, Gobancho, Chiyodx-ku, Tokyo 102-0076, Japan*

[3]*Department of Electrophysics, National Chiao Tung University, 1001 Ta Hsueh Road, Hsinchu 300, Taiwan*

[4]*Institute of Laser Engineering, Osaka University, 2-6 Yamada-oka, Suita, Osaka 565-0971, Japan*

**E-mail: kobayashi@ils.uec.ac.jp*

Ultrafast real-time vibrational dynamics in several J-aggregate systems were studied by femtosecond real-time spectroscopy using ultrashort pulses with durations shorter than 5 fs. Despite the weak exciton–phonon coupling in Frenkel exciton systems, coherent molecular vibrations are clearly observed in the time dependence of the transmission change of porphyrin J-aggregates. The experimental results are explained in terms of a transition dipole moment modulated by dynamic intensity borrowing from an intense B transition to a weak Q transition through vibronic interaction. The observed wave packet has been classified as a Herzberg–Teller type in contrast with the frequently studied conventional Franck–Condon type. In pseudo-isocyanine J-aggregates, intermolecular vibrations were observed in the lattice phonon coupled to the excitonic transition in J-aggregates and 11 vibrational modes were detected. We applied non-Condon-type exciton–phonon coupling in J-aggregates to explain the oscillatory signals. In addition, we discussed the influence of the excitation pulse duration and the laser spectrum. The vibrational amplitude was found to be proportional to the product of the integral of the absorbed spectrum and the one shifted by the vibrational mode frequency.

Keywords: Real-time vibrational spectroscopy, Isocyanine, Psuedoisocyanine, Porphyrin, Frenkel exciton, NOPA, OPA, Vibronic interaction, Exciton-vibration interaction, Vibrational wave packet, Multi-exciton state, Herzberg-Teller, Franck-Condon, Intensity borrowing

Introduction

The term aggregate is used to describe a collection of molecules that are generally self-assembled by electrostatic forces with a dimensionally reduced chain-like configuration in high-concentration solutions or in thin solid films.[1–5] The most obvious evidence for aggregate formation is the dramatic change in the absorption spectrum, which was independently discovered by Jelley and Scheibe in 1936.[3,4] They observed that the electronic absorption spectra of certain cyanine dyes changed dramatically during the increase of the dye concentration, and an unusually red-shifted sharp absorption band (J-band) compared with the one of the monomers was discovered at high concentration.

Usually, the self-assembled molecular aggregates can be classified into J- and H- type depending on whether the relative orientations of the induced transition dipoles of each molecules is ‘head-to-tail’ or ‘head-to-head’.[6] H-aggregates are distinguished from J-aggregates by an absorption band that is blue shifted relative to that of a single monomer.[6,7] It has been confirmed that the excited states in J-aggregates have characteristic of Frenkel exciton states, which are delocalized over an aggregate and caused by the coupling between the optical transitions of the individual molecule and the intermolecular dipole interaction between the monomers well oriented in the form of a chain.[8] The red-shift of the absorption peak is due to the negative sign of the electronic coupling between the individual molecules with smaller angle between the axis connecting the dipole centers of the neighboring molecules and the direction of the parallel dipoles of the components. The exciton band width is related to the coupling energy; the narrow linewidth was explained by an exchange narrowing effect.[8]

The optical properties of molecular aggregates have been extensively investigated ever since the discovery of J-aggregates[1–5] because they are expected to be intermediate between the macroscopic properties of bulk materials and the microscopic properties of isolated molecules. One-dimensional Frenkel-exciton model has been used to theoretically reproduce the linear absorption spectrum and to explain the effect of exciton–phonon coupling.[9–13] After that, the one-dimensional exciton model has subsequently been used to investigate the spectroscopic properties of many mesoscopic systems. In addition to obtaining valuable linear information from J-aggregates, the rapid development of nonlinear optical techniques for spectroscopy has resulted in many investigations of the nonlinear characteristics of molecular aggregates.[14–17] In 1993, Fidder and coworkers used a picosecond pump–probe technique to unambiguously observe the optical transition from a one-exciton band to a two-exciton band in pseudoisocyanine (PIC) aggregates for the first time.[14] Soon after this, Kobayashi *et al.* observed the same transition with a much higher temporal resolution by performing femtosecond time-resolved spectroscopy of J-aggregates. The decay time of the many exciton states and that of the biexcitonic annihilation process have been determined to be 200 fs and 1.5 ps, respectively.[15] Since then, ultrafast optical dynamics of excitons in J-aggregates have been intensively investigated by time-resolved absorption and fluorescence spectroscopies and several four-wave-mixing spectroscopies.[18–23] However, few studies have investigated the coupling between coherent molecular vibrations and excitons in J-aggregates in the real-time domain. This may be due to the weak interaction between the excitons and the molecular vibrations, and also due to the technical difficulty of directly measuring intramolecular vibrations.

Thanks to the recent progress of the ultrashort laser technique, laser pulse with pulse duration much shorter than the molecular vibrational period could be generated. These ultrashort laser pulses can be used to generate wave packets as simultaneous excitations of the vibrational manifold. The wave packets can be detected by femtosecond pump–probe real-time vibrational spectroscopy, which is a powerful tool for accurately analyzing relationships between structure and electronic properties with the goal of designing new molecular materials with improved behavior. It has the following advantages over conventional vibrational spectroscopies such as fluorescence line-narrowing and hole-burning spectroscopy, infrared absorption, and Raman scattering techniques:[24–26]

(1) Femtosecond pump–probe real-time vibrational spectroscopy can be used to simultaneously measure the electronic and vibrational dynamics of a system under the same conditions. Electronic relaxation and vibrational relaxation dynamics are generally measured separately. The former can be investigated by methods such as conventional pump–probe experiments, degenerate four-wave mixing, and pulsed optical pump–photoacoustic probe experiments. The latter has mainly been studied by time-resolved Raman spectroscopy and time-resolved infrared absorption spectroscopy.[24–26] However, femtosecond pump–probe real-time vibrational spectroscopy can observe the dynamics of vibrational modes by monitoring the time-dependent difference absorption represented by predominant periodic modulation in the transition probability change. Hence, vibrational dynamics can be studied in relation to the decay dynamics of electronic excited states utilizing the same measurement equipment under identical experimental conditions (e.g., laser power, temperature, and probe sensitivity).

(2) Resonance Raman signals are frequently swamped by the fluorescence signal, especially for highly fluorescent molecules. In contrast, real-time vibrational spectroscopy almost completely eliminates contamination or cross-talk effects of spontaneous fluorescence because the probe light used is much more intense and directional than the spontaneous fluorescence.

(3) Low-frequency modes can be easily studied by pump–probe methods as long as a few quanta of the modes lie within the width of laser spectrum with nearly constant phase. However, it is difficult to detect low-frequency modes by a Raman method due to the intense Rayleigh scattering of the excitation beam.[27]

(4) Real-time spectroscopy is sensitive to small instantaneous frequency changes so that it can detect changes of the molecular structure during relaxation process (e.g., formation of transition states).

(5) Since it simultaneously measures electronic and vibrational relaxation processes, the probe-dependent vibrational amplitude can be detected as a function of the pump–probe delay time. Therefore,

information about the vibrational phase can be obtained, which cannot be obtained by conventional Raman and infrared spectroscopies.

This chapter reviews recent research into several J-aggregate systems by femtosecond real-time vibrational spectroscopy using ultrashort pulses with durations shorter than 5 fs. The experimental configuration of real-time spectroscopy and details regarding the samples are described in Section 2. Section 3 discusses the application of sub-5-fs real-time spectroscopy to study exciton-vibration coupling in porphyrin J-aggregates. Despite the weak exciton–phonon coupling in Frenkel exciton systems, coherent molecular vibrations can be clearly observed in the time dependence of the transmission change. The experimental results are well explained by a new model that describes the modulated Q band transition dipole moment, which originates from the oscillator strength transfer from the intense B band to the weak Q band by dynamic intensity borrowing (DIB) through the vibronic interaction. The observed wave packet has been classified as a Herzberg–Teller type, in contrast with the frequently studied conventional Franck–Condon type. Section 4 discusses ultrafast exciton dynamics in PIC J-aggregates. Exciton–phonon coupling in the aggregates modulates electronic transitions. Despite exchange narrowing, as many as 11 vibrationally coupled modes are observed by ultrafast spectroscopy using few-cycle visible pulses. The origin of the molecular vibrations excited in PIC J-aggregates is explained in terms of non-Condon-type DIB. In addition, we discuss the different vibrational phase/amplitude distributions and vibronic coupling strengths obtained using different excitations.

Experimental Section

2.1 *Porphyrin J-aggregates*

In photobiology, molecular aggregates are one of the most extensively studied molecular systems because they play an essential role in energy or charge transfer in most photobiological processes, including photosynthesis.[28] Porphyrin aggregates are particularly important. They serve as models for artificial solar energy capture as in photosynthesis, namely the light-harvesting antenna chlorophyll system with a storage ring configuration and primary charge-separation systems in the photosynthesis reaction center.[29]

We have produced porphyrin aggregates using tetraphenylporphine tetrasulfonic acid (TPPS; Tokyo Kasei) and polyvinyl alcohol (PVA; Kanto Chemical) without further purification. 200 mg PVA was dissolved in 4 ml distilled water at ~100°C and then 20 mg TPPS was added to the hot solution. The TPPS concentration was 5×10^{-3} mol/L, which was sufficiently high for aggregate formation. A unidirectionally oriented film of porphyrin J-aggregates was prepared using a vertical spin-coating technique developed by our group.[30]

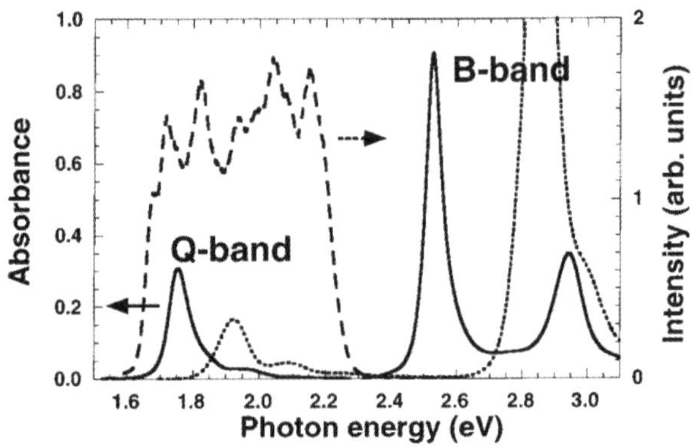

Figure 2.1. Linear absorption spectra of porphyrin J-aggregates (solid line) and monomer (dotted line). The dashed line shows the laser spectrum.

Figure 2.1 shows linear absorption spectra of the porphyrin J-aggregates and the monomer. Sharp absorption peaks at 2.53 eV (S_2 state; the Bx or Soret band) and 1.75 eV (S_1 state; Q_x band) respectively correspond to B and Q exciton bands, forming a quasi-two band Frenkel-exciton system. The B_x and Q_x bands have lower energies than the monomer bands (2.86 and 1.92 eV). They correspond to the second-lowest and the lowest singlet excited states (S_1 and S_2 states), respectively. The fine structure in the absorption bands indicates counterparts of the two bands, namely the B_y and Q_y bands, which have peaks at about 2.94 and 1.86 eV, respectively. The polarization dependence of the absorption indicates that the x and y components originate from the transition dipole moments, which are respectively parallel and perpendicular to the aggregate axis (Fig. 2.2).[31–33] One component of the aggregate is a protonated TPPS molecule, which has a nearly coplanar structure between the phenyl rings and the porphyrin ring. As shown in Fig. 2.2, the porphyrins are stacked. The sulfonic group of one TPPS unit is directly above the diacid charge of another, which contributes to the stability of the aggregate.

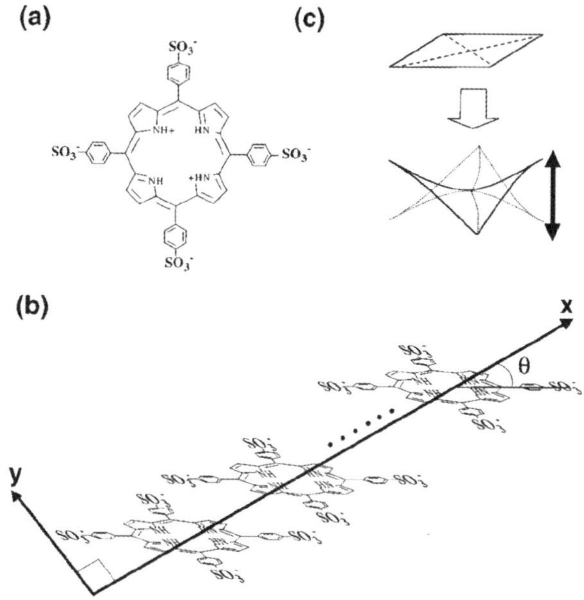

Figure 2.2. Structures of a porphyrin molecule (a) and J-aggregate (b). The x axis corresponds to a stacking direction. The porphyrin molecules make an angle θ with the x axis. (c) Structural diagram of the ruffling motion of the porphyrin molecule.

2.2 *Pseudoisocyanine J-aggregates*

Many experimental and theoretical studies have investigated the J-aggregates of cyanine dye molecules because of their physical importance as models for 1-D Frenkel excitons and because of their application as sensitizers.[9,10,14,34–45] We used one of the most typical cyanine dyes, PIC, to produce J-aggregates. To form J-aggregates of PIC, it is necessary to dissolve the molecule in a solvent with a high concentration. To obtain a sufficiently high concentration, 0.5 g PVA was dissolved in 4 ml pure hot water in a flask heated by an oil bath to a temperature slightly above 120°C. PIC powder (28 mg) was then added to the hot aqueous solution of PVA and it was completely dissolved by stirring the solution. Special care was taken to remove any scattering particles from the sample prior to spin coating. Spin-coated films were formed by casting the solution onto a spin coater rotating at 2000 rpm. The film thickness was estimated to be 15 μm from the film absorbance at the J-peak wavelength.

Figure 2.3 shows stationary absorption and fluorescence spectra of a film of the J-aggregates measured using conventional spectrometers. The stationary absorption spectrum has three peaks at 2.159, 2.337, and 2.492 eV and a shoulder at about 2.70 eV. The fluorescence spectrum has a peak at 2.152 eV and a shoulder around 2.02 eV. The Stokes shift is thus calculated to be 6.8 meV.

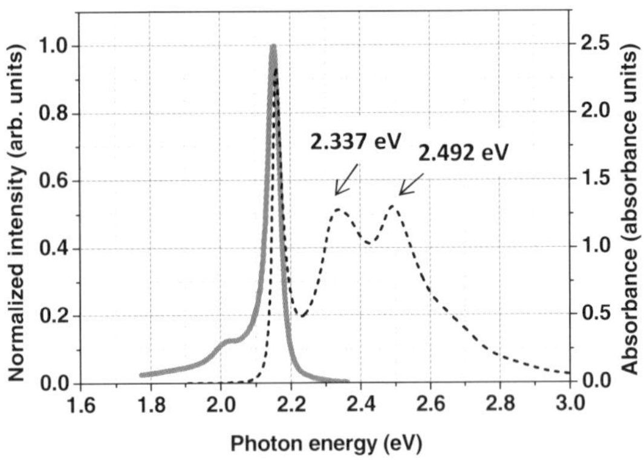

Figure 2.3. Stationary absorption (broken line) and fluorescence (dotted line) spectra of the sample of PIC J-aggregates.

2.3 *Sub-5-fs Real-time Pump–Probe Experimental Setup*

Several methods can be used to generate sufficiently short pulses to excite coherent vibration modes in molecules. Of them, noncollinear optical parametric amplifiers (NOPAs) have been attracting great interest as a novel method for generating ultrashort pulses. In our experiment, both pump and probe pulses were generated by a NOPA laser system, which was seeded by a white-light continuum (Fig. 2.4).[34,47,48] This NOPA system was pumped by a commercial regenerative amplifier (Spectra Physics, Spitfire; central wavelength: 800 nm; pulse duration: 50 fs; repetition rate: 5 kHz; average output power: 650 mW). The pulse duration of the NOPA output was compressed by a system composed of a pair of Brewster-angle prisms with apex angles of 68° and chirped mirrors (Layertec). A typical visible–near-infra red pulse with a pulse duration shorter than 5 fs had a spectral range extending from 1.6 to 2.3 eV. The pump pulse energy was 14 nJ and the photon density was 5.9×10^{14} photons/cm^2 in the porphyrin aggregate experiment. For PIC J-aggregates, two laser pulses were provided by the NOPA: one had a pulse duration of 4.7 fs and a wavelength range 500 to 750 nm and the other had a pulse duration of 8.8 fs and a wavelength range 530 to 750 nm; both pulses had pulse energies of about 40 nJ.

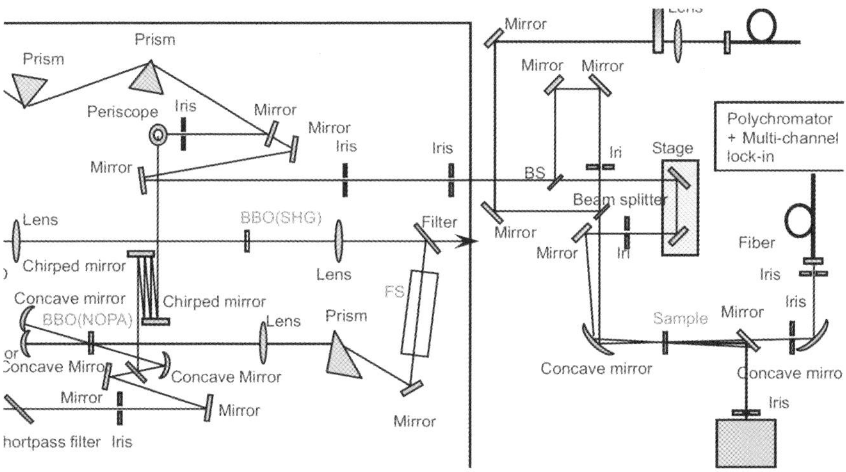

Figure 2.4. Block diagram of the apparatus of vibration real-time spectroscopy based on a pump-probe scheme. VND: variable neutral-density filter; FS: fused silica; BS: beam splitter.

Porphyrin aggregates were investigated by multichannel pump–probe spectroscopy by using a CCD camera with a polychromator. The reference and probe pulses were dispersed by a polychromator (Acton, 320i; 150 grooves/mm; 500 nm blaze) and simultaneously grabbed by a two-dimensional (2D) CCD camera (Roper Scientific, TE/CCD-1100-PB/UVAR; 1100×330 pixels) within 500 ms. The total system had a spectral resolution of about 4 nm. Two sets of 500 laser shots were accumulated under alternating excitation and non-excitation conditions;

this procedure was repeated 10 times for each delay time, which was controlled by an optical delay stage.

We increased the spectral resolution to 0.3 nm by employing a system consisting of a polychromator and a multichannel lock-in amplifier, which was utilized for pump–probe spectroscopy of J-aggregates formed using PIC (see Section 4). The signal was spectrally dispersed by a polychromator (Jasco, M25-TP) over 128 channels and only the spectral range from 2.09 to 2.25 eV (593 to 550 nm) was detected. Each channel was detected by an avalanche photodiode in conjunction with a lock-in amplifier locked onto the 2.5-kHz reference frequency set by an optical chopper, which was used to modulate the pump beam. The delay times between the pump and probe pulses were scanned with a step of 0.2 fs using the optical delay stage. One time-resolved spectrum was accumulated over 0.6 s. The integration time of the lock-in amplifier was set to 3 s. All measurements were performed at room temperature.

3. Ultrafast Real-time Spectroscopy in Porphyrin Aggregates

3.1 *Stationary Raman Spectrum*

Figure 3.1 shows the resonance Raman spectrum of the porphyrin aggregates under excitation of the B band (2.53 eV). Two vibrational frequency peaks are clearly observed at 244 and 316 cm^{-1}, which have also been detected by Ohno *et al.*[49] and Akins *et al.*[32] These groups assigned the two bands at 241 and 317 cm^{-1} to out-of-plane ruffling and doming modes, respectively.[49,50] They also found that these two bands are drastically enhanced by a factor of over 30 with respect to the corresponding two bands of the diacid monomer at 233 and 310 cm^{-1}. Similar enhancement has also been observed for the bands at 236 and 314 cm^{-1} of $TCPPH_4^{2+}$ (tetraphenylporphine tetracarboxyl acid) aggregates[32] and the band at 386 cm^{-1} of CuTOOPP ((5,10,15,20-tetrakis [4-(1-octyloxy)phenyl]-porphinato)copper-(II)) aggregates;[51] therefore, the enhancement is probably due to aggregate formation. Because aggregation formation is expected to influence the out-of-plane mode between adjacent porphinato macrocycles, it may also be necessary to account for intermolecular interactions.

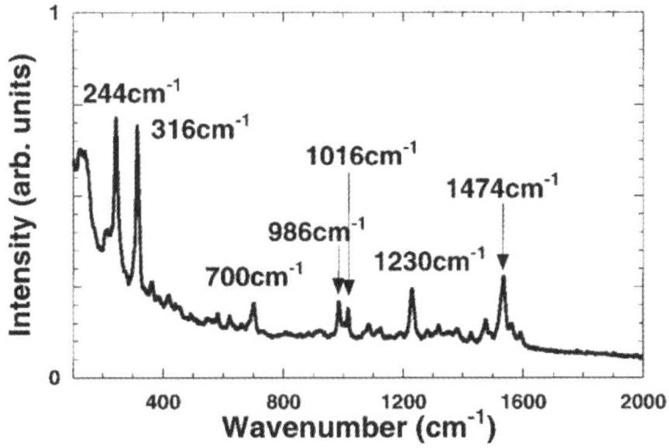

Figure 3.1. Resonance Raman spectrum of TPPS J-aggregates. The excitation wavelength is 488 nm. Reprinted with permission from H. Kano, T. Saito, and T. Kobayashi, *J. Phys. Chem. A* 2002, 106, 3445-3453. Copyright 2002, American Chemical Society.

3.2 *Dynamics of Excitonic States*

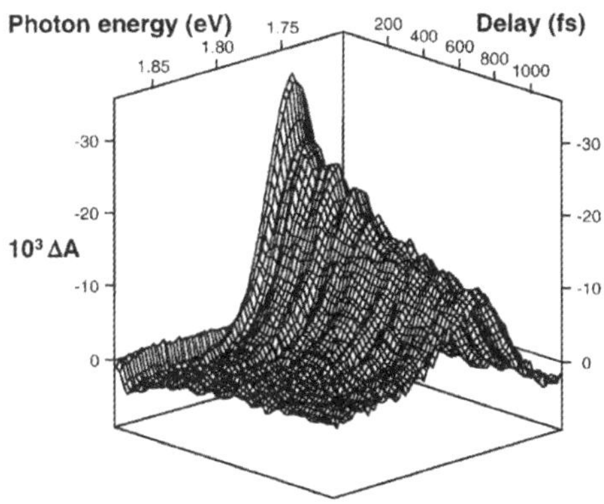

Figure 3.2. Three-dimensional plot of the time-resolved difference absorption spectrum of TPPS J-aggregates as functions of probe-photon energy and delay time. Reprinted with permission from H. Kano, T. Saito, and T. Kobayashi, *J. Phys. Chem. A* 2002, 106, 3445-3453. Copyright 2002, American Chemical Society.

Figure 3.2 shows a three-dimensional plot of the time-resolved difference absorption spectrum. It clearly indicates coherent oscillation together with a slow dynamics component. It is important to clarify that the S_2 exciton is not photoexcited in the excitation photon flux ($I = 1.2 \times 10^{29}$ photons/s cm^2) by the two-photon absorption process for the following reason. The two-photon absorption cross section for anthracene is typically of the order of 10^{-50} cm^4 s,[52] which is relatively large for aromatic molecules. Even if the two-photon absorption cross section of a porphyrin J-aggregate, $\sigma^{(2)}$, is assumed to be similarly large, the quantity $\sigma^{(2)}I = 1.2 \times 10^{-21}$ cm^{-2} will still be much smaller than the one-photon absorption cross section for a porphyrin J-aggregate (~10^{-15} cm^2). Hence, S_2 excitons can be neglected under the present experimental conditions.

Figure 3.3 shows the difference absorption spectrum as a function of delay time. The dominant feature that commonly appears in the traces is high-frequency oscillations superimposed on the underlying slow-dynamics transient components. The overall slow dynamics of the induced change of the absorbance (ΔA) by photoexcitation originates from the excited states of excitons. Following the excitation of J-aggregates, a negative ΔA is observed over the whole range of delay time from 1.71 to 1.78 eV. This is attributed to ground-state bleaching (BL) and photoinduced emission (PIE) of the S_1-exciton state. At probe energies higher than 1.78 eV, the ΔA signal becomes positive due to photoinduced absorption (PIA). In fact, because BL is around 1.85 eV, the ΔA signal overshoots and becomes negative around probe photon energy of 1.80 eV as the delay time increases. The positive ΔA signal is more clearly observed if only the red side of the Q band is photoexcited. Note that the BL, PIE, and PIA spectra overlap each other.

Measured decay traces were fitted to convoluted curves of the system response function using the following equation:

$$\Delta A(\hbar\omega) = A(\hbar\omega)\exp(-t/\tau) + B(\hbar\omega) \tag{3.1}$$

Figure 3.3(a) shows the results for A (dashed lines), B (dotted lines), and ΔA (solid lines). Near the Q band, the decay time is determined to be ~320 fs. As we checked the spectral profile of the two time components in Fig. 3.4, we found that the main difference between the 320 fs decay component ($A(\hbar\omega)$) and the constant component ($B(\hbar\omega)$) is the blue-shifted PIA signal, which only dominates in the 320 fs decay component. It is thus reasonable to attribute the blue-shifted PIA signal to transitions

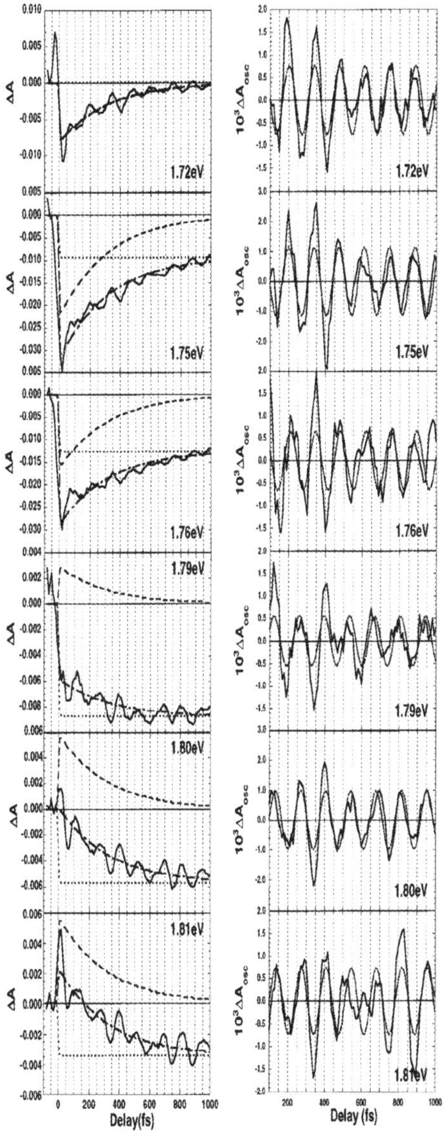

Figure 3.3. (a) Delay-time dependence of the induced absorption change of TPPS J-aggregates (solid line) and fitted results (dashed, dotted, and dot-dashed lines) at six probe-photon energies marked on the right; (b) Delay-time dependence of the oscillating component in a (solid line) and fitted result (dotted line). Reprinted with permission from H. Kano, T. Saito, and T. Kobayashi, *J. Phys. Chem. A* 2002, 106, 3445-3453. Copyright 2002, American Chemical Society.

to the multiexciton states [MES; (|n + 1, S_1>←|n, S_1>) (n = 1, 2, …)] and the 320 fs decay component to the fast relaxation process from higher MESs [|n, S_1> (n = 2, 3, …)] to the one-exciton state, which is expected to be faster than the decay of the S_1 state. A similar decay-time constant (200 fs) has been observed in PIC J-aggregates and it is also ascribed to relaxation from higher MES.[53] In addition, the constant component can be explained by several relaxations in the Q band and the lifetime of the Q exciton.

The slow relaxation process can be explained in the following way; excitonic transitions are classified into the following three components: BL and PIE due respectively to |1,S_1> ← |G> and |1,S_1> → |G> and PIA due to the transition to an MES [|n + 1, S_1> ← |n, S_1> (n = 1, 2, ...)]. Here, n in |n, S_1> denotes the number of S_1 excitons (Q excitons) in a single aggregate. Because of the small Stokes shift, BL is indistinguishable from PIE. Both BL and PIE give a negative signal in the induced absorption spectrum. On the other hand, the positive signal at a probe photon energy slightly higher than the peak of the Q band originates from PIA due to transitions such as |n+1, S_1> ← |n, S_1> (n = 1, 2, ...). Because a transient oscillation is observed only around the Q band, the signals due to PIA and BL/PIE are modulated. In addition, another PIA due to the transition to higher excited exciton states [|1, S_m> ← |1, S_1> (m = 2, 3, ...)] is also observed far above the Q band (E_{pr} >> 1.75 eV) and around E_{pr} < 1.71 eV. Here, |1, S_m> indicates the one-exciton state of a S_m exciton, namely the excited state of an exciton originating from an S_m excited state in a TPPS molecule.

3.3 *Coherent Molecular Vibrations*

Figure 3.5 shows the Fourier transformation of the oscillating component (Fig. 3.3(b)) obtained by subtracting the fitted slow-relaxation component from the trace. A 2D contour plot of the Fourier-power spectrum of the oscillating components is also shown in Fig. 3.6 as functions of the probe photon energy and oscillation frequency. The Fourier spectrum clearly shows an intense peak at 247±7 cm^{-1}. A peak is also observed around 247 cm^{-1} in the resonance Raman spectrum shown in Fig. 3.1; it is assigned to the ruffling mode. As Fig. 3.6 clearly shows,

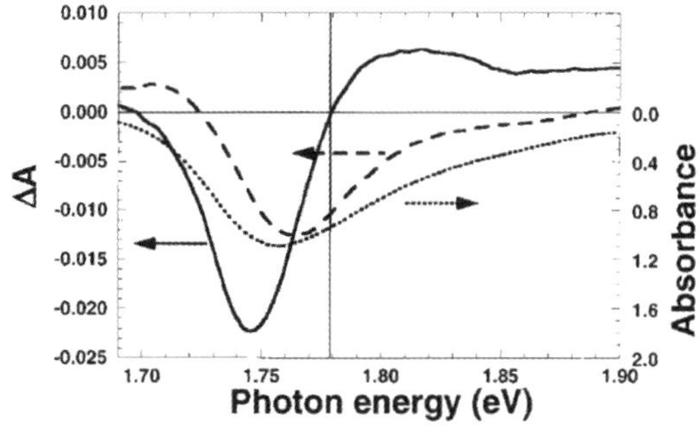

Figure 3.4. 320 fs decay (solid line) and constant (dashed line) components in the transient signal and the stationary absorption spectrum (dotted line). Reprinted with permission from H. Kano, T. Saito, and T. Kobayashi, *J. Phys. Chem. A* 2002, 106, 3445-3453. Copyright 2002, American Chemical Society.

the intensity of 247 cm^{-1} mode drastically decreases at probe photon energy of about 1.78 eV. It means that the sign of the oscillation amplitude reverses (i.e., a π-phase difference), which will be discussed in the following subsection.

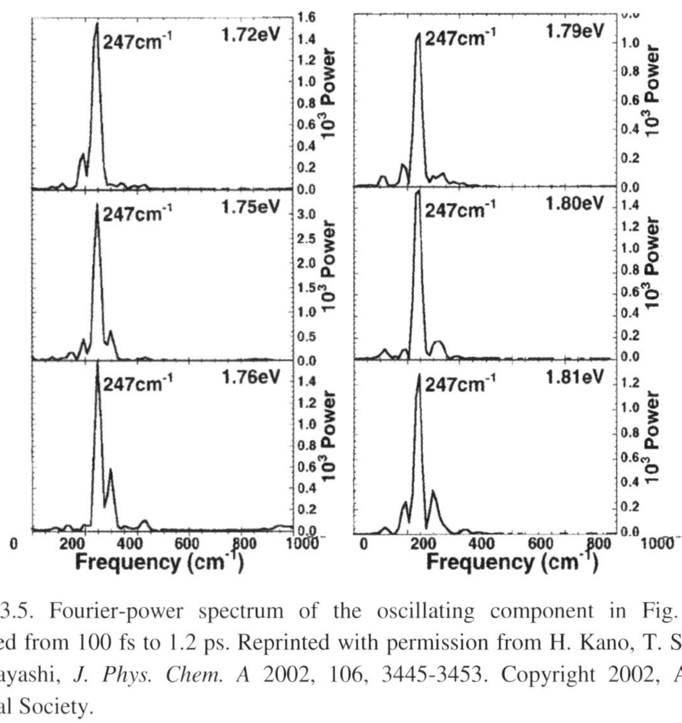

Figure 3.5. Fourier-power spectrum of the oscillating component in Fig. 3.3 (b) integrated from 100 fs to 1.2 ps. Reprinted with permission from H. Kano, T. Saito, and T. Kobayashi, *J. Phys. Chem. A* 2002, 106, 3445-3453. Copyright 2002, American Chemical Society.

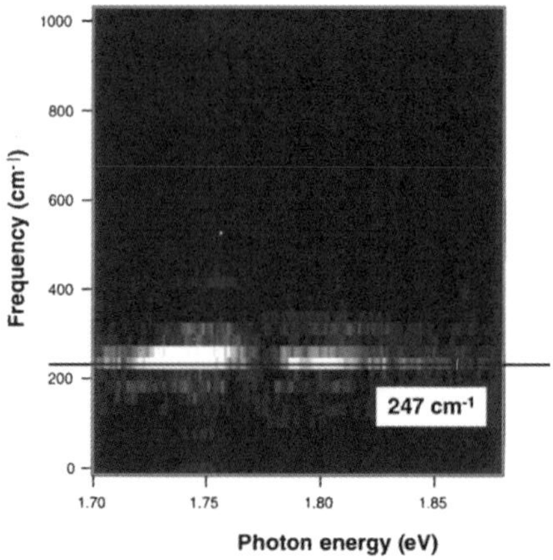

Figure 3.6. Two-dimensional contour plot of the Fourier power spectrum of the oscillating component of the induced absorption spectrum. The vertical and horizontal axes correspond to the probe-photon energy and oscillation frequency, respectively. Brighter regions indicate greater signal intensities. Reprinted with permission from H. Kano, T. Saito, and T. Kobayashi, *J. Phys. Chem. A* 2002, 106, 3445-3453. Copyright 2002, American Chemical Society.

3.4 *Phase and Amplitude Analysis*

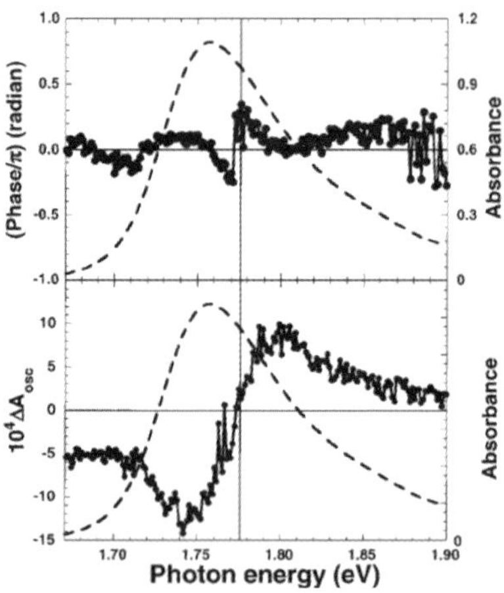

Figure 3.7. Probe-photon energy dependence of the phase (upper) and amplitude (lower) of the oscillation in the transient signals (dots). The oscillating component is fitted by a cosine function with a period of 135 fs. Reprinted with permission from H. Kano, T. Saito, and T. Kobayashi, *J. Phys. Chem. A* 2002, 106, 3445-3453. Copyright 2002, American Chemical Society.

Figure 3.7 shows the probe photon energy dependence of the phase and amplitude of the oscillation, which is obtained by fitting the oscillating component using a cosine function with a fixed frequency of 247 cm^{-1}. The continuous spectra of the phase and amplitude clearly show that the phase is constant throughout the Q band and that the sign of the amplitude is reversed around 1.78 eV. This result is consistent with the one obtained by performing a complex Fourier transformation. The probe photon energy at which the sign reverses is slightly higher than the peak of the Q band and located very close to the probe photon energy of the null absorbance change in the 320 fs decay component. Because the BL/PIE signal has the opposite sign to the PIA signal, the spectral profile of the amplitude indicates that the oscillations are in phase for BL/PIE and PIA. In other words, all the oscillations in the absolute value of the absorbance change over the whole spectral range are synchronous with each other and have a constant phase.

3.5 *Exciton–Vibration Interaction*

The monomer and aggregates have different vibronic interactions. Firstly, the Stokes shift of the fluorescence is quite small (less than 20 meV; i.e., smaller than 160 cm^{-1}). Secondly, the low-frequency Raman mode (244 and 316 cm^{-1}) is strongly enhanced. Finally, the transition from the one-exciton state to the two-exciton state is located at the probe-photon energy slightly higher than the stationary absorption peak corresponding to the exciton peak. As a result of these three differences, the spectral and temporal profiles exhibit characteristic properties that are inherent to the exciton system in J-aggregates.

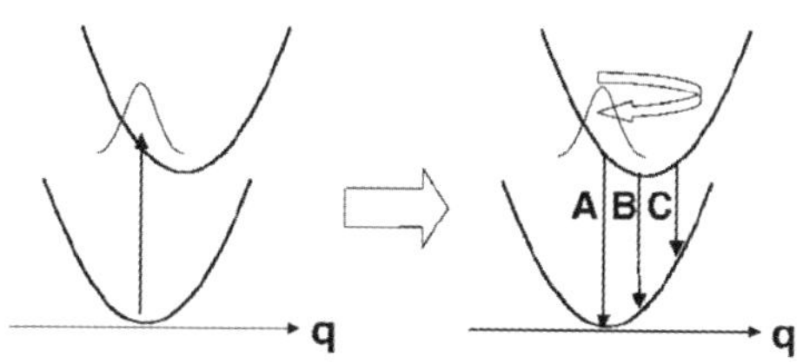

Figure 3.8. Wave packet motions on the excited state PES. The phases of the oscillation are 0, $\pi/2$, and $3\pi/2$ for A, B, and C, respectively.

The transient oscillation persists for a probe delay time longer than 1 ps, which is much longer than the relaxation time from the MES to the one-exciton state (~320 fs). Consequently, they are unlikely to be quantum beats between the different *n*-exciton states. It is also difficult to be explained in terms of the wave packet prepared on the ground-state potential energy surface (PES). Because the oscillation in the BL/PIE signal has an identical phase with that observed in the PIA range, which is uniquely ascribed to wave-packet motion on the excited-state PES, it is not reasonable to describe the observed oscillation signal by wave-packet motion on the ground-state PES.

Based on the above discussion regarding the vibrational phase, it seems likely that the wave-packet motion in the excited state could be used to interpret the time dependence of the absorbance change, as shown in Fig. 3.8. Then the oscillation has phase of 0, $\pi/2$, and π for the three probe photon energies at A, B, and C, respectively, which correspond to the probe photon energies of E_{abs}, E_{fluo}, and $2E_{fluo} - E_{abs}$ assuming wave packet motion on a harmonic PES. Here, E_{abs} and E_{fluo} are the peak photon energies of the absorption and fluorescence spectra, respectively. Therefore, the oscillation must exhibit a π phase difference between $E_{pr} = E_{abs} = 1.76$ eV and $E_{pr} = 1.72$ eV, which correspond to the positions labeled as A and C, respectively. However, a π phase difference is not observed between 1.76 and 1.72 eV in the present experiment. If the amplitude of the oscillation due to PIA is taken into consideration, the ratio of PIA to PIE must be 1.95±0.02 to explain the phase spectrum of the oscillation component; this is substantially larger than the theoretical estimation of 1.57 (see Ref. 54). Therefore, the vibrational signal cannot be explained only in terms of PIE from the excited-state wave packet. This model has two other problems: Firstly, the conventional wave packet model is based on simultaneous photoexcitation of the multi-vibrational quantum states, which depend on the Franck–Condon factor. However, the vibronic excitation of the Q band originates mainly from an Herzberg–Teller-type transition.[55] This is discussed in more detail below. Secondly, the signal intensity integrated over the whole spectral region of the Q band cannot be modulated in this model because the model only produces a spectral shift. However, as we show below, the integrated signal intensity shows non-negligible modulation, which cannot be explained by this model.

The above discussion is made under the assumption that the wavepacket is on the harmonic potential of the excited state or the ground state. The assumption of anharmonicity is not required since we are analyzing the Fourier transform of the real-time trace at each probe wavelength. This implies that only the harmonic component is selected from real-time traces. Even if anharmonicity is present, it can be restricted to overtones of the mode frequency. It may sometimes be difficult to detect because of the short vibration period of the overtones. Therefore, it does not affect the above discussion.

3.6 *Dynamic Intensity Borrowing*

In this subsection, we propose a new mechanism that induces coherent oscillation and modulation of the transition dipole moment of the Q band. This model is beyond the Condon approximation because the transition dipole moment does depend on the normal vibrational coordinate. The modulation frequency corresponds to the ruffling mode of 247 cm^{-1}. The critical aspect is the optical properties of the Q band, which is initially forbidden because of the symmetry of the electronic structure. But it becomes allowed due to the mixing of electronic properties with the B state by the configuration interaction (electron–electron correlation) and vibronic coupling.[55,56] Therefore, modulation of the Q transition can be explained by an additional intensity borrowing from the B band. We term this new mechanism *dynamic intensity borrowing* (DIB).

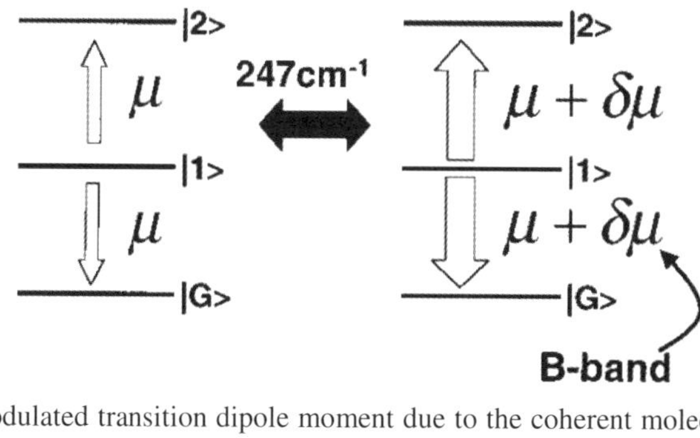

Figure 3.9. Modulated transition dipole moment due to the coherent molecular vibration. |G>, |1>, and |2> denote the ground state and the one- and two-exciton states, respectively. Reprinted with permission from H. Kano, T. Saito, and T. Kobayashi, *J. Phys. Chem. A* 2002, 106, 3445-3453. Copyright 2002, American Chemical Society.

Because of the small Stokes shift of the stationary fluorescence spectrum, the bottom of the excited-state PES is assumed to be located just above that of the ground-state PES. A wave packet is generated on the excited-state PES because of vibronic coupling and the small anharmonicity. The wave packet on the excited-state PES becomes narrower or broader with time evolution and no probe photon energy dependence is expected unless DIB occurs. DIB allows the oscillator strength to be modulated as a function of the normal coordinate, which results in coherent oscillation in the induced absorption signal, as shown in Fig. 3.9. The increase (or decrease) in the transition dipole moment is reflected by an increase (or decrease) in the signal intensity of both BL/PIE and PIA. Consequently, DIB can explain the synchronous oscillations for BL/PIE and PIA in terms of modulation of the transition dipole moment. If this model is valid, the total signal intensity associated with the Q exciton will also be modulated. This can be evaluated by integrating the signal intensity over the whole spectral region of the Q band, which is defined as $\int \Delta A(\hbar\omega)d\omega$ over the whole Q band. Figure 3.10 shows the signal intensity integrated from 1.65 to 1.91 eV as a function of the delay time. It clearly shows the non-negligible amplitude of the coherent oscillation that mainly has a frequency of 247 cm^{-1}. Since the spectral range from 1.65 to 1.91 eV is wide enough to sum all non-negligible oscillating signals due to the Q band, the oscillator strength of the Q exciton is also modulated by the 247 cm^{-1} frequency component.

As mentioned above, the modulation of the transition dipole moment depends on the normal coordinate. Hence, this model can also be called a "Herzberg–Teller-type" wave packet motion, in contrast with conventional "Franck–Condon-type" wave packet motion. In the following, we derive an analytical expression for the wave packet motion based on the coordinate-dependent transition dipole moment. In general, the vibronic coupling effect can be expressed approximately by expanding the transition dipole moment, $\hat{\mu}(\hat{q})$, by performing a Herzberg–Teller expansion of the ground- and excited-state wave functions. Here, $\hat{q}$ denotes the normal coordinate. In the present system, the vibronic coupling predominantly occurs between the Q and B bands; therefore, the following three wave functions are considered:

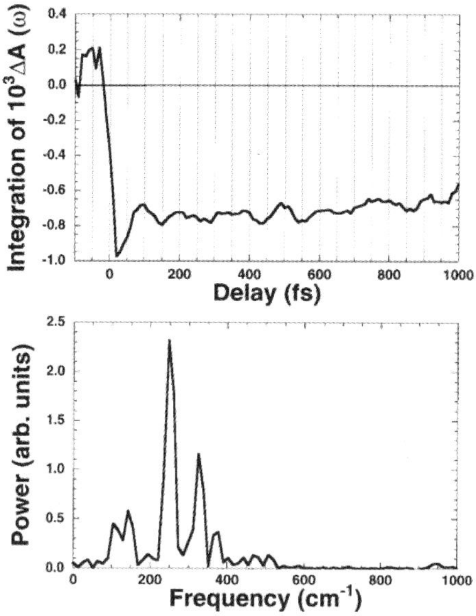

Figure 3.10. Integrated signal intensity of the induced absorption change from 1.65 to 1.91 eV (upper) and the Fourier power spectra of the oscillating component (lower). Reprinted with permission from H. Kano, T. Saito, and T. Kobayashi, *J. Phys. Chem. A* 2002, 106, 3445-3453. Copyright 2002, American Chemical Society.

$$|G(r,q)\rangle = |\psi_G(r)\chi_n^G(q)\rangle \tag{3.2}$$

$$|B(r,q)\rangle = \left(|\psi_B(r)\rangle - \frac{H_{vib}}{\Delta E_{BQ}} q |\psi_B(r)\rangle\right) |\chi_l^B(q)\rangle \tag{3.3}$$

$$|Q(r,q)\rangle = \left(|\psi_Q(r)\rangle + \frac{H_{vib}}{\Delta E_{BQ}} q |\psi_B(r)\rangle\right) |\chi_m^Q(q)\rangle \tag{3.4}$$

Here, $\psi_X(r)$ and $\chi^X(q)$ represent wave functions of the electron and nuclear for the *X* (= G, B, and Q) state, respectively, as a function of the electron and nuclear coordinates, *r* and *q*, respectively. In the nuclear wave functions, the vibrational quantum numbers are indicated as *n*, *l*, and m. The first term in Eq. (3.4) consists of the following two wave functions:

$$|\psi_Q(r)\rangle = |\psi_Q^0(r)\rangle + \alpha|\psi_B(r)\rangle \tag{3.5}$$

The first component is an optically forbidden state, whereas the second term is an optically allowed state that originates from the configuration interaction (CI) with the B band. The coefficient α represents the degree of CI. The matrix element of the transition dipole moment from the ground state to the Q band is given by

$$\langle Q(r,q)|\hat{\mu}(\hat{q})|G(r,q)\rangle$$
$$= \left\langle \chi_m^Q(q) \middle| \left(\langle\psi_Q(r)| + \frac{H_{\text{vib}}}{\Delta E_{\text{BQ}}} \langle\psi_{\text{B}}(r)|\hat{q} \right) \mu|\psi_G(r)\chi_n^G(q) \right\rangle$$

$$= \mu_{\text{QG}} \left\langle \chi_m^{\text{Q}} \middle| \chi_n^{\text{G}} \right\rangle \cong |\alpha|^2 \mu_{\text{BG}} \quad (\text{if } n = m = 0) \tag{3.6}$$

$$= \beta\mu_{\text{BG}} q_{mn} \quad (\text{if } n \neq m \text{ and } m \neq 0) \tag{3.7}$$

Here, $\mu_{QG} = \langle\psi_Q|\mu|\psi_G\rangle$, $\mu_{BG} = \langle\psi_B|\mu|\psi_G\rangle$, $\beta = H_{vib}/\Delta E_{BQ}$, and $q_{mn} = \langle\chi_m^Q|\mu|\chi_n^G\rangle$. Although multimode effects should be taken into account, we use the approximation $\langle\chi_0^Q|\chi_0^G\rangle \cong 1$, which means that all vibronic excitation is not caused by the Franck–Condon component but by the term in Eq. (3.7) associated with vibronic coupling. Although the symmetry affects the vibronic interaction and the coefficient, q_{mn}, the symmetry of the ruffling mode of the aggregate is not as well defined as that for the monomer. Accounting for the time evolution, the wave function, $|\psi(t)\rangle$, is expressed as follows by assuming an impulsive limit:

$$\begin{aligned} \left|\Psi(t)\right\rangle &\equiv \exp(-i\hat{H}t/\hbar)\hat{\mu}\left|G(r,q)\right\rangle \\ &\propto |\alpha|^2 \mu_{\text{BG}} \left|\psi_{\text{B}}\chi_0^{\text{Q}}\right\rangle + \beta\mu_{\text{BG}} \sum_{m\neq 0} q_{m0} \exp(-im\Omega t)\left|\psi_{\text{B}}\chi_m^{\text{Q}}\right\rangle \end{aligned} \tag{3.8}$$

Here, Ω corresponds to the vibrational frequency of 247 cm^{-1}. The transition from the excited-state wave packet, $|\Psi(t)\rangle$, to the ground state, $|G(r,q)\rangle$, is calculated as follows using Eq. (3.8):

$$\langle G(r,q)\,|\,\mu\,|\,\Psi(t)\rangle = |\alpha|^4 (\mu_{BG})^4 + |\beta|^2 (\mu_{BG})^2 \sum_{m\neq 0} |q_{m0}|^2 \exp(-im\Omega t) \qquad (3.9)$$

Because the ΔA signal is proportional to the square of Eq. (3.9), the wave packet can be created by the Q transition through vibronic coupling with the B transition. In other word, DIB corresponds to real-time observation of the vibronic coupling (vibronic intensity borrowing) between the Q and B bands. In a similar manner, the oscillator strength of the transition from the n-exciton state to the $(n+1)$-exciton state $[|n+1,S_1\rangle \leftarrow |n,S_1\rangle$ $(n = 1, 2, ...)]$ varies synchronously. Therefore, both the BL/PIE and PIA signals are modulated simultaneously, which agrees well with the observed experimental results.

Based on the DIB mechanism, the normalized modulation of the transition dipole moment is simply expressed as $\delta\mu_{QG} / \mu_{QG}$, where μ_{QG} represents the Q transition dipole moment of a TPPS molecule and $\delta\mu_{QG}$ indicates the change in μ_{QG} induced by the molecular vibration. Assuming all the transition dipoles in the molecules are parallel and to make an angle θ with the aggregate axis and the molecules are equidistant with a lattice constant a, the modulation of μ_{QG} also changes the intermolecular dipole–dipole interaction, J, to $J + \delta J$, which is given by

$$J_{QG} = \mu_{QG}{}^2(1-3\cos^2\theta)/\hbar a^3$$

$$\frac{\delta J_{QG}}{J_{QG}} = \frac{2\delta\mu_{QG}}{\mu_{QG}} + \frac{3\sin(2\theta)}{3\cos^2\theta - 1}\delta\theta - \frac{3}{a}\delta a. \qquad (3.10)$$

Because the transition energy from the ground state to the lowest one-exciton state is originally red shifted by $2J$ relative to that of the monomer, the modulation of J gives rise a peak shift of the J band in the transient absorption spectrum, which is one of the characteristic features of the exciton system. Below, the data are analyzed phenomenologically. The nonlinear spectrum, $S_{NL}(\omega)$, is assumed to be modulated as $S'_{NL}(\omega)$ by $\delta\mu_{QG}$ and $\delta(2J)$, which is given by

$$S'_{\mathrm{NL}}(\omega) = S_{\mathrm{NL}}(\omega - \delta\omega(t))(1 + \delta\alpha(t)), \tag{3.11}$$

where

$$\delta\alpha(t) = \delta\alpha\cos(\Omega t) \equiv \delta(\mu_{\mathrm{QG}}^2)/\mu_{\mathrm{QG}}^2\cos(\Omega t), \tag{3.12}$$

$$\delta\omega(t) = \delta\omega\cos(\Omega t) \equiv \delta(2J)\cos(\Omega t) \tag{3.13}$$

Here, $\delta\alpha$ corresponds to the amplitude modulation and $\delta\omega$ is the spectral shift in the nonlinear spectrum. The modulation frequency, Ω, corresponds to the 247 cm^{-1} ruffling mode. The present model assumes that both $\delta\alpha$ and $\delta\omega$ are independent of the probe photon energy, ω, and that all μ_{QG} and J are effectively modulated. Eq. (3.11) can be approximately expressed as

$$S'_{\mathrm{NL}}(\omega) \approx S_{\mathrm{NL}}(\omega) + \left(\delta\alpha S_{\mathrm{NL}}(\omega) - \delta\omega\frac{\mathrm{d}S_{\mathrm{NL}}(\omega)}{\mathrm{d}\omega}\right)\cos(\Omega t)$$
$$= S_{\mathrm{slow}}(\omega) + S_{\mathrm{osc}}(\omega)\cos(\Omega t). \tag{3.14}$$

Here, $S_{\mathrm{slow}}(\omega)$ is the slow dynamics component and $S_{\mathrm{osc}}(\omega)$ is the oscillating component.

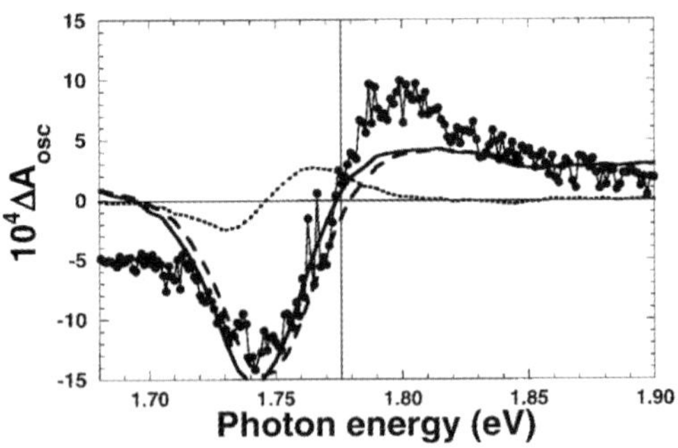

Figure 3.11. Simulation results for $\delta\alpha S_{NL}(\omega)$ (dashed line), $\delta\omega \cdot dS_{NL}(\omega)/d\omega$ (dotted line), and $S_{OSC}(\omega)$ in the DIB model. Reprinted with permission from H. Kano, T. Saito, and T. Kobayashi, *J. Phys. Chem. A* 2002, 106, 3445-3453. Copyright 2002, American Chemical Society.

Figure 3.11 shows the fitting results obtained using Eq. (3.14); they agree very well with the experimental data. In this simulation, the values of $\delta\alpha$ and $\delta\omega$ are determined to be $(6\pm1)\times10^{-2}$ and $(-8\pm4)\times10^{-4}$ eV, respectively. Hence, $\delta(\mu_{QG})/\mu_{QG}$ is calculated to be $(3\pm1)\times10^{-2}$. Assuming $\delta\theta = \delta\alpha = 0$, the dipole–dipole interaction of the Q band can be calculated to be $2J_Q = \delta\omega / \delta\alpha$ = 13±9 meV, which is much smaller than that calculated by energy shifts from the monomer to the J-aggregate in the stationary absorption spectrum (170 meV). Therefore, it is reasonable to conclude that the dipole–dipole interaction is not the only contribution to the energy shift in the stationary absorption spectrum. A Coulomb interaction between molecules caused by the aggregation could be one of the other mechanisms associated with this energy shift.

Below, we calculate the oscillator strength transfer from the B band to the Q band. The perturbed transition dipole moment of the Q band is described by

$$\mu'_{QG} = \mu_{QG} + \delta\mu_{QG} = \mu_{QG} + \frac{V_{DIB}}{\Delta E_{BQ}}\mu_{BG} \tag{3.15}$$

Here, V_{DIB} represents the interaction Hamiltonian due to DIB. From the stationary absorption spectrum of the monomer, ΔE_{BQ}, μ_{QG} / μ_{BG}, and $\delta\mu_{QG} / \mu_{QG}$ are calculated to be ~0.94 eV, ~0.32, and 3%, respectively. Therefore, V_{DIB} is calculated to be 9 meV (73 cm^{-1}) from Eq. 3.15, which is much smaller than the molecular vibrational frequency of 247 cm^{-1}. The modulated transition dipole moment can be expressed as

$$\mu'_{QG} = \mu_{QG} + \delta\mu_{QG} = (32 + 0.96)\times10^{-2}\,\mu_{BG} \tag{3.16}$$

Consequently, the modulation of the Q-transition dipole moment in the real-time spectrum is estimated to be about 0.96% of the B-transition dipole moment.

3.7 *Conclusion*

In this section, ultrashort laser pulses with durations shorter than 5 fs were used to study the coherent molecular vibration coupled with the Frenkel exciton in porphyrin J-aggregates using multichannel real-time spectroscopy. The ruffling mode with a molecular vibrational frequency of 247 cm^{-1} was clearly observed. The experimental result could not be explained by the conventional model of Franck–Condon-type wave-packet motion, but it could be explained by Herzberg–Teller-type wave-packet motion. Specifically, it could be well interpreted by a transition dipole moment modulated by DIB from the intense B transition to a rather weak Q transition through vibronic coupling.

Ultrafast Exciton Dynamics in Pseudoisocyanine Aggregates

4.1 *Electronic Excited-State Dynamics*

Figure 4.1 shows a 2D plot of the time-resolved difference absorption spectrum of PIC J-aggregates. It depicts both positive (i.e., absorbance increase) and negative (i.e., absorbance decrease)

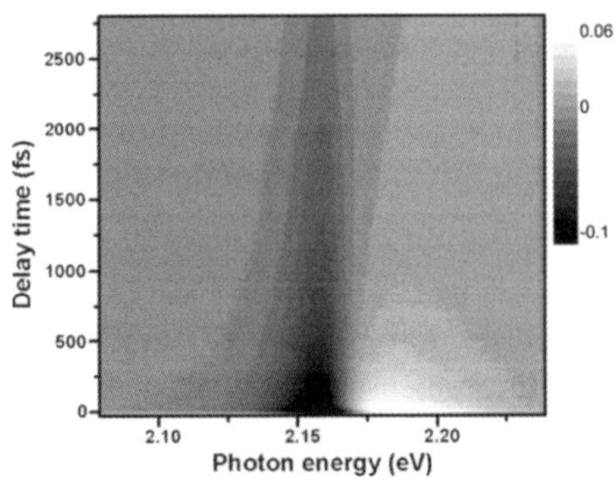

Figure 4.1. 2D real-time spectrum in the time region from – 30 to 2800 fs and in the spectral region from 2.07 to 2.24 eV.

changes. The 2D plot contains a high density of spectral- and time-dependent information. We first discuss the spectral and time domains separately. The time-resolved spectra shown in Fig. 4.2 can be decomposed into several contributions. The negative absorbance change (ΔA) around the peak of the J band (2.158 eV) is attributed to BL due to ground-state depletion and PIE from the one-exciton state. The negative BL and PIE signals overlap because of the small Stokes shift. The positive ΔA around 2.187 eV is due to the PIA from the n to $(n+1)$ exciton state,[14,15] where the integer n is the number of excitons in a single mesoaggregate in which coherent exciton is delocalized on a mesoscopic scale. The transition from the one-exciton state to the two-exciton state is expected to dominate weak pumping case.

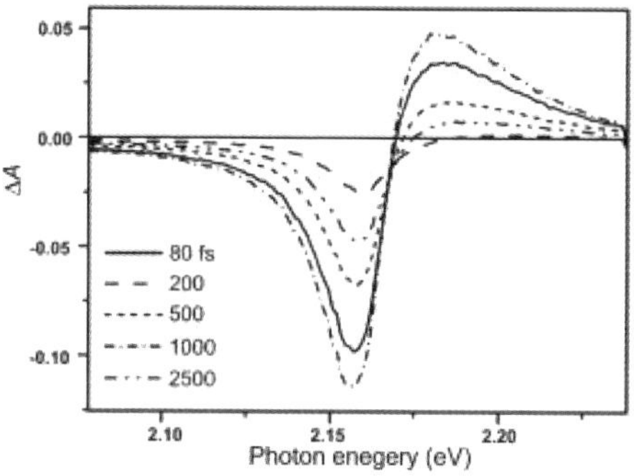

Figure 4.2. Averaged absorbance change spectra at delay times 80, 200, 500, 1000, 2500 fs over a 60-fs interval.

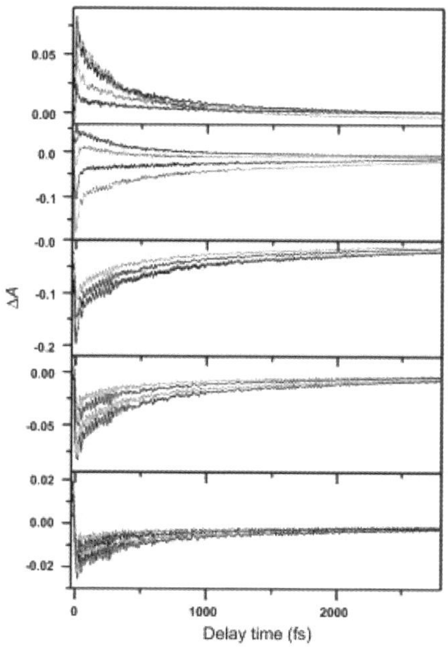

Figure 4.3. Time traces at 20 of typical probe-photon energies.

Figure 4.3 shows real-time traces for 20 probe photon energies plotted against the pump–probe delay time between −30 and +2800 fs. None of the traces can be fitted by a single exponential function but seem to have at least two exponential functions to be well fitted. They also have very fine oscillating features. Even after removing the oscillation features by smoothing, the time constants cannot be well fitted by single values since they depend on the probe photon energy. The observed complicated features can be phenomenologically fitted with the following function after removing the oscillating feature. The decay function is composed of four terms corresponding to processes 1–4:

$$\Delta A(t) = A_1\, e^{-t/\tau_1} + A_2 (e^{-t/\tau_2} - e^{-t/\tau_1}) + A_3 (e^{-t/\tau_4} - e^{-t/\tau_3}) + A_4 (1 - e^{-t/\tau_4}) \quad (4.1)$$

These processes have been extensively studied by several groups including ourselves and the main findings are given below[14,15].

Process 1 has a decay time constant τ_1 of the order of a few hundreds of femtoseconds, which corresponds to relaxation from the n-exciton state to the (n−1)-exciton state ($n \geq 3$). This process is usually dominated by relaxation from the three-exciton state to the two-exciton state. Process 2 has a growth time constant τ_1 of the order of a few hundred femtoseconds and a decay constant τ_2 = 2 ps. The time constant τ_2 corresponds to relaxation from the two-exciton state to the one-exciton state. Process 3 has a growth time constant τ_2 = 2 ps and a decay constant τ_3 = 5–10 ps. The decay constant corresponds to thermalization of the one-exciton state. The last process has a lifetime of about 300–500 ps and corresponds to relaxation of the thermal one-exciton state to the ground state.

In the case of τ_3, thermalization is induced by complicated mode coupling among intramolecular vibrational modes and interactions between intramolecular vibrational modes and lattice vibrations. Because thermalization cannot be well described by simple rate equations, it may be better characterized by a diffusion equation. Figure 4.4 shows two typical real-time traces together with expanded traces in the insets.

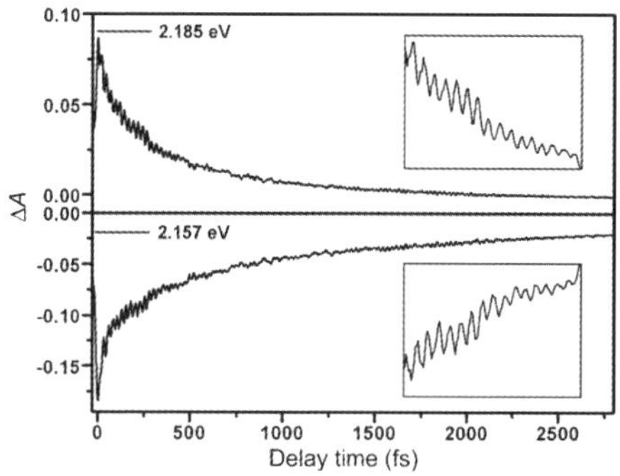

Figure 4.4. Time traces at peak photon energies. Inset shows enlargement of region surrounded by dashed box delay time from 100 to 500 fs.

4.2 *Vibrational Dynamics*

4.2.1 *FFT spectra and the origin of molecular vibrations*

To discuss the oscillating features in the real-time traces, fast Fourier transformation (FFT) is performed on the real-time probe photon energy versus delay time 2D data for probe delay times between 150 and 2967 fs. The resulting 2D FFT power spectrum is depicted in Fig. 4.5 as a function of the probe photon energy (horizontal axis) and the vibrational frequency (vertical axis). It contains some oblique lines that are due to interference between scattered pump and probe light. The 2D spectrum allows us to readily distinguish such false interference peaks from the true vibrational peaks.

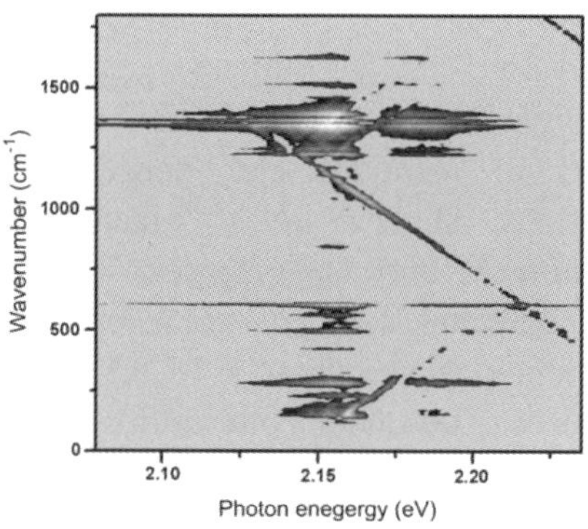

Figure 4.5. 2D FFT power spectrum. Wavenumber resolution is ≈ 10 cm^{-1} determined by the condition of FFT.

4.2.2 *FFT power spectra at a typical probe photon energy*

Figure 4.6 shows vertical slices of the two positive and negative peaks due to BL and PIE (2.157 eV) and IA (2.187 eV). Both spectra contain 11 prominent peaks at 152, 228, 282, 499, 607, 1225, 1247, 1344, 1366, 1388, and 1626 cm^{-1}. The insets show enlarged spectra of the three modes at 1344, 1366, and 1388 cm^{-1} in the dashed boxes, which are referred to as the triplet hereafter. The frequency resolution was limited to ~10 cm^{-1}. The arrows indicate false FFT peaks resulting from the above-mentioned interference.

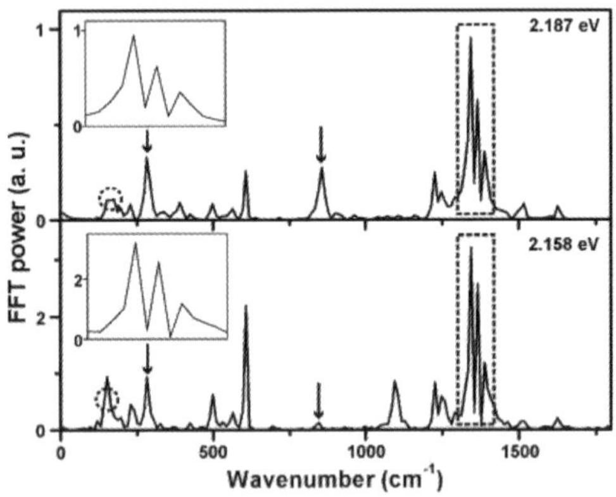

Figure 4.6. FFT power spectra of time trace at IA (top) and BL (bottom) peak photon energies 2.187 and 2.158 eV, respectively. Inset is enlarged portion of dashed box and mode at 152 cm^{-1} is pointed by dotted circle.

We first discuss the signals due to modes with frequencies over 100 cm^{-1}, which are assigned to resonant and nonresonant modes due to intramolecular vibrations. Akins *et al.*[57,58,59] measured stationary Raman spectra of PIC monomers and J-aggregates. They assigned the low-frequency modes of 223 and 278 cm^{-1} to resonance- and aggregation-enhanced exciton–phonon modes when a 17,100 cm^{-1} laser was used as the Raman pump source. The major contribution to the 223 cm^{-1} Raman band was assigned to the twisting mode, while the bending mode contributes the most to the 278 cm^{-1} band.[59] The vibrational modes in the present FFT spectra have corresponding bands in the stationary Raman spectra that lie within ±7 cm^{-1}, except for the band at 152 cm^{-1} (indicated by the dashed circle in Fig. 4.6). It is only possible to state that these corresponding bands coincide with the previously obtained Raman data to within the present frequency resolution of 10 cm^{-1}. The band at 152 cm^{-1} has probably not been well observed in Raman measurements because of the difficulty of detecting such a low-frequency mode against the intense background due to the Rayleigh peak. It has not been predicted by calculations either. Note that the 152 cm^{-1} band intensity for the probe photon energy of 2.158 eV (Fig. 4.6; bottom) is larger than the true value because a false interference peak overlaps with the band. The relative intensities of the present results differ remarkably from those in the stationary Raman spectra. First, the low-frequency modes at 228 and 282 cm^{-1} are enhanced less in the Raman spectra than other (mainly) higher-frequency modes. Second, the 282 cm^{-1} band is much stronger than the 228 cm^{-1} band, whereas the two bands have almost the same intensity in the Raman spectra. Third, of the triplet bands, the lowest band at 1344 cm^{-1} is the strongest in the present FFT spectra, whereas the middle band at 1366 cm^{-1} is the most intense in the Raman spectra.

One of the main reasons for these differences lies in the difference between the stimulated process in real-time spectroscopy and the spontaneous process in Raman measurements. The probe photon energy dependence of the molecular vibrational amplitude impulsively excited by 5-fs pulses can provide important information about vibronic coupling and mode assignment. We performed a comparative study of the amplitude spectrum excited by sub-5-fs pulses and the Raman excitation profile.[48,57–59] The vibrational amplitude profile obtained by the impulsive excitation provides information for the assignment of the vibrational mode. It was shown that there are several advantages of the former method over the Raman technique for the assignment of vibrational modes.[48]

Another possible reason for these differences is that in the above-mentioned Raman spectroscopy experiments[57,58,59] both monomeric and aggregated PIC were adsorbed onto a metal surface. Thus, the measurements were probably influenced by surface enhanced Raman scattering. In addition, there are fundamental differences between the present real-time spectroscopy and conventional stationary Raman spectroscopy. The spectrum of the present sub-5-fs laser pulses is so broad that it fully covers the exciton peak of PIC J-aggregates. This resonance condition differs from that of stationary Raman scattering in which the excitation laser is monochromatic and the degree of resonance varies with the laser wavelength. Ground-state vibrations are detected in stationary Raman spectra, whereas the real-time spectra wavepacket motions of both ground- and excited-states are detected in the FFT power spectrum (see Section 4.2.3). Even though the electronic absorption and fluorescence spectra of the J-aggregate system exhibit almost no vibrational progressions because of exchange narrowing[10], we have now been able to observe as many as 11 vibrationally coupled modes. This indicates the superiority of our ultrashort pulse multichannel real-time spectroscopy for detecting molecular vibrational modes coupled to excitonic transitions in systems such as J-aggregates and modes that couple to electronic transitions in molecular systems.

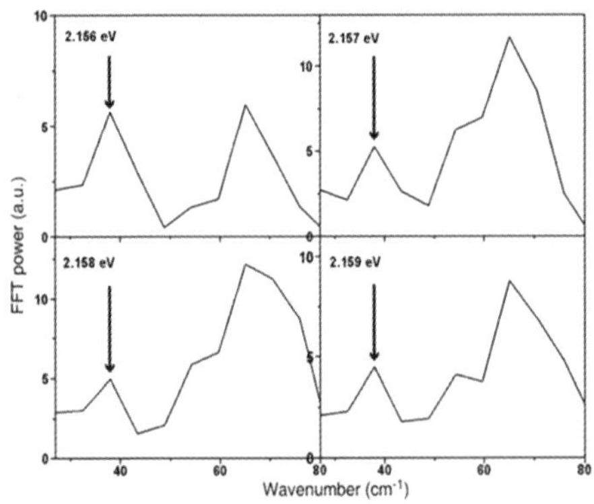

Figure 4.7. FFT power spectra of four probe-photon energies. The arrows indicate the positions of 38 cm^{-1}.

Next, we discuss the signals lower than 80 cm^{-1} showing in Fig. 4.7. These modes do not appear in Fig. 4.6 because high-pass filtering was applied to eliminate noise. Even though the signal-to-noise ratios of the lower-frequency power spectra are lower than those of the higher-frequency modes shown in Fig. 4.6, two modes are observed at 38 and 65 cm^{-1} at almost all probe photon energies. The mode at 65 cm^{-1} may be one of the bending modes of the whole PIC molecule, but it cannot be uniquely assigned to a specific bending mode. The 38 cm^{-1} mode can be assigned to the optical (lattice) phonon of the chain-structured J-aggregates, as discussed below. Kamalov *et al.*[44] discussed their experimental results for the temperature dependence of relaxation in benzimidacarbocyanine (BIC) aggregates in terms of temperature-assisted tunneling mediated by an optical phonon at 25–30 cm^{-1}. Since the molecular masses of BIC and PIC are respectively 779.5 and 327.4 u, the frequencies of the optical phonon can be calculated for PIC to be 39–46 cm^{-1} from the relation

$$\frac{\omega_{\mathrm{PIC}}}{\omega_{\mathrm{BIC}}} = \sqrt{\frac{m_{\mathrm{BIC}}}{m_{\mathrm{PIC}}}} \tag{4.2}$$

The present frequency of 38 cm^{-1} agrees well with that of the previous study.[44] Therefore, we have directly observed the intermolecular vibration of the lattice phonon, which had previously only been determined indirectly.

4.2.3 *FFT amplitude dependences on probe photon energy*

Figure 4.8 shows the FFT amplitude (top) and power (bottom) spectra of modulation signals due to molecular vibrations with frequencies of 228 and 282 cm^{-1}. Previous studies (including our own) have clearly demonstrated that the negative ΔA signal results from BL and PIE.[15] This implies that the wavepacket motions of both the ground- and one-exciton states contribute to the FFT power in the corresponding spectral regions. On the other hand, in the spectral range where ΔA is positive, the oscillatory signal is considered to be mainly due to IA. The relative amplitudes are larger on the negative ΔA side than on the positive side with respect to the absorbance changes. This is probably due to the above-mentioned fact that the relative sizes of contributions to the amplitude signals from the wavepackets of the exciton and ground states differ between the positive and negative signal regions. However, further study is required to clarify the cause of this difference.

As discussed in Sec. 3, if we apply the conventional Franck–Condon-type wavepacket model to explain the origin of this molecular vibration, the integrated FFT amplitude over the whole spectral region of IA is expected to be constant. However, coherent oscillations are present

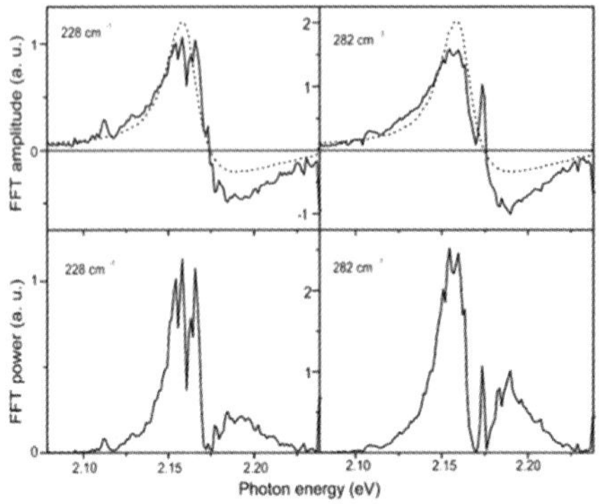

Figure 4.8. FFT power spectra of two low-wavenumber modes. FFT amplitude spectra (top, solid line) and the results of fittings (dashed line).

clearly in the integrated amplitude (data not shown). They cannot be explained in terms of the Franck–Condon-type wavepacket model, which is similar with the results obtained for porphyrin J-aggregates.

4.2.4 *Fourier amplitudes and phases of real-time traces*

As discussed in Subsection 3.6, the nonlinear spectrum, $S_{NL}(\omega)$, is phenomenologically assumed to be modified to $S'_{NL}(\omega)$:

$$\begin{aligned} S'_{NL}(\omega) &\cong S_{NL}(\omega) + \left(\delta\alpha S_{NL}(\omega) - \delta\omega \frac{dS_{NL}(\omega)}{d\omega} \right) \cos(\Omega t) \\ &\equiv S_{slow}(\omega) + S_{osc}(\omega)\cos(\Omega t) \end{aligned} \tag{4.3}$$

From the fitting coefficients $\delta\alpha$ and $\delta\omega$, we can estimate both $\delta\mu/\mu$ and $\delta(2J)$ by using Eqs. (3.12) and (3.13), respectively. The real-time spectrum used to fit the FFT amplitude spectrum was obtained by averaging the ΔA spectra in the time range 150 to 2967 fs. The two spectra at the top of Fig. 4.8 show the results of least-squares fitting (dashed line) two low-frequency modes (solid line). Each FFT amplitude spectrum was well fitted with only the $S_{NL}(\omega)$ spectrum without the $dS_{NL}(\omega)/d\omega$ spectrum. From this analysis, $\delta\mu/\mu$ for 228 and 282 cm^{-1} was estimated to be about 0.3 and 0.5%, respectively. One possible explanation for the absence of the $dS_{NL}(\omega)/d\omega$ contribution in the fit is an accidental cancellation of the modulation by synchronous modulation of the ground-state crystal shift. Another possible explanation is the accidental cancellation of the modulation by that of $\delta\theta$ or $\delta\alpha$ if they have opposite sign to $\delta\mu$. It is difficult to determine the origin of the negligible contribution of $dS_{NL}(\omega)/d\omega$ and further study is required to clarify it.

We recently discovered that the derivative-type contribution is simply due to the real part of the third-order susceptibility inducing a refractive index change that results in cross-phase modulation. This process does not actually exchange energy but does so virtually. The imaginary part of the susceptibility causing the energy exchange does not appear in the form of a derivative but in the form of a difference between the absorbed spectrum of the probe and the shifted absorbed spectrum by the relevant vibrational frequency to a higher or lower frequency, corresponding respectively to anti-Stokes and Stokes interactions,[60,61] as discussed in Subsection 4.3.2. The negligibly small derivative signal for J-aggregates is also considered to be due to the signal induced by a change in the refractive index that causes cross-phase modulation.

The fitting curves for both the 228 and 282 cm^{-1} bands are smaller than the FFT amplitude curves. This can be explained as follows. For photon energies less than 2.175 eV, the ΔA signal is caused by both BL and PIE. In both processes, wavepackets on both the ground- and excited-state potential curves contribute to the modulation. For photon energies larger than 2.175 eV, on the other hand, the ΔA signal is dominant due to IA. If the four contributions in the lower photon energy range (i.e., the ground-state wavepacket of BL and PIE and the excited-state wavepacket of BL and PIE) interfere destructively, then the relative sizes of the modulation in that region may be smaller than those at higher probe photon energies. There is another possible explanation for the smaller FFT amplitude in the smaller probe photon energy: IA can have a contribution from a transition other than that from the one- to two-exciton state, namely to the excited exciton. In this case, the properties of the transition from the ground state to the one-exciton state can be quite different from those of transition from the one-exciton state to the excited exciton state. The vibronic coupling strengths of the two transitions can then be quite different.

The Frenkel exciton system is well known to have very weak exciton–phonon coupling. We have observed real-time exciton–phonon coupling, which has resulted in modulation of electronic transitions associated with

BL, PIE, and IA. This is because of the enhanced transition probability, which is proportional to $N\mu^2$, where N is the number of molecules in the mesoaggregate.[62] In PIC J-aggregates, N was estimated to be ~55. The transition probability is hence enhanced by ~55 and any weak exciton–phonon coupling can induce a detectable change as large as $N\delta\mu$.

4.3 *Effect of Excitation Pulse Width and Spectrum*

4.3.1 *Excitonic dynamics in transient absorption spectra*

In this subsection, we describe a spectroscopic study of PIC using different excitation conditions from those mentioned in the previous subsection. The pump and probe pulses used are both generated by the same NOPA laser system in the two experiments. However, using a different alignment of the laser system, the spectrum of the NOPA output was tuned in the range 530–750 nm with a FWHM of 90 nm, as shown in Fig. 4.9. The Fourier-transform-limited pulse duration calculated from the spectrum was 6.6 fs. The pump and probe pulse intensities on the sample surface were 0.12 and 0.02 GW/cm^2, respectively. The signal in this experiment was spectrally dispersed using a polychromator (Jasco, M25-TP) over 128 channels from 2.09 to 2.25 eV (593 to 550 nm) with a spectral resolution of 0.3 nm.

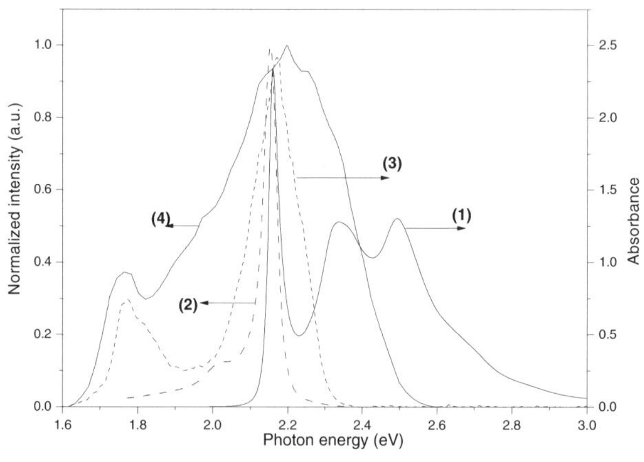

Figure 4.9. Stationary absorption (solid line (1)), fluorescence spectrum (dashed line (2)), laser spectrum (thin dashed line (3) for the one used in this subsection, and thin solid line (4) for the one in the former subsection) used in the two related experiments.

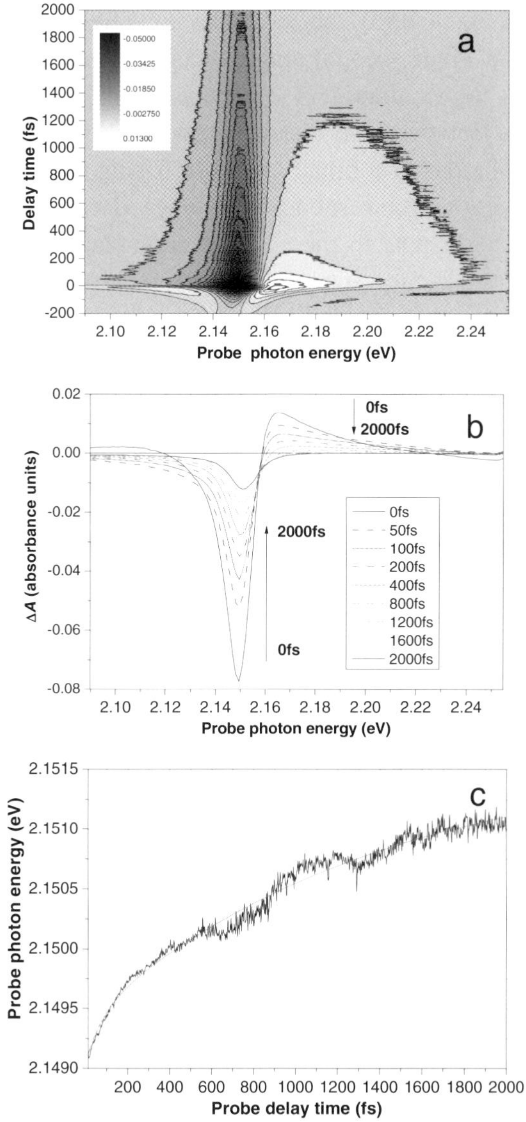

Figure 4.10. (a) Difference absorption spectrum plotted two-dimensionally against probe photon energy and pump-probe delay time. (b) Difference absorption spectra at nine different probe delay times integrated over 50fs delay time period with center delays indicated in the inset. (c) Probe delay time dependence of the photon energy of the negative peak of the difference absorption spectrum from 10 to 2000 fs (black line) and exponential fitting of the data (gray line).

As shown in Fig. 4.10(a), the zero delay time lies within a range of ±2 fs over nearly the whole spectral range, except at the edge of the range for photon energies higher than 2.24 eV, which implies that there is small frequency chirp. However, this does not affect the following discussion of the Fourier amplitudes of a broad spectrum. Using the same method, the values of the shorter (corresponding average decay time from the *n*-exciton state ($n \geq 3$) state to the $n = 2$ exciton state) and longer (decay time from the $n = 2$ exciton state to the $n = 1$ exciton state) time constants are respectively determined to be 140 ± 6 fs and 0.93 ± 0.03 ps, which are close to those reported in Section 4.1 (100 fs and 2 ps). The shorter decay time constant (about 140 fs) increases slightly with increasing photon energy. This can be explained by a change in the distribution in energy space during thermalization of the excitonic population. This change increases the average or the mass center of the transition energy of the induced absorption spectrum distributed over the corresponding exciton energy distribution. The differences in the time constants between the present case and previous results are considered to be due to cross talk between decaying components with similar spectra of the n=1 exciton and n>1 exciton states. The assignments of these time constants are also supported by the observation of the corresponding energy shifts during relaxation, as discussed below.

Figures 4.10(b) and (c) show that the negative peak located near 2.15 eV shifts at a rate of about 4×10^{-6} eV/fs (see Fig. 4.10(c)). The dynamics of this negative peak energy are expressed by

$$E = \Delta E_1 \exp(-t/\tau_{1e}) + \Delta E_2 \exp(-t/\tau_{2e}) + \Delta E_3 \tag{4.4}$$

The values of the parameters in this equation are determined to be $\Delta E_1 = -0.35\ meV$, $\Delta E_2 = -2.24\ meV$, $\Delta E_3 = 2.1517\ eV$, τ_{e1} = 84±9 fs, and τ_{e2} = 1.4±0.1 ps.

The values of these time constants are close to those obtained for the population decay time constants (τ_{a1} = 140±6 fs and τ_{a2} = 0.93±0.03 ps, respectively). The discrepancies between these two sets of values are considered to be due to crosstalk between the two components and the contribution of an even shorter relaxation time due to n>2 exciton states.

4.3.2 *Fourier amplitude spectra of the vibrational real-time spectra*

To analyze the oscillatory signals in the real-time traces, the slowly varying features in the pump–probe traces were subtracted. The fast Fourier transformation (FFT) of the pump–probe data was taken for probe delay times in the range 40 to 2000 fs. Figure 4.11 depicts the two-dimensional FFT amplitude spectrum plotted over the full range of probe photon energies investigated and as a function of the Fourier frequency. To illustrate the effect of the probe pulse wavelength on the Fourier power spectra, Figs. 4.12 shows the Fourier power spectra of the time-dependent absorbance changes at 2.160 and 2.140 eV, respectively. All the FT amplitude spectra obtained from the cross section of the 2D spectrum at specific probe photon energies contain many peaks. The spectra shown in Figs. 4.11 and 4.12 have 14 prominent peaks at 155, 232, 280, 428, 509, 602, 838, 1025, 1167, 1213, 1278, 1314, 1344, and 1392 cm^{-1} in the probe spectral range 2.14 to 2.16 eV. The frequency resolution for these peaks was limited to ~4 cm^{-1}.

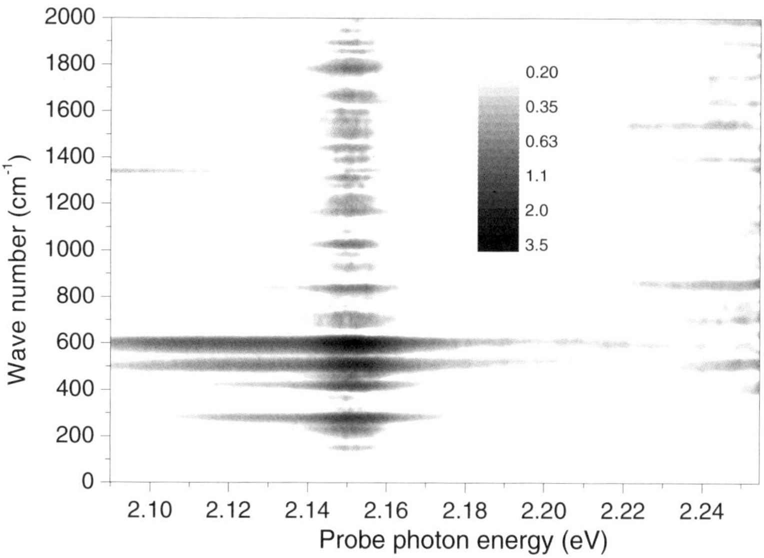

Figure 4.11. Two-dimensional contour map of Fourier amplitude plotted against probe photon energy and vibrational frequency. The Fourier amplitude is plotted on a logarithmic scale.

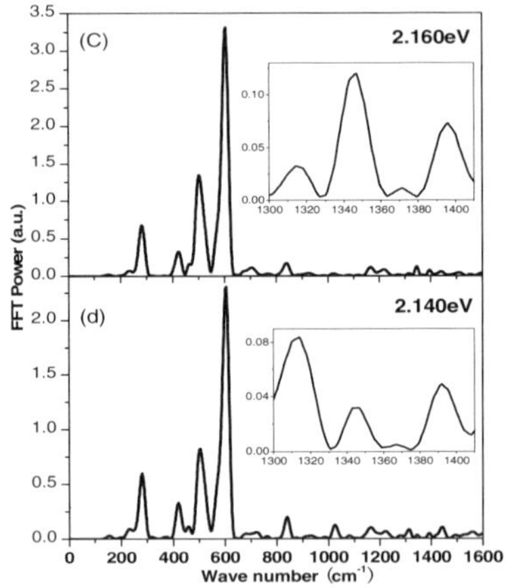

Figure 4.12. FFT power spectra of real-time traces probed at a photon energy of 2.160 eV (positive absorbance change) and of 2.140 eV (negative absorbance change). The enlarged vibrational frequencies between 1300 and 1400 cm^{-1} are shown in inset boxes.

Compared with Figs. 4.5 and 4.6 in Sec. 4.2.2, the most prominent difference between two experiment is the relative intensities of low-frequency modes (below 600 cm^{-1}) to higher frequency modes. They are higher in the present experiments but lower in the experiment described in Section 4.2. This is considered to be due to the difference in the laser spectra ($S(\omega)$). In both the ground and excited states, molecular vibrations are induced by the interaction between the two frequency components satisfying the relation, $\omega_v = \omega_1 - \omega_2$. The intensities of impulsively excited vibrational modes are approximately given by

$$\int S^{'}(\omega)\times S^{'}(\omega+\omega_v)d\omega. \tag{4.5}$$

Here, ω is the optical frequency of the probe laser, ω_v is the vibrational frequency, and $S'(\omega)$ is the absorbed photon energy distribution of the pump laser, which is given by

$$S^{'}(\omega) = S(\omega)\times(1-10^{-A(\omega)}) \tag{4.6}$$

Here, $S(\omega)$ is the pump laser spectral shape and $A(\omega)$ is the absorption spectrum of the sample. The frequency difference between the two spectral components in the integrand of Eq. (4.5) in the corresponding pair equals the associated vibrational mode frequency. Eq. (4.5) indicates summing all the pairs of spectral components to the mode with a frequency of ω_v in Table 4.1.

Table 4.1. Overlap of integral between absorbed spectrum and shifted absorbed spectrum at 280, 500, 600, 1344, and 1388 cm^{-1}.

	Sec. 4.2	Sec. 4.3
Shift amount(cm^{-1})	Integrated value(10^{-1})	Integrated value(10^{-2})
280	1.88	4.27
500	1.56	2.88
600	1.49	2.38
1344	1.28	0.22
1388	1.25	0.18

The Fourier amplitude $q(\omega_v)$ of the vibrational mode is given by

$$q(\omega_v : \omega_2 - \omega_1) \propto \int \chi^{(3)} E(\omega_2) E(\omega_2 - \omega_v)^* \, d\omega_2 \tag{4.7}$$

If we neglect the ω_2 dependence of the Raman susceptibility, which is assumed to be dependent on the vibration frequency, the squared amplitude of the vibration mode is given by

$$|q|^2 \propto \left|\chi^{(3)}\right|^2 \left|E(\omega_2)^2\right|^2 \left|E(\omega_2 - \omega_v)^2\right|^2 \tag{4.8}$$

Table 4.2 shows the calculated results for the integrals for the two experiments together with the values of $|\chi^{(3)}|^2$ for the 1344 cm^{-1} mode relative to the 280 cm^{-1} mode calculated from the observed FFT powers by using the overlap integral between the two corresponding components given by Eqs. 4.5 and 4.6. The ratio $|\chi^{(3)}|^2$(1344 cm^{-1})/$|\chi^{(3)}|^2$(280 cm^{-1}) is equal to 4.8±0.8 and 3.4±0.5 in the two experiments, respectively; they are in fair agreement with each other within the experimental error. This result indicates that the great difference in the Fourier power spectrum between the low-frequency components around 280 cm^{-1} and the high-frequency components around 1340 cm^{-1} is simply due to differences in the spectral components.

Table 4.2. Integrated value, FFT power, and relative χ^3 at 280 and 1344 cm^{-1} in L and S experiments.

	Frequency (cm^{-1})	Integrated value (10^{-1})	Normalized FFT Power ($280cm^{-1}$)	Relative $\lvert\chi^{(3)}\rvert^2$
Sec. 4.2	280	1.88	1	1
	1344	1.28	3.27	4.8
Sec. 4.3	280	0.43	1	1
	1344	0.02	0.18	3.3

4.3.3 *FFT amplitude dependences on probe photon energy*

Similar to the discussion in Section 4.2, the spectrum ($\delta\Delta A(\omega)$) of the difference absorbance is given by

$$\delta\Delta A(\omega) = \Delta A_{osc}(\omega) - \overline{\Delta A_{osc}(\omega)}$$
$$= \left(\frac{\delta(\mu^2)}{\mu^2}\Delta A(\omega) + \delta\omega\frac{d\Delta A(\omega)}{d\omega} + \delta\Delta\omega\frac{d^2\Delta A(\omega)}{d\omega^2} \right)\cos(\Omega t + \phi) \tag{4.9}$$

In the present study, we found that fitting only with the zeroth and first derivatives do not reproduce the amplitude spectra of many modes with sufficient accuracy. Hence, we analyze the amplitude spectra of many modes using the contributions of the zeroth, first, and second derivatives of the absorption spectrum of the ground state of J-aggregates.

Figure 4.13 shows the Fourier-amplitude spectrum of the oscillating component (black line) with the corresponding results (blue line) of fitting Eq. 4.9 to the 280 cm^{-1} mode. Table 3 lists the fitting parameters for several different vibrational modes. This fitting was performed using the ground-state absorption spectrum. Since a contribution from the wave-packet motion may modulate the excited-state absorption, we also fitted the probe photon energy dependence of the Fourier amplitudes using the zeroth, first, and second derivatives of the excited-state absorption spectrum and stimulated emission spectrum ($A(\omega)$) of J-aggregates. However, the fitting was very poor in this case. Therefore, modulation of the induced absorption and stimulated emission spectra is considered to make only small contributions to the analyzed spectral region of 2.13 to 2.20 eV.

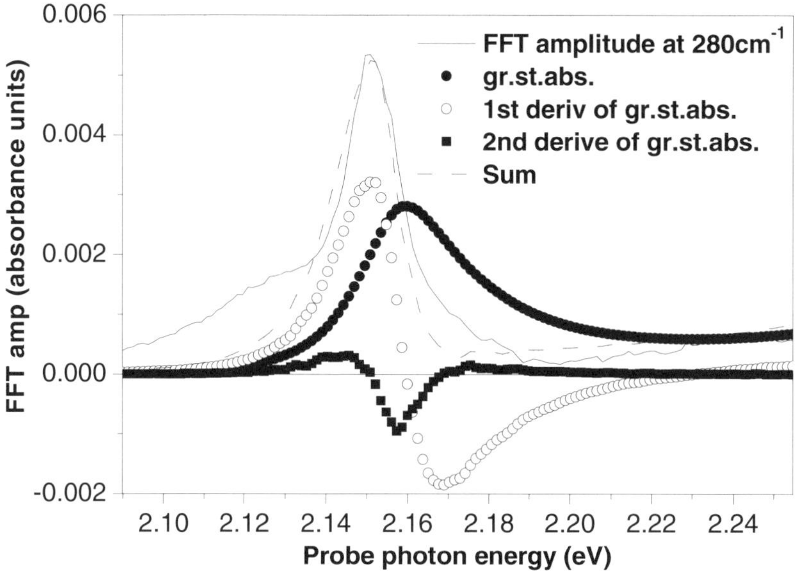

Figure 4.13. Fourier-amplitude spectrum of the oscillating component (solid line) with corresponding fitting results (dashed line) using Eq. 4.9 for 280cm^{-1} mode. The fitting curve is the sum of the ground state absorption spectrum (filled circle), its first derivative (open circle), and second derivative (square).

Table 4.3. Differences in the transition dipole moment, transition energy, and vibrational frequency between ground and exciton state and the initial phase of the molecular vibration.

Mode (cm^{-1})	$\|\delta\mu/\mu\|$ ($\times 10^{-3}$)	$\|\delta\omega\|$ (meV)	$\|\delta\omega/\omega_0\|$ ($\times 10^{-5}$)	$\delta\Delta\omega_v$ (meV)	$\|\delta\Delta\omega_v/\omega_v\|$ ($\times 10^{-3}$)	φ(rad)	Function
280	0.60	0.027	1.3	-0.21	6.1	-0.04π	+cos
428	0.19	0.008	0.37	-0.13	2.4	0.17π	+cos
509	0.38	0.016	0.73	-0.13	2.0	-0.33π	-sin
602	0.66	0.026	1.2	0.22	2.9	0.9π	-cos
838	0.08	0.003	0.15	-0.09	8.4	-0.55π	-sin

Although the fitting of the probe photon energy dependence of the vibrational amplitudes is not perfect, fitting of the probe photon energy dependence in the range 2.13 to 2.17 eV is relatively good. The difference in the dipole moment due to the non-Condon effect, the stabilization energy relevant to the Huang–Rhys factor, and the frequency change due to the change in the curvature of the potential curves between the ground and n=1 exciton states can be calculated from the values of the contributions of the zeroth, first, and second derivatives; Table 4.3 lists the obtained values. The value associated with the 280 cm^{-1} mode is quite different from the one given in Section 4.2. The reason for this difference is given as follows.

Different excitation conditions in terms of the excitation pulse spectra and duration will result in a different probe wavelength distribution of the vibrational amplitudes. In the present study, we selectively excited only the strongest sharp electronic transition, unlike in previous studies. Consequently, we observed some new vibrational modes (428, 509, 838, 1167, and 1543 cm^{-1}) that were not clearly visible in Section 4.2. In addition, the vibronic coupling strength may differ for the common modes observed in both experiments (e.g., the 280 cm^{-1} mode). In the present study, the coupling strength appears to be weaker than in the previous study because the oscillatory component is weaker. The FT amplitude is determined by the strength of the oscillatory component of the real-time traces, so the two experiments give quite different FT amplitudes. In addition, we only consider the contribution of the zeroth and first derivative in the previous study. We obtained the best estimation of 0.005 for the 280 cm^{-1} mode. We recently found that the second derivative also makes a significant contribution. This is another important finding of the present study, namely that the second derivative is important even though it is not generally considered (usually only first derivative is considered). As shown in Table 4.3, the first derivative gave the smallest contribution, which is consistent with our previous results. However, the contribution of the second derivative was of the same order as that of the zeroth derivative. The similarity between the zeroth and second derivatives near the absorption peak results in the zeroth derivative being over estimated in Section 4.2; its physical meaning is expressed in terms of the non-Condon effect. This is the reason for the discrepancy between the values obtained in two experiments for the 280 cm^{-1} mode (i.e., 0.005 and 0.0006). They are not identical due to the different experimental conditions (e.g., different excitation pulse durations and laser spectra).

4.4 *Conclusion*

Real-time coherent oscillations due to the vibrational modes of PIC J-aggregates pumped by sub-5-fs pulses were observed. We directly observed intermolecular vibrations in the lattice phonon coupled to the excitonic transition in J-aggregates. Eleven vibrational modes were extracted from the 2D FFT spectra of the real-time data. We detected the wavepacket motion both in the ground- and 1-exciton states separately according to their probe-photon energy. We used non-Condon-type exciton–phonon coupling in J-aggregates to explain the oscillatory signals and evaluated $\delta\mu/\mu$ for the vibrational modes at 228 and 282 cm^{-1} to be about 0.3 and 0.5%, respectively. In addition, the influences of the excitation pulse duration and the laser spectrum were discussed. We compared the Fourier power spectra of the real-time spectra observed with excitation pulse durations of 4.7 and 8.8 fs. The Fourier power spectrum was found to clearly depend on the pump laser spectrum. The intensity of the power spectrum was found to be proportional to the product integral $\int S'(\omega)\times S'(\omega+\omega_v)d\omega$ of the absorbed spectrum $S'(\omega)$ and its spectrum shifted by the vibrational mode frequency $S'(\omega+\omega_v)$.

Acknowledgements

The authors would like to thank Drs. Hideaki Kano and Kumiko Nishimura, and Mr. Tomohiro Hanamoto for their collaboration with the experiments described here. This work was supported by the Core Research for Evolutional Science and Technology (CREST) program and the International Cooperative Research Project (ICORP) program of the Japan Science and Technology Agency (JST), the National Science Council of the Republic of China, Taiwan (NSC 98-2112-M-009-001-MY3), and a grant from the Ministry of Education, Aiming for Top University (MOE ATU) Program at National Chiao-Tung University (NCTU). A part of this work was performed under the joint research project of the Institute of Laser Engineering, Osaka University under Contract No. B1-27.

References

1. J-agregates, ed. T. Kobayashi (World Scientific, Singapore, 1996).
2. E. E. Jelly, Nature 138 (1936) 1009.
3. E. E. Jelly, Nature 139 (1937) 631.
4. G. Scheibe, Angew. Chem. 50 (1937) 212.
5. H. von Berlepsch, and C. Böttcher, J. Phys. Chem. B 106 (2002) 3146.
6. M. Kasha, Radiation Res. 20 (1963) 55.
7. M. Kasha, H. R. Rawls, M. A. El-Bayoumi, Pure and Applied Chemistry: IUPAC,

Butterworths: London, (1965) 11, p371.
8. J. Franck, and E. Teller, J. Chem. Phys. 6 (1938) 861.
9. P. O. J. Scherer, and S. F. Fischer, Chem. Phys. 86 (1984) 269.
10. E. W. Knapp, Chem. Phys. 85 (1984) 73.
11. P. O. J. Scherer, E. W. Knapp, and S. F. Fischer, Chem. Phys. Lett. 106 (1984) 191.
12. E. W. Knapp, P. O. J. Scherer, and S. F. Fischer, Chem. Phys. Lett. 111(1984) 481.
13. F. C. Spano, J. R. Kuklinski, and S. Mukamel, Phys. Rev. Lett. 65 (1990) 211.
14. H. Fidder, J. Knoester, and D. A. Wiersma, J. Chem. Phys. 98 (1993) 6564.
15. K. Minoshima, M. Taiji, K. Misawa, and T. Kobayashi, Chem. Phys. Lett. 218 (1994) 67.
16. A. Tilgner, H.P Trommsdorff, J.M. Zeigler, and R.M. Hochstrasser, J. Chem. Phys. 96, (1992) 781.
17. S. de Boer, K.J. Vink and D.A. Wiersma, Chem. Phys. Lett. 137 (1987) 99.
18. M. A. Drobizheva, M. N. Sapozhnikova, I. G. Scheblykina, O. P. Varnavskya, M. Van der Auweraerb and A. G. Vitukhnovskya, Chem. Phys. 211 (1996) 455.
19. J. Moll, S. Daehne, J. R. Durrant, and D. A. Wiersma, J. Chem. Phys. 102 (1995) 6362.
20. E. Rousseau, M. Van der Auweraer, and F. C. De Schryver, Langmuir 16 (2000) 8865.
21. H. Kano, and T. Kobayashi, J. Chem. Phys. 116 (2002) 184.
22. S. Verma, A. Ghosh, A. Das, and H. N. Ghosh, J. Phys. Chem. B 114 (2010) 8327.
23. J. H. Lee, C. K Min, and T. Joo, J. Chem. Phys. 114 (2001) 377.
24. Springer Proceedings in Physics 4, Time-Resolved Vibrational Spectroscopy, edited by A. Laubereau and M. Stockburger, (Springer-Verlag, Berlin Heidelberg New York Tokyo, 1985).
25. Springer Proceedings in Physics 68, Time-Resolved Vibrational Spectroscopy V, edited by H. Takahashi, (Springer-Verlag, 1991).
26. Springer Proceedings in Physics 74, Time-Resolved Vibrational Spectroscopy VI, edited by A. Lau, F. Siebert, and W. Werncke, (Springer-Verlag, 1993).
27. A. Cupane, M. Leone, E. Unger, C. Lemke, M. Beck, W. Dreydrodt, and R. Schweitzer-Stenner, J. Phys.Chem. B. 102 (1998) 6612.
28. H. van Amerongen, L. Valkunas, and R. van Grondelle, Photosynthetic excitons (World Scientific, Singapore, 2000).
29. M. Y. Okamura, G. Feher, and N. Nelson, *Photosynthesis*, Govindjee, Ed., Academic Press: New York, 1982; pp 195-272.
30. K. Misawa, H. Ono, K. Minoshima, and T. Kobayashi, Appl. Phys. Lett. 63 (1993) 577.
31. J. M. Ribo, J. Crusats, J. A. Farrera, and M. L. Valero, J. Chem. Soc., Chem. Commun. (1994) 681.
32. D. L. Akins, S. Ozcelik, H. R. Zhu, and C. Guo, J. Phys. Chem. 100 (1996) 14390.
33. N. C. Maiti, S. Mazumdar, and N. J. Periasamy, J. Phys. Chem. 102 (1998) 1528.
34. C. G. Granqvist, O. Hunderi, Phys. Rev. B 16 (1977) 3513.

35. B. Kopainsky, W. Kaiser, Chem. Phys. Lett. 88 (1982) 357.
36. B. Kopainsky. J. K. Halllmeier, W. Kaiser, Chem. Phys. Lett. 87 (1982) 7.
37. Y. Wang, Chem. Phys. Lett. 126 (1986) 209.
38. V. Sundström, T.Gillbro, R. A. Gadonas, A. Piskarskas, J. Chem. Phys. 89 (1988) 2754.
39. F. C. Spano, S. Mukamel, Phys. Rev. A 40 (1989) 5783.
40. H. Fidder, J. Terpstra, D. Wiersma, J. Chem. Phys. 94 (1991) 6895.
41. J. Knoester, J. Chem. Phys. 99 (1993) 8466.
42. S. Kobayashi, F. Sasaki, Nonlinear Opt. 4 (1993) 305.
43. K. Misawa, K. Minoshima, T. Kobayashi, J. Raman Spectrosc. 26 (1995) 553
44. V.F. Kamalov, I.A. Struganova, and K. Yoshihara, J. Phys. Chem. 100 (1996) 8640.
45. N. Fukutake, T. Kobayashi, Chem. Phys. Lett. 356 (2002) 368
46. A. Shirakawa, and T. Kobayashi, Appl. Phys. Lett. 72 (1998) 147.
47. A. Shirakawa, I. Sakane, and T. Kobayashi, Opt. Lett. 23 (1998) 1292.
48. A. Baltuška, T. Fuji, and T. Kobayashi. Opt. Lett. 27 (2002) 306.
49. O. Ohno, Y. Kaizu, H. Kobayashi, J. Chem. Phys. 99 (1993) 4128.
50. B. Ren, Z. Q. Tian, C. Guo, D. L. Akins, Chem. Phys. Lett. 328 (2000) 17.
51. G. A. Shick, M. R. O'Grady, R. K. Tiwari, J. Chem. Phys. 97 (1993) 1339.
52. R. Pantell,; F. Pradere, J. Hanus, M. Schott, H. Pathoff, J. Chem. Phys. 46 (1967) 3507.
53. K. Minoshima, M. Taiji, K. Misawa, T. Kobayashi, Chem. Phys. Lett. 218 (1994) 67.
54. M. Burgel, d. Wiersma, and K. Duppen, J. Chem. Phys. 102 (1995) 20.
55. M. Perrin, M. Gouterman, C. L. Perrin, J. Chem. Phys. 50 (1969) 4137.
56. D. Dolphin, Ed. *The Porphyrins, volume III, Physical Chemistry, Part A*; Academic Press: New York, 1979.
57. D.L. Akins and J.W. Macklin, J. Phys. Chem. 93 (1989) 5999.
58. D.L. Akins, Y.H. Zhuang, H.-R. Zyu, and J.Q. Liu, J. Phys. Chem. 98 (1994) 1068.
59. C. Guo, M. Aydin, H.-R. Zyu, and D.L. Akins, J. Phys. Chem. B 106 (2002) 5447.
60. N. Ishii, E. Tokunaga, S. Adachi, T. Kimura, H. Matsuda, and T. Kobayashi, Phys. Rev. A 70, (2004) 023811.
61. T. Kobayashi and Z. Wang, IEEE J. Quantum Electron. 44 (2008) 1232.
62. T. Kobayashi, Mol. Cryst. Liq. Cryst. 314 (1998) 1.

CHAPTER 2

VIBRONIC COUPLING IN J-AGGREGATES

Frank C. Spano

Department of Chemistry, Temple University
Philadelphia, PA USA
E-mail: spano@temple.edu

Vibronic coupling in molecular aggregates leads to additional signatures for identifying J- and H-aggregates. In most π-conjugated molecules nuclear relaxation subsequent to electronic excitation occurs mainly along a symmetric vinyl-stretching mode with a frequency of 0.17 eV/$\hbar$. Analysis of the impact of intermolecular interactions on the mode progressions in absorption and photoluminescence leads to additional information about the nature of the excited states. In J-aggregates the ratio of the lowest two vibronic peak intensities in the absorption spectrum increases with exciton bandwidth W providing a direct measure of the latter. In addition, the ratio of the 0-0 and 0-1 line strengths in the photoluminescence spectrum provides a direct measure of the exciton coherence number, N_{coh}. N_{coh}. increases with W, and decreases with disorder and temperature.

Keywords: J-aggregate, exciton-vibrational coupling, polaron, exciton coherence

1. Introduction

The nature of electronic excitations in aggregates, films and crystals comprised of π-conjugated chromophores continues to be an area of significant current interest driven mainly by promising commercial applications including light-emitting diodes and solar cells.[1–4] Perhaps the most extensively studied collective organic systems are the so-called J-aggregates, which are typically composed of cationic dye molecules like pseudoisocyanine chloride (PIC). The formation of an intense red-shifted main absorption peak with aggregation[5,6] is a clear signature of exciton formation.[7,8] Although such aggregates were discovered many decades ago,[5,6] interest in their photo physical and transport properties has hardly waned, primarily because J-aggregates are ideal models with which to study excitons of reduced dimensionality.

The purpose of this contribution is to study in detail the impact of vibronic coupling on the absorption and emission spectra of J-aggregates. Unlike their inorganic counterparts, organic assemblies like J-aggregates are soft in the sense that energy and charge transport are accompanied by significant intramolecular nuclear rearrangements. Optical excitations are therefore composite particles involving electronic and vibrational degrees of freedom.[9–11] The intramolecular mode mainly responsible for geometric relaxation following electronic excitation in a great many conjugated chromophores like PIC involves a symmetric vinyl stretching mode (or closely-spaced cluster of modes) with energy $\approx$ 1400 cm^{-1}. Elongation along the mode coordinate subsequent to $S_0 \rightarrow S_1$ optical excitation is responsible for a nuclear reorganization energy of $\approx$ 0.2 eV. A seminal paper treating the impact of vibronic coupling on the absorption spectra of J-aggregates was presented by Scherer and Fischer[12] who accurately reproduced the measured PIC aggregate spectrum using variational theory as well as a full numerical analysis of the Hamiltonian matrix. More recently, several theoretical approaches have been advanced to better understand the impact of exciton-vibrational coupling involving the 1400 cm^{-1} mode on the optical response in various organic aggregates.[13–22] By appreciating how the Franck-Condon (FC) progression is distorted in going from the isolated molecule to a molecular assembly one can derive important information about molecular packing, the exciton bandwidth, the nature of disorder and the exciton coherence length. In addition, one can probe the structure of the fundamental excitations responsible for absorption and emission. Particularly useful in this respect is the multi-particle basis set originally introduced by Philpott.[23] The basis set allows one to obtain essentially exact steady-state absorption and emission spectral profiles using a Holstein-like Hamiltonian[11] to represent exciton-vibrational coupling.

From a purely electronic perspective the impact of molecular aggregation on the optical response was worked out early on by Kasha.[7,8] When the sign of the resonant electronic (excitonic) coupling is negative,[24] as for a pair of rod-shaped molecules in a “head-to-tail” orientation (with their transition dipole moments oriented along the long molecular axes), the main absorption peak shifts to lower energies and fluorescence is enhanced.[25,26] These are J-aggregates. Conversely, in H-aggregates the couplings are positive, resulting in a spectral blue-shift. H-aggregates occur when "side-by-side" orientations dominate and are subradiant at low temperatures.[27,28]

When vibronic coupling is included the FC progressions characterizing absorption and emission are affected differently by J- and H- aggregation. The contrasting ways in which the progressions are distorted away from a Poissonian distribution in these extreme aggregate types is the subject of a recent review article.[29] In this chapter we focus mainly on J-aggregates. Although generally recognized as systems in which vibronic coupling is not a dominant factor (witness the negligible Stokes shift between absorption and emission in many J-aggregates) subtle effects do arise and can be exploited to obtain a great deal of information about the excitons involved in absorption and emission. For example, the main (0-0) peak characterizing the photoluminescence (PL) spectrum of J-aggregates is accompanied by its first side-band (0-1), even in the limit of strong excitonic coupling.[30] The ratio of the 0-0 and 0-1 line strengths provides a direct means for determining the coherence length of the emitting exciton, even in the presence of static and dynamic disorder.[30]

Emission characteristic of J-aggregates can be found in more complex morphologies such as the herringbone packing arrangements found in polyacene crystals[31,32] and oligothiophene and oligophenylene vinylene crystals.[27,28,33] In such systems oscillator strength is distributed over the bottom and top of the exciton band, so they are not purely H- or J-aggregates. Nevertheless, the *b*-polarized and *ac*-polarized components of the PL spectrum can be understood in terms of J- and H-aggregate emission, respectively.[29,30]

In the next section we will present an analysis of the vibronic excitations in molecular aggregates based on a Holstein-like Hamiltonian. Absorption and PL spectra for disorder-free J-aggregates at T = 0K are considered in Section 3 and the effects of diagonal disorder and increasing temperature are taken up in Section 4. The penultimate Section deals with the exciton coherence size and its relationship to the PL spectrum. Our findings are summarized in the final section.

2. Excitonic Polarons

Electronic excitations in J-aggregates are generally classified as Frenkel excitons (or excitonic polarons), comprised of a vibronically excited central molecule surrounded by vibrationally – but not electronically – excited neighbors. The latter are geometrically distorted, due to elongation along one (or more) symmetric vibrational coordinate(s). The situation is analogous to a bowling ball on a mattress as shown in Fig. 2.1. The ball and the spring directly underneath correspond to the vibronically excited molecule while the neighboring springs correspond to the surrounding vibrationally excited molecules. As the bowling ball traverses the mattress the compression field of the underlying springs travel with it, in exact analogy to the Frenkel polaron considered here (with the exception that excited molecules are elongated and not compressed along the vibrational coordinate.) Delocalization loosely corresponds to the range over which the ball travels.

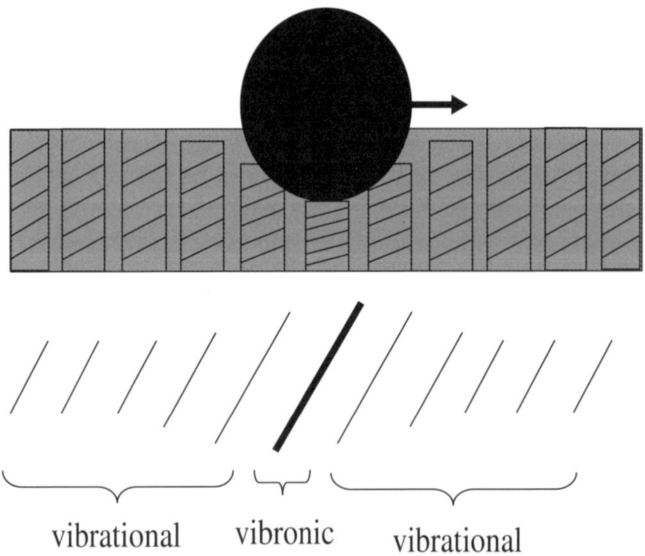

Figure 2.1. Analogy of the motion of a Frenkel polaron in a linear aggregate with a bowling ball on a mattress. Sticks represent rod-shaped molecular chromophores. Stick length represents symmetric mode elongation.

In order to account quantitatively for exciton-vibrational coupling we have assumed a simple model in which the nuclear potentials corresponding to the ground (S_0) and electronically excited (S_1) molecular states are shifted harmonic wells of identical frequency, ω_0. The shift is quantified by the Huang-Rhys (HR) factor, λ^2, which is approximately unity for the 1400 cm^{-1} mode in most organic chromophores. To describe the collective excitations we employ the multi-particle basis set[14,23] as depicted in Fig. 2.2. A single-particle excitation, $|n,\tilde{v}>$, consists of a vibronically excited chromophore at site n with $\tilde{v}$ excited state quanta in the S_1 nuclear potential, with all other molecules electronically and vibrationally unexcited. A vibronic /vibrational *pair* excitation, denoted $|n,\tilde{v};n',v'>$, is a two-particle state. In addition to a vibronic excitation at n, this state includes a *vibrational* excitation at $n'(\neq n)$ with $v'(\geq 1)$ quanta in the S_0 potential, as demonstrated schematically in Fig. 2.2. Such states are necessary for describing the spatial extent of the vibrational distortion field surrounding the central vibronic excitation, characterized by the polaron radius. Within the two-particle approximation (TPA) the αth eigenstate of the aggregate Hamiltonian can be expanded in one- and two-particle states;

$$|\psi^{(\alpha)}>=\sum_{n,\tilde{v}} c_{n,\tilde{v}}^{(\alpha)}|n,\tilde{v}>+\sum_{n,\tilde{v}}\sum_{n',v'} c_{n,\tilde{v};n'v'}^{(\alpha)}|n,\tilde{v};n',v'>. \quad (2.1)$$

The above expansion is highly accurate for obtaining the optical response in organic assemblies where the singlet exciton bandwidth ranges from 0 to 1 eV and where the nuclear relaxation energy, $\lambda^2\omega_0$, is approximately 0.2 eV. In this parameter range three-particle contributions have negligible effect on absorption and emission.[14,34] Hence, we can obtain an essentially exact solution to the polaron's optical response for relatively large aggregates.

The optical response from the excitons in (2.1) derives from the following basic properties: (i) Only one-particle states are optically allowed from the vibrationless ground state. Two-particle states are forbidden via one- (and two-) photon excitation. Hence, only one-particle states contribute to 0-0 emission. (ii) Both one- and two-particle states radiatively couple to the electronic ground state with one or more vibrational excitations. Thus, side-band emission involves both one- and two-particle states. (iii) When disorder is absent and periodic boundary conditions apply, the exciton wave number k becomes a good quantum

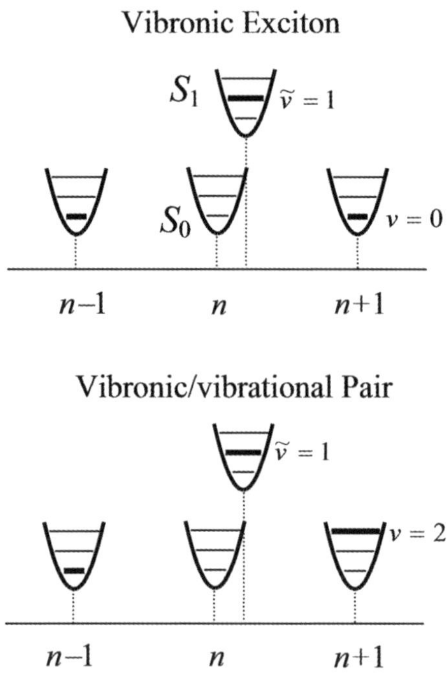

Figure 2.2: Examples of the fundamental excitations in ordered organic assemblies. Thick lines represent occupied levels. The vibronic (single-particle) excitation is $|\, \boldsymbol{n}, \tilde{\boldsymbol{v}} = 1 >$ while the vibronic/vibrational pair (two-particle state) is $|\, \boldsymbol{n}, \tilde{\boldsymbol{v}} = 1; \boldsymbol{n}+1, \boldsymbol{v} = 2 >$. Reprinted with permission from Ref. 29. Copyright 2011 American Chemical Society.

number. Only nodeless excitons with $k = 0$ can be optically excited from the vibrationless ground state. In J- aggregates the $k = 0$ exciton resides at the band bottom while the dark $k = \pi$ exciton with N-1 nodes resides at the top of the band. The order is reversed in H-aggregates.

The coefficients involved in the expansion (2.1) can be evaluated by diagonalizing the aggregate Hamiltonian H. The diagonal elements of H consist of the energies of the localized states: taking $\hbar = 1$ we have $\omega_{0-0} + D + \tilde{v}\omega_0 + \Delta_n$ for the energy of the one-particle state, $\left|n, \tilde{v}\right\rangle$ and $\omega_{0-0} + D + (\tilde{v} + v)\omega_0 + \Delta_n$ for the energy of the two-particle state, $\left|n, \tilde{v}; n', v'\right\rangle$. Here, $\omega_{0\text{-}0}$ is the gas-phase 0-0 molecular transition energy corresponding to the lowest optically allowed transition (normally, $S_0 \rightarrow S_1$), D is the gas-to-crystal shift due to nonresonant intermolecular interactions, Δ_n represents a disorder-induced change in the transition energy at site n, and ω_0 is the aforementioned energy of the symmetric intramolecular vibration. The off-diagonal elements of H are represented by the usual excitonic Hamiltonian,

$$H_{ex} = \sum_{m,n} J_{mn} |m\rangle\langle n| \tag{2.2}$$

where $|n\rangle$ represents a pure (S_1) electronic excitation at site n (with all other molecules electronically unexcited) and J_{mn} is the excitonic coupling between the mth and nth molecules. The off-diagonal matrix elements of H connecting one-particle states allow for resonant energy transfer in the conventional Frenkel exciton theory. However, H_{ex} also allows for resonant transfer amongst two-particle states, as well as the coupling between one- and two-particle states (and generally, between n- and (n+1)-particle states).[14,23,28] Because of the latter, the one-photon oscillator strength, which is confined entirely to one-particle excitations, becomes increasingly "borrowed" by two-particle states as the excitonic coupling increases and the wave functions take on an increasing mixed one- and two-particle character. All of the aforementioned matrix elements of H_{ex} necessarily involve vibrational overlap integrals which depend on the HR factor.[16] The Hamiltonian as described above is exactly equivalent to the Holstein Hamiltonian[11] when the latter is represented in the one- and two-particle basis. Finally, in all that follows we invoke periodic boundary conditions.

3. Absorption and Emission in Disorder-free J-aggregates

In what follows we consider a linear array of N chromophores (see Fig. 2.1) with nearest neighbor coupling ((J_0<0) only and periodic boundary conditions. All molecules are oriented in a common direction so that the transition dipole moments are all aligned. The J-aggregates are *ideal* in the sense of containing a single molecule per unit cell with no disorder present (Δ_n = 0). We further assume the temperature to be low enough to effectively discount thermally excited emission. Finally, the number of chromophores N is chosen to be large enough (N = 20) to minimize finite-size effects.

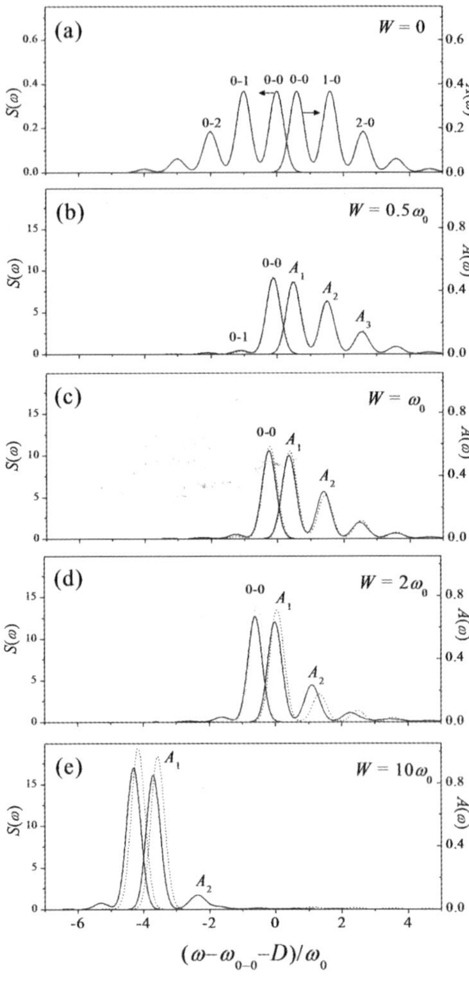

Figure 3.1. Calculated absorption and photoluminescence spectra for linear J- aggregates containing $N = 20$ molecules with nearest-neighbor-only coupling $J_0 < 0$. The line width, $\sigma_{hom} = 0.3\omega_0$, is taken for all spectra. In all panels $\lambda^2 = 1$ and $\omega_0 = 0.17$ eV (1400 cm^{-1}.) The exciton bandwidth $W = 4|J_0|$ increases from top to bottom. Solid spectra are evaluated using one- and two-particle states while dotted spectra use only one-particle states. The 0-0 PL band and A_1 absorption band overlap (no Stokes shift). To better distinguish them the absorption spectrum has been blue-shifted by $0.5\omega_0$. Note the change in vertical scale for emission beyond the non-interacting chromophore limit in (a).

We begin with absorption. Figure 3.1 shows calculated spectra for increasing exciton bandwidth, $W = 4|J_0|$, evaluated from the expression,

$$A(\omega) = \frac{1}{N\mu^2}\sum_{\alpha}|\langle G|\hat{\boldsymbol{M}}|\psi^{(\alpha)}\rangle|^2\,\Gamma(\omega-\omega^{(\alpha)}). \qquad (3.1)$$

$\hat{\boldsymbol{M}}$ is the aggregate electronic transition dipole moment operator,

$$\hat{\boldsymbol{M}} = \boldsymbol{\mu}\sum_{n}\{|g\rangle\langle n| + |n\rangle\langle g|\},$$

with |*g*> representing the pure electronic ground state of the aggregate (all molecules unexcited) and with $\boldsymbol{\mu}$ representing the $S_0 \rightarrow S_1$ molecular transition dipole moment. The sum in Eq. (3.1) runs over all dipole- allowed transitions from the vibrationless ground state $|G\rangle$ to the excitons $|\psi^{(\alpha)}\rangle$ with energies $\omega^{(\alpha)}$. $\Gamma(\omega)$ is the homogeneous line shape function, taken to be a peak-normalized Gaussian, $\Gamma(\omega) = \exp(-\omega^2/\sigma_{\text{hom}}^2)$, in Fig. 3.1. The parameters defining the Hamiltonian *H* are given in the Figure caption. We note that the spectra in Fig. 3.1 differ slightly from those published in Ref. [29] since here we are taking periodic boundary conditions as opposed to the open boundary conditions employed in Ref.[29] (Slightly different homogeneous line widths were also used).

The absorption spectra in Fig. 3.1 consist of vibronic progressions involving the symmetric vinyl stretching mode. For the isolated molecule in Fig. 3.1(a) the intensity of the $n \leftarrow 0$ transition ($n = 0,1,\ldots$) scales as the FC factor, $\lambda^{2n}\exp(-\lambda^2)/n!$. For the value $\lambda^2 = 1$ used in the figure, the 0-0 and 1-0 peak intensities are therefore equal. In the aggregate absorption spectra the peak labeled A_{n+1} correlates to the *n*-0 molecular transition. A_{n+1} is only slightly red-shifted from the *n*-0 molecular peak when the exciton coupling is relatively weak $(W < \lambda^2\omega_0)$ as in Fig. 3.1(b) and (c). (The gas-to-crystal red-shift |*D*|, is set to zero in the figure in order to focus on excitonic shifts.) Fig. 3.1 shows spectra calculated from only one-particle states (dotted) and more accurate spectra calculated using one- and two-particle states (solid). We have

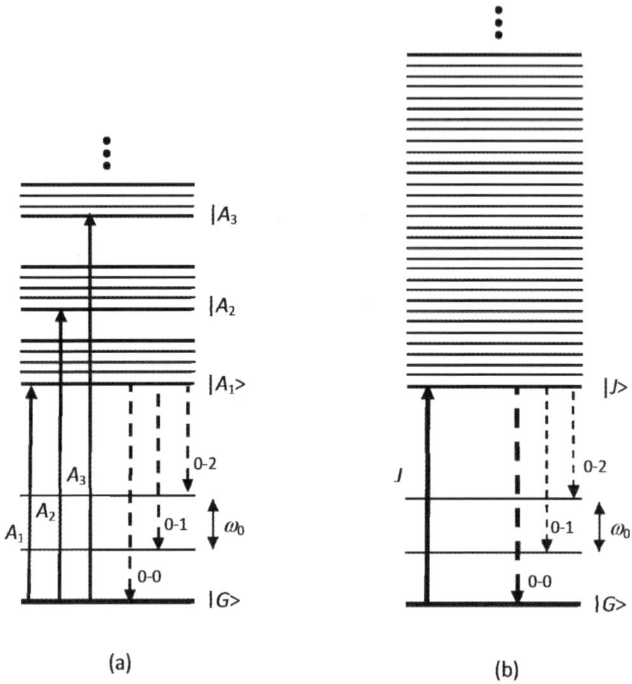

Figure 3.2: Approximate energy level diagrams corresponding to the weak (a) and strong (b) exciton coupling regimes in ideal J-aggregates. In (a) the states $|A_n\rangle$ are nodeless ($k = 0$) excitons while the upper level in each vibronic band is a $k = \pi$ exciton. Three and higher phonon states in the ground electronic state are not shown. Note that the bandgap is not to scale. Reprinted with permission from Ref. 29. Copyright 2011 American Chemical Society.

also calculated spectra with up to three-particle states but even for the highest W in Fig. 3.1, three-particle states contribute only small changes (<10%) to the spectrum.

As W increases A_1 increases (while A_3 decreases) relative to A_2. In addition, the vibronic peaks become irregularly spaced, with the separation between A_1 and A_2 slightly exceeding ω_0.[35,36] Such aggregation-induced spectral distortions also occur in H-aggregates, although the peak intensities show the opposite trend with increasing W.[29] The latter behavior has been observed in P3HT thin films[37–39] and loosely-packed carotenoid aggregates[36,40] such as the lutein diacetate H-aggregates.

In the weak excitonic coupling regime the inclusion of two-particle states has practically no effect on the spectral line shape (Figs.3.1 *(b)* and *(c)*) indicating that the states responsible for absorption remain localized (small radius) polarons. However, as W increases into the intermediate coupling regime, two-particle states cause significant deviations, indicating the involvement of larger radius polarons. The deviations arise from enhanced one- and two-particle mixing induced by the matrix elements of H_{ex} which connect one- and two-particle states. In the strong coupling regime (Fig. 3.1(e)) the absorption spectrum is dominated by an intense red-shifted peak. The effect is clearly observed in a variety of cationic dye (J-) aggregates[26] beginning with the seminal works of Scheibe,[5] Jelley[6] and Cooper[41] and is the primary signature of J-aggregation. Note, however that a higher-energy side band, captured correctly only at the two-particle level, is also evident. In the large N limit (not shown) we find the absorption spectral shape converges, with a dominant A_1 peak, and a less intense (but broader) phonon side-band.

The redistribution of oscillator strength towards the lowest energy A_1 peak in J-aggregates can be understood in the weak coupling regime from the energy level diagram in Fig.3.2(a). Treating H_{ex} as a perturbation, the zero-order excitations which absorb light are mainly delocalized single-particle excitons. The small admixture of two-particle excitations has little effect on absorption.[42] each exciton is characterized by a wave vector quantum number, k. To zero order, excitons are organized into vibronic bands with the $\tilde{v}$ th band ($\tilde{v} = 0,1,\ldots$) consisting of all single particle excitons with $\tilde{v}$ excited state vibrational quanta:

$$|k,\tilde{v}\rangle^{(0)} = N^{-1/2}\sum_n \exp(ikn)\,|n,\tilde{v}\rangle \tag{3.2}$$

Only the nodeless $k = 0$ states are optically allowed from the vibrationless ground state, $|G\rangle$.

In J-aggregates the $k = 0$ exciton in the vth vibronic band, $|A_{v+1}\rangle$, resides at the bottom of the band – see Fig. 3.2 *a*. The oscillator strength redistribution observed in Fig.3.2*(b),(c)* is primarily a result of first-order interband coupling between the (zeroth-order) $k = 0$ vibronic excitons in *different* bands (see Eq. (4)), and to a much lesser extent on the coupling between the single- and two-particle excitations as detailed in Refs.15 and 35. The result is an increasing value of R_{abs} with exciton bandwidth

(W) in J-aggregates, where $R_{abs} \equiv I_{A_1}/I_{A_2}$ is the ratio of the oscillator strengths in the A_1 and A_2 bands. In the weak coupling regime R_{abs} is given by,[15,35,36]

$$R_{abs} = \frac{(1-0.48\tilde{J}_{k=0}/\omega_0)^2}{(1+0.146\tilde{J}_{k=0}/\omega_0)^2} \quad \lambda^2=1,\ W << \lambda^2\omega_0 \tag{3.3}$$

where the excitonic shift of the k=0 exciton is generally given by,

$$\tilde{J}_{k=0} = 2\sum_{s>0} J_{n,n+s}. \tag{3.4}$$

Here $J_{n,n+s}$ is the coupling between a chromophore at n and one at $n+s$. Accordingly, the exciton bandwidth is $W = |\tilde{J}_{k=0} - \tilde{J}_{k=\pi}|$. In the special case of nearest-neighbor coupling only (J_0) we have simply, $\tilde{J}_{k=0} = 2J_0$ and $W = 2|\tilde{J}_{k=0}|$. Eq.(3.3) shows that R_{abs} increases with W in J-aggregates since $\tilde{J}_{k=0} < 0$. Interestingly, the effect is exactly opposite in H-aggregates $(\tilde{J}_{k=0} > 0)$, where R_{abs} decreases with increasing W.[29] Hence, a comparison of R_{abs} between isolated and aggregated molecules provides a reliable test for H- or J- type aggregation.[36] Moreover, R_{abs} allows one to measure the k = 0 interaction sum, which is especially valuable in systems which lack Davydov splitting.

It is quite interesting that perhaps the most celebrated J-aggregates, those formed from PIC in solution, show H-aggregate behavior in the early dimerization stages.[43] One clearly sees from the absorption spectrum in Ref. 43 a marked decrease in R_{abs} upon dimerization relative to the monomer in solution. (However, in the large aggregate phase the packing geometry changes leading to J-aggregates. Here, the A_1 peak completely dominates and R_{abs} is much greater than the monomer value.) In H-aggregates, the attenuation of R_{abs} has been used to monitor self-assembly in perylene diimide aggregates;[44,45] elucidate the packing arrangement in carotenoid aggregates;[36] and extract the exciton bandwidth and average conjugation lengths in P3HT π-stacks.[15,35,38] The ratio has also been shown to correlate crystal quality with charge mobilities in P3HT thin film transistors.[37,46]

Figure 3.1 shows that as W increases into the intermediate and strong excitonic coupling regimes the main absorption peak continues to red-shift. In the strong exciton coupling regime ($W >> \lambda^2\omega_0$) oscillator strength is mainly concentrated in a single (A_1) peak due to absorption by a nearly free $k = 0$ exciton with energy $\approx \omega_{0-0} + D + \lambda^2\omega_0 + 2J_0$. This exciton, labeled as $|J\rangle$ in Fig. 3.2(b), is created with virtually no change in the ground state nuclear coordinates — there is essentially no vibronic relaxation subsequent to the vertical excitation because the excitation resonantly jumps to a neighbor before nuclear relaxation can occur. In this regime the wave functions are approximately B.O. products of free excitons and phonons,[7,47] where the latter are delocalized 1400 cm^{-1} vibrational excitations defined by a wave vector q. Hence, the approximate (zero-order) wave functions are,

$$|J\rangle^{(0)} = \frac{1}{\sqrt{N}}\sum_n |n\rangle \otimes |vac\rangle \qquad W >> \lambda^2\omega_0 \tag{3.5}$$

where $|vac\rangle$ indicates the vacuum state for all phonons.

We now turn to the photoluminescence spectrum at low temperature. At $T = 0$ K aggregate emission proceeds from the lowest excited state $|em\rangle$ with transition energy ω_{em} in accordance with Kasha's rule, which assumes that the exciton lifetime is sufficiently long compared to the inverse spectral diffusion (relaxation) rate induced primarily by exciton — lattice phonon scattering. At 0K the "reduced" emission profile takes the form of a vibronic progression,

$$S(\omega) = \sum_{v_t=0,1,2,\ldots} I^{0-v_t}\Gamma(\omega - \omega_{em} + v_t\omega_0). \tag{3.6}$$

The spectrum in Eq. (3.6) is simplified by excluding the cubic frequency dependence as well as the frequency-dependent index of refraction in order to focus on the impact of aggregation on the dimensionless $0 \rightarrow v_t$ line strengths,

$$I^{0-v_t} = \frac{1}{\mu^2}\sum_{T(v_t)} |< em|\hat{\boldsymbol{M}}|T(v_t)>|^2 \tag{3.7}$$

The terminal states in the $0 - v_t$ transition, $|T(v_t)\rangle$, form a degenerate set of electronic ground states, in which a total of v_t vibrational quanta are distributed over N molecules. Since $\Gamma(0) = 1$, the peak vibronic intensities in $S(\omega)$ directly reflect the line strengths. In the limit of noninteracting molecules the line strengths are the single-molecule FC factors, $I^{0-v_t} = \lambda^{2v_t} \exp(-\lambda^2)/v_t!$.

Since the absorption origin is also the emission origin in J-aggregates (see Fig.3.2) there is no Stokes shift and the emission band red-shifts along with the main absorption band with increasing W — see Fig. 3.1*(b)*-*(e)*. The emitting exciton has wave vector $k = 0$ making the 0-0 emission allowed and enhanced by a factor of N compared to an isolated molecule. The dimensionless intensity of the 0-0 peak is given by,

$$I^{0-0} = \frac{1}{\mu^2} |\langle em | \hat{\boldsymbol{M}} | G \rangle|^2, \tag{3.8}$$

which reduces to the simple form,[30]

$$I^{0-0} = FN \tag{3.9}$$

where F is a generalized FC factor ranging from $\exp(-\lambda^2)$ in the weak exciton coupling limit, to approximately unity in the strong exciton coupling limit where $|em\rangle$ approaches the nodeless wave function in Eq.(3.5). The N-fold enhancement arises from a concentration of oscillator strength in the $|em\rangle \rightarrow |G\rangle$ transition and leads to superradiant decay rates.[14,25,33,47,48] More generally, 0-0 emission in J-aggregates is enhanced by the coherence number, N_{coh} ($<N$) which is the number of molecules over which the wave function is coherently spread, as determined by static and dynamic disorder (see Section 5).

The 0-1 peak, observed as a small shoulder in Fig.3.1, *increases* with increasing exciton bandwidth and arises from a *constructive* interference between the one- and two-particle coefficients of the emitting exciton. In the general case, the 0-1 intensity is given by[30]

$$I^{0-1} = F\lambda^2 \tag{3.10}$$

Hence, in the limit of very large W where the BO wave functions are accurate zero-order approximations the 0-1 intensity converges to the value of λ^2,[49] considerably larger than the single molecule value of $\lambda^2 e^{-\lambda^2}$. More importantly, the 0-1 intensity is *not* coherently enhanced by N (or, more generally, by N_{coh}). Figure 3.3 shows that despite a four-fold increase in N the 0-1 sideband intensity remains essentially constant.

The insensitivity to N is true for the higher replicas as well. (If extended excitonic couplings are included then W initially increases with N. This contributes a slight increase to *all* band intensities through the F factor in Eqs. (3.9) and (3.10). However, as W eventually converges, the sidebands stabilize with N, in contrast to the 0-0 intensity which continues to increase with N according to Equation (3.9). Equation (3.9) breaks down only when the aggregate length approaches an optical wave length.)

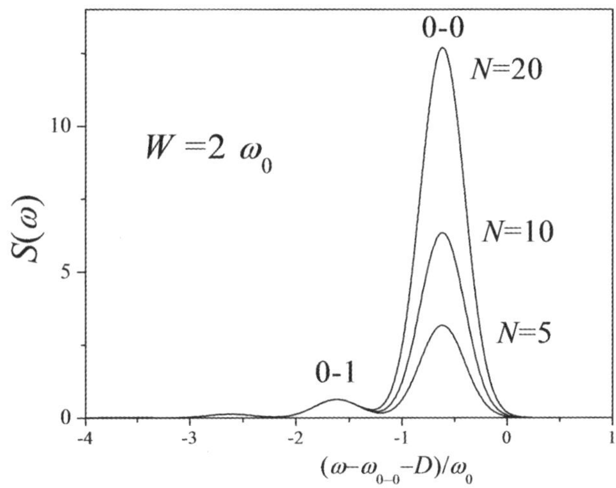

Figure 3.3: Calculated emission spectrum for various size linear J-aggregates with nearest-neighbor coupling, $J_0 = -0.5\omega_0$ and with HR factor, λ^2=1. Periodic boundary conditions are assumed. The ratio of the 0-0 to 0-1 peak intensities is exactly N/λ^2.

Equations (9) and (10) lead to a very simple yet extremely useful result for the ratio of the 0-0 to 0-1 line strengths:

$$R_{em} \equiv I^{0-0} / I^{0-1} = N / \lambda^2 . \tag{3.11}$$

The relationship was shown to be exact in Ref. 30 and is valid over the entire range of exciton-phonon coupling, from the weak to strong coupling regimes. It also remains valid for extended excitonic couplings as long as periodic boundary conditions are maintained. The validity of Eq.(3.11) is evident in Fig. 3.3, where the ratio is exactly N (since $\lambda^2=1$) for all three cases shown. As we discuss further in Section 5 when disorder is included or the temperature is raised Eq.(3.11) maintains its simple form, but with N replaced by the coherence number.

The accuracy of Eq.(3.11) is more readily demonstrated using a basis set which includes (at least) two-particle states. For example, including only one-particle states in Fig. 3.1(d) results in $R_{em} \approx 70$, over three times greater than the value, $R_{em} \approx 20$, obtained using the two-particle approximation. The latter is in excellent agreement with the value of 20 from Eq. (3.11). The situation is more dramatic with increased exciton bandwidth: in Fig. 3.1(e), the ratio increases to approximately 700 under the single particle approximation, but to only 22.3 under the two-particle approximation.

Figure 3.1 shows that R_{em} jumps immediately to large values compared to the monomer when W is activated. This is due to our taking T = 0 K and assuming emission from only the lowest energy exciton. For small but physically attainable temperatures, emission can be considered as originating from the lowest energy exciton only when $4\pi^2 F\,|J_0|/N^2 >> kT$, i.e. when the energy separation between the two lowest excitons (with k = 0 and 1) exceeds kT. For aggregates at a nonzero temperatures, R_{em} will smoothly increase with increasing $|J_0|$ until the aforementioned inequality is satisfied. During this transition thermally activated excitons will also contribute to the emission. We consider such effects in the next section.

4. Static Disorder and Thermal Effects

In this Section we focus on the effect of site-energy disorder and increasing temperature on J-aggregate absorption and emission. The impact of disorder on the absorption spectral line width including motional narrowing has already been covered in great detail and will only be briefly dealt with here.[50,51] We also omit any discussion surrounding the temperature dependent homogeneous line width.

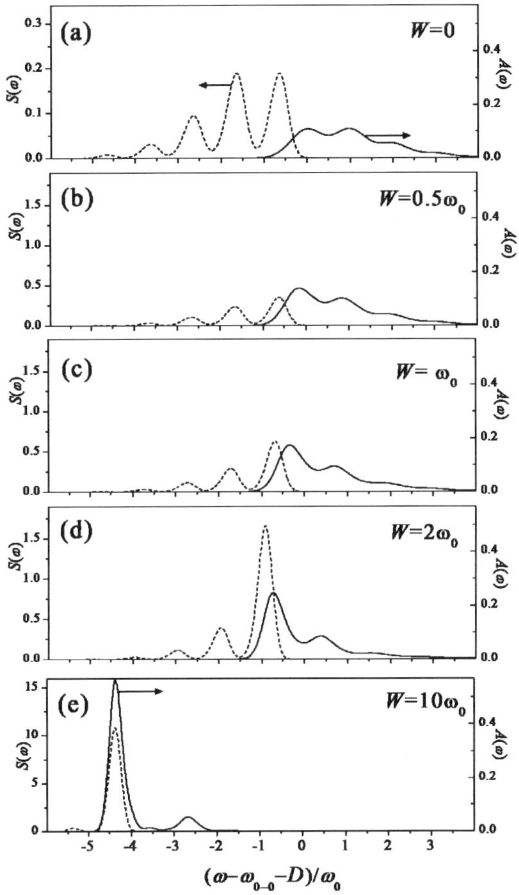

Figure 4.1 Calculated absorption (solid) and emission (dashed) spectra for disordered linear J- aggregates containing N=20 molecules with nearest-neighbor-only coupling J_0 and with $\sigma = 0.5\omega_0$ (see Eq.(4.1)). The spectra were obtained by averaging Eq.(3.6) over 10^4 configurations using a homogeneous line width of $0.15\omega_0$. In addition, $\lambda^2 = 1$ and ω_0= 0.17 eV (1400cm^{-1}.) The exciton bandwidth $W = 4|J_0|$ increases from top to bottom. All one- and two-particle states are included. Note the change in the PL vertical scale following a).

4.1 *Static Disorder*

We first demonstrate that introducing site disorder does not seriously impact the accuracy of Eq.(3.3) as long as the disorder does not cause line broadening significantly beyond the vibronic spacing ω_0. Figure 4.1 shows absorption and emission spectra for several values of exciton bandwidth for aggregates with substantial site disorder. We averaged Eq.(3.6) over 10^4 configurations of transition frequency offsets, Δ_n, chosen randomly from a Gaussian distribution,

$$P(\Delta_n) = (\sqrt{\pi}\sigma)^{-1} \exp(-\Delta_n^2 / \sigma^2) \tag{4.1}$$

with inhomogeneous (1/e) line width, $2\sigma = \omega_0$. As the figure shows the absorption spectral ratio R_{abs} is practically the same as that obtained in the homogenous case of Fig.3.1 (Due to the overlapping vibronic lines, R_{abs} should be evaluated using the spectral *areas* of the A_1 and A_2 peaks). Hence Eq.(3.3) remains valid for disordered aggregates. This is because the main effect of introducing site disorder is to mix exciton states within the *same* vibronic band causing band broadening but no significant effect on the band spectral area. Mixing between states in different bands is minimal as long as σ is less than ω_0.

Due to motional narrowing, the absorption line width in Fig. 4.1 drops significantly with increasing excitonic bandwidth. For example, in the non-interacting limit (Fig. 4.1(a)) the (1/e) full width is 2σ but drops to only 0.76σ in Fig. 4.1(e) when the excitonic coupling is strong. Interestingly, the emission line width is significantly less than σ even when there are no excitonic interactions, as is clearly observed in Fig. 4.1(a). This has nothing to do with motional narrowing but simply reflects the distribution of the *lowest* energy state in a chain of N chromophores where each chromophore is assigned a detuning according to Eq. (4.1).[16,52]

Figure 4.1 shows that in the presence of disorder the emission ratio R_{em} increases significantly with exciton bandwidth, reflecting an increase in the exciton coherence number, N_{coh}. R_{em} also decreases with increasing disorder when the exciton bandwidth is held constant (compare Fig. 3.1 and 4.1), reflecting a diminishing value of N_{coh}. Figure 4.2(a) further demonstrates how the emission vibronic progression for J-aggregates changes with increasing disorder. The demise in R_{em} is almost entirely due to the decrease in the 0-0 emission line strength, which further coincides with a diminishing coherence number, N_{coh}. This is also indicated by a strong reduction in the (super) radiative decay rate which scales with $\omega_{em}^3 I^{0-0}$ when the 0-0 peak is dominant. In the limit of extreme disorder $S(\omega)$ approaches the broad, isolated-molecule spectrum with $R_{em} \rightarrow 1/\lambda^2$.

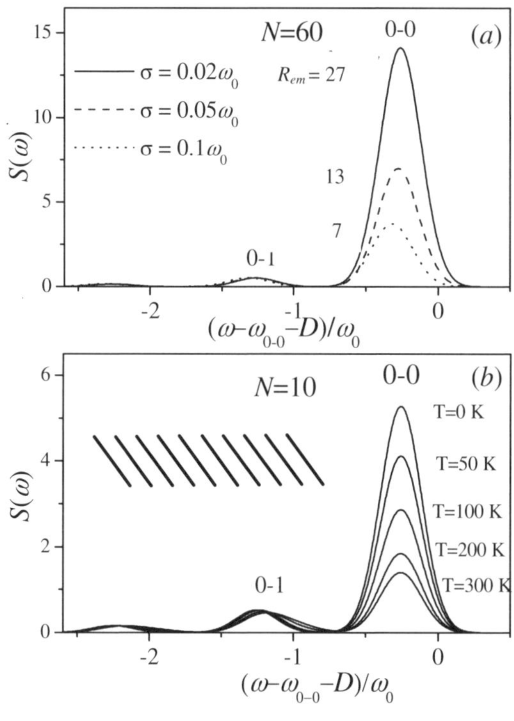

Figure 4.2 (a) Reduced PL spectrum for an ensemble of N=60 linear J-aggregates for several values of disorder width σ. The spectra were obtained by averaging Eq.(3.6) over 10^4 configurations using a homogeneous line width of $0.2\omega_0$. (b) Reduced PL spectrum for an ensemble of N=10 linear J-aggregates at several temperatures. In a) and b) the exciton bandwidth is $W=4|J_0|=\omega_0$ (=1400 cm^{-1}) and $\lambda^2=1$. All spectra were calculated using the TPA. Reprinted with permission from Ref. 30. Copyright 2011 American Chemical Society.

It is interesting to observe the contrasting ways the 0-0 emission responds to disorder in J and H-aggregates. Whereas in J-aggregates the 0-0 line strength is diminished by disorder in H-aggregates exactly the opposite behavior is observed: the 0-0 emission is enhanced by disorder.[29] This arises because in H-aggregates the emitting state is a k = π exciton; in a defect-free aggregate symmetry forbids 0-0 emission. However, as disorder is introduced symmetry is broken and 0-0 emission is no longer forbidden. Hence as disorder increases (and N_{coh} shrinks) the emission spectral ratio R_{em} increases (decreases) in H- (J-) aggregates. Recently, N_{coh} was derived from the measured value of R_{em} in MOPV4[52,53] and P3HT[16] H-aggregates. Detailed analysis of the emission spectral shape and Stokes shift using spatially correlated disorder also allowed for a crude determination of the incoherent exciton migration length — the extent to which energy can hop between coherent domains.[16] In marked contrast, the sidebands are almost entirely insensitive to coherence.[16,31,47]

4.2 *Thermal Effects*

We next consider thermal effects on the intrinsic exciton emission. The analysis is simplified by omitting thermally activated trap states and possible competition from excimer emission. We focus entirely on the effect of temperature on the line strengths, and omit the T-dependent broadening due to scattering with lattice phonons. The spectra in Fig.4.2(b) were evaluated by averaging over a Boltzmann distribution of excitons.

As observed in Fig.4.2(b) increasing temperature leads to attenuated 0-0 emission in J-aggregates. (The opposite behavior is observed in H-aggregates.[29]) The demise of the 0-0 emission with increasing T derives from the fact that thermally excited excitons with $k \neq 0$ cannot emit to the vibrationless ground state, but *can* emit to vibrationally excited electronic ground states (phonons) which satisfy momentum conservation. The latter process proceeds with an emission probability that is only weakly dependent on k. The overall effect can also be described in terms of a thermal coherence size, N_{coh}, with $I^{0-0} = FN_{coh}$. N_{coh} has an activated temperature dependence when N or T are small enough so that kT is smaller than the exciton level separations, but turns over to,[30,48,54]

$$N_T \approx 1 + (4\pi\hbar\omega_c / kT)^{d/2} \qquad (4.2)$$

in the thermodynamic limit when N exceeds N_T. Here, d is the aggregate dimension (d = 1,2,3) and ω_c is the curvature the exciton band-bottom.[48,54] Eq. (4.2) was derived using a parabolic approximation for the exciton band near $k = 0$. The "1" was inserted afterwards to correct for the high-temperature limit where N_T reduces to unity. Eq. (4.2) shows that in J-aggregates superradiance can be strongly quenched with increasing T.[48,54]

The theory also predicts temperature-independent side-band line strengths as observed in Fig. 4.2(b). In particular, $I^{0-1} \approx F\lambda^2$, so that the coherence number can be obtained directly from R_{em}. The general features of the temperature dependence of $S(\omega)$ are not limited to ideal *J*-aggregates, but are present in any morphology in which the 0-0 transition is dipole allowed, including the herringbone packing found in oligothiophene,[28,48,54] anthracene[31] and tetracene[32] thin films. Finally, when temperature is increased in *disordered* aggregates the effect on R_{em} is qualitatively similar to that in disorder-free aggregates.[16,54] In this case, R_{em} responds to a coherence number N_{coh} defined by both the magnitude of disorder and temperature as discussed in the next Section.

5. Exciton Coherence Number

The (emitting) exciton coherence number, N_{coh} indicates the number of coherently connected chromophores involved in emission. To quantify this for a linear J-aggregate we utilize the spatial coherence function[30]

$$C^{(em)}(s) = \langle \Psi^{(em)} \left| \sum_n B_n^\dagger B_{n+s} \right| \Psi^{(em)} \rangle, \qquad (5.1)$$

where $B_n^\dagger \equiv |n; vac\rangle\langle g; vac|$ is the "exciton" creation operator corresponding to the nth monomer. The coherence number for a disordered J-aggregate at a temperature T is then found by averaging the coherence function over a thermal distribution of emitters as well as many realizations of disorder. We have,

$$N_{coh} \equiv \frac{1}{\langle C^{(em)}(0) \rangle_{T,C}} \sum_s \langle C^{(em)}(s) \rangle_{T,C}\ , \text{ J-agg} \qquad (5.2)$$

where $\langle\cdots\rangle_{T,C}$ indicates a thermal and configurational average. For a perfectly ordered aggregate at $T = 0$ K the coherence number is equal the number of chromophores comprising the aggregate, $N_{coh} = N$. When static or dynamic disorder is introduced the coherence number is less than N.

Figure 5.1 shows how N_{coh} varies with N for disordered aggregates at T = 0 K for the values of σ encountered in Fig. 4.2(a). Figure 5.1(b) shows how the coherence number varies with N for perfectly ordered aggregates at several temperatures corresponding to Fig. 4.2(b). In Figures 5.1(a) and (b) the PL spectral ratio, R_{em}, is also plotted, showing that to a very good approximation the following relationship holds,

$$R_{em} = N_{coh} / \lambda^2 . \qquad \sigma, k_b T << \hbar\omega_0 \tag{5.3}$$

Eq. (5.3) reduces to Eq. (3.11) in the limit of no disorder and T = 0 K. Both Fig. 5.1(a) and (b) show that the coherence number is initially equal to N but eventually levels off to a N-independent value. In Fig. 5.1(b), the leveling-off behavior is characteristic of the thermodynamic limit where the number of exciton states within k_bT of the band-bottom is sufficiently large. In this limit the coherence number N_{coh} is equal to N_T in Eq.(4.2). It is important to note that the accuracy of Eq.(5.3) strongly depends on inclusion of two-particle states. For even modest exciton bandwidths, inclusion of only one-particle states will predict values of R_{em} several times greater than the correct value given by Eq. (5.3).

Equation (5.3) shows that the coherence number can be readily extracted from the PL spectral line shape. This method is an alternative to the more conventional means of obtaining N_{coh} from the superradiant decay rate or an analysis of the nonlinear absorption line shape.[55,56] One important advantage of the PL technique is that R_{em} is insensitive to nonradiative decay processes, which serve only to attenuate the PL spectrum uniformly. R_{em} is, however, sensitive to self-absorption which must be carefully eliminated by using optically thin samples or collecting front face emission.[39] Finally, intrinsic exciton emission may require early time collection,[31,32] before low-energy traps (chemical impurities and structural defects[14]) can be populated.

6. Conclusions and Outlook

The manner in which the FC progression based on the ubiquitous vinyl stretching mode in conjugated chromophores is distorted upon aggregation allows one to determine the exciton bandwidth, the exciton coherence number, and the parameters describing disorder. In J- aggregates, the ratio, R_{abs}, of the oscillator strengths of the first two vibronic peaks in the absorption spectrum increases with exciton bandwidth W thereby providing a measure of the latter, even in the presence of disorder. In the PL spectrum, the 0-0/0-1 line strength ratio, R_{em}, decreases with increasing disorder and rising temperature and is a direct measure of the (emitting) exciton coherence number, N_{coh}. The behavior of the absorption and emission spectral ratios in H-aggregates is exactly the opposite.[29] These contrasting behaviors establish a basis for analyzing the photophysical response in more complex morphologies. For example, in packing arrangements with multiple molecules in a unit cell, oscillator strength generally resides at the top and bottom of the

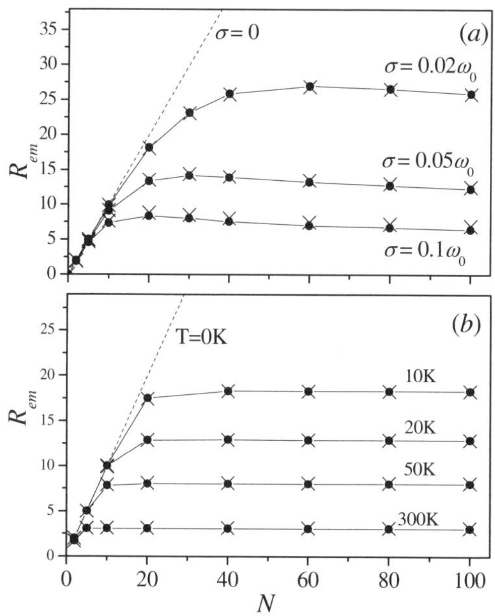

Figure 5.1: Rem (solid dots) and Ncoh (x's) as functions of N for several values of disorder width, σ in (a) and temperature in (b). Ncoh was evaluated using Eq.(5,s), and averaging over 10^4 configurations of disorder in a) and over a Boltzmann distribution of emitters in b). In both cases the exciton bandwidth is W=4|J0|=ω0 (=1400 cm-1) and λ^2=1. The line strengths in Rem were determined using the TPA. Reprinted with permission from Ref. 30. Copyright 2011 American Chemical Society.

exciton band, leading to optical properties which contain elements of both ideal J- and H-aggregate behavior, depending upon the polarization. In oligo-phenylene vinylene (OPV*n*) and oligothiophene (OT*n*) crystals[28] most of the oscillator strength resides near the top of the band, yielding a dominant H-like *ac*-polarized absorption profile. Emission from the weakly allowed band-bottom (lower Davydov component) contains both *ac*- and *b*-polarized components, resembling ideal H- and J-aggregate emission, respectively. In particular, the *b*-polarized 0-0 transition intensity decreases with increasing disorder and rising temperature in a manner similar to the J-aggregate portrayed in Fig.5.1.[48,54,57] However, the 0-0 transition is weak in OPV*n* and OT*n* aggregates since it is sourced by the difference of the two (nearly parallel) molecular transition dipoles in the unit cell, requiring a threshold[13,28] in N_{coh} to realize superradiance.[33] This is not the case in polyacene crystals[31,32] where the *b*-polarized lower Davydov component possesses substantial oscillator strength, resulting in dominant b-polarized emission. The PL spectral line shapes of anthracene and tetracene[31,32] strongly resemble those of linear J-aggregates: the (early time) spectra contain a pronounced 0-0 peak followed by a much weaker side-band (0-1). Moreover, with increasing temperature the 0-0 peak significantly attenuates, in marked contrast to the side-bands intensities which are practically temperature-independent.

In linear *J*-aggregates N_{coh} can be deduced directly from, R_{em} via Eq.(5.3). This simple relationship also applies to herringbone aggregates with two molecules per unit cell as long as R_{em} is evaluated using the polarization components of the 0-0 and 0-1 line strengths which correspond to the polarization of the lower Davydov component (from which emission originates).[30] In OPV*n* and OT*n* (*n* even) herringbone aggregates the lower Davydov component is *b*-polarized. Thus, the ratio of the *b*-polarized 0-0 and 0-1 line strengths gives directly the exciton coherence number:[30]

$$I_b^{0-0} / I_b^{0-1} = N_{coh} / \lambda^2. \tag{6.1}$$

Moreover, since only the 0-0 line strength is coherently enhanced in the vibronic progression, the mixed ratio, I_b^{0-0} / I_{ac}^{0-1} is also proportional to the coherence number, although the relationship is more complicated than Eq.(6.1).[14,30,48,54] A more dramatic example is the intrinsic emission fom thin films of anthracene[31] and tetracene[32] in which the lower (*b*-

polarized) Davydov component is strongly allowed, in contrast to OT*n* and OPV*n*. For early times following excitation (before traps can be populated), the PL spectral shape consists of an intense (*b*-polarized) 0-0 origin followed by a much weaker (*b*-polarized) 0-1 sideband — exactly what is expected for ideal J-aggregates. The ratio R_{em} is approximately four at T = 4K.[32] Assuming an HR factor of unity, Eq.(6.1) predicts a coherence number of also 4-5.[30]

The 0-1 line strength in the PL spectra of J-aggregates is very sensitive to the inclusion of two-particle states; excluding them leads to very large errors in R_{em} even for relatively weak excitonic couplings. In J-aggregates the 0-1 line strength is due to a constructive interference between one- and two-particle emission pathways[29,30] (in H-aggregates there is a destructive interference[47]). Hence, in disorder-free aggregates the 0-1 line strength rises from $\lambda^2\exp(-\lambda^2)$ for very weak excitonic coupling to λ^2 for strong excitonic coupling, a factor of roughly three for the physically relevant value of $\lambda^2 = 1$. An ideal background-free measure of the two-particle contribution — and therefore the polaron radius - is provided by circularly polarized luminescence in chiral aggregates, where the side-band dissymmetry scales directly with the two-particle coefficients.[53]

Finally, the coupling to additional lower-energy vibrational modes beyond the symmetric stretching mode can lead to interesting inter-mode synergistic effects.[49,58,59] Because low-energy modes ($\lesssim$ 500 cm^{-1}) are better described under the B.O. approximation, a more robust theory for the photophysical response of J-aggregates and organic assemblies in general involves a polaron-phonon basis set.[60] We are currently exploring further the intricacies of multiple mode coupling

Acknowledgements

FCS is supported by the NSF, grant DMR-0906464.

References

1. S. R. Forrest, Nature **428**, 911-918 (2004).
2. S. Gunes, H. Neugebauer, and N. S. Sariciftci, Chem. Rev. **107**, 1324-1338 (2007).
3. R. H. Friend, R. W. Gymer, A. B. Holmes, J. H. Burroughes, R. N. Marks, C. Taliani, D. D. C. Bradley, D. A. D. Santos, J. L. Bredas, M. Logdlund, and W. R. Salaneck, Nature **397**, 121-128 (1999).

4. A. J. Heeger, Reviews of Modern Physics **73**, 681-700 (2001).
5. G. Scheibe, Kolloid-Zeitschrift **82**, 1-14 (1938).
6. E. E. Jelley, Nature **138**, 1009-1010 (1936).
7. M. Kasha, Radiation Research **20**, 55-70 (1963).
8. E. G. McRae and M. Kasha, J. Chem. Phys. **28**, 721-722 (1958).
9. Toyozawa, Prog. Theor. Phys. **26**, 29 (1961).
10. Y. B. Levinson and E. I. Rashba, Rep. Prog. Phys. **36**, 1499-1565 (1973).
11. T. Holstein, Ann. Phys. **8**, 325-342 (1959).
12. P. O. J. Scherer and S. F. Fischer, Chem. Phys. **86**, 269-283 (1984).
13. F. C. Spano, Chem. Phys. Lett. **331**, 7-13 (2000).
14. F. C. Spano, J. Chem. Phys. **116**, 5877-5891 (2002).
15. F. C. Spano, J. Chem. Phys. **122**, 234701 (2005).
16. F. C. Spano, J. Clark, C. Silva, and R. H. Friend, J. Chem. Phys. **130**, 074904 (2009).
17. S. C. J. Meskers, R. A. J. Janssen, J. E. M. Haverkort, and J. H. Wolter, Chem. Phys. **260**, 415-439 (2000).
18. A. Stradomska and P. Petelenz, J. Chem. Phys. **130**, 094705 (2009).
19. A. Eisfeld and J. S. Briggs, Chemical Physics **281**, 61-70 (2002).
20. A. Eisfeld and J. S. Briggs, Chem. Phys. **324**, 376-384 (2006).
21. M. Hoffmann and Z. G. Soos, Phys. Rev. B **66**, 024305 (2002).
22. J. Seibt and V. Engel, Chem. Phys. **347**, 120-126 (2008).
23. M. R. Philpott, J. Chem. Phys. **55**, 2039-2054 (1971).
24. This assumes that the wave function phases are chosen so that the in-phase combination corresponds to transition dipole moments pointing in a common direction.
25. H. Fidder, J. Knoester, and D. A. Wiersma, Chem. Phys. Lett. **171**, 529-536 (1990).
26. *J-aggregates*, edited by T. Kobayashi (World Scientific, Singapore, 1996).
27. D. Oelkrug, H.-J. Egelhaaf, J. Gierschner, and A. Tompert, Syn. Met. **76**, 249-253 (1996).
28. F. C. Spano, Ann. Rev. Phys. Chem. **57**, 217-243 (2006).
29. F. C. Spano, Acc. Chem. Res. **43**, 429-439 (2010).
30. F. C. Spano and H. Yamagata, J. Phys. Chem. B **115**, 5133–5143 (2011).
31. T. S. Ahn, A. M. Muller, R. O. Al-Kaysi, F. C. Spano, J. E. Norton, D. Beljonne, J. L. Bredas, and C. J. Bardeen, J. Chem. Phys. **128**, 054505 (2008).
32. S.-H. Lim, T. G. Bjorklund, F. C. Spano, and C. J. Bardeen, Phys. Rev. Lett. **92**, 107402 (2004).
33. F. Meinardi, M. Cerminara, A. Sassella, R. Bonifacio, and R. Tubino, Phys. Rev. Lett. **91**, 247401-247404 (2003).
34. A. Stradomska and P. Petelenz, J. Chem. Phys. **131**, 044507 (2009).
35. F. C. Spano, Chem. Phys. **325**, 22-35 (2006).
36. F. C. Spano, J. Am. Chem. Soc. **131**, 4267-4278 (2009).

37. J. Clark, J. F. Chang, F. C. Spano, R. H. Friend, and C. Silva, Applied Physics Letters **94**, 163306 (2009).
38. J. Clark, C. Silva, R. H. Friend, and F. C. Spano, Phys. Rev. Lett. **98**, 206406 (2007).
39. P. J. Brown, S. D. Thomas, A. Kohler, J. S. Wilson, J.-S. Kim, C. M. Ramsdale, H. Sirringhaus, and R. H. Friend, Phys. Rev.B **67**, 064203-064201-064216 (2003).
40. F. Zsila, Z. Bikadi, Z. Keresztes, J. Deli, and M. Simonyi, J. Phys. Chem. B **105**, 9413-9421 (2001).
41. W. Cooper, Chem. Phys. Lett. **7**, 73-77 (1970).
42. In order to simplify the discussion we omit the zero-order mixing between degenerate one- and two-particle states which leads to small spectral splittings not resolvable in Fig.3.
43. B. Kopainsky, J. K. Hallermeier, and W. Kaiser, Chem. Phys. Lett. **83**, 498-502 (1981).
44. A. D. Shaller, W. Wang, H. Y. Gan, and A. D. Q. Li, Ang. Chem. Int. Ed. **47**, 7705-7709 (2008).
45. V. Dehm, Z. J. Chen, U. Baumeister, P. Prins, L. D. A. Siebbeles, and F. Wurthner, Organic Letters **9**, 1085-1088 (2007).
46. J. F. Chang, J. Clark, N. Zhao, H. Sirringhaus, D. W. Breiby, J. W. Andreasen, M. M. Nielsen, M. Giles, M. Heeney, and I. McCulloch, Phys Rev. B **74**, 115318 (2006).
47. F. C. Spano, J. Chem. Phys. **118**, 981-994 (2003).
48. F. C. Spano, J. Chem. Phys. **120**, 7643-7658 (2004).
49. Z. Zhao and F. C. Spano, J. Phys. Chem. C **111**, 6113-6123 (2007).
50. E. W. Knapp, Chem. Phys. **85**, 73-82 (1984).
51. J. Knoester, J. Chem. Phys. **99**, 8466-8479 (1993).
52. L. van Dijk, P. A. Bobbert, and F. C. Spano, J. Phys. Chem B **113**, 9708-9717 (2009).
53. F. C. Spano, S. C. J. Meskers, E. Hennebicq, and D. Beljonne, J. Amer. Chem. Soc. **129**, 7044-7054 (2007).
54. F. C. Spano, Phys. Rev. B **71**, 235208 (2005).
55. M. Dahlbom, T. Pullerits, S. Mukamel, and V. Sundstrom, J. Phys. Chem. B **105**, 5515-5524 (2001).
56. L. D. Bakalis and J. Knoester, J. Phys. Chem. B **103**, 6620-6628 (1999).
57. F. Meinardi, M. Cerminara, S. Blumstengel, A. Sassella, A. Borghesi, and R. Turbino, Phys. Rev. B **67**, 184205-184201-184205-184206 (2003).
58. L. Silvestri, S. Tavazzi, P. Spearman, L. Raimondo, and F. C. Spano, J. Chem. Phys. **130**, 234701 (2009).
59. F. C. Spano and L. Silvestri, J. Chem. Phys. **132**, 094704 (2010).
60. Z. Zhao and F. C. Spano, J. Chem. Phys. **122**, 114701 (2005).

CHAPTER 3

HIERARCHAL STRUCTURE OF LIGHT-HARVESTING PORPHYRIN AGGREGATES

Jeanne L. McHale

Department of Chemistry
Washington State University
Pullman, WA, USA 99164-4630
E-mail: jmchale@wsu.edu

The structural and spectroscopic properties of self-assembled light-harvesting chromophores are discussed. In photosynthetic organisms, light-harvesting aggregates are optimized for absorption of particular wavelengths of the solar spectrum and for energy transfer to the photosynthetic reaction center. Artifical light-harvesting aggregates composed of porphyrins provide model systems for understanding how the hierarchal structure of these assemblies determines their optical properties. Resonance Raman and scanning probe microscopy of porphyrin aggregates reveal the structural features and intermolecular interactions which control this assembly.

Keywords: Light-harvesting, chlorophyll and bacteriochlorophyll aggregates, porphyrin aggregates, resonance Raman spectroscopy, excitonic coupling.

1. Introduction

Nature provides us with remarkable examples of how the assembly of molecules which are strong absorbers of light can lead to large perturbations to their optical and electronic properties. The same quantum mechanical quantity which results in strongly colored molecules, the transition dipole moment, is also responsible for perturbations to the optical spectra of these molecular assemblies. The optoelectronic properties of noncovalently bonded light-harvesting aggregates (LHAs) depend strongly on the relative positions and orientations of the chromophores, and excitonic coupling can lead to spectral-shifts, line narrowing, energy transfer, and enhanced radiative relaxation. In photosynthetic organisms, nature takes advantage of this transition-dipole coupling in the form of organized assemblies of chlorophyll derivatives which are optimized for the absorption of specific wavelengths of visible light. LHAs do the job of rapidly funneling absorbed light energy to the reaction center where electron transfer initiates the process of photosynthesis.

Better understanding of how the supramolecular structure of chromophore aggregates determines their optical and electronic properties will permit the properties of excitonically coupled LHAs, such as excited state delocalization, energy transfer, superradiance, photoconductivity, and photosensitization, to be exploited for devices. The ultimate application of self-assembled chromophores might be the imitation of photosynthesis, i.e., the use of biomimetic chromophore assemblies to split water and make fuel using solar energy. Synthetic porphyrin derivatives are convenient mimics of natural chlorophylls and bacteriochlorophylls found in photosynthetic plants and bacteria, respectively, and the aggregation properties of porphyrins have been widely studied. In this chapter, we review recent work using spectroscopy and imaging to understand how the supramolecular structure of porphyrin aggregates determines their optical properties. We begin with a brief overview of what is known about natural LHAs and what remains to be learned about them. We begin with an example of a porphyrin aggregate discovered in our lab for which the structure and optical spectra depend on the nature of the counterion. We then proceed to an admittedly incomplete review of the vast literature on self-assembled porphyrin aggregates, particularly as related to their hierarchal structure, with an emphasis on the widely studied but still incompletely understood aggregate of 5,10,15,20-tetra(4-sulfanotophenyl)-porphyrin (TSPP), which aggregates in aqueous solution upon protonation of the pyrrolic nitrogens to form the diacid, H_2TSPP^{2-}. Finally, recent work from the author's lab and that of our collaborators is presented, leading to some surprising conclusions about the structure of the TSPP aggregate. The key questions to be addressed are the following. How does the structure of the chromophore determine that of the aggregate? What kinds of intermolecular interactions drive the self-assembly? Finally, what is the role of the environment, be it solvent or substrate in *in vitro* studies or other components of the membrane in *in vivo* systems, in determining supramolecular structure and optical spectra of light-harvesting assemblies?

2. Light-Harvesting Aggregates in Photosynthetic Organisms

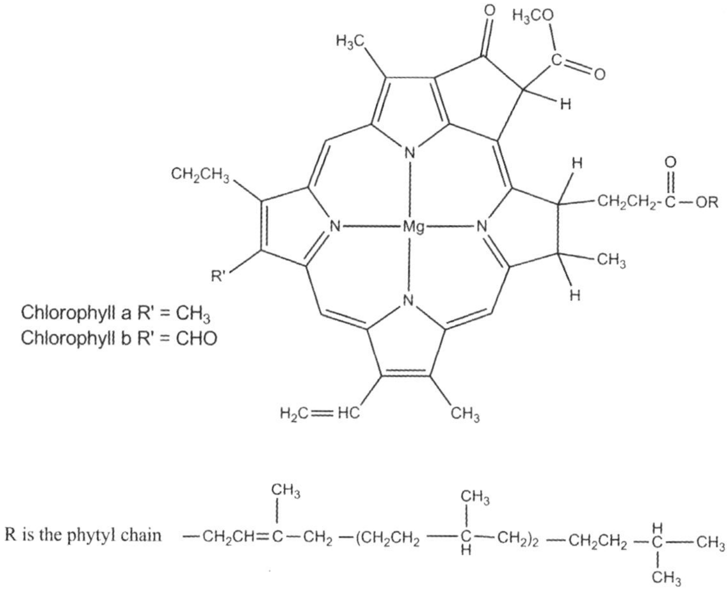

Figure 2.1 Structure of chlorophyll a and chlorophyll b

Nature optimizes the structure of LHAs, also called antenna complexes, in photosynthetic organisms in order to maximize the collection of light and to funnel energy to the reaction center.[1–3] In plants, chlorophyll molecules perform the daunting task of converting about 1 photon/sec of incident sunlight into a turnover rate greater than 100 s^{-1} at the reaction center. An astonishing 99% of the chlorophyll content of green plants is devoted to light harvesting, not electron transfer![4] The structures of chlorophyll a and b are shown in Fig. 2.1. The basis for both structures is a magnesium substituted tetrapyrrole (more precisely, a 17,18-dihydroporphyrin macrocycle) with an additional five-membered ring external to the macrocycle, a long phytyl chain which imparts hydrophobic character, and various substituents. The characteristic blue and red absorption bands of chlorophyll correlate to the B and Q bands of the more symmetric porphyrin derivatives discussed below; however, a complete understanding of the optical spectrum of chlorophyll derivatives is still an active are of research.[5,6] A relatively minor difference in the substituents, i.e. the replacement of the methyl group of chlorophyll a (Chl a) by the formyl group of Chl b results in favorable spectral shifts which leave a window of transparency only from about 500 to 600 nm when both pigments are present, a gap which may be covered by accessory pigments such as carotenoids.

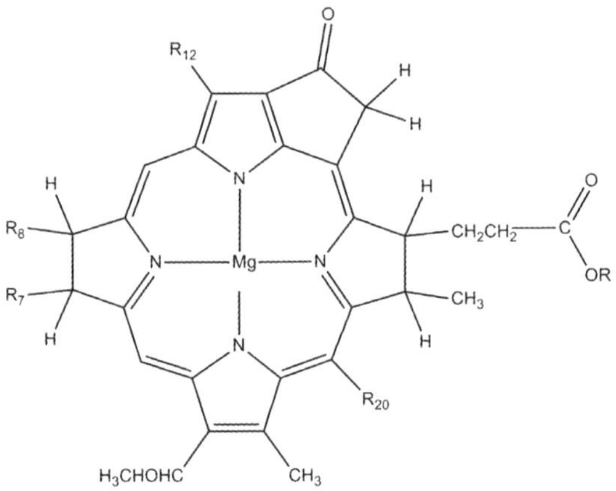

BChl c, $R_7 = CH_3$, $R_8 = C_2H_5$, $R_{12} = CH_3$, $R_{20} = CH_3$

BChl d, $R_7 = CH_3$, $R_8 = C_2H_5$, $R_{12} = CH_3$, $R_{20} = H$

BChl e, $R_7 = CHO$, $R_8 = C_2H_5$, $R_{12} = C_2H_5$, $R_{20} = CH_3$

Figure 2.2 Structures of bacteriochlorophylls found in chlorosomes. There is some variation in the R_8 substituents, which may be ethyl, propyl, isobutyl or neopentyl. The R group on the ester function may be farnesyl, stearyl, phytyl, or other long alkyl chain.

Bacteriochlorophyll molecules found in photosynthetic bacteria have very similar structures to those shown in Fig. 2.1. In both bacteriochlorophyll a (BChl a) and BChl b, the R′ group is methyl. In BChl a, the ethenyl group of Fig. 2.1 is replaced by an acetyl group, and the $-CH_2CH_3$ group is replaced by the ethylidene group $=CH-CH_3$ which is doubly bonded to the pyrrole ring. The result is a red-shift that extends the absorption into the near-infrared. The BChl c, d, and e molecules, on the other hand (Fig. 2.2) are the major components of chlorosomes, which are rod-shaped aggregates found in green photosynthetic bacteria. It is significant that these latter variants lack the methoxycarbonyl group found in the other chlorophylls and bacteriochlorophylls. This group presents steric-hindrance which inhibits π-stacking of adjacent chromophores.[7] The chlorosomes of green bacteria differ from other LHAs in that they lack protein scaffolding, and indeed the BChl pigments isolated from bacteria have been shown to self-assemble *in vitro* resulting in optical properties similar to those of intact chlorosomes.[8–10] Even in cases where a protein scaffold and associated pigments are essential to the organization of the LHAs *in vivo*, functional protein-free and lipid-free light harvesting complexes can also self-assemble *in vitro*.[11–13]

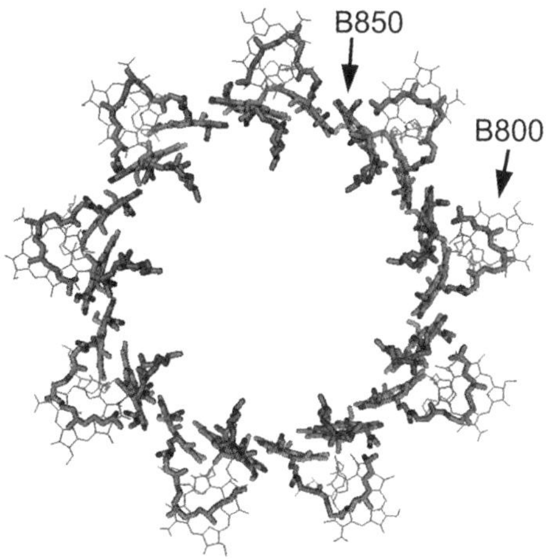

Figure 2.3 BChl molecules in the LH2 complex from *Rps. acidophila* according to the crystal structure of Cherezov *et al.* in Ref. 16.

Photosynthetic bacteria provide two examples of LHAs that may be mimicked by assemblies of synthetic porphyrins, the rod-shaped chlorosomes of green photosynthetic bacteria and the cyclic assemblies of bacteriochlorophylls found in purple photosynthetic bacteria. In contrast, the chlorophylls found in the light-harvesting complexes of plants and algae are not as regularly arranged. In the case of purple photosynthetic bacteria, X-ray crystallography has revealed beautiful ring-shaped LHAs[14,15] such as the LH2 complex from *Rps. acidophila* depicted in Fig. 2.3[16] and drawn using coordinates from the Protein Data Bank. In these complexes, BChl a molecules and carotenoids are noncovalently bound to membrane apoproteins to form a complex with n-fold rotational symmetry, where n is typically 8 or 9. BChl a molecules are arranged in two rings. One is a ring of $2n$ BChl a molecules having their mean molecular planes perpendicular to the plane of the cyclic assembly, referred to as B850 according to the corresponding peak in the absorption spectrum. A second ring of n B800 molecules arranged coplanar to the ring of the complex leads to a peak at 800 nm. The 800 nm and 850 nm absorptions are red-shifted relative to the spectrum of monomeric BChl a. A larger red-shift in the case of the ring of B850 molecules results from greater excitonic coupling from transition dipole moments arranged parallel to the direction of aggregation, i.e., tangential to the ring of the complex. The B800 molecules, in contrast, are arranged with their transition dipoles perpendicular to the direction of aggregation (radially) and intermolecular separations are larger than for the B850 ring, leading to a smaller red-shift.

While detailed structures of light-harvesting complexes from purple bacteria have been determined from X-ray crystallography, the internal structure of the chlorosomes from green bacteria is not known with certainty. The overall morphology of these LHAs is rod-shaped with dimensions of approximately 20 by 50 by 200 nm, and spiral, lamellar, and tubular assemblies have been proposed.[13,17] Chlorosomes are comprised of BChls c, d, and e, which unlike chlorophyll and BChl a and b have two reduced pyrrole rings as shown in Fig. 2.2. Resonance Raman studies using model self-assembled compounds have shed light on the nature of the intermolecular interactions which control the supramolecular structure of chlorosomes.[18] Proposed structures are based on evidence for hydrogen bonding between the carbonyl group of the external five-membered ring and the hydroxyl group of an adjacent molecule. The –OH group acting as a donor in the hydrogen bond is diametrically opposed to the carbonyl acceptor group along the direction of the Q_y transition moment. The hydroxyl group is also coordinated to the central magnesium atom, controlling the interactions in the perpendicular direction. But how do these near-neighbor arrangements lead to the overall ellipsoidal shape of chlorosomes imaged by electron microscopy? We shall see that the same question is echoed in studies of synthetic porphyrin aggregates described below.

3. Porphyrin Aggregates

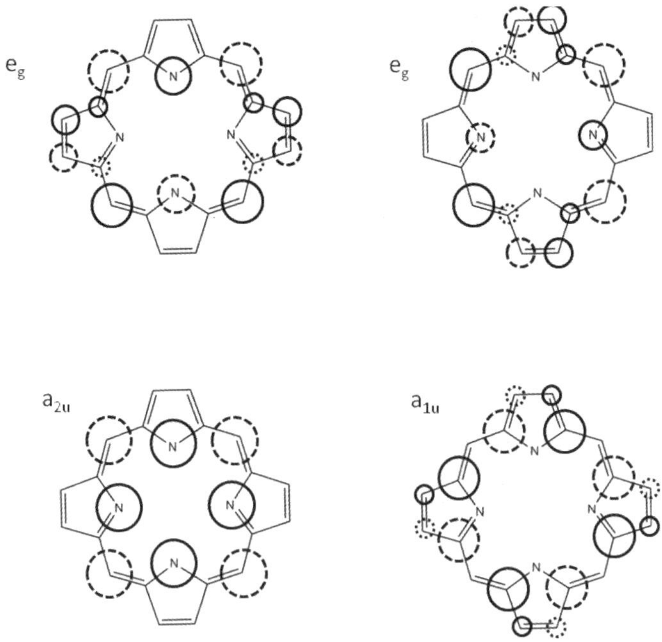

Figure 3.1 Symmetries of the frontier orbitals of porphyrin in D_{4h} symmetry.

Aggregation of porphyrins, particularly in aqueous solution, is well-known.[19] Such aggregates do not always lead to clearly-defined excitonic coupling via transition moments, but rather to spectral shifts and increased spectral width which are merely the result of intermolecular interactions in heterogeneous agglomerates. We are more interested here in systems which exemplify excitonic effects, i.e., excited electronic states which are delocalized over a number of molecules through coupling of transition moments. Though porphyrins are closely related to their chlorophyll cousins, their higher symmetry presents additional considerations in applying simple theories[20,21] to interpret spectral shifts of excitonic states. As shown in Fig. 3.1, the porphine ring which is the basis for porphyrins has four-fold symmetry in the diacid form or when a metal atom occupies the core, but in the free base form the symmetry is lower. (As will be discussed below, protonation of the pyrrole nitrogens may result in symmetry lowering, but that is neglected here.) Under four-fold symmetry, the S_1 (Q band) and S_2 (B or Soret band) excited levels are doubly degenerate. In D_{4h} symmetry, Gouterman's four-orbital model[22] explains the origin of these transitions using excited states derived from configurations based on the nearly-degenerate a_{1u} and a_{2u} HOMOs and the doubly degenerate $e_{g,x}$, $e_{g,y}$ LUMO. Since the $a_{1u}e_g$ excited configuration is nearly degenerate with $a_{2u}e_g$, strong configuration interaction leads to in-phase and out-of-phase linear combinations of these configurations which correspond to the B and Q excited states, respectively. The ground to S_2 transition is thus more intense than the ground to S_1 because of constructive interference of the transition moments for the $\left|a_{1u}(1)e_g(2)\right\rangle$ and $\left|a_{2u}(1)e_g(2)\right\rangle$ configurations in the former case, compared to destructive interference in the case of the out-of-phase combination. When the symmetry is lower than D_{4h}, the weighting of the two excited configurations is not equal (i.e., when θ in Eqs. 3.1 below is not 45°), the cancellation of the two transition moments is incomplete and the Q band gains intensity at the expense of the B band. The y-polarized excited states, for example, are represented by:

$$\left|B_y\right\rangle = \cos\theta\left|a_{2u}(1)e_{g,y}(2)\right\rangle + \sin\theta\left|a_{1u}(1)e_{g,x}(2)\right\rangle$$
$$\left|Q_y\right\rangle = \sin\theta\left|a_{2u}(1)e_{g,y}(2)\right\rangle - \cos\theta\left|a_{1u}(1)e_{g,x}(2)\right\rangle \tag{3.1}$$

The x-polarized excited states are obtained by swapping the x and y subscripts in Eqs. 3.1 and interchanging the plus and minus signs. The two components of the doubly-degenerate LUMO, $e_{g,x}$ and $e_{g,y}$, account for the perpendicularly polarized in-plane transition moments corresponding to each of the S_1 and S_2 excited states. The result is a strong Soret band in the blue region and a weaker Q band in the red, both of which are transitions to in-plane polarized doubly degenerate excited states.

Fig. 3.2 illustrates the influence of symmetry on the visible absorption spectrum of 5,10,15,20-tetra (4-sulfonatophenyl)-porphyrin (TSPP) in its monomeric form. Protonation of the pyrrole nitrogens converts the two-fold symmetric free base form to the more symmetric diacid, accompanied by structural changes in the porphyrin core and rotation of the phenyl rings. The result is a shifting of both the Soret and Q bands to the red. The influence of symmetry is more apparrent in the Q band, which shows distinct vibrational sub-bands. In four-fold symmetric porphyrins such as the TSPP diacid, the (0,0) and (0,1) components of

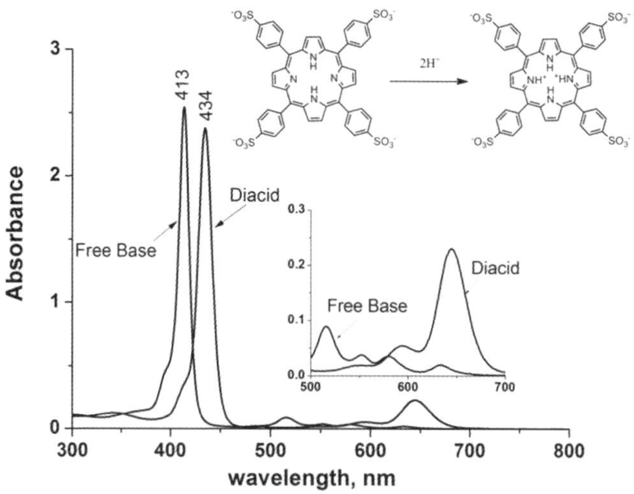

Figure 3.2 Structure and optical spectrum of TSPP in its free base and diacid forms. The inset at the bottom shows the vibrationally-resolved Q-bands.

the Q band are resolved, while the occurrence of four peaks in the Q band region: $Q_x(0,0)$, $Q_x(0,1)$, $Q_y(0,0)$ and $Q_y(0,1)$, is diagnostic of less than four-fold symmetry and is observed for the free base. Note that these may be approximate symmetries owing to the orientation of the phenyl rings and the sulfonato groups. Using group theoretical considerations, Gouterman showed how perturbations of certain symmetries (namely, the A_{1g}, A_{2g}, B_{1g}, and B_{2g} irreducible representations of D_{4h}) can result in configuration mixing which leads to increase of the Q band intensity at the expense of the B-band.[22] A ruffled, saddled, or bowed distortion of the porphyrin macrocycle is thus capable of causing an increase in the Q band intensity. This sort of perturbation is revealed in Fig. 3.2 where the increased intensity of the Q band of the diacid results from nonplanar distortion of the porphyrin ring on protonation, owing to steric crowding of the hydrogens in the porphyrin core. Another effect of protonation is a rotation of the phenyl rings to become more nearly coplanar with the mean porphyrin plane,[23] which also contributes to the overall red-shift owing to increased π-electron delocalization.

The existence of two or more excited levels in porphyrin monomers results in additional considerations when treating excitonic coupling in porphyrin aggregates. The much larger transition dipole moment of the B-band results in stronger exciton coupling than for the Q-band, but

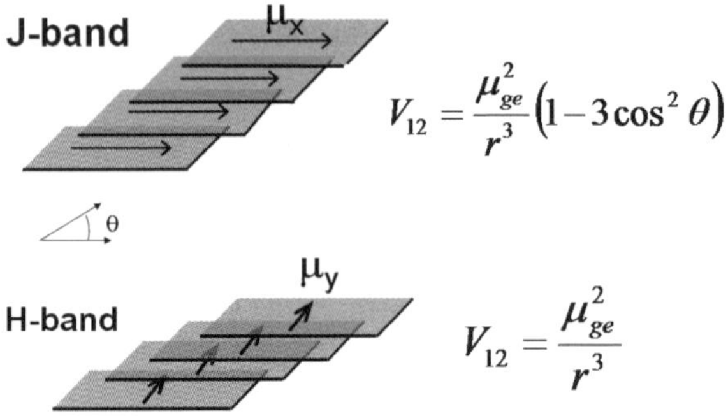

Figure 3.3 Formation of red-shifted (J-band) and blue-shifted (H-band) exciton transitions from coupling of transition dipole moments which are respectively parallel and perpendicular to the direction of aggregation.

symmetry-lowing perturbations in the local environment of the molecule in the aggregate can lead to B-Q intensity borrowing.[24] There is also the possibility that aggregation facilitates excitonic coupling of the B and Q excited states.[25] At the simplest level of theory, i.e., the near-neighbor coupling of point transition dipoles, the existence of an in-plane polarized degenerate excited state means that a single aggregate structure, such as the widely invoked staircase model of Fig. 3.3, leads to both red-shifted ("J-band") and blue-shifted ("H-band") transitions. The transition dipole coupling between molecules 1 and 2 in the aggregate is given by:

$$V_{12} = \frac{\mu_{ge}^2}{r^3}\left[\hat{u}_1 \cdot \hat{u}_2 - 3\left(\hat{u}_1 \cdot \hat{r}\right)\left(\hat{u}_2 \cdot \hat{r}\right)\right] \tag{3.2}$$

where μ_{ge} is the magnitude of the transition dipole moment, $\hat{u}_{1,2}$ is a unit vector in the direction of the dipole moment for molecule 1 or 2, r is the distance between the molecular centers, and $\hat{r}$ is a unit vector in the direction of aggregation. The sign of the coupling V_{12} determines the spectral shift. For the arrangement of transition dipole moments shown in Fig. 3.3, the two components of the Soret band transition moment can be taken to be parallel (μ_x) or perpendicular (μ_y) to the chain of the J-aggregate. Taking θ to be the angle between the transition dipole moment μ_x and $\hat{r}$, it is clear that the sign of V_{12} is negative for the coupling of the dipole moments parallel to the chain, provided that

$\cos^2\theta > 1/3$ (θ < 54.7°). This leads to a red-shifted J-band. The transition moments μ_y on the other hand are perpendicular to $\hat{r}$, leading to a positive value of V_{12} and a blue-shifted H-band. Thus a degenerate excited electronic state may give rise to more than one excitonic transitions in a single aggregate structure.

Another important aspect of LHAs of porphyrins, and perhaps also those of chlorophylls, is that nonplanarity of the monomer units may be significant. For example, the bacteriochlorophylls which comprise the antenna complexes of *Rhodopseudomonas acidophila* are bowed by about 7 to 10°,[26] and experiments show that the Mg atom of chlorophyll is about 0.4 Å above the mean porphyrin plane.[5] The flexibility of tetrapyrrole macrocycles may be biologically important.[27] It has been proposed, for example, that different degrees of distortion for the two BChl b molecules of the special pair of the *R. viridis* reaction center correlate to differences in their electronic structure and result in unidirectional electron transfer.[28] The tetraphenylporphyrin (TPP) derivatives discussed further below aggregate in acidic aqueous solution where they exist in their diacid forms. Protonation of TPP leads to significant structural perturbations, namely, nonplanarity of the porphyrin core, such as saddled or ruffled geometries.[29,30] At the same time, the peripheral phenyl rings, which are perpendicular to the porphryin plane in the free base, rotate on protonation to become more nearly coplanar with the mean porphyrin plane in the case of the diacid.[23] This leads to a red-shift in the B and Q band transitions resulting from the enhanced conjugation of the porphyrin and phenyl π systems. These distortions have important consequences for the self-assembly of porphyrin diacids because they preclude the formation of the linear staircase "J-aggregate" model depicted in Fig. 3.3. In addition, they may explain our surprising observation of porphyrin aggregates for which the structure and spectra depend on counterion.

3.1 *Counterion Dependence*

An accidental discovery in our lab initiated our interest in the unexpected phenomenon of counterion-dependent porphyrin aggregates.[31,32] In a study of interfacial electron transfer by *meso*-tetrakis (p-carboxyphenyl)porphyrin (TCPP) adsorbed on TiO_2, we noticed that the optical spectrum of this porphyrin in aqueous acid was different in the presence of Cl^- and NO_3^- counterions, as shown in Fig. 3.4. The monomer diacid H_2TCPP^{2+} can be observed in acidic ethanol and has a Soret band maximum at 440 nm. In acidic aqueous solution at pH below about 1, on the other hand, aggregation leads to counterion-dependent spectral shifts as shown in Fig. 3.4. In aqueous HNO_3, the Soret band splits into an H-band at 406 nm and a J-band at 467 nm, while in aqueous HCl, only an H-band is observed, at 417 nm. In either case the H-band is not significantly exchange-narrowed. Though dimerization of the free base form of TCPP had been previously reported,[33] protonation of the pyrrole nitrogens of TCPP, which occurs below a pH of about 5, leads to eventual precipitation, hindering spectroscopic studies of the extended aggregates. In contrast to the more widely studied aggregate of the diacid of TSPP, the weaker acid strength of the carboxylic acid group as compared to the sulfonic acid group results in a net positive charge for the diacid form, H_2TCPP^{2+}. Thus it is not too surprising that negative counterions would be involved in the structure of the TCPP aggregate. Interestingly, we also observed spectral signatures of a TCPP dimer, independent of counterion, at moderately low pH. Using aqueous polyvinyl alcohol (PVA) to impart kinetic stability to the aggregate solutions, we were able to measure the resonance Raman spectrum as a function of pH and counterion, as shown in Fig. 3.5. Both aggregates display a resonance light scattering spectrum in the Soret band region (see Refs. 31 and 32), which is diagnostic of a delocalized excited electronic state, but a stronger signal is observed in the presence of nitrate counterion in the vicinity of the J-band, indicating a greater degree of excitonic coupling. This is consistent with the decreased spectral width of this band in the absorption spectrum which indicates greater coherence for the J-band excited state.

The optical spectrum of the TCPP aggregate in aqueous nitric acid, shown in Fig. 3.4, bears much resemblance to that of the TSPP aggregate at low pH. Formation of the monomer diacid (which forms readily without aggregating in acidic ethanol solution) results in a red-shift of the Soret band from 414 nm for free base, TCPP, to 440 nm for the diacid, H_2TCPP^{2+}, accompanied by collapse of the Q-bands and increase in their relative intensity. This is characteristic of reduced symmetry and enhanced nonplanarity of the macrocycle. On further decreasing the pH to below one, the Soret band splits into a red-shifted J-band and a blue-shifted H-band, accompanied by further increase and red-shifting of the Q-band. The J-band shows a degree of exchange narrowing, suggestive of a coherence number of about 4, while the H-band is not significantly

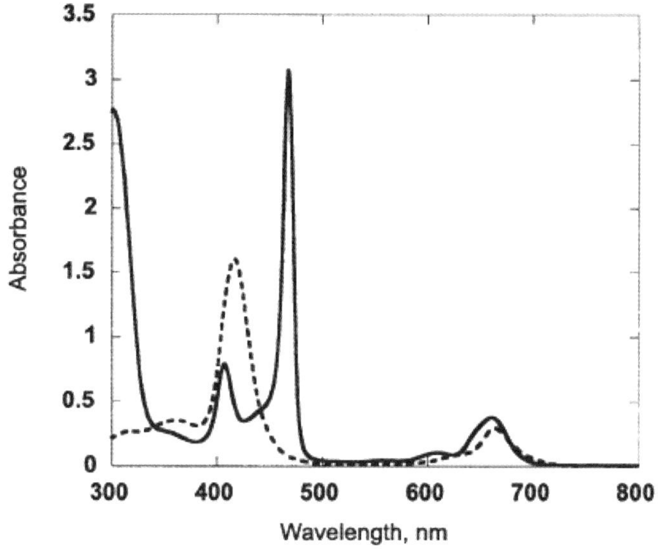

Figure 3.4 Optical spectrum of meso-tetrakis(4-carboxyphenyl)porphyrin (TCPP) aggregates in aqueous HCl (dashed line) and aqueous HNO_3 (full line), both at pH ~ 1. Reprinted with permission from Choi, M.Y.; Pollard, J. A.; Webb, M. A.; McHale, J. L. *J. Am. Chem. Soc.* **2003**, *125*, 810-820. Copyright 2003, American Chemical Society.

narrower than the Soret band of the monomer. As discussed further below, the unequal intensities of the H- and J-bands presents a challenge to simple exciton theory, if both arise from a slipped near-neighbor arrangement as assumed by the staircase model. In the presence of Cl^- counterions, only an H-band is seen in the Soret band region, suggesting an arrangement in which neighboring porphyrins are stacked like plates. In the pH range between about 1 and 4.5, a counterion-independent spectrum is observed in which the Soret band is split into two components at 404 and 444 nm, which we supposed to be a dimer in which hydrogen bonding between the pyrrole nitrogens and the carboxylic acid group holds the two monomers together in a perpendicular arrangement.[31] Based purely on the positions of the shifted Soret peaks of the aggregate, we proposed that the near-neighbor arrangement of porphyrins is characterized by a slip angle (θ in Fig. 3.3) of zero in the case of chloride counter ion, while a slip angle of 42° leads to the predicted H- and J-bands in the presence of nitrate.

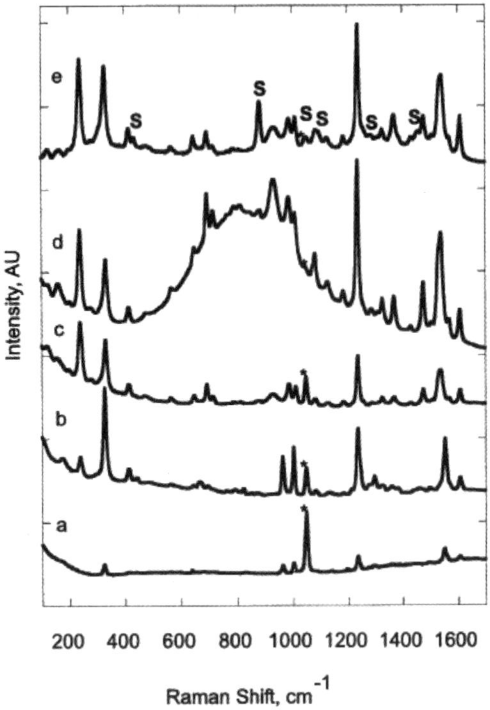

Figure 3.5. Resonance Raman spectra of 50 μM TCPP, excited at 457.9 nm. a) Free base monomer in water at pH 10, b) diacid monomer in water at pH 1.2, c) Aggregate in aqueous HCl at pH ~ 1, d) aggregate in aqueous HNO_3 at pH ~ 1, and e) diacid monomer in acidified ethanol. "S" marks solvent lines of ethanol. Reprinted with permission from Choi, M.Y.; Pollard, J. A.; Webb, M. A.; McHale, J. L. *J. Am. Chem. Soc.* **2003**, *125*, 810-820. Copyright 2003, American Chemical Society.

Figure 3.5 shows the resonance Raman spectra of the TCPP free base, diacid monomer, dimer, and aggregate in the presence of Cl^- and NO_3^-. Aggregation does not greatly perturb any of the Raman frequencies of H_2TCPP^{2+}, but it does result in enhancement of low-frequency vibrations. Especially in the presence of strong excitonic coupling, resonance enhancement of low-frequency modes appears to be a signature of aggregation.[34–36] The most intense modes in resonance Raman spectroscopy are those for which the excited state geometry is displaced along the corresponding normal coordinate. In other words, enhanced vibrational modes are those which mimic the geometry change of the molecule in the excited state. Thus the basis for enhancement of low-frequency modes, which are likely out-of-plane distortions of the macrocycle, is that they couple strongly to the delocalized excited state by modulating the interchromophore separation.

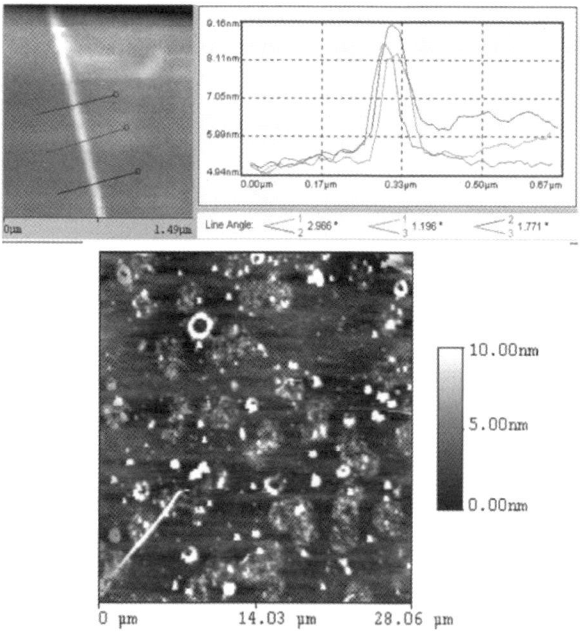

Figure 3.6. AFM images of TCPP aggregates deposited on silica from aqueous HNO_3 (top, showing lone nanorod and line analysis) and from aqueous HCl (bottom, showing wheels, spirals and a stray nanorod.) Reprinted with permission from Doan, S. C.; Shanmugham, S.; Aston, D. E.; McHale, J. L. *J. Am. Chem. Soc.* **2005**, *127*, 5885-5892. Copyright 2005, American Chemical Society.

Atomic force microscopy was performed on the TCPP aggregates deposited on silica from PVA-free aqueous HNO_3 and HCl.[32] As shown in Fig. 3.6, TCPP aggregates deposited from aqueous HCl result in images which show wheels ("nanodonuts") ranging from about 2 to 4 nm in height to as much as 15 nm. The diameters of the wheels range from about 200 to 2000 nm. Some of these structures are imaged as spirals or incomplete wheels. These structures might be consistent with stacks of H_2TCPP^{2+} molecules hydrogen-bonded to chloride counterions through the protonated pyrroles. This sort of interaction was reported by Rosa et al. in Ref. 29 based on the influence of halide ions on the optical spectrum of the tetraphenylporphyrin diacid cation, H_2TPP^{2+}, revealing that the saddling of the porphyrin ring was influenced by the identity of the halide. If this hydrogen bonding is weak and nonspecific, variations in the near neighbor geometry could predispose stacks of H_2TCPP^{2+} and intercalated counterions to curl around with different radii, resulting in a range of nanodonut diameters. A small angle between adjacent porphyrin planes in these nanodonuts rationalizes the existence of only a blue-shifted Soret band, the red-shifted exciton being forbidden in this case. The dimension of an individual porphyrin is about 2 nm, comparable to the smallest height of the wheels.

Also shown in Fig. 3.6 (top) is the AFM image of a nanorod of the H_2TCPP^{2+} aggregate deposited from aqueous HNO_3. These nanorods are 3 to 4 nm in height, reminiscent of what is observed for the optically similar aggregate of H_2TSPP^{2-}.[37,38] A rod-shaped structure is satisfying in that it seems qualitatively in accord with the staircase model for both the H_2TSPP^{2-} and H_2TCPP^{2+}/NO_3^- aggregates, and accounts for the presence of both H- and J-bands arising from excitonic transitions polarized respectively perpendicular and parallel to the aggregate chain. However, the question remains as to why the nanorods are taller than the dimensions of the monomer, which is only ~ 2 nm for both porphyrins.

Evidently, the simple linear staircase model for the aggregate is insufficient to explain the AFM images, and is inconsistent with the known nonplanarities of porphryin diacids. In the next section, we focus on the more well-studied H_2TSPP^{2-} aggregate, considering the influence of porphyrin nonplanarity and evidence for structural order on different length scales.

3.2 *Evidence for hierarchal structure in aggregates of TSPP*

The well-known aggregate of TSPP that forms in acidic aqueous solution has been probed by a variety of structural and spectroscopic techniques. Figure 3.7 shows the splitting of the Soret band of the diacid into H- and J-band components in the aggregate. Note also the increase in intensity and further red-shift of the Q band on aggregation. While the existence of red- and blue-shifted components of the aggregate Soret band are consistent with the staircase model, the apparent weaker intensity and greater breadth of the H-band compared to the J-band is not predicted by simple theories. However, inclusion of electron-vibrational coupling in the exciton model may be able to account for differences in the lineshapes of the H- and J-bands.[39,40]

Flow-induced linear dichroism of the TSPP aggregate revealed the J-band and H-band to be polarized parallel and perpendicular to the flow direction, respectively.[41] However, the polarization of the perpendicular component is quite weak and incomplete polarization of the J-band was interpreted to result from imperfect alignment with the flow. For some porphyrins, notably 5-phenyl-10,15-20-tris(4-sulfonatophenyl)porphyrin, vortex-induced circular dichroism (CD) has also been reported and associated with helical tapes imaged in AFM. The helicity of the tapes and the sign of the CD signal can be reversed by swirling in

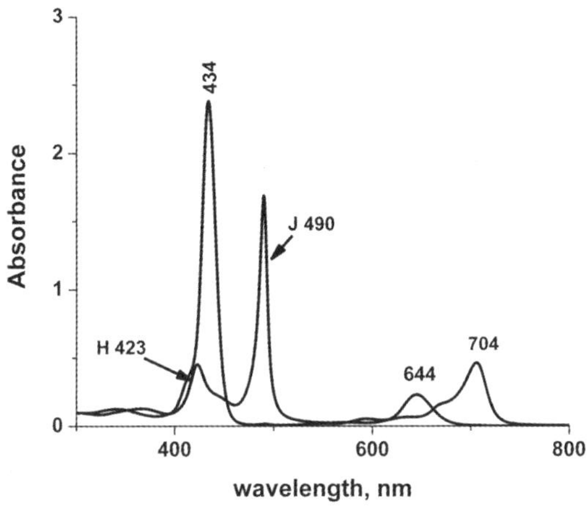

Figure 3.7 Absorption spectrum of H_2TSPP^{2-} monomer diacid, with Soret band peak at 434 nm, and the spectrum of aggregate thereof, with H- and J-bands at 423 and 490 nm, respectively.

opposite directions.[42,43] In the case of the tetrakis-substituted compound highlighted in this chapter, TSPP, chiral sign induction has sometimes been observed in the presence of a chiral template,[44,45] though in other studies a weak CD signal has been reported to result from vortex stirring in the absence of template.[41,46]

Light-scattering data for the TSPP aggregate in solution are suggestive of fractal structure.[47] Micali et al.[48] interpreted their dynamic light scattering data in terms of large (1-1.5 μm), medium (100-200 nm) and small (3-6 nm) aggregates, the latter having an aggregation number between 6 and 32. Small-angle X-ray scattering[49] and electron microscopy[50] both reveal a hollow tube with a ~2 nm shell thickness. A model of a flattened tube formed from rolling up a tape derived from the staircase model was used in Ref. 38 to explain the nanorod images observed by AFM. Vlaming et al.[50] assumed a helical aggregate structure obtained by imagining a planar sheet of overlapping porphyrins which can be rolled into a nanotube. They used the tilt angles of the molecular planes as adjustable parameters to optimize agreement of the calculated optical spectrum with experiment. The resulting small tilt angles would not, on their own, account for a 2 nm shell thickness, however, and this discrepancy was attributed to the possibility that waters of hydration contribute to the shell thickness revealed by electron microscopy.

Any structural model of the TSPP aggregate must account for the following puzzling features: the existence of an isosbestic point in the normalized concentration-dependent absorption spectrum,[24] and a wide range of reported coherence numbers.[31,51–53] An isosbestic point can also be observed in a solution of fixed H_2TSPP^{2-} concentration when the ionic strength is varied.[31] This is suggestive of an equilibrium between two different absorbing species and would not be expected if the aggregate forms in a sequential manner; i.e., monomer → dimer → trimer, etc., if each step results in perturbations to the optical spectrum. This may be evidence that the initial step in aggregation is the formation of a smaller assembly (*N*-mer) within which the excitonic coupling is strong. These *N*-mers might be more weakly coupled to one another on assembly into the larger aggregate resulting in smaller spectral perturbations. Perhaps related to this, the coherence number N_c, which is the number of molecules which share the delocalized excited state, has been variously reported in the literature. Using the ratio of the linewidth of the monomer diacid absorption band to that of the aggregate J-band

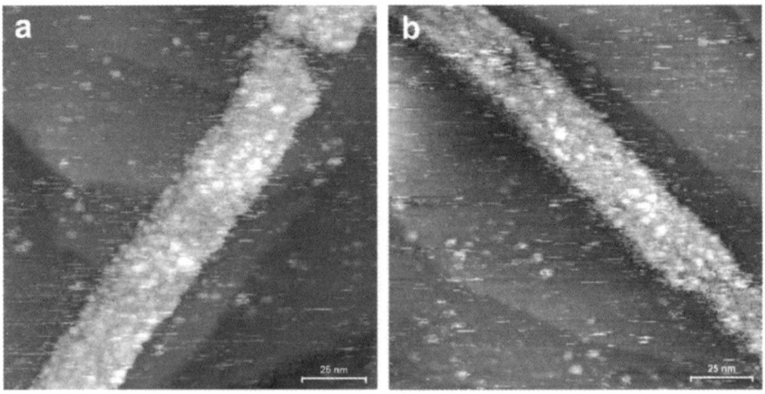

Figure 3.8 Constant current STM images of two different samples of TSPP aggregates deposited on Au(111) from a 5 μM solution of TSPP in 0.75 M HCl. Reprinted with permission from Friesen, B. A.; Nishida, K. A.; McHale, J. L.; Mazur, U. *J. Phys. Chem. C* **2009**, *113*, 1709-18. Copyright 2009, American Chemical Society.

results in $N_c \approx 11$.[31] Taking thermal broadening of the absorption band into account, Koti et al.[51] obtained $N_c \approx 16$. Maiti et al.[52] used fluorescence anisotropy to find $N_c = 22$, while excited state polarizability determined from Stark spectroscopy gives N_c as large as 500.[53] It is possible that these variations derive from experimental conditions which favor different degrees of hierarchal assembly, or that different experiments view the exciton on different timescales.

We recently began a collaboration with the group of Ursula Mazur to use scanning tunneling microscopy (STM) to image the aggregates of TSPP deposited on Au(111).[54] The results, shown in Fig. 3.8, were quite surprising. In agreement with AFM, the aggregates are imaged as nanorods about 4 nm in height and 25 to 27 nm in width. The STM images show that the lengths of the nanorods vary, being as long as microns, and bundling and twinning are often seen. However, higher resolution STM images also reveal ~6 nm diameter disks as constituents of the nanorods. The heights of these disks, 1-2 nm, are consistent with the dimension of the monomer. On repeated probing with the STM tip, these disks begin to separate, and indeed the gold surface is littered with lone disks in some images. Occasional images reveal an intact nanotube, such as the one shown in Fig. 3.9(b), but in most cases the tubes are flattened and imaged as rods. The ~4 nm height of the flattened nanotubes is in good agreement with the ~2 nm shell thickness of nanotubes reported by Vlaming et al. from cryo-electron microscopy.[50]

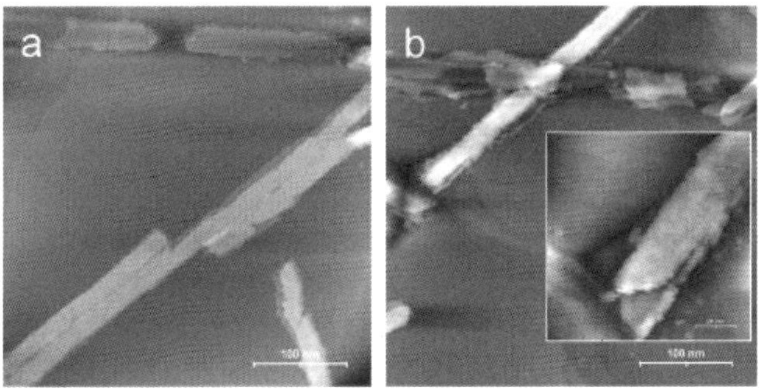

Figure 3.9. Large scale constant current images of TSPP aggregates on Au(111) showing (a) collapsed double layer nanorods and (b) a fractured nanotube . Reprinted with permission from B. A. Friesen, C. C. Rich, U. Mazur, and J. L. McHale, *J. Phys. Chem. C*, **2010**, *114*, 16357-16366. Copyright 2010, American Chemical Society.

We hypothesize that the 6 nm disks are cyclic N-mers driven to assemble by electrostatic and π-stacking interactions. If planar molecules are assumed, the near-neighbor slipped geometry of the staircase model permits the negative charges on the peripheral sulfonato groups to overlap the positive charges from the protonated pyrroles, leading to the conventional cartoon for a J-aggregate. However, protonation of tetraphenylporphyrin derivatives results in diacids which are distorted from planarity by as much as 33°,[55] thus the formation of curved aggregates is expected. A cyclic N-mer is favored when the distortion angle α is an integral divisor of 360°. As a working hypothesis, we assumed the 6-nm disks to be 16-mers arranged as shown in Fig. 3.10, where $N = 16$ is consistent with a 21° distortion. The monomers are assembled on the ring to maximize the interaction of opposite charges on adjacent molecules. Note that we do not resolve individual molecules within the aggregate; the assumption that N is 16 is consistent with a

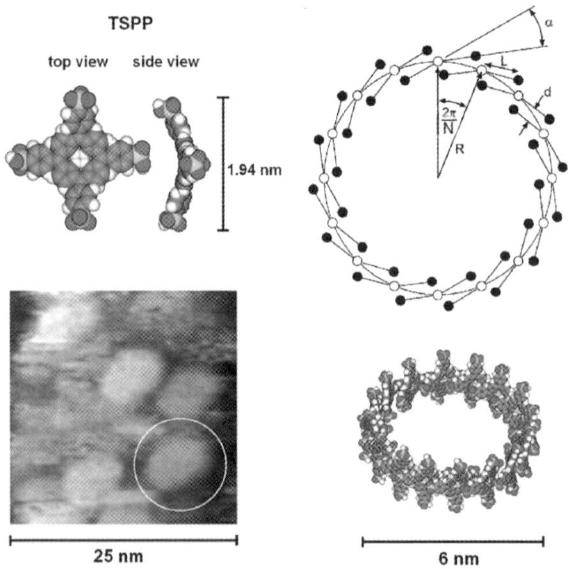

Figure 3.10 Proposed model for 6-nm diameter disks imaged in STM (circled in white on lower left) as cyclic N-mers, driven to assemble in a closed circle by a nonplanar distortion by angle $\alpha = 2\pi/(N+1)$. A distortion by 21° is consistent with the formation of a cyclic 16-mer assembled to permit favorable electrostatic interactions of the positively charged porphyrin centers (represented by open circles in the cartoon on the upper right) with the negative peripheral charges from the SO_3^- groups (filled circles). Reprinted with permission from Friesen, B. A.; Nishida, K. A.; McHale, J. L.; Mazur, U. *J. Phys. Chem. C* **2009**, *113*, 1709-18. Copyright 2009, American Chemical Society.

reasonable estimate of the nonplanarity, the size of the disks, and the smallest coherence numbers reported. Our model does not explain the absence of a hole in the middle of the disk. Counterions and water are expected to contribute to the tunneling current, and water is known to play an important structural role.[56]

The formation of a helical nanotube from cyclic 16-mers is envisioned in Fig. 3.11, and explained further in Fig. 3.12. In a fashion analogous to the depiction of a carbon nanotube, we imagine that the 2-nm thick disks are first packed into a two-dimensional hexagonal array, then rolled along a circumference vector for which the length (51 nm) is consistent with the twice the width of the flattened tubes and a nanotube radius of 8.2 nm. The angle between the planes of adjacent disks in the helix is 140°, and is perhaps a consequence of hydrogen bonding involving the hydrated sulfonato groups. The thickness of the nanotube shell, about 2 nm, corresponds to the size of the porphyrin and is consistent with images from cryo-electron microscopy.[50] Our model accounts for the apparent ~4 nm height of the flattened nanotubes when imaged in both AFM and STM. The helical nanotube model also provides an explanation for vortex-induced circular dichroism. In the next section, the validity of the proposed internal structure of the aggregate will be explored by examining how well it corresponds with polarized resonance Raman data.

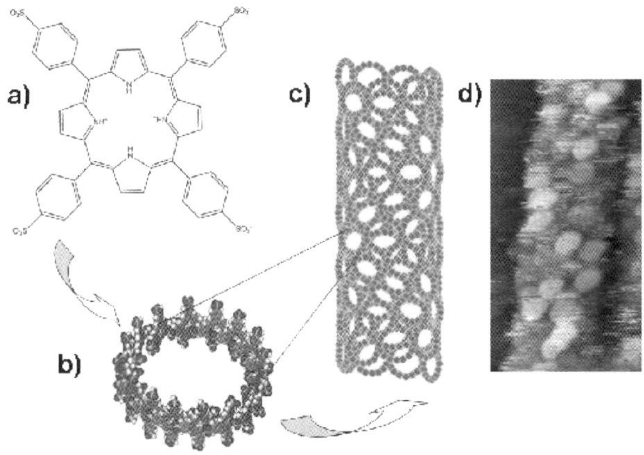

Figure 3.11. Formation of a helical nanotube from the diacid of TSPP (a), which first assembles into a cyclic 16-mer (b), then a helical nanotube, (c) where the 16-mers are represented as strings of beads. The helical nanotube model captures the structural features of the flattened nanotubes as imaged by STM, shown in (d). The drawings in (b) and (d) are adapted with permission from Friesen, B. A.; Nishida, K. A.; McHale, J. L.; Mazur, U. *J. Phys. Chem. C* **2009**, *113*, 1709-18. Copyright 2009, American Chemical Society.

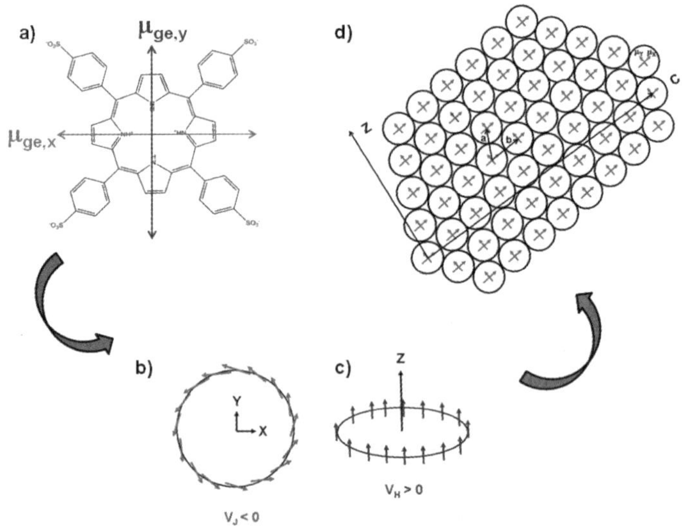

Figure 3.12. Formation of a two-dimensional hexagonal pattern of 6 nm disks. (a) The transition dipole moments of the TSPP diacid monomer are shown. The molecules are assembled into a cyclic 16-mer in which the *x*- and *y*-components of the transition dipole moments are respectively parallel (b) and perpendicular (c) to the plane of the ring, leading to red- and blue-shifted excitons. The former is polarized in the plane of the ring (X,Y) and the latter is perpendicularly polarized (Z). These rings (representing 6-nm diameter disks which are 2 nm thick) are assembled in a hexagonal closed-packed array as shown in (d), where the in-plane components of the J-band transition moments of the cyclic 16-mer are shown. The helical nanotube (see Fig. 3.11) is constructed by rolling the sheet along the circumference vector **C**. **Z** is the tube axis. **a** and **b** are basis vectors of length 6 nm.

3.3 *Resonance Raman Spectroscopy of TSPP Aggregates*

Like many other porphyrin aggregates, those of TSPP form in aqueous solution at high concentration, low pH, and high ionic strength. When the aggregate is deposited on a substrate for scanning probe microscopy studies, it is reasonable to ask whether the imaged structures are the same as those in solution. Figure 3.13 compares resonance Raman (RR) spectra of the TSPP aggregates in solution and deposited on Au(111), excited at 488 nm and thus resonant with the J-band. Clearly, the Raman spectrum is very similar for the two samples. As noted previously, the strong enhancement of the low-frequency modes at 240 and 314 cm^{-1} is characteristic of excitonic coupling. However, our polarized resonance Raman data raised some new questions. When a Raman spectrum is gathered at an excitation wavelength which is resonant with a single electronic level, the depolarization ratio ρ of a mode depends on the symmetry of the mode and the degeneracy of the resonant state. In free porphyrins, resonance with the doubly degenerate Soret band, for example, results in $\rho = 1/8$ for totally symmetric modes, and $\rho = 3/4$ for nontotally symmetric modes. Note that depolarization ratios of 1/8 are observed for totally symmetric modes of the free base porphyrin in the Soret-band excited Raman spectrum, owing to negligible splitting of the this band despite two-fold symmetry. Generally, the totally symmetric modes are the strongest because, in the absence of a change in symmetry on excitation, the excited state displacements are along totally symmetric modes only. When the resonant state is nondegenerate, we expect $\rho = 1/3$ for enhanced totally symmetric modes. Simultaneous resonance with more than one excited electronic level leads to depolarization ratios which depend on excitation frequency and deviate from the simple ratios expected for resonance with a single excited electronic level. Note that these considerations apply to solution phase systems of randomly ordered molecules or aggregates.

Returning to the standard cartoon for a J-aggregate in terms of a linear staircase, the presence of H- and J-bands correlates to complete lifting of the Soret band degeneracy, and one expects $\rho = 1/3$ when resonance Raman spectra are acquired at wavelengths within either the H- or J-band. However, in our cyclic *N*-mer model, the H- and J-bands correspond to transitions polarized respectively perpendicular and parallel to the plane of the circle, as shown in Fig. 3.12. The J-band excited state of an isolated cyclic *N*-mer would be doubly degenerate with mutually perpendicular in-plane transition moments. Thus, neglecting further coupling, among the *N*-mers, the J-band excited resonance Raman spectrum would give $\rho = 1/8$. Surprisingly, the two strongest lines in the J-band excited RR spectrum of the aggregate, at about 240 and 315 cm^{-1}, revealed depolarization ratios close to 0.5. Note also the slight changes in relative intensity and frequency of these two bands on going from solution to Au(111). Though these low-frequency modes have been referred to in the literature as ruffling and doming vibrations,[57] respectively, their precise assignment is uncertain. Both modes are red-shifted in D_2O which strongly suggests they relate to out-of-plane activity of the pyrrolic hydrogens.[58]

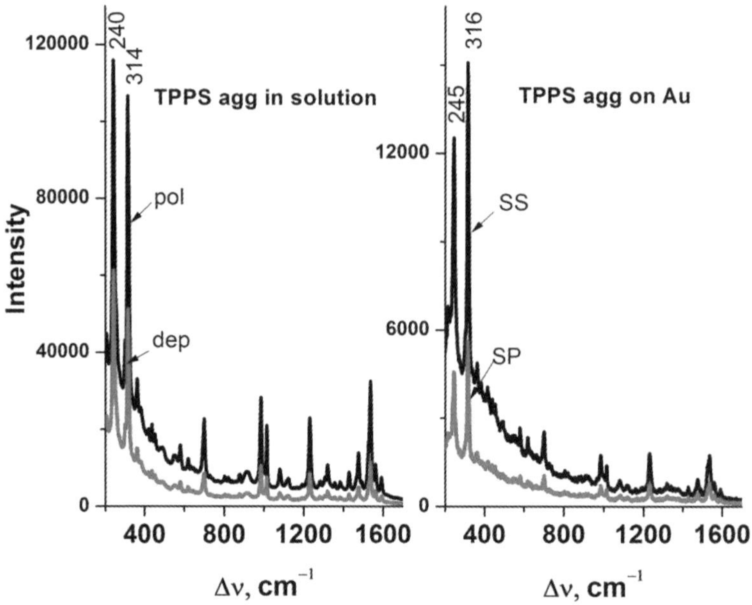

Figure 3.13 Polarized resonance Raman spectra of TSPP aggregates in solution (left) and deposited on Au(111) (right). Adapted with permission from Friesen, B. A.; Nishida, K. A.; McHale, J. L.; Mazur, U. *J. Phys. Chem. C* **2009**, *113*, 1709-18. Copyright 2009, American Chemical Society.

However, assembly of cyclic *N*-mers into a helical nanotube leads to further excitonic coupling, albeit weaker than that within the rings. The degenerate J-band of an isolated *N*-mer splits into two allowed excitons, separated by several cm^{-1} when the *N*-mers are assembled into the nanotube. One of these excitons is nondegenerate and polarized along the axis of the nanotube, while the other is a doubly degenerate level polarized in the orthogonal direction. In this model, the Raman polarizability tensor of a mode on resonance with the J-band would have $\alpha_{xx} = \alpha_{yy} \neq \alpha_{zz}$. As shown in Ref. 24, the occurrence of the nonstandard value of $\rho \approx 0.5$ can be rationalized within this picture. Note that we are using lower case subscripts *x*, *y*, *z* to denote the aggregate (molecular) coordinate system, where *z* is the long axis of the nanotube.

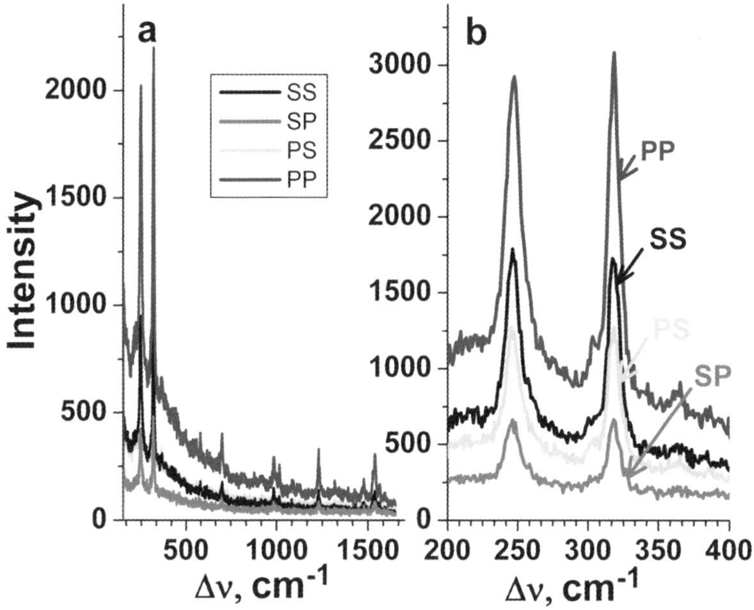

Figure 3.14. Resonance Raman spectrum of TSPP aggregates on Au(111) excited at 488 nm, resonant with the J-band. "S" and "P" denote perpendicular and parallel polarization with respect to the scattering plane, respectively, where the first letter pertains to the exciting radiation and the second letter to the scattered radiation. The spectrum in a) shows the stronger relative intensity of the low-frequency modes, shown in detail in b). Reprinted with permission from B. A. Friesen, C. C. Rich, U. Mazur, and J. L. McHale, *J. Phys. Chem. C* **2010**, *114*, 16357-16366. Copyright 2010, American Chemical Society.

For further insight, we turned to the RR spectrum of the aggregate on Au(111) as a function of the polarization of the incident and scattered light, which may be polarized either perpendicular ("S") or parallel ("P") to the plane defined by the propagation vectors of the incident and scattered beams. Figure 3.14 shows the results for the four possible combinations of incident and scattered light polarizations; PP, SS, PS, and SP, where the first and second letter denote the polarization of the incident and scattered light, respectively. The average intensity ratios PP:SS:PS:SP of the two low-frequency modes at 245 and 316 cm^{-1} are 4.9:3.0:2.0:1.0 We used the surface selection rules derived by Moskovits[59,60] to interpret our results:

$$SS \propto \left|\alpha_{XX}(1+r_s)(1+r_s')\right|^2$$

$$PS \propto \left|\alpha_{XY}(-1+r_p)(1+r_s')\cos\phi + \alpha_{XZ}(1+r_p)(1+r_s')\sin\phi\right|^2$$

$$SP \propto \left|\alpha_{YX}(1+r_s)(1-r_p')\cos\phi' + \alpha_{ZX}(1+r_s)(1+r_p')\sin\phi'\right|^2$$

$$PP \propto \Big|\left[\alpha_{YY}(1-r_p)\cos\phi + \alpha_{YZ}(1+r_p)\sin\phi\right](1-r_p')\cos\phi' + \left[\alpha_{ZY}(-1+r_p)\cos\phi + \alpha_{ZZ}(1+r_p)\sin\phi\right](1+r_p')\sin\phi'\Big|^2 \quad (3.3)$$

In Eqs. (3.3), ϕ is the angle between the propagation direction of the incident beam and the surface normal, while ϕ' is that between the scattered beam and the surface normal. α_{XX}, α_{XY}, etc. are the lab frame components of the polarizability tensor. The relative intensities in Eq. (3.3) also depend on the Fresnel coefficients for S and P polarization evaluated at the wavelength of the incident (r_s, r_p) and scattered (r'_s, r'_p) light. To use the above equations, we found the lab-frame components of the polarizability in terms of the molecule frame components by assuming nanotubes which lie flat on the gold surface with random angle of rotation about the surface normal. We also assumed a diagonal polarizability tensor consistent with expectations for a totally symmetric vibration. We found that we could account for the relative intensities of the four polarized J-band excited RR spectra with the assumption that $|\alpha_{yy}/\alpha_{zz}| \approx 0.06$ and $|\alpha_{xx}/\alpha_{zz}| \approx 0.4$, where x and y are both perpendicular to the nanotube long axis, with x taken to be the direction normal to the surface. The two perpendicular components of α should be equal in the nanotube, but on flattening on the gold surface the x and y directions are no longer equivalent. If we assumed instead a linear aggregate with a J-band polarized completely along the long axis, we would predict intensity ratios in poor agreement with experiment, and the solution phase depolarization ratio would be 1/3 which is also in disagreement with experiment. Assuming $\alpha_{xx} = \alpha_{yy} \neq \alpha_{zz}$ in solution, where the nanotubes are assumed to be intact, ratios of $|\alpha_{xx}/\alpha_{zz}| = |\alpha_{yy}/\alpha_{zz}| \approx 0.14$ result in a depolarization ratio of 0.5 in agreement with experiment, in fair agreement with the average of $|\alpha_{yy}/\alpha_{zz}|$ and $|\alpha_{xx}/\alpha_{zz}|$ determined from the data for the nanotubes on Au(111). Thus polarized RR data suggest that the J-band is a composite of two transitions: one is a transition to a long-axis polarized, nondegenerate excited state and the other is a transition to a doubly degenerate excited state polarized in the perpendicular direction. The estimated coherence length N_c derived from comparing the linewidths of the monomer Soret band and aggregate J-band would thus be underestimated as a result of this spectral overlap, and indeed the resulting value of $N_c \approx 11$ is smaller than the numbers obtained from other experimental techniques.

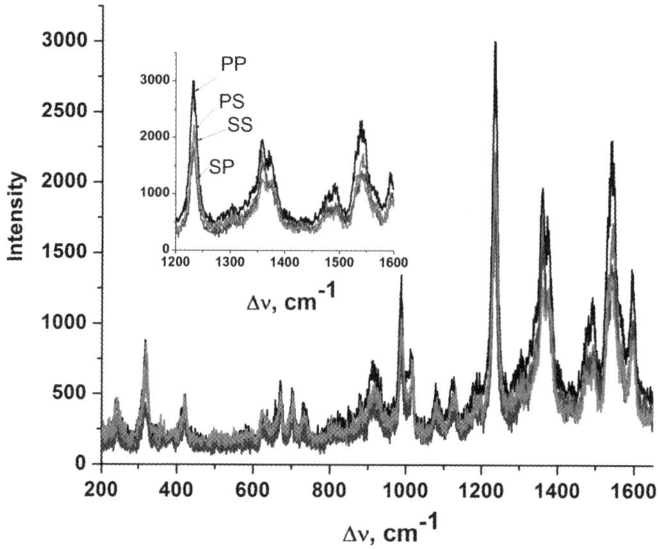

Figure 3.15. H-band excited resonance Raman spectrum of TSPP aggregates obtained at 413 nm on Au(111) as a function of incident and scattered light polarization. The survey scan shows the relatively weaker relative intensity of the low-frequency modes, compared to the J-band excited Raman data of Fig. 3.14. The inset shows the 1200–1600 cm^{-1} region on an expanded scale. Reprinted with permission from B. A. Friesen, C. C. Rich, U. Mazur, and J. L. McHale, *J. Phys. Chem. C* **2010**, *114*, 16357–16366. Copyright 2010, American Chemical Society.

Excitation at a wavelength that is near resonance with the H-band gives a completely different story. As shown in Fig. 3.15, the low-frequency modes that dominate the J-band RR spectrum are rather weak compared to the skeletal modes in the H-band RR spectrum, for both the solution phase aggregate (not shown) and the one deposited on gold.[24] Consistent with the connection between coherence and activity of low-frequency modes, we also found the resonance light scattering signal in the vicinity of the H-band to be weak compared to that from the J-band.[24] An attempt to use the surface selection rules of Eqs. (3.3) to interpret the relative intensities PP:SS:PS:SP of the polarized surface RR data, under the assumption of a diagonal polarizability tensor, were unsuccessful. Indeed, the relative intensities for the different polarizations are more similar to one another in the case of H-band excitation. This may suggest several features of the H-band which distinguish it from the J-band. The H-band transition moment may not be well-aligned with respect to the nanotube, such that the approach used to transform lab-frame to molecule-frame polarizabilities in the case of J-band excitation would not apply. This too could be a logical consequence of reduced H-band coherence, but still begs the question as to how the H-band came to be less intense than the J-band. Another consideration is the possible occurrence of off-diagonal components of the Raman tensor, which cause the problem of obtaining tensor components from relative intensities to be underdetermined. Such off-diagonal components result from vibronic (i.e., non-Born-Oppenheimer) coupling of nearby excited electronic states and result in enhancement of non-totally symmetric vibrational modes. In Ref. 24, we speculated that the reduced intensity of the H-band results from vibronic coupling of the B_y component of the monomer Soret band excited state with the Q_x excited state, where x and y here are directions parallel and perpendicular to the direction of aggregation, respectively. In the lowered site symmetry inside the aggregate, the B_x-Q_y vibronic coupling is not necessarily equal to the B_y-Q_x coupling. If the latter is larger, then the perpendicularly polarized "H-band" B_y loses intensity and shifts to the blue while the Q_x band red-shifts and increases in intensity, in accord with experiment. There is evidence that the red-shifted Q-band is polarized along the long-axis of the aggregate,[41,50] despite its apparent lack of coherence as indicated by resonance light scattering.

While these arguments for reduced H-band intensity seem reasonable, they are by no means a definitive explanation. In fact, splitting of the Soret band on aggregation into a weaker blue-shifted component and a stronger red-shifted one is observed in a number of different porphyrin systems,[61] and could derive from general symmetry considerations for the slipped near-neighbor arrangement expected for J-aggregates, or from difference in electron-vibrational coupling in the two bands.[39,40]

Another possibility is that the flexibility of the porphyrin diacid enables further nonplanar distortions in the aggregate. This might account for the increased intensity and red-shift of the Q-band on aggregation as shown in Fig. 3.7. It is conceivable that further distortion, beyond that associated with the formation of the diacid, accompanies aggregation in order to optimize the electrostatic or hydrogen-bonding interactions. Yet another possibility is that excitonic coupling of the H-band with the Q-band perturbs the frequencies and intensities. In the next section, we consider whether simple exciton theory, subject to the constraints of our structural model, can explain the observed optical and resonance Raman spectra.

3.4 *Theory and Spectroscopy of Excitons in a Helical Nanotube*

We pursue a model in which a cyclic 16-mer of H_2TSPP^{2-} is considered to be the basis for the 6 nm diameter disks imaged in STM. The excitonic coupling within the 16-mer is calculated using a point-dipole model for the transition moment, treating only near-neighbor interactions. The choice of $N = 16$ is consistent with the expected electrostatic and π-stacking forces for macrocycles which are distorted from planarity by 21°. We have obtained evidence from STM that the diacid monomer adopts a saddled configuration,[62] and the optical spectrum of the aggregate is suggestive of further distortion upon self-assembly. In the following, we refer to these cyclic *N*-mers as "disks". Further excitonic coupling among the disks will result from treating them as supermolecules with transition dipole moments amplified by a factor of $N^{1/2}$. These 6-nm disks are assumed to be 2 nm thick, in keeping with the porphyrin dimension, and are envisioned to pack in a two-dimensional hexagonal close-packed array which is rolled into a nanotube in a fashion similar to the construction of a carbon nanotube, as shown in Fig. 3.12. Larger inter-disk separations as compared to the couplings of adjacent molecules within the same disk will result in much weaker excitonic splittings despite the larger transition moments. Thus the excitonic coupling within the cyclic *N*-mer will largely decide the optical spectrum with further small perturbations encountered when the disks are assembled into nanotubes.

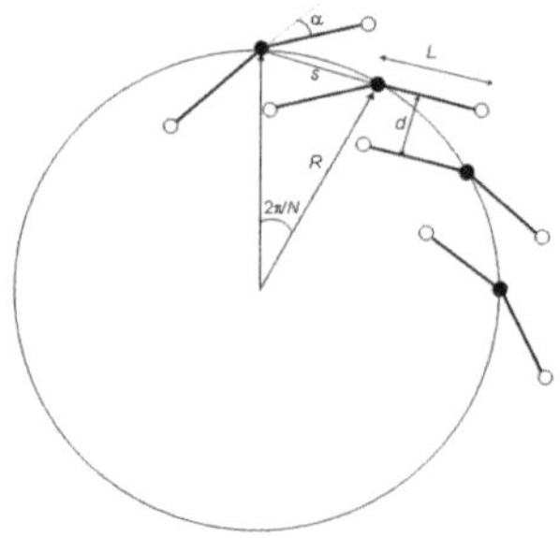

Figure 3.16 Geometrical constraints for the construction of a cyclic N-mer of radius R. The monomer dimension is $2L$ and the distortion from planarity is $\alpha = 2\pi/(N + 1)$.

Figure 3.16 shows the geometrical constraints for the construction of a cyclic N-mer in which the porphyrin diacids, depicted side-on, are distorted from planarity by an angle $\alpha = 2\pi/(N + 1)$ in order to form a closed circle. Typical values of α in the range 20–30° correspond to N from 17 to 12. The interplanar separation of the neighboring planes might be expected to be on the order of $d = 0.35$ nm in keeping with π-stacking, but this is difficult to realize in the distorted macrocycles. In order to optimize the electrostatic interaction of the positively charged porphyrin core and the peripheral negative charges, we require $d^2 + L^2 = s^2$, where s is the distance between the centers of neighboring molecules. The width of the monomer before distortion is taken to be $2L = 2$ nm, and the radius of the disk is $R = 3$ nm. Assuming $N = 16$ and $\alpha = 21°$ leads to an interplanar separation d of about 0.6 nm. The splitting of the Soret band on aggregation depends strongly on this distance as well as the orientation, which is different for the parallel and perpendicular components of the transition moment. The near-neighbor coupling V, in cm^{-1}, is

$$V_{J/H} = \frac{\mu_{ge}^2}{hcs^3}\left[\hat{u}_1 \cdot \hat{u}_2 - 3\left(\hat{u}_1 \cdot \hat{s}\right)\left(\hat{u}_2 \cdot \hat{s}\right)\right] \tag{3.4}$$

where $\hat{u}_1$ is a unit vector in the direction of the in-plane (J) or out-of-plane (H) transition moment of molecule 1 and similarly for molecule 2, s is the distance between neighboring molecules and $\hat{s}$ a unit vector in that direction. The limitations of the point-dipole model are recognized[63] (for example, the dependence of the coupling strength of Eq. (3.4) is a very sensitive function of the presumed intermolecular distance), but we shall adopt it for simplicity to see if the qualitative features of the optical spectrum can be captured.

Upon assembly into the disk, the x component of the Soret-band transition moment will be taken to be parallel to the plane of the ring, while the y component is perpendicular. The coupling strength of Eq. (3.4) is negative for the coupling of the x-components of the transition dipole moments, leading to a red-shifted J-band, and positive for the coupling of the y-components leading to a blue-shifted H-band. In this section, we use lower case letters x, y and z for the molecule frame coordinates and upper case letters X, Y and Z for the coordinate system of the cyclic N-mer. As a first approximation, we take the x and y components of the transition dipole moment to be $\mu_{ge,x} = \mu_{ge,y} \approx 12$ Debye, in accord with the integrated intensity of the Soret band. However, the nonplanarity of the molecule could result in different values of the transition moment in the directions parallel and perpendicular to the disk. While some of the reduced intensity of the H-band could result from this effect, we begin by considering $\mu_{ge,x} = \mu_{ge,y}$. The assumption of a16-mer results in a near-neighbor separation of $s = 1.2$ nm and coupling strength on the order of $\mu_{ge}^2 / hcs^3 \approx 450\ \text{cm}^{-1}$.

We consider next the implications of the cyclic N-mer according to exciton theory.[64] The excitonic wavefunction is written as a superposition of basis states consisting of products of the single-molecule eigenfunctions in which the n-th molecule is in the excited electronic state and all others are in the ground electronic state:

$$|n\rangle \equiv |g_1\rangle|g_2\rangle \ldots |e_n\rangle \ldots |g_{N-1}\rangle|g_N\rangle \tag{3.5}$$

The periodicity of the ring is satisfied by wavefunctions of the form

$$\psi_k = \frac{1}{\sqrt{N}} \sum_{n=1}^{N} e^{i2\pi kn/N} |n\rangle \tag{3.6}$$

where for even N, $k = 0, \pm 1, \pm 2, \ldots N/2$. With the assumption that all but near-neighbor couplings are zero, Eq. (3.6) leads to exciton energies given by

$$E_k = \langle k|H|k\rangle = \sum_{n,m} \left[e^{i2\pi k(n-m)/N} \langle n|H|m\rangle \right] = E_M + 2V \cos\left(\frac{2\pi k}{N}\right) \tag{3.7}$$

where E_M is the excited state energy of the monomer and V is the near-neighbor coupling strength. Thus the excitonic coupling splits the N degenerate excited states into $N/2 + 1$ levels which are all doubly degenerate except for the lowest and highest energy levels.

The frequencies of the H and J band transitions are found using

$$\begin{aligned}\tilde{\nu}_J &= E_B / hc + 2V_J \cos(2\pi k / N) \\ \tilde{\nu}_H &= E_B / hc + 2V_H \cos(2\pi k / N)\end{aligned} \tag{3.8}$$

where E_B/hc is the frequency of the uncoupled Soret (or "B") band, which is 23,000 cm^{-1} (434 nm) in the absence of site shifts in the local environment of the porphyrin in the N-mer. We wish to estimate the coupling strengths V_J and V_H, however, the nonplanarity of the molecule introduces uncertainty in the direction of the transition moment. We consider that the component μ_{ge} in the plane of the disk is parallel to the line segments which are exterior to the circle of Fig. 3.16. These planes make an angle $\varphi = \cos^{-1}(L/s) - \pi/N \approx 21°$ with respect to local tangent and an angle of $\varphi + \beta \approx 32°$ with respect to the line connecting the centers, where $\beta = \pi/N$ is the angle between the local tangent and the polygon edge s connecting adjacent molecular centers. Taking molecule 1 to be the one at 12 o-clock then gives $\hat{u}_1 \cdot \hat{s} \approx 0.85$ and $\hat{u}_2 \cdot \hat{s} \approx 1$. Since $\hat{u}_1 \cdot \hat{u}_2 = \cos(2\pi / N + 1) = 0.93$, this results in the coupling strength for the J-band: $V_J / hc = 450[0.93 - 3(0.85)(0.98)] = -706$ cm^{-1}. As shown below, the $k = \pm 1$ component of the J-band exciton manifold is the only allowed transition, leading to the prediction of a doubly degenerate J-band at 21,700 cm^{-1} (460 nm). Experimentally, the J-band is observed at about 490 nm, so the calculation greatly underestimates the coupling strength. Similarly, the predicted H-band, for which the transition moments are perpendicular to the plane of the disk and only the $k = 0$ exciton is allowed, should be blue shifted from the monomer transition by 900 cm^{-1}, leading to a predicted H-band at 420 nm in good agreement with experiment. Aside from the discrepancies in the individual wavelengths, the predicted splitting of about 2300 cm^{-1} is smaller than the experimental value of 3400 cm^{-1}. Attempting to bring the predicted values in better alignment with experiment by assuming a site shift would increase the agreement with experiment for the J-band at the expense of that for the H-band. Similarly, while the assumption of a shorter distance s would make the splitting larger, it is hard to justify on the basis of the structural data, since the porphyrin nonplanarity forces a greater separation between neighboring molecules. Also, the evidence for B-Q coupling as the origin of the intensity loss of the H-band further implies a blue-shift in the H-band in addition to that caused by exciton coupling. It is evident that a higher level of theory will be needed to account for these details.

The structural distortion which we propose to be the basis for formation of cyclic *N*-mers presents problems for the application of simple exciton theory, since we don't know the in-plane and out-of-plane components of the transition dipole. One could imagine that the distortion which makes the formation of a closed ring possible also transfers oscillator strength from perpendicular to the parallel component of the transition moment, potentially explaining the greater shift and larger intensity of the J-band relative to the H-band. Given the flexibility of tetraphenylporphyrin diacids, it is possible that the appropriate monomer transition frequency to use in the calculation should be red-shifted from the value for the free monomer, explaining the close overlap of the H-band with the absorption of residual monomer. Though it is impossible to experimentally determine the value of the site shift of the monomer, there is evidence from similar porphyrin derivatives and BChl that puckering of the macrocycle can be associated with red-shifts as large as 70 nm.[65,66] Further perturbations may also result from excitonic coupling of the B and Q transitions in the aggregate. Despite these uncertainties, the assumption of a cyclic *N*-mer appears to be in better qualitative agreement with structural and spectroscopic data than the conventional staircase model.

The assumption of cyclic symmetry limits the activity of the *N* exciton states. The intensity of the transition from the ground state $|g\rangle \equiv |g_1\rangle|g_2\rangle \cdots |g_N\rangle$ to the *k*-th exciton state is proportional to the square of the transition dipole moment, which in turn depends on the vector sum of the molecular transition moments.

$$\mu_{gk} = \frac{1}{\sqrt{N}} \sum_n e^{i2\pi kn/N} \vec{\mu}_n \tag{3.9}$$

Equation (3.9) predicts that only the $k = 0$ exciton state in the H-band manifold is allowed, while the allowed J-band transition is to the doubly degenerate $k = \pm 1$ excitonic state. For the J-band, $\vec{\mu}_n = \langle g_n | \vec{\mu} | e_n \rangle$ has components in the X and Y directions, while for the H-band the transition moment is in the Z direction. Using the identity:

$$\sum_{n=1}^{N} e^{i2\pi kn/N} = N\delta_{k0} \tag{3.10}$$

the X and Y components of the allowed J-band excitons are found as summarized in Table 3.1, where θ is the angle between the transition moment and the local tangent

Table 3.1. Components of the J-band transition moment in the cyclic N-mer.

k	$\mu_{k,X}$	$\mu_{k,Y}$
+1	$\frac{\sqrt{N}\mu_{ge}}{2} e^{i\theta} e^{i2\pi/N}$	$\frac{\sqrt{N}\mu_{ge}}{2} (-i)e^{i\theta} e^{i2\pi/N}$
−1	$\frac{\sqrt{N}\mu_{ge}}{2} e^{-i\theta} e^{-i2\pi/N}$	$\frac{\sqrt{N}\mu_{ge}}{2} (i)e^{-i\theta} e^{-i2\pi/N}$

The total intensity of the J-band is proportional to $N\left|\mu_{ge}\right|^2$ and is equally shared by the $k = +1$ and $k = -1$ excitons. It is convenient to take linear combinations of the $k = \pm 1$ excitons to generate excited states for which the transition is polarized in the X and Y directions, respectively, in which case the absolute values of the X and Y-components of the transition moment are given by $\sqrt{N/2}\mu_{ge}$ and are of course equal to one another. Coupling of transition moments perpendicular to the plane of the ring gives rise to the H-band. For the perfect alignment pictured in Fig. 3.12(c), $\mu_{1Z} = \mu_{2Z} = \cdots \mu_{NZ} = \mu_{ge}$. Thus Eq. (3.9) gives a nonzero transition moment only for $k = 0$:

$$\mu_{kZ} = \sqrt{N}\mu_{ge}\delta_{k0} \tag{3.11}$$

In the absence of further assembly, the resonance Raman spectrum of the cyclic N-mer would reveal totally symmetric modes with depolarization ratios of 1/8 for J-band excitation and 1/3 for H-band excitation. While the strongly enhanced modes in the H-band excited RR have $\rho \approx 1/3$, the low-frequency modes in the J-band RR have $\rho \approx 0.5$. To understand this, we turn to the construction of the helical nanotube and its consequences for further coupling of the *X*,*Y* and *Z* components of the transition moments of the cyclic *N*-mers.

The in-plane (*X*,*Y*) and out-of plane (*Z*) components of the disk transition moments are respectively parallel and perpendicular to the 2D array of Fig. 3.12. The tube axis **Z** is perpendicular to the circumference vector **C**. Bold face letters **X,Y,Z** are used below to denote the coordinate system of the nanotube. The nanotube is constructed by rolling the sheet to overlap the head and tail of the circumference vector. The basis vectors **a** and **b** of length 6 nm can be used to construct the 2D lattice. To achieve a flattened tube with a width on the order of 26 nm (approximately half the circumference), we take $\vec{C} = 9\vec{b} - \vec{a}$ to get a circumference of 51 nm, consistent with a tube radius *R* of 8.2 nm. The *X* axes of the disks make an angle of δ = 5.8° with respect to **C** and the helical angle φ is

$$\phi = 2\pi \frac{\vec{b} \cdot \vec{C}}{C^2} = \frac{2\pi}{8.59} = 41.9^\circ \qquad (3.12)$$

The angle φ is the angle of rotation about the **Z**-direction from one disk center to the adjacent one, projected onto the **XY** plane of the nanotube. The coordinates of the center of the *n*-th disk are then $\mathbf{X}_n = R\cos[(n-1)\varphi]$, $\mathbf{Y}_n = R\sin[(n-1)\varphi]$ and $\mathbf{Z}_n = (n-1)a\sin\delta$. The helical angle corresponds to an angle of 138° between the planes of neighboring rings.

This helical nanotube structure was used as the basis for calculating the matrix elements of the transition dipole coupling. Since experimental evidence suggests a much larger degree of coherence for the J-band than for the H-band, we considered only the excitonic coupling of the transition dipole moments in the plane of the disks:

$$H_{nm} = \frac{\mu_J^2}{hcr^3}\left[\hat{u}_n \cdot \hat{u}_m - 3\left(\hat{u}_n \cdot \hat{r}\right)\left(\hat{u}_m \cdot \hat{r}\right)\right] \qquad (3.13)$$

for $n \neq m$. The distance *r* in Eq. 3.13 is the separation between the centers of the *n*-th and *m*-th disks. The zero-order energies of the allowed disk excitons constitute the diagonal elements of this matrix. Our initial efforts to obtain the exciton energies and transition moments for the helical nanotube included both the H- and J-band transitions of the rings as zero-order states, keeping the transition moments as defined above:

$\mu_Z = \sqrt{N}\mu_{ge}$ for the H-band (perpendicular to the nanotube axis) and $\mu_X = \mu_Y = \sqrt{N/2}\,\mu_{ge}$ for J-band (in the plane of the nanotube surface). Consistent with the theory of helical excitons[67] the ring H-band transition moments lead to allowed transitions which are polarized perpendicular to the long axis of the nanotube, and the J-band transition moments give both parallel and perpendicular components which arise from the components of the disk transition dipoles which are respectively parallel to the long axis and to the circumference of the nanotube. However, it was clear that such a calculation could not lead to different intensities for the final H- and J-bands. Further, our experimental evidence[24] argues against a strongly delocalized H-band excited state, justifying its neglect in the treatment of the excitonic coupling within the nanotube. Efforts to include the Q band transition moments, which like the Soret band lead to parallel (J-band) and perpendicular (H-band) moments of the disk, did lead to increased intensity of the Q-band at the expense of both the H- and J-bands, but the calculated H- and J-band transitions remained equal in intensity. Since the Q-band leads to a very small RLS signal, it was considered reasonable to omit the Q-band from the calculation as well.

The calculation of the J-band excited states of the nanotube is very simple in the limit of an infinite length tube.[67] In this case, the component of the transition dipole moment in the nanotube **Z** direction leads to an allowed transition to the $k = 0$ exciton only, while the tangential component of the transition moment results in a doubly degenerate allowed exciton with $k = \pm 1$ polarized perpendicular to **Z.** The in-plane J-band transition moments of the rings can be aligned to point in the longitudinal and circumferential directions, leading to two predicted J-band transitions of the nanotube. One is polarized along the nanotube **Z** direction with transition strength of $N\mu_{ge}^2/2$ per ring. There is also a doubly degenerate transition polarized in the **X** and **Y** directions with each component contributing a transition strength of $N\mu_{ge}^2/4$. This conclusion is independent of the details of the nanotube geometry and certainly one could envision other choices for the circumference vector **C** of Fig. 3.12 which would also be consistent with structural data. However, in all cases this model leads to a diagonal polarizability tensors for which $\alpha_{xx} = \alpha_{yy} \neq \alpha_{zz}$, in intact nanotubes, consistent with our observations for the J-band excited polarized resonance Raman spectra.

To find the energies of the parallel and perpendicular J-bands of the nanotube, we used the geometry of Fig. 3.12 to calculate the matrix elements for a nanotube consisting of 720 disks (11,520 porphyrins), corresponding to a nanotube length of about 440 nm, then used Matlab to find the eigenvectors and eigenvalues. We used a zero-order energy E_J/hc = 20,500 cm^{-1} (488 nm) for the diagonal elements of the matrix, i.e., the J-band energy in the isolated ring. The resulting energies for the parallel and perpendicular components of the J-band in the nanotube were found to be 20,360 cm^{-1} (491 nm) and 20,480 cm^{-1} (488 nm). Each of the two J-band energies of the nanotube is predicted to be slightly red-shifted from its value in the isolated cyclic *N*-mer, more so for the exciton polarized parallel to the nanotube axis. Thus the observed J-band is a superposition of two closely spaced transitions at the frequencies $\tilde{\nu}_{Ja}$ and $\tilde{\nu}_{Jb}$, and the components of the Raman tensor for the fundamental (0 → 1) transition of a totally symmetric mode in the nanotube frame are:

$$\alpha_{xx} = \alpha_{yy} \propto \sum_{v} \left|\mu_{ge,x}\right|^2 \left[\frac{\langle 0|v\rangle\langle v|1\rangle}{\tilde{\nu}_{Ja} + \tilde{\nu}_{vib} - \tilde{\nu}_0 - i\Gamma_e} \right] \tag{3.14a}$$

$$\alpha_{zz} \propto \sum_{v} \left|\mu_{ge,z}\right|^2 \left[\frac{\langle 0|v\rangle\langle v|1\rangle}{\tilde{\nu}_{Jb} + \tilde{\nu}_{vib} - \tilde{\nu}_0 - i\Gamma_e} \right] \tag{3.14b}$$

where $\tilde{\nu}_{vib}$ is the frequency of the mode, $\tilde{\nu}_0$ is the excitation frequency, Γ_e is the dephasing rate, and the numerator contains the Franck-Condon factors connecting the excited vibronic states $|v\rangle$ to the initial, $|0\rangle$, and final, $|1\rangle$, vibrational states within the ground electronic state. The *a*-component of the J-band is taken to be the higher frequency one, polarized perpendicular to the nanotube axis. If the dephasing rates, Franck-Condon factors and deviation from resonance were all equal for both the *a* and *b* components, this would result in $\alpha_{xx} = \alpha_{yy} = 0.5\alpha_{zz}$. The experimentally derived ratios $|\alpha_{xx}/\alpha_{zz}|$ and $|\alpha_{yy}/\alpha_{zz}|$ are smaller than 0.5, an observation consistent with the use of an excitation wavelength closer to resonance with the parallel polarized transition.

The model accounts for the following experimental features. Using resonance Raman spectroscopy of the aggregate in solution and deposited on Au(111), excited at 488 nm, we have found that the relative intensities of strongly-enhanced low-frequency modes as a function of polarization of the incident and scattered light can be accounted for by a diagonal Raman tensor with a larger component in the longitudinal than the transverse directions. This implies a larger value of the transition moment in the nanotube **Z** direction than in each of the two perpendicular directions, in agreement with the helical exciton model. We have also recently obtained Raman depolarization ratios at several wavelengths within the J-band and find depolarization ratio dispersion which is consistent with the composite nature of the J-band.[68] In addition, the imperfect polarization of the J-band in a flow-induced linear dichroism experiment[41,50] is explained. The model also explains the observation of circular dichroism in the presence of chiral substances,[69,70] which would result in a preference for a right- or left-handed helix. We also note that less-than-perfect isosbestic points reported by us[24] and by others[34] can be understood in our model, since the formation of an initial *N*-mer causes pronounced changes in the optical absorption spectrum followed by more subtle spectral shifts when the rings assemble into nanotubes. Such a scenario has also been envisaged based on light scattering data, which led to the conclusion that the initial aggregation which determines the spectrum is followed by the formation of a loose network.[47] Finally, widely disparate values of the reported coherence number for the TSPP aggregate could result from experimental conditions which perturb the weak coupling among the rings in the nanotube, in addition to the nature of the experimental probe. While the interactions which hold the porphyrins together in the cyclic *N*-mers are presumed to be electrostatic in nature, we suggest that weaker interactions, perhaps water-mediated hydrogen bonds, hold the rings together in the nanotube. The failure to resolve individual disks in all STM images of TSPP nanotubes may indeed be a consequence of the strong excitonic coupling between them. As noted in our previous work,[54] probing by the STM tip tends to perturb the disks and liberate them from the network, and lone disks are observed in the vicinity of these perturbed tubes. Studies of the influence of the medium on the hierarchal structure of the aggregate are thus of great interest and are currently being pursued in our laboratory.[68]

4. Conclusions

The structure and optical spectra of porphyrin derivatives provide insight into the nature of excitonic couplings that result from self-assembly. A key contribution of our work, which has emphasized derivatives of tetraphenylporphyrin diacids, is that porphyrin flexibility and nonplanarity can direct the hierarchal structure of light-harvesting aggregates. Previously puzzling counterion dependence of aggregate morphology and optical spectra of the TCPP diacid aggregate may derive from the influence of the counterion on the nonplanarity of the porphyrin, or from intercalation of counterions in between stacked porphyrins. High resolution STM images of the previously studied TCPP aggregates will help to address this question. Another interesting question that would benefit from theoretical study is whether aggregation is accompanied by further distortion of the porphyrin macrocycle as suggested by the changes the optical spectrum of TSPP diacid on aggregation. The nature of the intermolecular forces which cause the cyclic *N*-mers to assemble into nanotubes needs further study, and we are pursuing the solvent dependence of the optical spectra and scanning probe microscopy images of this aggregate.

Acknowledgments

The support of the National Science Foundation through grant CHE-0848511 is gratefully acknowledged. The author wishes to acknowledge the fruitful contributions of graduate students Chris Rich and Ben Friesen and our collaborators Prof. Ursula Mazur and Prof. K. W. Hipps. I am also grateful to former students Myong Yong Choi and Stephanie Doan for their early contributions to this work. Thanks are due to my colleague Prof. Jim Satterlee for his help with preparing Fig. 2.3, and to Dr. Fritz Knorr for his critical reading of the manuscript.

References

1. G. McDermott, S. M. Prince, A. A. Freer, A. M. Hawthornthwaite-Lawless, M. Z. Papiz, R. J. Codgell, and N. W. Isaacs, *Nature* **374**, 517 (1995).
2. X. Hu and K. Schulten, *Phys. Today* **50**, 28 (1997).
3. S. M. Prince, M. Z. Papiz, A. A. Freer, G. McDermott, A. M. Hawthornthwaite-Lawless, R. J. Codgell and N. W. Isaacs, *J. Mol. Biol.* **268**, 412 (1997).

4. D. G. Nichols and S. J. Ferguson, *Bioenergetics*3 (Academic Press, San Diego, 2002).
5. J. Linnanto and J. Korppi-Tommola, *PhysChemChemPhys* **8**, 663 (2006).
6. H. L. Kee, C. Kirmaier, Q. Tang, J. R. Diers, C. Muthiah, M. Taniguchi, J. K. Laha, M. Ptaszek, J. S. Lindsey, D. F. Bocian and D. Holten, *Photochem. Photobiol.* **83**, 1125 (2007).
7. T. Oba and H. Tamiaki, *Photosynth. Res.* **74**, 1 (2002).
8. J. Chiefari, K. Griebenow, N. Griebenow, T. S. Balaban, A. R. Holzwarth and K. Schaffner, *J. Phys. Chem.* **99**, 1357 (1995).
9. Z.-Y. Wang, T. Kadota, M. Kobayashi, A. Kasuya and T. Nozawa, *J. Phys. Chem. B* **108**, 15422 (2004).
10. Y. Saga, S. Akai, T. Miyatake and H. Tamiaki, *Bioconjugate Chem.* **17**, 988 (2006).
11. F. Franci, G. Venturoli, D. Oesterhelt and J. Köhler, *FEBS Lett.* **546**, 345 (2003).
12. A. M. Van Oijen, M. Ketelaars, J. Koohler, T. J. Aartsma and J. Schmidt *Science* **285**, 400 (1999).
13. V. Huber, M. Lysetska and F. Würthner, *Small* **3**, 1007 (2007).
14. J. Deisenhofer, O. Epp, K. Miki, R. Huber and H. Michel, *Nature* **318**, 618 (1985).
15. J. P. Allen, G. Geher, T. O. Yeates, H. Komiya and D. C. Rees, *Proc. Natl. Acad. Sci. USA* **84**, 6162 (1987).
16. V. Cherezov, J. Clogston, M. Z. Papiz and M. Caffrey, *J. Mol. Biol.* **337**, 1605 (2006).
17. J. M. Linnanto and J. E. I. Korppi-Tommola, *Photosynth. Res.* **96**, 227 (2008).
18. P. Hilderbrandt, H. Tamiaki, A. R. Holzwarth and K. Schaffner, *J. Phys. Chem.* **98**, 2192 (1996).
19. W. I. White, *The Porphyrins*; Dolphin, D.; Ed. (Academic: NewYork, 1979) Vol V.
20. E. W. Knapp, *Chem. Phys.* **85**, 73 (1984).
21. M. Kasha, H. R. Rawls and M. A. El-Bayoumi, *Pure Appl. Chem.* **11**, 371 (1965).
22. M. Gouterman, *J. Chem. Phys.* **30**, 1139 (1959).
23. A. Stone and E. B. Fleischer, *J. Am. Chem. Soc.* **90**, 2735 (1968).
24. B. A. Friesen, C. C. Rich, U. Mazur and J. L. McHale, J. L. *J. Phys. Chem. C* **114**, 16357 (2010).
25. J. Zimmerman, U. Siggel, J. H. Fuhrhop and B. Röder, *J. Phys. Chem. B* **107**, 6019 (2003).
26. A. Freer, S. Prince, K. Sauer and M. Papiz, A. Hawthornthwaite-Lawless, G. McDermott, R. Cogdell and N. W. Isaacs, *Structure* **4**, 449 (1996).
27. B. Cheng, O. Q. Munro, H. M. Marques and W. R. Scheidt, *J. Am. Chem. Soc.* **119**, 10732 (1997).
28. J. Deisenhofer and H. Michel, *EMBO J.* **8**, 2149 (1989).
29. A. Rosa, G. Ricciardi, E. J. Baerends, A. Romeo and L. M. Scolaro, *J. Phys. Chem. A* **107**, 11468 (2003).
30. R. Rubires, J. Crusats, Z. El-Hachemic, T. Jaramillo, M. López, E. Valls, J.-A. Farrera and J. M. Ribó, *New J. Chem.* **23**, 189 (1999).

31. M. Y. Choi, J. A. Pollard, M. A. Webb and J. L. McHale, *J. Am. Chem. Soc.* **125**, 810 (2003).
32. S. C. Doan, S. Shanmugham, D. E. Aston and J. L. McHale, *J. Am. Chem. Soc.* **127**, 5885 (2005).
33. R. F. Pasternack, P. R. Huber, R. Boyd, G. Engasser, I. Francesconi, E. Gibbs, P. Fosella, G. Cerio Ventura and L. de C. Hinds, *J. Am. Chem. Soc.* **94**, 4511 (1972).
34. D. L. Akins, H.-R. Zhu and C. Guo, *J. Phys. Chem.* **98**, 3612 (1994).
35. D. M. Chen, T. He, D.-F. Cong, Y.-H. Zhang and F. C. Liu, *J. Phys. Chem. A* **105**, 3981 (2001).
36. B. Ren, Z.-Q. Tian, C. Guo and D. L. Akins, *Chem. Phys. Lett.* **328**, 17 (2000).
37. A. D. Schwab, D. E. Smith, C. S. Rich, E. R. Young, W. F. Smith and J. C. de Paula, *J. Phys. Chem. B* **107**, 11339 (2003).
38. R. Rotomskis, R. Augulis, V. Snitka, R. Valiokas and B. Liedberg, *J. Phys. Chem. B.* **108**, 2833 (2004).
39. F. C. Spano, *Acc. Chem. Res.* **43**, 429 (2010).
40. A. Esifeld and J. S. Briggs, *Chem. Phys.* **324**, 376 (2006).
41. O. Ohno, K. Kaizu and H. Kobayashi, H. *J. Chem. Phys.* **99**, 4128 (1993).
42. J. M. Ribó, J. Crusats, F. Sagués, J. Claret and R. Rubires, *Science* **292**, 2063 (2001).
43. Z. El-Hachemi, O. Arteaga, A. Canillas, J. Crusats, J. Escudero, R. Kuroda,. T. Harada, M. Rosa and J. M. Ribo, *Chemistry – A European Journal* **14**, 6438 (2008).
44. N. Micali, V. Villari, M. A. Castriciano, A. Romeo, L. M. Scolaro, *J. Phys. Chem. B* **110**, 8289 (2006).
45. L. Liu, Y. Li and M. Liu, *J. Phys. Chem. C* **112**, 4861 (2008).
46. S. Jiang, L. Zhang and M. Liu, *Chem. Commun.* 6252 (2009).
47. P. J. Collings, E. J. Gibbs, T. E. Starr, O. Vafek, S. Yee, L. A. Pomerance and R. F. Pasternack, *J. Phys. Chem. B* **103**, 8478 (1999).
48. N. Micali, F. Mallamace, A. Romeo, R. Purrello and L. M. Scolaro, *J. Phys. Chem. B* **104**, 5897 (2000).
49. S. C. M. Gandini, E. L. Gelamo, R. Itri and M. Tabak, *Biophys. J.* **85**, 1259 (2003).
50. S. M. Vlaming, R. Augulis, M. C. A. Stuart, J. Knoester and P. H. M. van Loosdrecht, *J. Phys. Chem. B* **113**, 2273 (2009).
51. A. S. R. Koti, J. Taneja and N. Periasamy, *Chem. Phys. Lett.* **375**, 171 (2003).
52. N. C. Maiti, M. Ravikanth, S. Mazumdar and N. Periasamy, *J. Phys. Chem.* **99**, 17192 (1995).
53. T. Katsumata, K. Nakato, T. Ogawa, K. Koiki and T. Kobayashi, *Chem. Phys. Lett.* **477**, 150 (2009).
54. B. A. Friesen, K. A. Nishida, J. L. McHale and U. Mazur, *J. Phys. Chem. C* **113**, 1709 (2009).
55. M. J. Hamor, T. A. Hamor and J. L. Hoard, *J. Am. Chem. Soc.* **86**, 1938 (1964).
56. J. M. Ribo, R. Rubires, Z. El-Hachemi, J.-A. Farrera, L. Campos, G. L. Pakhomov and M. Vendrell, *Mat. Sci. Eng. C* **11**, 107 (2000).
57. H. Kano, T. Saito and T. Kobayashi, *J. Phys. Chem. A* **106**, 3445 (2002).

58. D. L. Akins, S. Özçelik, H.-R. Zhu and C. Guo, *J. Phys. Chem.* **100**, 14390 (1996).
59. M. Moskovits, *J. Chem. Phys.* **77**, 4408 (1982).
60. M. Moskovits and J. S. Suh, *J. Phys. Chem.* **88**, 5526 (1984).
61. H. Donker, R. B. M. Koehorst and T. J. Schaafsma, *J. Phys. Chem. B* **109**, 17031 (2005).
62. B. A. Friesen, B. Wiggins, J. L. McHale, U. Mazur and K. W. Hipps, *J. Am. Chem. Soc.* **132**, 8554 (2010).
63. G. R. Fleming and G. D. Scholes, *Adv. Chem. Phys.* **132**, 57 (2006).
64. S. E. Dempster, S. Jang and R. J. Silbey, *J. Chem. Phys.* **114**, 10015 (2001).
65. K. M. Barkigia, L. Chantranupong, K. M. Smith and J. Fajer, *J. Am. Chem. Soc.* **110**, 7566 (1988).
66. E. Gudowska-Nowak, M. D. Newton and T. Fajer, *J. Phys. Chem.* **94**, 5796 (1990).
67. C. Didraga, J. A. Klugkist, J. Knoester, *J. Phys. Chem. B* **106**, 11474 (2002).
68. C. C. Rich and J. L. McHale, *Phys. Chem. Chem. Phys.*, accepted for publication. DOI: 10.1039/C2CP23363B.
69. L. Zhang and M. Liu, *J. Phys. Chem. B* **113**, 14015 (2009).
70. Z. El-Hachemi, C. Escudero, O. Orteaga, A. Canillas, J. Crusats, G. Mancini, R. Purrello, A. Sorrenti, A. S'Urso and J. M. Ribó, *Chirality* **21**, 408 (2009).

CHAPTER 4

THE MORPHOLOGIES OF MOLECULAR CYANINE DYE AGGREGATES AS REVEALED BY CRYOGENIC TRANSMISSION ELECTRON MICROSCOPY

Hans v. Berlepsch* and Christoph Böttcher†

Forschungszentrum für Elektronenmikroskopie, Institut für Chemie und Biochemie, Freie Universität Berlin, Fabeckstraße 36a, D-14195 Berlin, Germany
E-mail: h.v.berlepsch@fzem.fu-berlin.de and christoph.boettcher@fzem.fu-berlin.de

Cyanine dyes are known since more than 70 years. While the optical properties have early been explained by molecular excitons, detailed structural data of aggregates in solution were missing for a long time. Here we review the morphological structures of cyanine dye aggregates revealed recently by cryogenic transmission electron microscopy. Soft two-dimensional molecular layers, quasi one-dimensional threadlike and tubular aggregates were detected. Often the structures appear to be hierarchically organized to form more complex superstructures.

Keywords: Molecular aggregates, cyanine dyes, spectroscopic properties, cryogenic transmission electron microscopy.

1. Introduction

The study of molecular dye aggregates has a long history, starting in the 1930s with the pioneering works of Jelly[1] and Scheibe[2] on the spectral sensitizer pseudoisocyanine (PIC). For concentrated aqueous solutions they observed a new and sharp optical absorption band, which is shifted to longer wavelengths with respect to the long-wavelength absorption band of the monomers and is accompanied by an intense and nearly resonant fluorescence emission. In honor of their discoverers assemblies showing such a behavior are named J- or Scheibe-aggregates and the novel red-shifted absorption band is called the J-band. Scheibe has early attributed the peculiar spectroscopic behavior of the aggregates to a stacking of the chromophores into polymer-like assemblies. Moreover, from fluorescence quenching experiments he concluded[3] that an optical excitation should be able to migrate along the assemblies over many thousands of molecules.

Since the discovery of J-aggregation a lot of different dyes have been synthesized that are able to form molecular aggregates. Among them are also dyes, which show a hypsochromic shift of its absorption band upon aggregation. The corresponding assemblies have been termed H-aggregates.

Due to their peculiar spectroscopic properties J-aggregates have become commercially important as spectral sensitizers in the photographic industry[4] and they have a large potential for applications in the field of opto-electronics and nonlinear optics.[5]

Today it is generally accepted that the optical spectra of molecular aggregates can be explained by Frenkel´s model of molecular excitons,[6] which is based on dipole-dipole coupling between the transition dipole moments of neighboring molecules. Therefore, the aggregate spectra strongly depend on the spatial arrangement of the dye molecules within the assembly. The first structural models were also suggested by Scheibe.[7] For dye aggregates adsorbed on surfaces he proposed three alternative two-dimensional dye arrangements as schematically shown in Fig. 1.1: a brickstone-work (a), a ladder (b), and a staircase arrangement (c). For aggregates in solution, however, he suggested one-dimensional variants with a molecular arrangement analogous to the two-dimensional structures shown in (e)–(g).

The first quantitative simulations of optical spectra were performed by Kasha[8] and Kuhn.[9,10] They found, that the excitonic interaction depends on the orientation of the transition dipoles relative to the aggregate axis. Thus, large molecular slippage ($\alpha > 54.7°$) results in a hypsochromic shifted H-band while small slippage ($\alpha < 54.7°$) results in a bathochromic shifted J-band. In the case of oblique transition dipoles

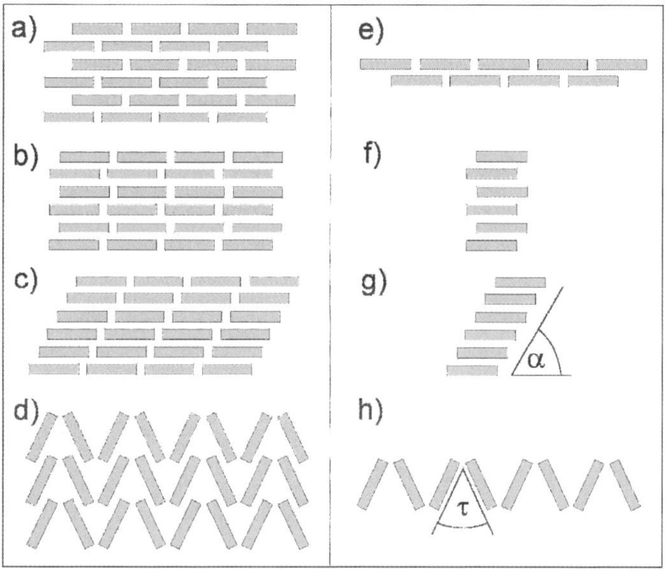

Figure 1.1 Conceivable arrangements of dye molecules in two-dimensional (a–d) and quasi one-dimensional assemblies (e–h).

(tilt angle $\tau \neq 0$) Davydov splitting leads to two exciton states that are perpendicularly polarized, a simplified model is the so-called "herringbone" type arrangement as schematically drawn in Fig. 1.1(d) and (h). Although these models explain specific experiments very well they are often insufficiently confirmed by experimental structure data.

With the development of scanning probe microscopy in the 1990s structural data of dye assemblies formed at interfaces became available. These studies verified the proposed sheet- or thread-like morphology of many of these aggregates and in some cases revealed even their molecular stacking pattern.[11–19] A few dyes form lyotropic liquid-crystalline phases (chromonic phases) and techniques like NMR spectroscopy, polarized-light optical microscopy, confocal microscopy, and small-angle x-ray scattering could be successfully applied to characterize the corresponding mesophases.[20–23] The characterization of individual aggregates dispersed in solution turned out to be more difficult as they are usually of "soft" nature. They usually do not organize in crystalline packing and therefore x-ray crystallographic structure analysis cannot be used. Indirect methods such as light scattering, however, help to characterize size, shape and aggregation number of supramolecular dye assemblies.[24,25] Additionally, fluorescence microscopy under linearly polarized light excitation allows one to determine the alignment of molecules within the aggregates.[26–28] Though a lot of data were obtained, which broadened the knowledge of the supramolecular architecture of aggregates the exact molecular organization of nearly all aggregates remained still vague.

Motivated by the success of cryogenic transmission electron microscopy (cryo-TEM) in the characterization of self-assembled amphiphilic aggregates, polymers and biological systems[29–33] we started to investigate the microstructure of dye aggregates.[34] This technique provides direct high resolution images of the aggregates in their native environment that are free of the drying artifacts due to the specific sample preparation technique. Although the technique does not generally resolve the aggregates internal molecular organization, it often gives valuable quantitative information on the nanometer scale that can be used to improve theoretical structure models.[35–38] In the following we present typical structural data of different supramolecular carbocyanine dye assemblies, which have been characterized by the cryo-TEM technique.

2. Techniques

2.1 *Cryogenic Transmission Electron Microscopy*

Transmission electron microscopy (TEM) is one of the most useful techniques in the study of self-assembled nanostructures in fluid systems. Because the contrast of organic materials for electrons is rather low heavy metal compounds have been introduced[39] as staining additives to enhance the contrast by making use of electron scattering absorption effects. However, the negative staining technique never got entirely rid of the stigma to promote the formation of artifacts through chemical interfering and the final sample drying step.

The drying and staining preparation technique is connected with serious limitations. The changing concentration on drying, dehydration, selective adsorption to certain parts of the specimen, or chemical reactions taking place can lead to drastic changes in the microstructure. Though the negative staining technique is still widely used in inorganic and supramolecular chemistry, structural biology or material science there is only one early TEM study to characterize dye aggregates.[40] These authors published the first micrographs of mesoscopic H-aggregates of a thiacarbocyanine dye grown from aqueous solution and prepared by the staining technique.

The drawbacks of staining and drying can be overcome by using thermal fixation of ultrathin sample layers (100–200 nm). Ultra-fast cooling of the aqueous suspension to a non-crystalline solid state, a process called vitrification, guarantees a low vapor pressure and a freezing of motion. The thermally fixed sample may then be directly examined in the microscope. The necessary contrast for the imaging of objects with low inherent (related to mass density and atomic number) contrast makes use of the phase contrast arising from the coherent interference of the scattered and transmitted electron beams. The theory of image formation shows that the phase contrast is modulated by the so-called phase contrast transfer function (CTF),[41] which originates from aberrations of the microscopic lenses. By defocusing of the objective lens it is possible to enhance the contrast for a certain band of reciprocal spacings, but due to the oscillating character of the CTF contrast reversions occur at the zeros of the function and limit the interpretable optical resolution of the microscope. In practice for every specimen the optimum defocus must be determined. For the dye aggregates we have usually chosen a defocus of 0.9 μm, corresponding to a first zero of the CTF at 1.8 nm (acceleration voltage: $U_B = 100$ kV, spherical aberration: $C_s = 2$ mm).

An important source of artifacts that can result in misinterpretations of TEM images is the sensitivity of the organic material to electron-beam induced radiation damage. Microscopy at low specimen temperature (–175°C) and usage of a minimal electron exposure on recording the micrographs ("low-dose mode") have proven to reduce the radiolysis effects to a negligible level. All the methodology and instrumentation has been developed in the 1980s[42,43] and cryo-TEM is becoming today a standard technique available in many laboratories.[29–33]

The most critical step in cryo-TEM is the specimen preparation. Samples are prepared at room temperature by placing a droplet (~ 5 μL) of the solution on a commercially available perforated carbon film. The electrical conducting carbon film reduces charge induced image distortions by drawing off the applied electrons from the surface. In addition, the carbon film can be used for the microscope alignment and focusing procedures. To enable a perfect spreading of the sample droplet on the grid its surface must be made wettable, which is accomplished by a glow-discharge treatment (60 s treatment at 8 W) shortly before the sample preparation. The excess fluid is blotted off by a sheet of filter paper to create an ultrathin layer (typical thickness of 100 nm) of the solution spanning the holes of the carbon film. Then the grids are immediately vitrified in liquid ethane at its freezing point (–184°C) using a special plunging device. The vitrified samples are transferred under liquid nitrogen into the transmission electron microscope (Philips CM12) using a cryoholder and -stage (Model 626 from Gatan).

2.2 *Image Processing and 3D-Reconstruction*

High-resolution details of cryo-TEM images are often hidden in the noise caused by electron optical effects such as inelastic electron scattering, CTF oscillation, incoherence of the illumination, charging of the sample *etc*. The signal (S) to noise (N) ratio in the individual images is quite unfavorable (a typical ratio is in the order of $S^2/N^2 = 0.001$) and one has to use methods where the ratio can be improved in favor of the signal. If a sufficient number of structurally identical objects (e.g. protein molecules, helices, micelles *etc.*) is available image processing methods following the so-called "single particle approach"[44] can be used to overcome the limitations. Such methods were refined in the recent decade for the successful processing and three-dimensional structure determination of proteins from noisy TEM images. The currently available software packages IMAGIC,[45] SPIDER,[46] EMAN[47] or IMIRS[48] and others can be used to perform the relevant calculations.

The basic idea of the single particle 3D-reconstruction is to use the "projection-slice theorem" to construct a three-dimensional model of the object through a series of two-dimensional projected images. Each TEM image is a 2D-image of the object projected through different angle. The Fourier transforms of these images generate a set of slices which can be interpolated to build up a complete Fourier transform of the object. The inverse Fourier transform is then used to build up the final 3D-model of the object.

The reconstruction of a helical object is a special case inasmuch as we need only a single projection to describe the full three-dimensional volume.[49] Provided we know the helical repeat (pitch) it makes no difference if we follow the helix along its long axis or we revolve around the helix axis at 360°. Therefore, dividing the long axis of the helix over the pitch distance into 360 segments (i.e. twice the length of the repetitive motifs of the projection images as determined by the complementary surface characterization) gives a complete description of the helix. A set of 360 motifs each shifted about one segment along the fiber axis therefore is the structural equivalent of a stepwise 1° movement around the helix. We now have to combine each segment with the corresponding space angles (Euler angles) and we obtain the full description of the helix in exactly the way as a tomographic series (where the angular relation of the subsequently recorded images is known) can be used for the determination of a three-dimensional volume.

3. Cyanine Dyes, Sample Preparation and Spectroscopic Characterization

3.1 *Chemical Structure of Cyanine Dyes*

Within the group of dyes that spontaneously form supramolecular assemblies in solution, cyanine dyes are of outstanding importance. This is due to the unusually high polarizability of the π-electrons along the polymethine backbone in the ground state, which leads to dispersion forces (van der Waals forces) between neighboring molecules and gives rise to substantial attractive forces for aggregation.[50] Examples of cyanine dyes, which were investigated by cryo-TEM, are listed in Fig. 3.1. Besides PIC, CD-1, and ICG we studied several differently substituted derivatives of the 5,5´,6,6´-tetrachlorobenzimidacarbocyanine chromophore (TBC). TTBC ($n = 1$, $m = 1$, $R_1 = R_2 = H$) and TDBC ($n = 1$,

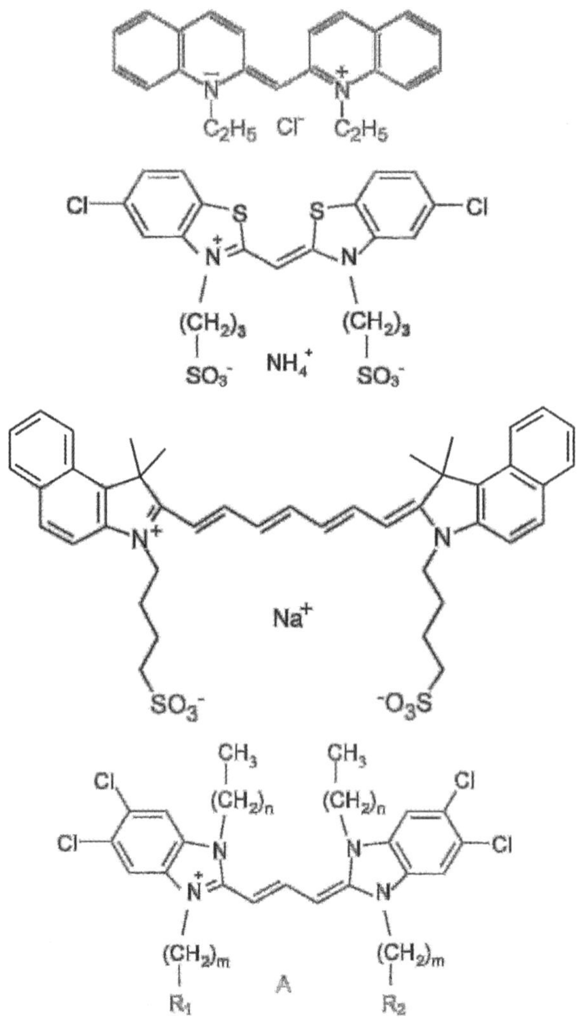

PIC-Cl

CD-1

ICG

TBC

n	m	R_1	R_2	A	Abbreviation
1	2	H	H	Cl^- or I^-	**TTBC**
1	4	SO_3^-	SO_3^-	Na^+	**TDBC**
7	3	COOH	COOH	Br^-	**C8O3**
7	4	COOH	COOH	Br^-	**C8O4**
7	2	SO_3^-	SO_3^-	Na^+	**C8S2**
7	3	SO_3^-	SO_3^-	Na^+	**C8S3**

Figure 3.1. Structural formulas of cyanine dyes that have been investigated by cryo-TEM.

m = 4, $R_1 = R_2 = SO_3^-$) are well-known representatives of the series of TBC dyes that form J-aggregates in aqueous solution.[51,52] Starting from this molecular framework S. Dähne and coworkers[53,54] synthesized a series of novel compounds abbreviated C8O3, C8O4, C8S2, and C8S3 (cf. Fig. 3.1) by systematically varying the N-alkyl-substituents. Some of these dyes are now commercially available from FEW chemicals.[55] The peculiar properties of the polymethine backbone causing strong dispersion forces are combined with structural elements typical for amphiphilic molecules with hydrophobic and hydrophilic substituents. Solvophobic interactions gain significant influence here and compete with the electrostatic repulsion forces of the charged substituents and the van der Waals forces. The balance of all forces leads to assemblies characterized by the typical spectroscopic features of J-aggregates but novel particularly interesting supramolecular aggregation behavior and aggregate architectures. Because of these features this class of dyes was termed[56] amphi-PIPEs (amphiphiles with *p*igment *i*nteraction *p*erforming *e*nergy migration).

3.2 *Sample Preparation*

In general, prior to the structural characterization all dye solutions were thoroughly characterized by absorption, fluorescence, and circular dichroism spectroscopy as well as fluorescence microscopy. These measurements were necessary to evaluate the optimal experimental conditions under which the dyes form stable aggregates suitable for cryo-TEM investigations.

First, the water solubility of the dye has been checked. While PIC-Cl, CD-1, and ICG are highly soluble in pure water at millimolar concentrations, the TBC-derivatives TDBC, C8O4 and C8O3 gave homogeneous solutions with maximum formation of J-aggregates only in the presence of 1 to 10 mM of NaOH.[34,57] Sometimes stirring for several days was necessary for complete dissolution of the dye.

It is well known, that the nature of the counterion can markedly influence the solubility of the dye in water. Chloride salts of TTBC and PIC,[58] e.g., are much better water-soluble than the respective iodide salts. In the case of poor water solubility we used a so-called "solvent mediated route" instead. Here the dye was first completely dissolved in an organic solvent, such as DMF, DMSO or MeOH and then added to water for aggregation. However, even this preparation protocol was not suitable for all systems. In some cases the dye precipitated before aggregation,[58] or the evolving aggregates became instable after several hours and precipitated.

A remarkable effect was observed for the amphi-PIPEs. When methanolic dye stock solutions were added to water new types of aggregates evolved. This effect can be traced back to the amphiphilic nature of the alcohol, which adsorbs at the aggregate-solvent interface thereby disturbing the packing of dye molecules. Because of the strong impact of surface-active additives on structure formation these effects were carefully investigated. Several non-chiral and chiral alkanols, ethylene glycol, poly(vinyl alcohol), and different surfactants were systematically tested with respect to their effect on the spectroscopic properties, the aggregates morphology and stability.[59–64] For aqueous C8S2 solutions containing methanol a partial phase diagram could be constructed.[65] The amphi-PIPEs are very sensitive to all changes in the molecular packing[66] and therefore the concept of combined spectroscopic and structural investigations was proven to be extremely successful. On the other hand, the results showed, that care has to be taken in the preparation of samples and the interpretation of data.

3.3 *Absorption Spectra*

Normalized absorption spectra of the dyes listed in Fig. 3.1 were measured under different solvent conditions and are presented in Fig. 3.2. The monomer spectra commonly recorded in pure methanol are given as reference to prove that aggregation occurred. The most conspicuous features of the aggregate spectra in aqueous solutions will be briefly discussed in the following.

Fig. 3.2(a) presents the absorption spectra of PIC-Cl. The monomer spectrum (long dashes) corresponds to a single vibrational progression of three lines, which are strongly broadened by coupling to other modes of the monomer or the solvent.[67] The aggregate spectrum (solid line) is characterized by the extremely narrow J-band at 572 nm (full-width at

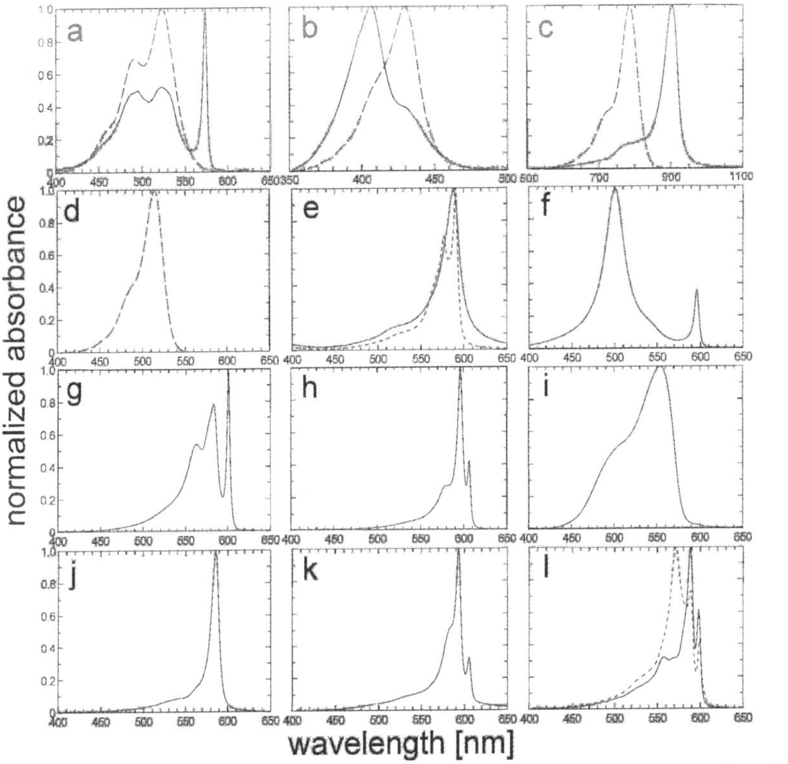

Figure 3.2. Normalized absorption spectra of cyanine dyes. (a) PIC-Cl; dashed: 2.35×10^{-5} M in H_2O; solid line: 4.1×10^{-4} M in 200 mM NaCl solution. (b) CD-1; dashed: 1.0×10^{-3} M in MeOH; solid line: 6.77×10^{-4} M in H_2O. (c) ICG; dashed: 1.0×10^{-3} M in MeOH, solid line: 7.0×10^{-3} M in H_2O after 3 months. (d) TTBC-Cl; 1.0×10^{-3} M in MeOH. (e) TTBC-I; solid line: 7.0×10^{-4} M in 1 mM NaOH after 2 hours, [MeOH] = 16 wt%; dashed: 7.0×10^{-4} M in 1 mM NaOH after 1 d, [DMSO] = 5 wt%; (f) TTBC-Cl; 1.0×10^{-4} M in H_2O. (g) C8O3; 5.0×10^{-5} M in 10 mM NaOH. (h) C8O3; 2.5×10^{-5} M in 10 mM NaOH, [SDS] = 2.5×10^{-5} M. (i) C8O3; 2.35×10^{-5} M in 10 mM NaOH after 3 months, [TTAB] = 2.8×10^{-5} M. (j) C8O4; 8.5×10^{-6} M in 1mM NaOH. (k) C8S3; 3.4×10^{-4} M in H_2O after 7 days. (l) C8S3; solid line: 4.2×10^{-4} M in H_2O after 10 minutes, [MeOH] = 19.5 wt%; dashed: 4.2×10^{-4} M in H_2O after 17 days, [MeOH] = 19.5 wt%.

half-maximum: fwhm = 180 cm^{-1}) and at least three additional bands at shorter wavelengths[68] whose intensity and peak locations are known to strongly depend on the dye concentration. They may be attributed to H-band contributions and residual monomers.[69] Though this particular dye is known since about 75 years the precise arrangement of molecules within the aggregates is still unexplained.

The monomer spectrum of the thiacarbocyanine dye CD-1 is again typical for cyanine dyes but shows one small vibrionic shoulder only. CD-1 represents a dye, which forms H-aggregates at low concentration (6.77×10^{-4} M), while a J-band is absent (Fig. 3.2(b)). At high concentration, and in particular in the presence of salt, however, H- and J-aggregates coexist.[26,70]

The NIR-dye ICG shows the typical monomer spectrum with one vibrionic sub-band. The aggregate spectra of this dye are strongly time- and concentration-dependent.[71] In Fig. 3.2(c) a spectrum is shown obtained at high concentration (7.0×10^{-3} M) after three months of storage. The J-band located around 900 nm is very strong and fairly narrow (fwhm = 580 cm^{-1}). The spectrum shows a remarkable short-wavelength absorption tail that is characterized by a clear shoulder at the position of the monomer band (0–0 transition) and still noticeable absorption at the position of the H-band (697 nm), i.e. monomers, H- as well as J-aggregates coexist.[72]

The monomer spectrum of the TTBC chloride in methanol is presented in Fig. 3.2(d). Identical spectra with a 0–0 transition peaked at around (515–520) nm and a first vibrionic satellite band around 490 nm are found for all derivatives of the TBC family. This behavior contrasts with the diversity of the aggregate spectra (Figs. 3.2(e)–(l)) obtained for the various derivatives under different experimental conditions. The exciting diversity in the aggregate spectra is the expression of a certain degree of freedom in the molecular packing arrangements. Such strong dependence has not been expected[34] and is obviously due to the delicate balance of driving forces, which govern the particular aggregate structures, such as the dispersion forces due to the delocalized π-electron systems, the electrostatic forces due to the ionized groups, the formation of hydrogen bonds or the activity of the hydrophobic effect.

In spite of the complexity of the aggregate spectra some general features or recurring spectral types may be identified. Fresh TTBC iodide solutions prepared by the "solvent mediated route" from a methanolic stock solution reveal a single J-band (Fig. 3.2(e), solid line). It is the classical spectrum observed for many J-aggregating dyes (e.g. Figs. 3.2(c) and (j)), but here its bandwidth is comparatively large (fwhm = 710 cm^{-1}). Interestingly, when prepared from DMSO the J-band is split into two components (dashed line). The reason is not fully clear, but decreasing absorbance and the appearance of precipitating material within a few days after preparation indicate that the aggregates formed under these conditions are not long-term stable. Probably different types of J-aggregates coexist or merge into one another, while another part of the dye is lost due to precipitation or crystallization.

Investigations on the TTBC chloride salt revealed a completely different absorption behavior. A respective aggregate spectrum is shown in Fig. 3.2(f). Due to the high water solubility of TTBC-Cl the solutions could be prepared by the "direct route" in 1mM NaOH. The measured absorption spectrum consists of a relatively broad and strong H-band and an accompanying weak and narrow J-band.[52] The same type of spectrum has been found for C8S2[65] and certain thiacarbocyanines[73,74] and is commonly explained by Davydov splitting.[6] J-aggregates of the classical single-banded type are never observed for TTBC-Cl.

Surprisingly, the same type of two-banded Davydov split spectrum is obtained for certain preparations of TTBC iodide as well.[75] For a solution prepared from DMSO we observed such a spectrum after a long-term storage of several months. Thus, the above discussed changes in the shape and intensity of the J-band obviously reflect only the beginning of a couple of structure transformation processes. The observed precipitation is obviously not complete within a few days, but a part of the dye remains in solution and forms later the novel type of aggregate.

The marked differences in the spectroscopic behavior of the iodide and chloride salts of TTBC reveal a close interrelation between the dye´s solubility and aggregation behavior. From absorption spectroscopy it is to be expected that at least two different morphologies exist.

The absorption spectrum of the C8O3 aggregates is shown in Fig. 3.2(g). The spectrum consists of four single bands,[34] where the longest-wavelength band is extremely narrow (fwhm = 110 cm^{-1}). By polarized absorption spectroscopy it was shown[56,61,76] that all sub-bands belong to the exciton spectrum of a tubular aggregate. The fluorescence emission is in resonance with the lowest energy transitions.[61] The addition of surface-active compounds into aqueous C8O3 solutions leads to changes of the spectral properties. Drastic changes have been observed by the addition of ionic surfactants. For example, the presence of the anionic sodium dodecyl sulfate (SDS) leads to a novel three-banded absorption spectrum (Fig. 3.2(h)) with the middle band becoming the most prominent. Complexation with the cationic trimethyl-tetradecylammonium bromide (TTAB) on the other hand yields stable colloidal solutions characterized by an absorption spectrum that is shown in Fig. 3.2(i). The spectrum looks like a red-shifted monomer absorption spectrum. It suggests that the chromophores within these aggregates are only weakly coupled and/or strongly disordered.[62] The formation of ionic complexes between oppositely charged surfactant and C8O3 dye ions is a remarkable finding, because in the case of all other TBC derivatives complete precipitation by the formation of crystals is observed.[63]

The absorption spectrum of C8O4 aggregates is shown in Fig. 3.2(j). The spectrum is quite simple, exhibiting one single narrow red-shifted J-band (fwhm = 310 cm^{-1}) and a weak second maximum at around 540 nm. The very different behavior of C8O4 compared to C8O3 is remarkable, as the molecular differences are small. C8O4 has only one more methylene group in each of the carboxylate-terminated substituents.

The absorption spectrum of an aggregated C8S3 solution prepared by the "direct route" in pure water is given in Fig. 3.2(k). Three bands can be discerned similar to the C8O3-SDS spectrum, but the peak positions differ. When prepared by the "solvent mediated route" from methanol, however, a completely different spectrum composed of five single bands is obtained immediately after preparation,[64] cf. Fig. 3.2(l) (solid line). This finding impressively demonstrates how the preparation conditions affect the aggregation behavior of C8S3. The spectral effect may in principle be understood by the formation of mixed C8S3-MeOH aggregates. Methanol is a weak amphiphile that adsorbs preferentially at the aggregate-solvent interface. Thereby the packing geometry of the aggregated chromophores is modified resulting in the changes of the optical spectra. Kinetic effects become visible if larger amounts of methanol are present and indicate further structural transformations. This can be seen in the dashed spectrum plotted in Fig. 3.2(l), which slowly evolved within 17 days of storage. Here, a new strong band at around 570 nm emerged, which indicates the formation of a second type of J-aggregates.

A highly complex spectroscopic behavior was also observed for the second sulfonate-substituted dye C8S2.[65] Here again two types of spectra evolved depending on the preparation conditions. One spectrum is characterized by a split J-band, the other by a strong H-band and an accompanying weak and narrow J-band (Davydov split spectrum).

4. Aggregate Structures

4.1 *PIC — Threadlike J-Aggregates*

In a relatively narrow concentration range between 1 and 10 mM PIC-Cl solutions show an exorbitant increase of the J-bands absorbance that is accompanied by the appearance of strong viscoelasticity.[2] Both effects are due to the explosive growth of J-aggregates that form a dense network of fibers.[12,68] To optimize the imaging conditions it appeared useful to add 200 mM of sodium chloride, which enhances J-aggregation and thus lowers the threshold concentration above which aggregation occurs. In this way a looser network of fibers is formed and individual J-aggregates (elementary fibers) can be visualized.[77] Figure 4.1 shows such a typical loose network, which is characterized by the appearance of single fibers and fiber bundles. The diameter of the smallest elementary fiber is 2.3 ± 0.2 nm (cf. the inset to Fig. 4.1) and agrees well with that obtained for the J-aggregates of PIC-Cl in the absence of salt[68] or for solutions of PIC-I. The persistence length of single J-aggregates is at least on the order of 100 nm. These aggregates are generally many hundreds of nanometers long and fiber ends are very rarely found,[77] indicating large end-cap energy. This is presumably caused by the highly anisotropic arrangement of molecules within the threadlike aggregates, which results in high free energy costs for a modified packing at the ends of aggregates. Although the sandwich-like packing of dye molecules into quasi one-dimensional strands (Fig. 1.1(e)) is a very common structure model,[58,68,78] the exact molecular architecture in the fiber is still unknown. Under the assumption of a space filling arrangement

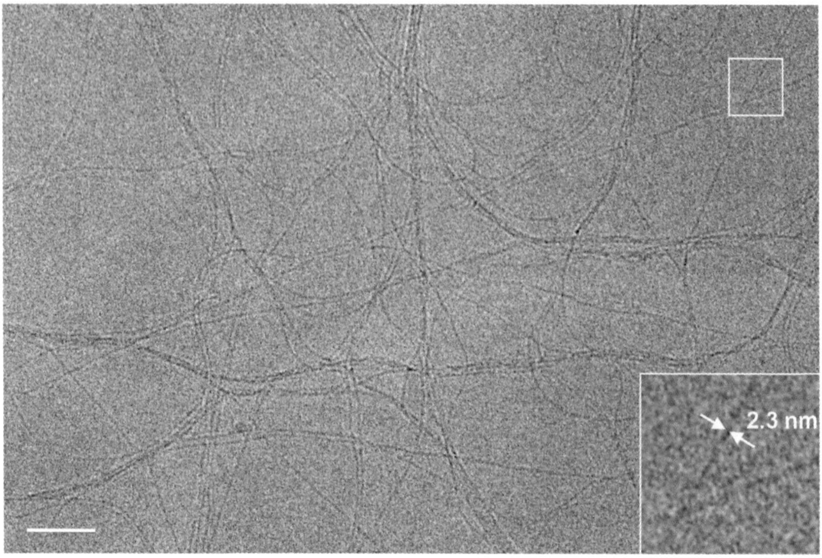

Figure 4.1. Cryo-TEM image of 4.1×10^{-4} M PIC-Cl solution in the presence of 200 mM NaCl. The inset shows the boxed area at high magnification. Bar = 50 nm.

approximately six "unit strands" are needed to completely fill a cylinder with a cross-sectional diameter of ~ 2.3 nm.

4.2 *Mesoscopic H-Aggregates*

H- and J-aggregates often coexist in dye solutions and a strict discrimination of their morphologies is not possible. Moreover, it is a widely held opinion that H-aggregates of cyanine dyes consist of a few molecules only and thus they should be too small to be satisfyingly characterized by electron microscopy. If the particles reach mesoscopic size, TEM becomes the method of choice. Therefore it is not surprising, that only a few structural studies were reported on H-aggregates.[27,40,70,72]

In the case of the dye CD-1, however, individual samples could be prepared, which contained H- or J-aggregates exclusively according to their absorption spectra. This separation was possible by selecting a suitable dye concentration (cf. Fig. 3.2(b)) or by adding salt.[70] Only negligible amounts of monomers coexisted in the respective solutions. Aggregates of mesoscopic size were detected in each solution. The

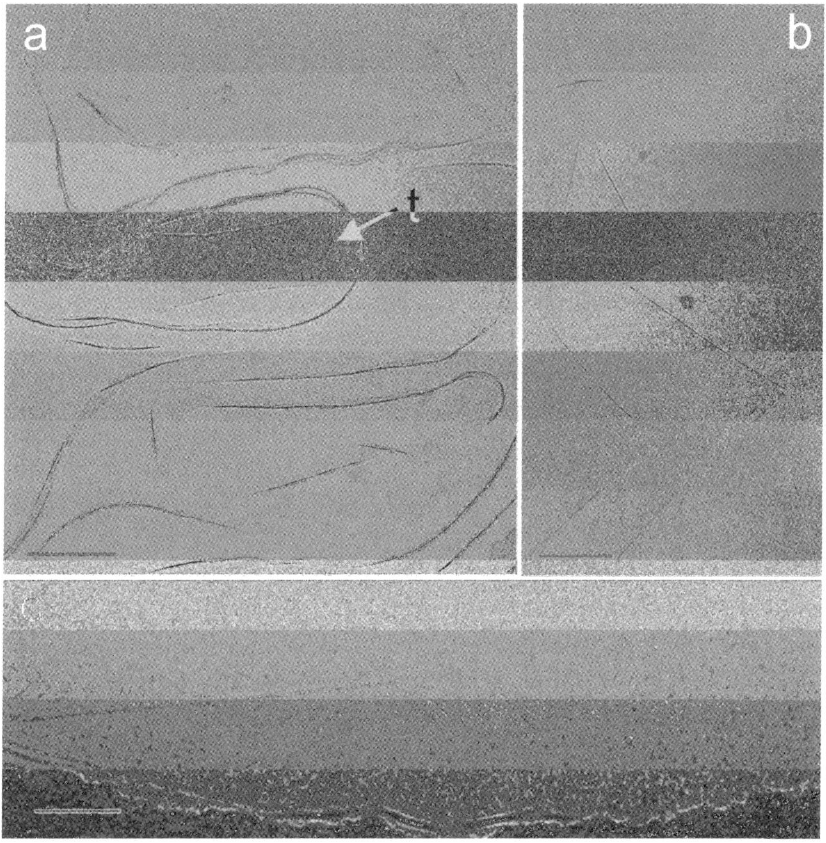

Figure 4.2. Cryo-TEM images reveal mesoscopic H-aggregates. (a) 6.77×10^{-4} M CD-1 solution. Ribbon-like and tubular (t) structures may be discerned. Bar = 100 nm. (b) For comparison are shown the J-aggregates formed in a fresh 3.12×10^{-4} M aqueous CD-1 solution containing 177×10^{-3} M NaCl. Bar = 100 nm. (c) H- and J-aggregates coexist in a 7.0×10^{-3} M ICG solution aged for 3 months. The dark spots represent small H-aggregates. The dark straight lines emerge from folds of sheet-like J-aggregates. Bar = 50 nm.

structures of the H-aggregates are shown in Fig. 4.2(a). Partially folded ribbon-like aggregates dominate, but also short single-layered tubular structures (t) can be seen. In top view orientation the ribbons look smooth and unstructured. Folds appear as dark lines and allow to directly measure the layer thickness of 2.0 ± 0.5 nm. This value corresponds to the dimension of a monomolecular layer. Horizontally oriented tubes (with respect to the incident electron beam direction) can be discerned by their dark walls and the diffuse low-contrast open ends (perpendicular to the tubes axes). Also vertically embedded tubes were sometimes found and could be unambiguously identified by their circular cross-section.[70]

A micrograph revealing the J-aggregates formed in a fresh salt-containing CD-1 solution is reproduced in Fig. 4.2(b) for comparison. Compared to the narrow and rather ribbon-like H-aggregates the J-aggregates typically appear as extensive sheets.

In the case of ICG, H- and J-aggregates could not be separated into individual solutions. The aggregate absorption spectrum displayed in Fig. 3.2(c) is dominated by a strong J-band. A representative cryo-TEM micrograph is reproduced in Fig. 4.2(c). The J-aggregates are sheet-like and possess very low contrast. They can be identified at best from folds, which appear as straight dark lines. The additional dark spots of 2.1 ± 0.5 nm diameter and the short threadlike assemblies of variable length with a diameter of 2.5 ± 0.5 nm represent H-aggregates. This classification is based on the observation that their number density increases with the strength of the H-band.[72] The estimated size of about 2 nm for the smallest particles points to oligomers consisting of a few molecules, the larger assemblies are presumably composed of linearly stacked molecules (Figs. 1.1(f), and (g)).

4.3 *Monolayered Sheet-like and Tubular J-aggregates*

A representative micrograph of a fresh TTBC-I solution prepared by the "solvent mediated route" from a DMSO dye stock solution is presented in Fig. 4.3(a), the respective split absorption spectrum of the solution is shown in Fig. 3.2(e) (dashed line). The cryo-TEM image reveals two types, of aggregates. The dominating morphology consists of tubular aggregates. A line scan perpendicular to a tubule's long axis (inset to Fig. 4.3(a)) yields a value of about 8.2 nm for the outer diameter. The wall thickness can be estimated from the density profile, which gives an average width of about 2 nm, which is in agreement with a monolayered wall architecture. The tubules usually reach lengths in the micrometer range but also a few shorter and wider tubes (giant tubes: gt) are found.

The second morphological type is of sheet-like geometry. It is the most frequently observed structure type of J-aggregates (detected for

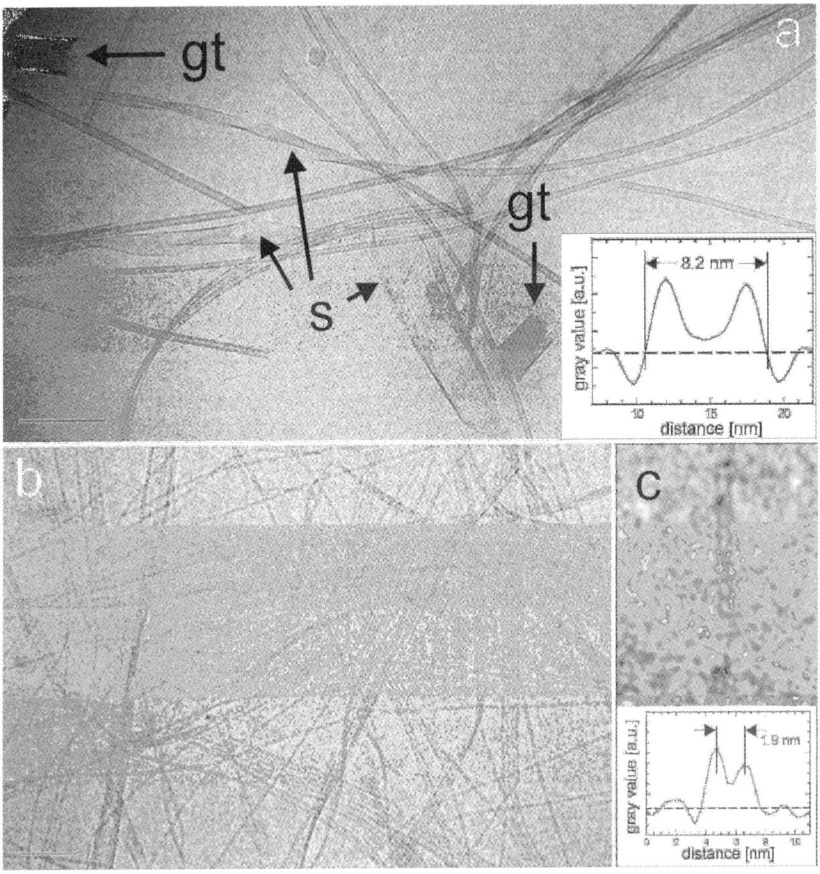

Figure 4.3. Cryo-TEM images of the different types of J-aggregates formed by TTBC. (a) A 1.0×10^{-4} M TTBC-I solution in 1 mM NaOH showed 1.5 hours after its preparation from a DMSO dye stock tubular and sheet-like (s) J-aggregates. The inset shows a line scan across a tubular aggregate. Bar = 100 nm. (b) Threadlike J-aggregates of a 5.4×10^{-4} M TTBC-Cl solution in 1 mM NaOH after six months of storage. Bar = 50 nm. (c) At high magnification the tubular architecture of the threadlike aggregates becomes visible. The inset shows a line scan across a thread-like aggregate.

TDBC,[34] CD-1,[70] ICG,[72] IR-140[72] or 3,3´-bis(4-sulfobutyl)-5,5´-diphenyl-9-ethyloxacarbocyanine betaine sodium salt) and also found for the H-aggregates of CD-1.[70] In the presented case of TTBC-I the two-dimensional layers (s) are extended in one direction and also twisted and hence can be more accurately described as twisted ribbons. The thickness of all these layers is in the order of ~ 2 nm as estimated from the folds of sheets representing direct images of their cross-section. A more accurate value can be extracted from small-angle neutron scattering measurements, as our former study of TDBC showed.[34] The thickness value points to a monolayered arrangement of molecules, whose long axes lie parallel to the layer plane supporting a brickwork-layer model as schematically drawn in Figs. 1.1(a)–(c).

It might be a case that the two types of aggregates possess different absorption bands, which produce an apparently split spectrum (Fig. 3.2(e), dashed line) in superposition, but it cannot unequivocally be experimentally proven. It appears even improbable because another sample, which was prepared from methanolic stock solution and displaying an unstructured single J-band (Fig. 3.2(e), solid line), revealed both morphologies as well.

We were very surprised by the observation that within several months of storage the J-band dominated absorption spectrum of the TTBC-I solution transformed into a typical Davydov-split spectrum consisting of a strong and broad H-band and an accompanying weak J-band as was obtained for the TTBC-Cl solutions immediately after the start of aggregation (Fig. 3.2(f)). Cryo-TEM revealed a novel and expectedly uniform type of aggregate in either case that is characterized by ~ 3 nm thin thread-like assembly. A representative micrograph is shown in Fig. 4.3(b) for TTBC-Cl. Similar to the PIC-Cl aggregates the single fibers often form bundles. At high magnification, however, a pattern of very fine lines becomes visible in the TTBC-Cl fiber bundles and even in isolated individual fibers, which is an indication of the aggregates tubular character (Fig. 4.3(c)). A line scan across a single fiber is added as inset to the figure and reveals a dent density profile of the tubular assemblies. The tube walls are spaced 1.9 nm apart, a value that agrees with the approximate length of a single dye molecule. Recently[75] the split absorption spectrum was ascribed to a flat two-dimensional herringbone arrangement of molecules (Fig. 1.1(d)). The present structural data suggest, however, a more likely tubular structure consisting of a rolled-up herringbone layer.

4.4 *Double-layered J-Aggregates of Amphi-PIPEs*

The TBC derivatives with attached octyl side chains, i.e. C8O3, C8O4, C8S2, and C8S3, introduce the amphiphilic structure element to the family of self-assembling dyes. As expected the octyl substituents try to avoid the contact with the surrounding aqueous solvent by forming bilayers, whereas the charged substituents point towards the solvent. The process of self-assembly is not exclusively driven by the hydrophobic effect, however, but certainly also by the stacking interactions of the chromophores.

Figure 4.4 shows cryo-TEM micrographs for the C8O3 system. The majority of aggregates are composed of several single strands that are twisted to form superhelices (Fig. 4.4(a)).[34] Total lengths up to several micrometers and diameters ranging from about 10.0 ± 1.0 nm for a single strand up to several tens of nanometers for a superhelix are found. Fig. 4.4(b) shows two single strands with well defined walls indicating a tubular architecture. A line scan perpendicular to the aggregate axis reproduced in Fig. 4.4(c) reveals a double-layered density profile for the ~ 3.6 nm wide walls. The double-layered tubular architecture is the structural origin of the complex absorption spectrum[35,38] showing four bands (Fig. 3.2(g)). Several cryo-TEM images for different superhelix diameters, numbers and arrangements of single strands have been analyzed.[34,79] The comparison with theoretically simulated projection images showed that the thick aggregates are composed of single tubules closely packed together and twisted around the superhelix axis in a rope-like manner. The nature of the intermolecular forces keeping the tubular strands together is not fully clear. We speculated that hydrogen bonds between neighbored partially protonated strands are formed and stabilize the complex superstructure. The helicity is an expression of molecular chirality. Since the C8O3 molecules are planar and non-chiral in solution, it must be assumed that the observed optical activity[80] is caused by a chiral packing structure of molecules spontaneously induced during self-assembly in solution. The exact molecular structure could not be determined up to now, however.

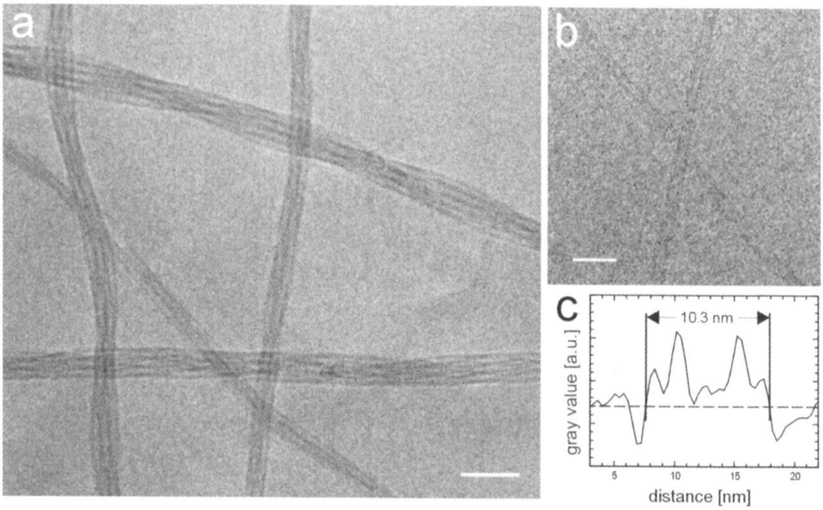

Figure 4.4. Cryo-TEM micrographs of double-layered tubular C8O3 J-aggregates. (a) Helically twisted bundles of tubular aggregates from 2.8×10^{-4} M C8O3 solution in 10 mM NaOH. Bar = 50 nm. (b) Cryo-TEM image of individual tubules. Bar = 30 nm. (c) The line scan across the single tubule reveals its double-layered wall architecture.

A surprising finding was, that increasing the acido-alkyl chain length by only one additional methylene group (C8O4) gives a completely different picture. A simple single-banded absorption spectrum (Fig. 3.2(j)) and a completely different supramolecular morphology was found here.[34] Cryo-TEM (Fig. 4.5) revealed exclusively striated rodlike assemblies of micrometer length. The stripes originate from the stacking of individual ribbons, if oriented parallel with respect to the incident electron beam direction (cf. inset in Fig. 4.5). Between 5 up to 10 individual ribbons build up such multilayered assemblies. The quantitative interpretation of stripes is problematic, because in defocused images of multilayers differences in the scattering potentials give rise to Fresnel fringe formation at the interfaces of adjacent layers.[34,70] The periodicity of the Fresnel fringes reproduces the periodicity of the layers, but the individual widths of bright and dark areas in the image do not provide direct structural information, as they are a function of the defocus value chosen for imaging. The repetition period of stripes for a vertically oriented aggregate section (inset to Fig. 4.5) is 4.0 ± 0.3 nm

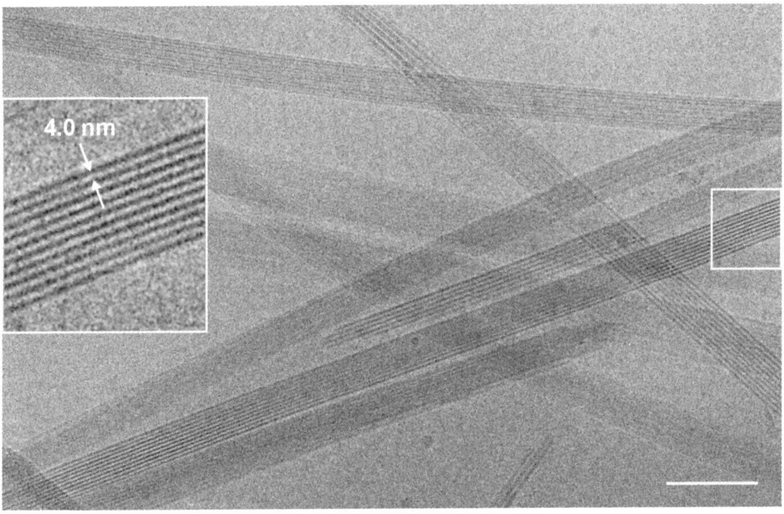

Figure 4.5. Cryo-TEM image of a 2.0×10^{-4} M C8O4 solution in 1 mM NaOH showing only slightly twisted rod-like assemblies. The boxed area reveals the multilayered architecture of the assembly at high magnification. Bar = 100 nm.

and can be interpreted as the thickness of an individual ribbon. The value corresponds fairly to the wall thickness found for C8O3 tubules, which points again to a double-layered architecture of the individual ribbons. The constituting chromophore monolayers are not resolved, however. The layered supramolecular architecture of aggregates and the absence of a marked twist are consistent with the simple absorption spectrum and missing optical activity. Hydrogen bonds could again be suspected as driving forces responsible for the stacking of layers.

The sulfonate-substituted dye C8S3 exhibits similarities to both acido-alkyl derivatives. Fresh solutions are commonly characterized by broad absorption spectra consisting of several sub-bands, but also spectra composed of a single band have been observed under certain experimental conditions.[35,64] Thus, it was not very surprising, to observe both tubular and ribbon-like structures. The peculiarity of this dye is its sensitivity to the conditions of sample preparation and long-term aging.

Prepared by the "direct route" C8S3 forms isolated double-layered tubules (Fig. 4.6(a)) of 17.4 nm diameter (Fig. 4.6(c)), which produce the absorption spectrum composed of three bands displayed in Fig. 3.2(k).

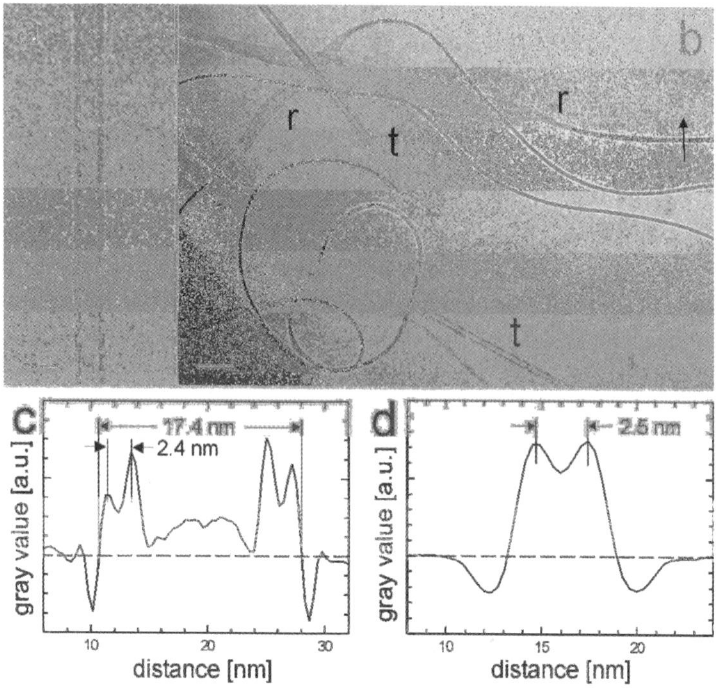

Figure 4.6. Cryo-TEM images of C8S3 J-aggregates. (a) Tubular aggregate formed via the "direct route" 9 days after preparation in pure water. Bar = 20 nm. (b) 4.2×10^{-4} M C8S3 solution prepared via the "solvent mediated route" from a MeOH dye stock after 17 days of storage. Tubular (t) and ribbon-like (r) aggregates coexist. [MeOH] = 19.5 wt%. Bar = 50 nm. (c) Cross-sectional density profile of the tubule shown in (a). (d) Line profile of a double-layered ribbon-like aggregate along the arrow (in b).

When prepared by the "solvent mediated route" from a methanolic stock solution a much broader absorption spectrum with at least five single bands was measured (Fig. 3.2(l)) that changed over the time. After 17 days of storage ribbon-like (r) and tubular (t) aggregates were detected to coexist, as demonstrated by Fig. 4.6(b). Fresh solutions contained only tubules, while ribbons became visible after the appearance of a novel peak in the absorption spectrum (Fig. 3.2(l)). The appearance of ribbon-like aggregates can at least in this particular case be followed by absorption spectroscopy.

The diameter of the tubules in the presence of methanol is 13.0 ± 0.5 nm. The double-layered wall architecture of both C8S3 tubules and ribbons appears frequently well resolved. The line scan across a vertically oriented individual ribbon plotted in Fig. 4.6(d) gives a total width of roughly 4 nm and for the distance between the density maxima constituting the chromophore layers a value of about 2.5 nm, which is equivalent to the corresponding value for the wall of tubules (Fig. 4.6(c)). For ribbons at oblique orientation in vitrified ice a pattern of fine lines becomes often visible (Fig. 4.6(c)). This pattern is not a direct view of the aggregates molecular architecture but probably related to interference effects (Moiré pattern).

The evolution of the two different morphologies is related to the presence of methanol. Methanol is a weak amphiphile that modifies the interaction between the dye molecules by its adsorption at the aggregate-solvent interface. Methanol thereby destabilizes the tubules and produces ribbon-like aggregates, which grow in length and width during several weeks of storage.[64] Multiple stacked ribbons (i.e. consisting of more than two bilayers) as are typical for C8O4 were not observed for C8S3. The structural transformations can be explained by energetic reasons. The driving force is the competition between the free energy gain from the face-to-face attraction between bilayers and the penalty due to the edge energy.[81] When stacking of ribbons is favorable, such as in the case of the C8O4, large edge energy can be tolerated. For C8S3 aggregates, stacking seems most unfavorable and the edge energy promotes growth of ribbons in length and width and eventually their closure into giant tubes.[64]

Individual tubules and multiple stacked ribbons have also been detected for the dye C8S2.[65] Moreover, hybrid-aggregates combining both structural motifs within one individual entity were found here. The appearance of the different morphologies correlates with changes observed in the absorption spectra. Noticeable are the differences in the packing of molecules within the various ribbon-like J-aggregates deduced from the corresponding absorption spectra. While the single J-band in the case of C8O4 and C8S3 points to a simple brickwork-layer arrangement (Figs. 1.1(a)–(c)), the Davydov-split spectrum of C8S2 predicts a packing with two dye molecules per unit cell (e.g., model (d) in Fig. 1.1).

4.5 *Surfactant-induced Structural Transitions*

Ionic surfactants like SDS or TTAB induce drastic changes in the absorbance of C8O3 J-aggregates (Figs. 3.2(h) and (i)). The reason is the strong disturbance of the dye molecules´ original packing geometry through the incorporation of a chemically different amphiphile. Also the morphology changes drastically. In the presence of SDS the twisted rope-like C8O3 aggregates are replaced by single-walled and stiff tubules with an outer cross-sectional diameter of 15.0 ± 1.0 nm and typical bilayer wall thickness of about 4.3 nm (Fig. 4.7(a)).[62] After some days of storage most of the tubules transformed into multilamellar tubes of micrometer length and approximately 70 nm diameter. At high magnification the constituent chromophore monolayers are visible as a pattern of fine lines with a periodicity of 2.05 nm (inset to Fig. 4.7(a)). The giant tubes are composed of multiple concentrically stacked bilayers. SANS measurements on samples containing deuterated SDS proved that the surfactant was indeed incorporated into the dye aggregates. CD measurements indicated that the new structures are optically inactive, which correlates with the absence of helical structural features.

While the C8O3-SDS system still showed the split J-band typical for bilayered tubules, a broad and unstructured monomer-like spectrum appeared when TTAB was added instead (Fig. 3.2(i)). Cryo-TEM (Fig. 4.7(b)) revealed spherical mono- and multilamellar vesicles with bilayer profiles of 4.1 ± 0.2 nm thickness. The small undulations reflect shape fluctuations and indicate a high flexibility of the membrane. A reduced effective charge due to the formation of anion-cation C8O3-TTAB complexes might explain the effect.

The addition of SDS to aggregated C8O4 solutions leads to a separation of the stacks of bilayer ribbons into individual ribbons (Fig. 4.7(c)). Because dye and surfactant head groups are equally charged the adsorption of SDS should increase the effective surface charge and therefore the repulsive electrostatic forces between adjacent bilayers. The repulsive interaction becomes obviously large enough to overcome the

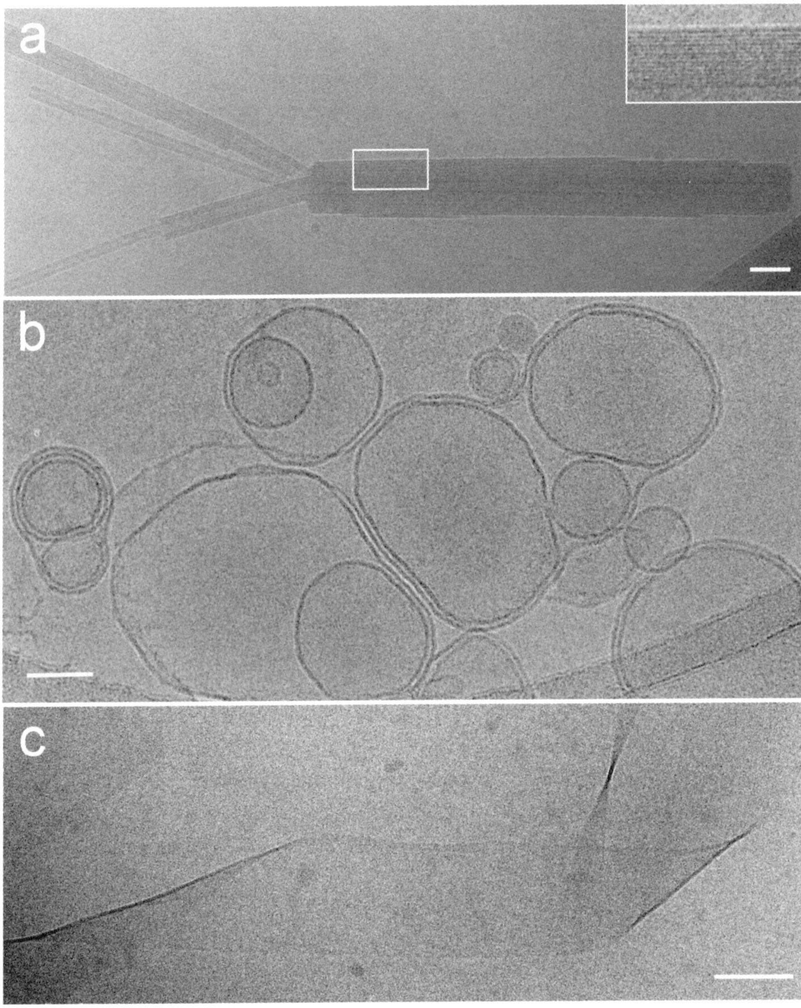

Figure 4.7. Surfactant-induced changes in the morphology of J-aggregates. (a) 2.0×10^{-4} M C8O3 solution in 10 mM NaOH to which one molar equivalent of the anionic surfactant SDS has been added. Giant tubes are formed within one day of storage. The boxed area reveals the multilayered architecture of the wall of tube at high magnification. The repetitive period of the layers is 2.05 nm. Bar = 50 nm. (b) Cryo-TEM image of a 2.0×10^{-4} M C8O3 solution in 10 mM NaOH to which 3.5 equivalents of the cationic surfactant TTAB had been added one day before. Uni- and multilamellar vesicles are seen. The estimated wall thickness is ~ 4.1 nm. Bar = 50 nm. (c) Cryo-TEM image of a 3.0×10^{-4} M C8O4 solution in 1 mM NaOH showing single bilayer ribbons after the addition of one equivalent of SDS. Folds appear as dark lines and allowed to estimate an effective bilayer thickness of ~ 4.2 nm. Bar = 100 nm.

attractive hydrogen-bonding forces between the neighboring and partially protonated bilayers in the stack, resulting in bilayer unbinding. The absorbance remains nearly unaffected by the presence of SDS.[63] This is in strong contrast to the behavior of the tubular C8O3 aggregates. A plausible reason could be packaging constraints due to the differences in the curvature of the assemblies. The high curvature in tubules compared to more planar (and probably more densely packed) ribbons gives rise to sensitive changes in the dipole arrangement.

4.6 *3D-Reconstruction of Tubular Aggregates*

In the case of tubular C8S3 aggregates prepared by the "solvent mediated route" from methanol prominent ultrastructural details are detectable in the raw data (Fig. 4.8(a)) offering a chance for a more detailed structural differentiation using image processing techniques.

By the translational and rotational alignment of structurally identical motifs and subsequent summing the statistically distributed noise present in all individual images is averaged out and the structural information identical over all images is added. 180 of the repetitive motifs along the tubule axis were clipped from several micrographs (e.g., from Fig. 4.8(a) and several other micrographs more) followed by centering and subsequent alignment. By summing the aligned images an almost noise-free image as shown in Fig. 4.8(b) is obtained which shows structural details with a visibly improved contrast.

It should be noted that in the present case all motifs are structurally identical (irrespective of the fiber orientation) and reveal a cross-like pattern which is indicative for a repetitive pattern expected from a helical supramolecular structure. The helical pitch and the handedness of the helix can, however, not been identified from the projection images obtained by the cryo-TEM method. For this purpose a complementary technique is necessary which characterizes the assemblies´ surface structure. Shadowing experiments where the surface of the specimen were decorated by platinum/carbon particles (by evaporation under an oblique angle in a high vacuum chamber) allowed for a surface characterization and gave a left-handed helical tubule with a pitch of 12.8 nm.

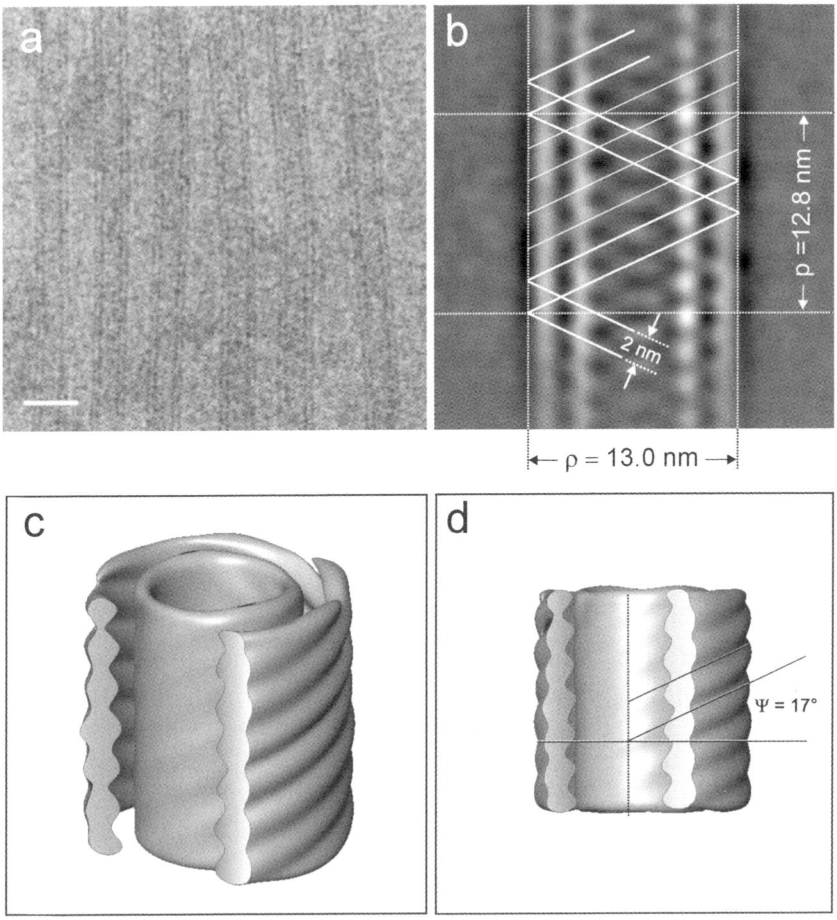

Figure 4.8. 3D-reconstruction of the tubular C8S3 aggregate. (a) Representative cryo-TEM image of tubular J-aggregates prepared via the "solvent mediated route" from MeOH. [C8S3] = 2.0×10^{-4} M. [MeOH] = 10 wt%. Bar = 20 nm. (b) Averaged image calculated from 180 extracted and aligned individual motifs taken from (a). (c) Tilted side view of the reconstructed aggregate. For a better view on the inner layer of the chromophores one quarter of the outer layer was removed. (d) Reconstruction of the aggregate in side view with the pitch angle Ψ indicated.

Following the long axis of the 13.0 nm wide tubule centro-symmetric cross-like striations at a regular distance of 2.0 nm are most notable. The tubule wall is characterized by a typical bilayer density profile at 3.6 nm with an undulating repeat of 2.0 nm. These geometrical relations suggests that the aggregate is composed of 2.0 nm wide bilayered ribbons helically enwrapping a central probably water filled channel.

In Figs. 4.8(c) and (d) the reconstructed helical tubule (using the software package IMAGIC[45]) is shown where the three-dimensional appearance of the structure is graphically obtained by the use of a surface rendering algorithm implemented in AMIRA 5.0 software (**Visage Imaging, Inc.,** San Diego, CA, USA) which gives a vivid description of the spatial organization of the supramolecular assembly.

According to the obtained three-dimensional structure (Fig. 4.8(d)), bilayered strands of ~ 3.6 nm thickness and ~ 2.0 nm width surround the tube with an outer diameter of $\rho = 13.0$ nm in a helical manner, where the repeat distance of one helical turn (pitch) is $p = 12.8$ nm and the inclination angle is $\Psi = 17°$. At this geometrical conditions six parallel bilayered strands are necessary to form a closed helical cylinder per helical turn. For better visibility of the inner monolayer one quarter of the outer layer was removed. Fig. 4.8(c) shows a slightly tilted side view of the aggregate structure reconstruction. The two constituent chromophore layers are clearly visible. Because the pitch of the outer and inner strands must be identical for one helical repeat, the respective pitch angles differ. Hence, an inherent twist between the inner and outer layers is required, but not directly apparent from the projection image in Fig. 4.8(b).

The essential problem to be solved is to find a quantitative molecular packing model that is consistent with the boundary conditions imposed by the presented 2.0 nm resolution 3D-reconstruction and which explains the excitonic band structures of the aggregate. One can expect, that the chromophors will be stacked within each layer plane-to-plane, whereas the octyl chains are intercalating within the 3.6 nm bilayer due to the hydrophobic effect. The geometrical dimensions of the chromophore are approximately $2.0 \times 0.75 \times 0.4$ nm^3, where the 2.0 nm extension is the molecular width measured along the axis connecting the two planar dichlorobenzimidazole moieties. The most simple model, which assumes the helical ripples of ~ 2 nm width to be represented by face-to-face stacked chromophors of 2 nm width, can be ruled out, however, because this type of cofacial stacking would lead to a hypsochromically shifted absorption band, characteristic of H-aggregate structures. As all other molecular dimensions remain well below the resolution achieved in the reconstruction there are still degrees of freedom to speculate about the exact molecular orientation.

Essential aspects of these structural investigations were the starting point of a quantitative interpretation of the linear absorption spectra of the tubular C8S3 J-aggregates prepared via the "direct route", i.e. in the absence of methanol.[35] The array of chromophores forming the aggregate was modeled by wrapping a two-dimensional brickwork layer onto a cylindrical surface. To simulate the double-layered structure two of these cylindrical aggregates were concentrically stacked into each other where the radii of chromophore layers differ by the thickness of the intercalated layer of octyl chains. A quite similar model[38] provided a basis for understanding the nonlinear optical spectra of these[82] and related tubular J-aggregates.[76,83–87]

5. Summary

Cyanine dye aggregates have been known since more than 70 years. While their optical properties have early been explained by the theory of molecular excitons, detailed structural data of aggregates in solution were missing for a long time. The situation clearly changed in the last 15 years when a number of novel analytical methods became available. Here we reviewed the morphological structures of cyanine dye aggregates that were revealed by cryogenic transmission electron microscopy. The two-dimensional organization of chromophors into soft molecular layers has proved to be a widely spread building principle of many J-aggregates, but also quasi one-dimensional threadlike and tubular aggregates were detected. Often the structures appeared to be hierarchically organized to form more complex superstructures. The morphological investigations should be valuable in understanding the molecular architecture of the dye assemblies.

Acknowledgments

The authors thank Prof. S. Dähne for introducing them into the fascinating field of J-aggregates. We are further grateful to Dr. S. Kirstein for insightful discussions. Financial support was obtained from grants of the Deutsche Forschungsgemeinschaft, DFG (KI544/3-1, DA287/5-1, and SfB448).

References

1. E. E. Jelly, *Nature* **138**, 1009 (1936).
2. G. Scheibe, *Angew. Chem.* **50**, 51 (1937).
3. G. Scheibe, A. Schöntag and F. Katheder, *Naturwissenschaften* **27**, 499 (1939).
4. T. Tani, *Photographic Sensitivity; Theory and Mechanisms* (Oxford University Press, New York, 1995).
5. T. Kobayashi, *J-Aggregates* (World Scientific, Singapore, 1996).
6. A. S. Davydov, *Theory of molecular excitons* (Plenum Press, New York, 1971).
7. G. Scheibe, *Angew. Chem.* **52**, 631 (1939).
8. M. Kasha, H. R. Rawls and M. A. El-Bayoumi, *Pure Appl. Chem.* **11**, 371 (1965).
9. V. Czikkely, H. D. Försterling and H. Kuhn, *Chem. Phys. Lett.* **6**, 11 (1970).
10. V. Czikkely, H. D. Försterling and H. Kuhn, *Chem. Phys. Lett.* **6**, 207 (1970).
11. L. Wolthaus, A. Schaper and D. Möbius, *Chem. Phys. Lett.* **225**, 322 (1994).
12. D. A. Higgins and P.F. Barbara, *J. Phys. Chem.* **99**, 3 (1995).
13. L. Dähne, J. Tao and G. Mao, *Langmuir* **14**, 565 (1998).
14. R. W. Owens and D. A. Smith, *Langmuir* **16**, 562 (2000).
15. G. Janssens, F. Touhari, J. W. Gerritsen, H. van Kempen, P. Callant, G. Deroover and D. Vandenbroucke, *Chem. Phys. Lett.* **344**, 1 (2001).
16. M. Vacha, M. Furuki, L. S. Pu, K. Hashizume and T. Tani, *J. Phys. Chem.* **B105**, 12226 (2001).
17. C.-H. Tian, D.-J. Liu, R. Gronheid, M. Van der Auweraer and F. C. De Schryver, *Langmuir* **20**, 11569 (2004).
18. T. E. Kaiser, V. Stepanenko and F. Würthner, *J. Am. Chem. Soc.* **131**, 6719 (2009).
19. D. M. Eisele, J. Knoester, S. Kirstein, J. P. Rabe and D. A. Vanden Bout, *Nature Nanotechnol.* **4**, 658 (2009).
20. T. K. Attwood, J. E. Lydon and F. Jones, *Liquid Crystals* **1**, 499 (1986).
21. G. J. T. Tiddy, D. L. Mateer, A. P. Ormerod, W. J. Harrison and D. J. Edwards, *Langmuir* **11**, 390 (1995).
22. W. J. Harrison, D. L. Mateer and G. J. T. Tiddy, *J. Phys. Chem.* **100**, 2310 (1996).
23. H. Stegemeyer and F. Stöckel, *Ber. Bunsenges. Phys. Chem.* **100**, 9 (1996).

24. P. J. Collings, E. J. Gibbs, T. E. Starr, O. Vafek, C. Yee, L. A. Pomerance and R. F. Pasternack, *J. Phys. Chem.* **B103**, 8474 (1999).
25. B. Neumann, K. Huber and P. Pollmann, *Phys. Chem. Chem. Phys.* **2**, 3687 (2000).
26. H. Yao, M. Omizo and N. Kitamura, *Chem. Commun.* , 739 (2000).
27. H. Yao, K. Domoto, T. Isohashi and K. Kimura, *Langmuir* **21**, 1067 (2005).
28. A. N. Lebedenko, G. Ya. Guralchuk, A. V. Sorokin, S. L. Yefimova and Y. V. Malyukin, *J. Phys. Chem.* **B110**, 17772 (2006).
29. Y. Talmon, *Ber. Bunsenges. Phys. Chem.* **100**, 364 (1996).
30. H. Cui, T. K. Hodgdon, E. W. Kaler, L. Abezgauz, D. Danino, M. Lubovsky, Y. Talmon and D. J. Pochan, *Soft Matter* **3**, 945 (2007).
31. J. Frank, *Three-Dimensional Electron Microscopy of Macromolecular Assemblies* (Oxford University Press, New York, 2006).
32. H. Friedrich, P. M. Frederik, G. de With and N. A. J. M. Sommerdijk, *Angew. Chem. Int. Ed.* **49**, 7850 (2010).
33. C. Schalley, *Analytical Methods in Supramolecular Chemistry*, Second extended edition (Wiley-VHC, Weinheim), in preparation.
34. H. von Berlepsch, C. Böttcher, A. Ouart, C. Burger, S. Dähne and S. Kirstein, *J. Phys. Chem.* **B104**, 5255 (2000).
35. C. Didraga, A. Pugžlys, P. R. Hania, H. von Berlepsch, K. Duppen and J. Knoester, *J. Phys. Chem.* **B108**, 14976 (2004).
36. S. Ganapathy, G. T. Oostergetel, P. K. Wawrzyniak, M. Reus, A. Gomez Maqueo Chew, F. Buda, E. J. Boekema, D. A. Bryant, A. R. Holzwarth and H. J. M. de Groot, *Proc. Natl. Acad. Sci. USA* **106**, 8525 (2009).
37. S. M. Vlaming, R. Augulis, M. C. A. Stuart, J. Knoester and P. H. M. van Loosdrecht, *J. Phys. Chem.* **B113**, 2273 (2009).
38. J. Sperling, A. Nemeth, J. Hauer, D. Abramavicius, S. Mukamel, H. F. Kauffmann and F. Milota, *J. Phys. Chem.* **A114**, 8179 (2010).
39. S. Brenner and R. W. Horne, *Biochim. Biophys. Acta.* **34**, 103 (1959).
40. E. S. Emerson, M. A. Conlin, A. E. Rosenoff, K. S. Norland, H. Rodriguez, D. Chin and G. R. Bird, *J. Phys. Chem.* **71**, 2396 (1967).
41. L. Reimer, *Transmission Electron microscopy. Physics of Image Formation and Microanalysis*, Springer Series in Optical Sciences, Vol. 36, Second edition (Springer-Verlag, Berlin, 1989).
42. M. Adrian, J. Dubochet, J. Lepault and A. W. McDowall, *Nature* **308**, 32 (1984).
43. J. Dubochet, M. Adrian, J.-J. Chang, J.-C. Homo, J. Lepault, A. W. McDowall and P. Schultz, *Quart. Rev. Biophysics* **21**, 129 (1988).
44. D. DeRosier and A. Klug, *Nature* **217**, 130 (1968).
45. M. van Heel, G. Harauz, E. V. Orlova, R. Schmidt and M. Schatz, *J. Struct. Biol.* **116**, 17 (1996).
46. J. Frank, M. Radermacher, P. Penczek, J. Zhu, Y. Li, M. Ladjadi and A. Leith, *J. Struct. Biol.* **116**, 190 (1996).
47. S. J. Ludtke, P. R. Baldwin and W. Chiu, *J. Struct. Biol.* **128**, 82 (1999).

48. Y. Liang, E. Y. Ke and Z. H. Zhou, *J. Struct. Biol.* **137**, 292 (2002).
49. C. Boettcher, H. Stark and M. van Heel, *Ultramicroscopy* **62**, 133 (1996).
50. S. Dähne, *Science* **199**, 1163 (1978).
51. A. H. Herz, *Photogr. Sci. Eng.* **18**, 323 (1974).
52. H. Hada, C. Honda and H. Tanemura, *Photogr. Sci. Eng.* **21**, 83 (1977).
53. U. De Rossi, J. Moll, M. Spieles, G. Bach, S. Dähne, J. Kriwanek and M. Lisk, *J. Prakt. Chem.* **337**, 203 (1995).
54. A. Pawlik, A. Ouart, S. Kirstein, H.-W. Abraham and S. Dähne, *Eur. J. Org. Chem.*, 3065 (2003).
55. FEW Chemicals GmbH, Industriepark Wolfen-Thalheim, P.O.Box 1340, D-06756 Wolfen, Germany.
56. C. Spitz, J. Knoester, A. Ouart and S. Dähne, *Chem. Phys.* **275**, 271 (2002).
57. I. A. Struganova, H. Lim and S. A. Morgan, *J. Phys. Chem.* **B106**, 11047 (2002).
58. E. Daltrozzo, G. Scheibe, K. Gschind and F. Haimerl, *Photogr. Sci. Eng.* **18**, 441 (1974).
59. H. von Berlepsch, S. Kirstein and C. Böttcher, *Langmuir* **18**, 7699 (2002).
60. H. von Berlepsch, S. Kirstein and C. Böttcher, *J. Phys. Chem.* **B107**, 9646 (2003).
61. H. von Berlepsch, S. Kirstein, R. Hania, C. Didraga, A. Pugžlys and C. Böttcher, *J. Phys. Chem.* **B107**, 14176 (2003).
62. H. von Berlepsch, C. Böttcher, A. Ouart, M. Regenbrecht, S. Akari, U. Keiderling, H. Schnablegger, S. Dähne and S. Kirstein, *Langmuir* **16**, 5908 (2000).
63. H. von Berlepsch, M. Regenbrecht, S. Dähne, S. Kirstein and C. Böttcher, *Langmuir* **18**, 2901 (2002).
64. H. von Berlepsch, S. Kirstein, R. Hania, A. Pugžlys and C. Böttcher, *J. Phys. Chem.* **B111**, 1701 (2007).
65. H. von Berlepsch, S. Kirstein and C. Böttcher, *J. Phys. Chem.* **B108**, 18725 (2004).
66. H. v. Berlepsch, K. Ludwig, S. Kirstein and C. Böttcher, *Chem. Phys.* **385**, 27 (2011).
67. H. Mustroph, K. Reiner, J. Mistol, S. Ernst, D. Keil and L. Hennig, *ChemPhysChem* **10**, 835 (2009).
68. H. von Berlepsch, C. Böttcher and L. Dähne, *J. Phys. Chem.* **B104**, 8792 (2000).
69. A. Eisfeld and J. S. Briggs, *Chem. Phys.* **324**, 376 (2006).
70. H. v. Berlepsch, E. Brandenburg, B. Koksch and C. Böttcher, *Langmuir* **26**, 11452 (2010).
71. M. L. J. Landsman, G. Kwant, G. A. Mook and W. G. Zijlstra, *J. Appl. Physiol.* **40**, 575 (1976).
72. H. v. Berlepsch and C. Böttcher, *J. Photochem. Photobiol. A: Chemistry* **214**, 16 (2010).
73. S. Kirstein and H. Möhwald, *Adv. Mater.* **7**, 460 (1995).
74. I. G. Scheblykin, M. A. Drobizhev, O. P. Varnavsky, M. van der Auweraer and A. G. Vitukhnovsky, *Chem. Phys. Lett.* **261**, 181 (1996).
75. B. Birkan, D. Gülen and S. Özçelik, *J. Phys. Chem.* **B110**, 10805 (2006).

76. A. Pugžlys, P. R. Hania, C. Didraga, J. Knoester and K. Duppen, *Solid State Phenom.* **97–98**, 201 (2004).
77. H. von Berlepsch and C. Böttcher, *J. Phys. Chem.* **B106**, 3146 (2002).
78. D. M. Coles, A. J. H. M. Meijer, W. C. Tsoi, M. D. B. Charlton, J.-S. Kim and D. G. Lidzey, *J. Phys. Chem.* **A114**, 11920 (2010).
79. S. Kirstein, H. von Berlepsch, C. Böttcher, C. Burger, A. Ouart, G. Reck and S. Dähne, *ChemPhysChem* **1**, 146 (2000).
80. U. De Rossi, S. Dähne, S. C. J. Meskers and H. P. J. M. Dekkers, *Angew. Chem. Int. Ed.* **35**, 760 (1996).
81. J. V. Selinger, M. S. Spector and J. M. Schnur, *J. Phys. Chem.* **B105**, 7157 (2001).
82. A. Pugžlys, R. Augulis, P. H. M. van Loosdrecht, C. Didraga, V. A. Malyshev and J. Knoester, *J. Phys. Chem.* **B110**, 20268 (2006).
83. J. M. Womick, S. A. Miller and A. M. Moran, *J. Phys. Chem.* **B113**, 6630 (2009).
84. J. M. Womick, S. A. Miller and A. M. Moran, *J. Phys. Chem.* **A113**, 6587 (2009).
85. A. Nemeth, F. Milota, J. Sperling, D. Abramavicius, S. Mukamel and H. F. Kauffmann, *Chem. Phys. Lett.* **469**, 130 (2009).
86. F. Milota, J. Sperling, A. Nemeth, D. Abramavicius, S. Mukamel and H. F. Kauffmann, *J. Chem. Phys.* **131**, 054510 (2009).
87. F. Milota, J. Sperling, A. Nemeth and H. F. Kauffmann, *Chem. Phys.* **357**, 45 (2009).

CHAPTER 5

INTERACTION BETWEEN AGGREGATES OF CYANINE DYES AND BIOMOLECULES

Qian Shang, Junfeng Xiang, Qianfan Yang, Yalin Tang*

Beijing National Laboratory for Molecular Sciences (BNLMS), Center for Molecular Sciences, State Key Laboratory for Structural Chemistry for Unstable and Stable Species, Institute of Chemistry, Chinese Academy of Sciences (ICCAS), Beijing, 100190, P. R. China

**E-mail: tangyl@iccas.ac.cn*

Aggregates states and chiralities of cyanine dye are sensitive to the change of solution environment. This property can be utilized in different fields, especially in fields of biology. Due to the distinct change of spectroscopy properties resulting from the interaction between aggregates of cyanine dye and biomolecules, aggregates of cyanine dye become a excellent probe in recognition of different biomolecules. This chapter mainly focuses on the interaction between aggregates of cyanine dyes and biomolecules, as well as the application in recognition of aggregates of cyanine dyes to different biomolecules.

Keywords: Cyanine dyes; aggregates; bimolecule; interaction.

1. Introduction

Supramolecular chemistry and nanotechnology, together with their application in the construction of functional assemblies and devices, have merged into a challenging field. Among them, constructing chiral supramolecules have been of great importance for decades as these

* Corresponding author.

molecules have exhibited interesting properties for a wide range of applications.[1-7] Compared with achiral supramolecules, chiral supramolecules show interesting feature which makes them a unique class of materials.[8] They have attracted great attention on the scientific community for a long time because of their applications in technology, such as recording information,[9,10] antitumor reagent[11,12] and biological test reagent.[13,14] With the discovery of the novel photophysical and photochemical properties, cyanine dyes have been applied in numerous added fields, such as nonlinear optics,[15–18] biological fluorescent detection,[19,20] biomedicine[21] and photodynamic therapy of tumor.[22]

The generic cyanine dyes consist of two nitrogen centers, one of which are positively charged and are linked by a conjugated chain of an odd number of carbon atoms to the other nitrogen.[23] This prominent feature has been regarded as “Push-pull” alkenes and also forms the basis of the polymethine dyes, which contain the streptopolymethine unit as the chromophore. Depending upon the charge of the streptomethine unit, these dyes are classified as follows: Cationic streptopolymethine-cyanine and hemicyanine dyes (**a**), Anionic streptopolymethine-oxonol dyes (**b**), Neutral streptopolymethine-Merocyanine dyes (**c**), Zwitterionic squaraine-based cyanine dyes (**d**). The feasible construction of cyanine dyes with various functional groups has lead to the widespread applications of cyanine dyes in various fields.

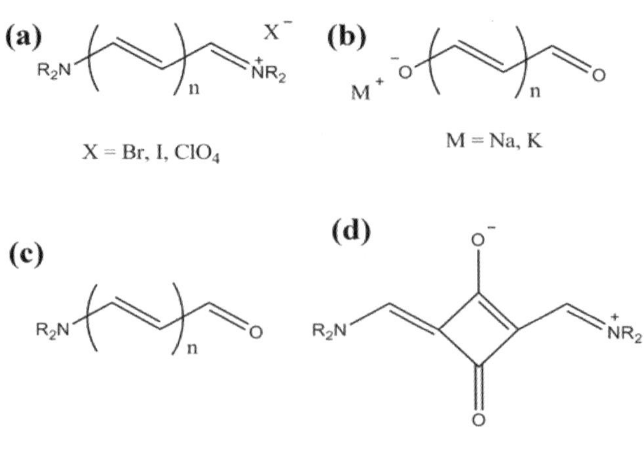

The aggregation of dyes in solution or at the solid-liquid interface is a frequently encountered phenomenon in dye chemistry owing to strong intermolecular van der Waals-like forces between the molecules. Due to its unique characters, such as flexible design of molecule structure and varied states of aggregate, aggregate of cyanine dye attracts researchers' attention greatly. Until today, many researchers have focused on manipulating self-assembly process or assembled aggregates of cyanine, in order to obtain cyanine aggregates with specific chiralities.[24,25]

The aggregates in solution exhibit distinct changes in the absorption band as compared with those of the monomeric species. Based on the spectral shifts, various aggregation patterns of the dyes have been proposed. The bathrochromically shifted J-bands[26] (J for Jelly, one of the first researchers who investigated these shifts) and hypsochromically shifted H-bands (H for hypsochromic) of the aggregates have been explained in terms of molecular exciton coupling theory. Generally, both J- and H-aggregates are composed of parallel dye molecules stacked plane-to-plane and end-to-end, and form two-dimensional dye crystals. The angle between the line of centers of a column of dye molecules and the long axis of any one of the parallel molecules is called the 'angle of slippage', α. Large molecular slippage ($\alpha > 32°$) results in a bathochromic shift (J-aggregates), and small slippage ($\alpha < 32°$) results in a hypsochromic shift[27] (H-aggregate). Extensive studies on J- and H-aggregates have resulted in the proposal that these aggregates exist as a one-dimensional assembly in solution that could be in a brickwork, ladder, or staircase type of arrangement.[28–30]

The behaviors of cyanine dyes are known to be sensitive to the environment. The change of conditions, for example adding inorganic salts, surfactants, polymers, polyelectrolytes and introduction of meso-substituent,[31] can affect the aggregate of cyanine dyes, inducing the transformation among monomer, dimer, H-aggregates and J-aggregates.[32] Among them, the concentration of inorganic salt is one of the important factors affecting aggregation of the cyanine dye. Generally the higher the concentration of inorganic salts is, the easier the formation of aggregate is and the more stable the aggregate is.

Koti *et al.*[3] reported that ammonium ion concentration controlled the spectroscopic aggregation number (coherence length) and physical size of the J-aggregates of anionic mesotetrakis (4-sulphonatophenyl) porphyrin (H_4TPPS^{2-}). Chibisov *et al.*[34] reported that the kinetics of J-aggregation of 3,3′-di(c-sulfopropyl)-5-methoxy-4′,5-benzo-9-ethylthiacarbocyanine in aqueous solution in the presence of salts. The rate of formation of J-aggregates increases with both dye and salt concentrations, which is the highest for trivalent metal ions, moderate for divalent and smallest for monovalent ions and independent of the type of anion.

Xiang JF *et al.*[35] reported the effects of NaCl on the aggregation of two typical thiacarbocyanine dyes (3,3′-di(3-sulfopropyl)-4,5,4′,5′-dibenzo-9-phenyl-thiacarbocyanine triethyl ammonium salt (PTC) and 3, 3′-di(3-sulfopropyl)-4,5,4′,5′-dibenzo-9-methyl-thiacarbocyanine triethyl ammonium salt (MTC)) (Chart 1.1) in aqueous solution by using absorption spectroscopy, fluorescence spectroscopy,[36] and so on. Inorganic salts promote the formation of J-aggregates[32,37] of cyanine dye in water by increasing its effective dielectric constant and reducing the electrostatic repulsion between dye anions. ^{1}H-NMR spectra of two dyes also shows that the Na^+ ions penetrate into the J-aggregates and replace the counterion (triethylammonium ions) in two dyes, which accelerates the aggregation of the two dyes in the presence of NaCl.

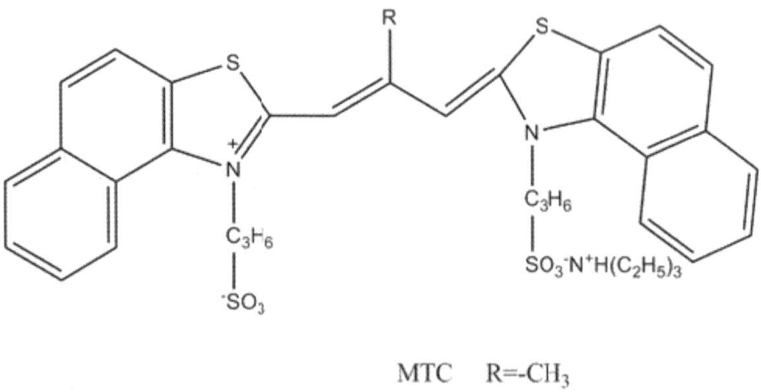

MTC R=-CH_3

PTC R=

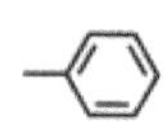

Chart 1.1 The structure of MTC and PTC.

Structural conversion between J-aggregate and H-aggregate of MTC is realized by adjusting K^+ together with 18-crown-6 ether using molecular recognition technique[38, 39]. In the presence of KCl, K^+ bound to two adjacent dye anions has a high positive ionic character than that of triethylammonium ion, which reduce the electrostatic repulsion between dye anions and thus facilitate the J-aggregation[35]. In the presence of 18-crown-6 ether[40], K^+ is isolated from the aggregate molecules, the driving force of neutralization of the electrostatic repulsion between dye anion due to K^+ would be reduced, and thus facilitate the H-aggregation.

Tang YL and coworkers also investigated[41] the effects of metal dication chloride on J-aggregation of two cyanine dyes (MTC and PTC). The promotion efficiency on J-aggregation of the dye molecule and the magnitude of effective coherence length of the J-aggregate are found to be dependent on the dyes' structure and the ability of coordination of dye molecule to metal ions.

Because the formation of J-aggregate of cyanine dyes is sensitive to solution condition, the mode of assembly could be regulated and chiral inducement of J-aggregates of cyanine dyes could be realized by adjusting solution condition which will greatly expand the application of J-aggregates of cyanine dyes. Recently, with the development of life science and molecular biology, the macromolecules in creatures become attractive owing to their unique 3D structures and irreplaceable physiological functions. All kinds of macromolecules resulted from biological evolution, rather than artificial design, open up a novel avenue for material science research. The unique 3D structure of the macromolecules offers chemists endless novel design strategies and perfect molecular models. Undoubtedly, bringing novel structural and functional properties into supramolecular assembly provides a short cut to develop advanced materials. This review mainly focuses on the interaction between aggregates of cyanine dyes and biomolecules, as well as the application in recognition of different biomolecules by aggregates of cyanine dyes.

2. Interaction Between Aggregates of Cyanine Dyes and Biomolecules

Biomacromolecules, such as nucleic acid[25] and proteins,[42] are chiral and hence chirality can be imposed on the interaction between aggregates of cyanine dyes and biomolecules to induce negative or positive chirality[2, 24, 25, 43, 44] of aggregates of cyanine dyes. Recently, many researchers focus on assembling chiral supramolecule from achiral cyanine dye by these biomacromolecules. Among the biological substances, the investigation of the interaction between DNA and aggregates of cyanine dyes is particularly interesting.

2.1 *Nucleic Acid*

DNA plays an important role in the transmission of human genetic information. DNA has a unique double-helical structure, which is one of the most interesting supramolecular architectures. Furthermore, it has a large amount of phosphate groups and ribose units on its helical backbone and shows the chirality and typical properties of a polyanion. In general, the J-aggregates formed by cyanine dyes can be induced to show chirality when dye molecules self-assemble on the unsymmetrical or chiral environment. Due to the particular double-helical structure with helical grooves of its surface, DNA is a so good template that many dye molecules can interact with it by groove binding mode and exhibit induced chirality.[1, 45, 46]

Wang *et al.*[25] reveal that a tricationic cyanine dye spontaneously assembles into a helical J-aggregate within the minor groove of DNA to form H- and J-aggregates by UV-vis and circular dichroism (CD) spectroscopy. The type of formed aggregate is highly dependent on several factors, for example, DNA length, dye concentration, temperature, and ionic strength. The strong dependencies on these factors can be attributed to the highly cooperative nature of J-aggregation and the critical role of electrostatics in the assembly of the aggregate. The aggregate likely consists of dimeric units assembled in an offset, face-to-face orientation within the minor groove of the DNA.

A model is proposed to relate H- and J-aggregates structures by translation of one monomer from a given dimer along the floor of the minor groove. The offset introduced by this translation reduce the intensity of the transition to the upper state but allow transition to the lower state, accounting for the observed red-shift in the absorption spectrum (Fig. 2.1). Translation of one dye monomer from a given dimer necessarily requires the adjacent monomer from the next dimer in the groove, leading to the observed cooperativity.

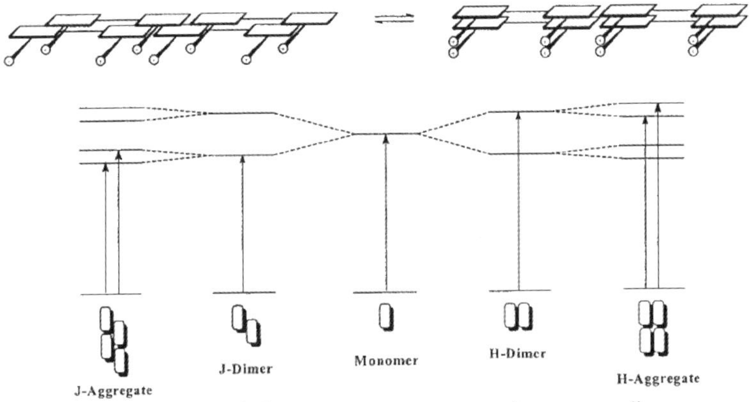

Figure 2.1. Proposed H- and J-aggregate structures and corresponding energy level diagram illustrating allowed electronic transitions for monomer, dimer, and extended aggregate structures. (Copyright: American Chemical Society)

Hannah *et al.*[47] describe the relationship between supramolecular assembly and dye structure and DNA sequence. Binding of one dye in the minor groove significantly enhances the binding of a second one to form a cofacial dimer. Dimerization causes the groove to widen beyond its natural width, which facilitates the assembly of additional dimers directly adjacent to the first dimmer. Termination occurs when the aggregate reaches the end of the template or when the sequence changes to one that does not support dimerization. Figure 2.2 shows a molecular model of three $DiSC_2(5)$ (Chart 2.1) dimers assembled in the minor groove of DNA and reflects the intimate association between the dye molecules and the DNA.[48] In this model, the sulfur atoms on the dye project into the minor groove, while the ethyl substituents project out of the groove. This orientation allows the curvature of the dye to more closely match that of the groove.

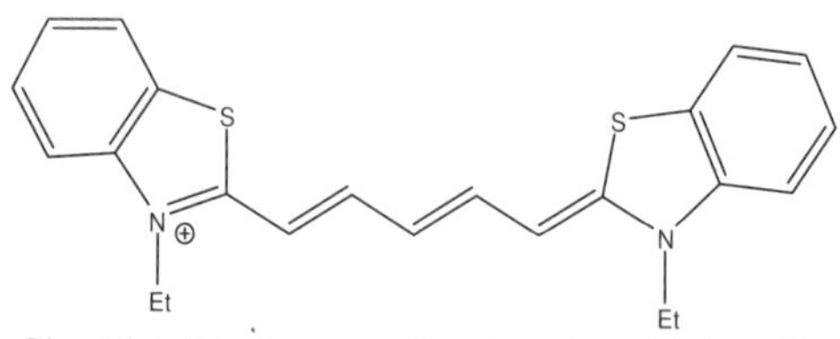

Chart 2.1 Structures of Cyanine Dyes $DiSC_2(5)$

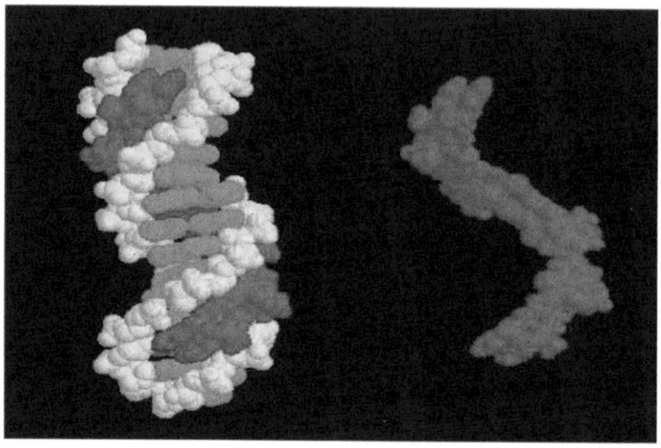

Figure 2.2. Molecular model showing an aggregate of three $DiSC_2(5)$ dimers aligned end-end in the minor groove of a DNA template. The DNA is removed in the figure on the right to emphasize the right-handed helical morphology of the dye aggregate. (Copyright: American Chemical Society)

In water solution, 'free' carbocyanine dye 3-methyl-2-[3-methy l-2-(3-methyl-2,3-dihydro-1,3-benzothiazole-2-iliden)-1-butenyl]-1,3-benzothiazole-3-il iodide (Cyan βiPr) (Chart 2.2) stays mainly in monomeric due to spatial impediments of an isopropyl fragment from Cyan βiPr molecule hindering the formation of the H-aggregates, while in the presence of DNA the dye molecules form J-aggregates. Ogul'chansky *et al.*[49] suppose that the Cyan βiPr molecules bind with DNA in a monomeric form and mainly via the groove binding mode. By increasing the dye concentration the Cyan βiPr molecules form J-aggregates[50] in the DNA grooves. Meanwhile an increase of ionic strength of solution leads to the release of dye molecules from DNA grooves and prevents J-aggregates formation.

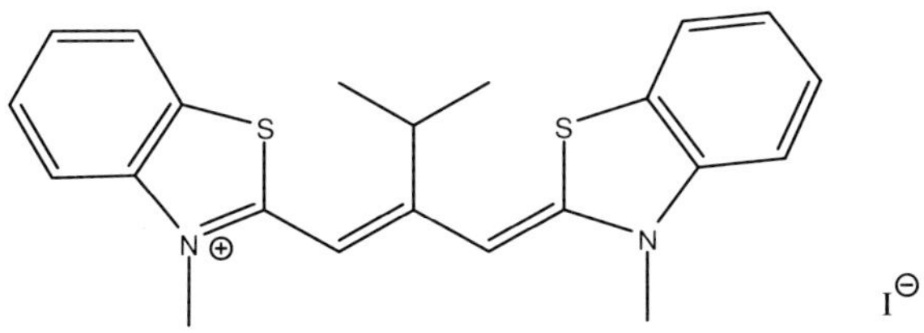

Chart 2.2 Structure of the Cyan βiPr carbocyanine dye.

In the presence of different types of double stranded DNA, Cyan βiPr molecules represent different form. The absorption spectra of the Cyan βiPr solutions with the double-stranded poly(dA/dT) and poly (dGC/dGC) polynucleotides are given in Fig. 2.3.

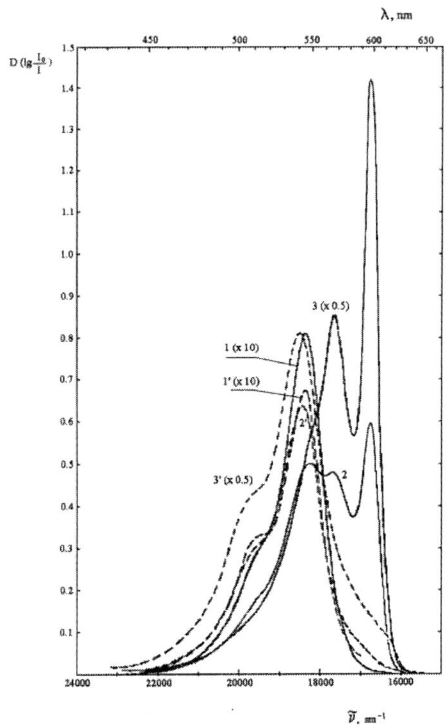

Figure 2.3 Absorption spectra of the CyanβiPr buffer solutions in the presence of poly (dA/dT) (c.1–3, —) and poly (dGC/dGC) (c.1′–3′,---). Concentrations of polynucleotides $C_{poly}=1.2\times10^{-4}$ M b. CyanβiPr concentrations: 1,1′, 10^{-6} M; 2,2′, 10^{-5} M; 3,3′, 3×10^{-5} M. (Copyright: Elsevier)

2.2 *Protein*

Proteins and polypeptides play an important role in constitution of human tissues and organs, which is the key class of biological macromolecules present in all organisms. The vital physiological activities are accomplished with the help of them. The growth of body, update of issues, repair of hurt issues are from the contribution of proteins. To be able to perform their biological function, proteins fold into one or more specific spatial conformations, driven by a number of non-covalent interactions such as hydrogen bonding, ionic interactions, Van Der Waals forces, and hydrophobic packing. All proteins are polymers of amino acids. Each protein polymer consists of a sequence of 20 different L-α-amino acids.

Owing to regular molecular orientation and novel physiological function, natural biomacromolecules, such as proteins, are considered as excellent template for fabricating and regulating chiral cyanine aggregates. However, natural biomacromolecules have longer peptides chains containing several of amino acids which constitute complex advanced construction, as a result, it is difficult to construct and adjust functional supramolecules by template design, because self-assemble of cyanine dye templated by such biomacromolecules is affected by many factors such as non-covalent bonds and steric hindrance.

2.2.1 *Synthetic peptides*

By designing sequence of amino acids, natural biomacromolecules could be imitated by synthesizing peptides which have shorter peptide chains and simple structures. Yang QF etc investigated[51] the structure and chirality of cyanine dye in the presence of small peptides with different length. Three peptides with different chain length (K4, K5, and K6) (Chart 2.3) were designed to interact with cyanine dye MTC. The three peptides could affect the transformation of dye assembled together with their chirality.[52] Different orientation of the side chains of the peptides owing to different chain length plays very important roles in dye aggregation.

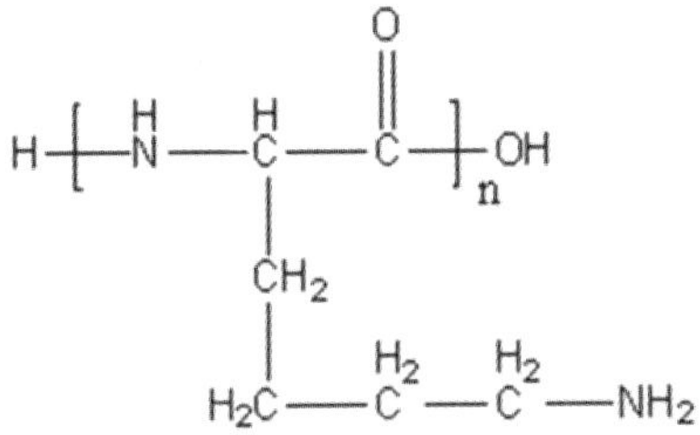

Peptides	K4	K5	K6
n	4	5	6

Chart 2.3 Molecular Formulas of the Peptides K4, K5, and K6

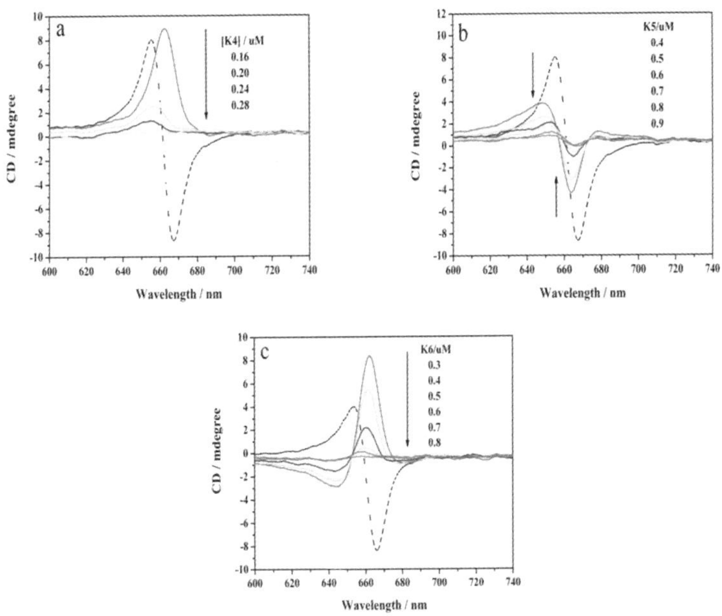

Figure 2.4 The CD spectra of 5 μM MTC J-band with different concentrations of the peptides in PBS. (a) K4, (b) K5, (c) K6. The dash line is the CD signal of MTC J-aggregates assembled in PBS without the peptides. (Copyright: American Chemical Society)

The chain length of the peptides may influence the peptides' conformation which affects the interaction of the dye molecules among aggregates, so H- and J-aggregates exhibit different chiralities in the presence of peptides with slightly different chain length. And molecular modeling method is adopted to prove above experimental results.

The peptides with short chain do not exhibit any well-ordered secondary structure as natural proteins do, so they do not have pocketlike interspace to accommodate a single dye molecule as proteins do.

2.2.2 Natural protein

Zhang YZ etc[52] reported an approach to construct a chiral supramolecule of achiral J-aggregates of PTC templated by human serum albumin (HSA) and to alter the chirality of the supramolecule by varying either the ratio of HSA to PTC or the solution pH value. The high concentration of HSA results in the reversion of the induced chirality of the J-aggregates, which is significantly different to previous reports[53] (Fig. 2.5(b)). The complex behavior of induced chirality of J-aggregates might be ascribed to the binding of J-aggregates to the α-helix and random-coil conformation of HSA. The induced optical activity of the J-aggregates shows significant pH dependence, which can be attributed to conformational changes of HSA at different pH values (Fig. 2.6). It opens new promising ways towards the design of efficient, chiral supramolecular devices.

In phosphate buffer solution (PBS) due to the salt effect,[32] PTC is in the form of J-aggregates. With addition of HSA to the phosphate buffer solution of PTC, a new peak of monomeric PTC binding to HSA appears. (Fig. 2.5(a)) CD measurements show that the introduction of HSA results in the J-band exhibiting a bisignated CD signal.[54]

Tatikolov *et. al.*[55] investigated non-covalent interactions between polymethine dyes of various types (cationic and anionic thiacarbocyanines as well as anionic oxonols and tetracyanopolymethines) and HSA by means of absorption, fluorescence and CD spectroscopies. Cationic thiacarbocyanines exhibit the lowest K and very weak CD signals, which points to a location of the dyes far from chirality centers and high affinity binding sites (possibly in superficial area of HSA) (Dye 1-4). Anionic thiacarbocyanines with sulfonate groups show much stronger association with HSA (and more intense CD signal) due to their bipolar structure and interaction with both positive and negative charges of HSA (Dye 5-8). Anionic oxonols strongly bind to HSA (due to tentative contribution of hydrogen

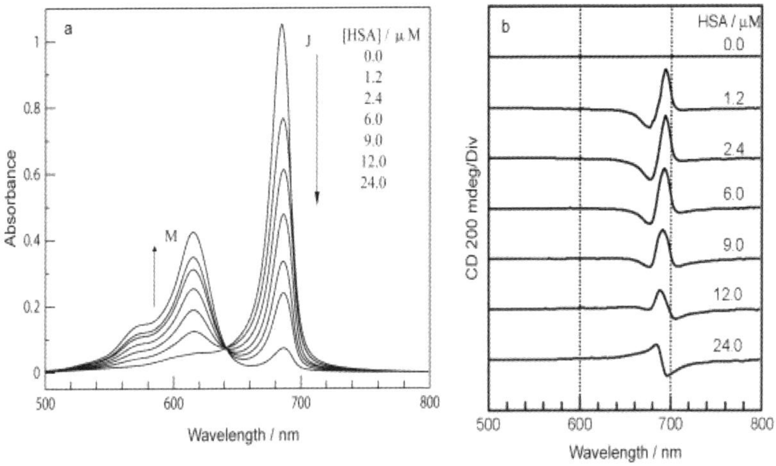

Figure 2.5 Absorption (a) and CD (b) spectra of 4.8 μM PTC induced by HSA at pH 7.20. (Copyright: Wiley)

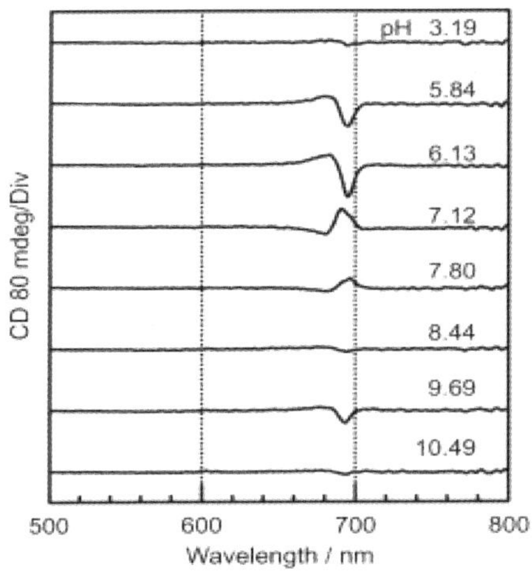

Fig. 2.6. CD spectra of 4.8 μM PTC induced by 4.8 μM HSA at different pH values. (Copyright: Wiley)

bonding), showing the highest K and CD bands (Dye 9-13) (Fig. 2.7). Interaction of meso-substituted anionic thiacarbocyanines with HSA results in cis→trans isomerization and, as a consequence, an appearance and a steep rise of dye fluorescence, which indicates strong electrostatic field in the corresponding binding sites.

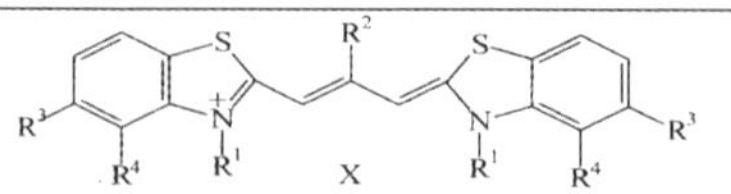

Dye	R^1	R^2	R^3	R^4	X
DTC (1)	C_2H_5	H	H	H	I^-
2	$(CH_2)_2OH$	H	H	H	Cl^-
3	$(CH_2)_2OH$	CH_3	H	H	Cl^-
4	n-C_6H_{13}	H	H	H	Br^-
5	$(CH_2)_3SO_3^-$	H	H	H	$C_5H_5NH^+$
6	$(CH_2)_3SO_3^-$	C_2H_5	OCH_3	H	$C_5H_5NH^+$
7	$(CH_2)_3SO_3^-$	C_2H_5	–CH=CH–CH=CH–		$C_5H_5NH^+$
8	$(CH_2)_3SO_3^-$	NH_2	H	H	NH_4^+

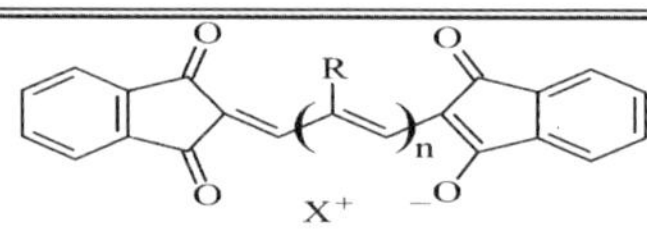

Dye	R	n	X^+
9	H	1	K^+
10	F	1	$(CH_3)_2NH_2^+$
11	CH_3	1	K^+
12	H	2	$(C_2H_5)_3NH^+$
13	H	0	K^+

Chart 2.4. The structure of dye 1-13(Copyright: Elsevier)

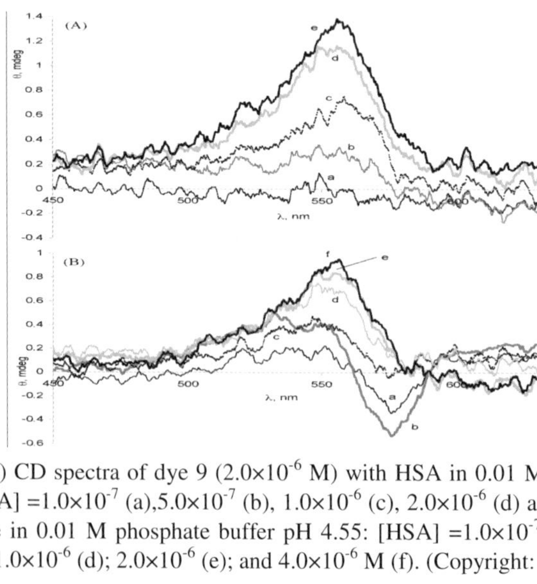

Figure 2.7 (A) CD spectra of dye 9 (2.0×10^{-6} M) with HSA in 0.01 M phosphate buffer pH 6.95: [HSA] =1.0×10^{-7} (a),5.0×10^{-7} (b), 1.0×10^{-6} (c), 2.0×10^{-6} (d) and 4.0×10^{-6} M (e). (B) The same in 0.01 M phosphate buffer pH 4.55: [HSA] =1.0×10^{-7} (a); 2.0×10^{-7} (b); 5.0×10^{-7} (c); 1.0×10^{-6} (d); 2.0×10^{-6} (e); and 4.0×10^{-6} M (f). (Copyright: Elsevier)

The complexation is accompanied by the appearance of a positive band in the CD spectrum (Fig. 2.7(a)). Aggregation of dye 9 on HSA at low pH is confirmed by the change of the CD signal shape, which is biphasic (a couple of positive and negative bands) at relatively small HSA concentrations when the aggregation takes place, and transforms into the uniform (positive) band at higher [HSA] when the aggregates decompose into the bound monomer (Fig.2.9(b)). While the uniform CD band can be produced by the monomeric dye molecules, the biphasic CD spectra are known to characterize dye aggregates on biomacromolecules (as a result of exciton coupling of the adjacent dye molecules).[25,43,56]

3. The Application of Aggregates of Cyanine Dyes

In solution or at the solid-liquid interface, cyanine dyes tend to aggregate. As well as with the change of environment, they are prone to transform between different states. The unique property of cyanine dyes is applied in different aspects especially in the field of biology. In the last part, we reviewed the performance of different biomolecules, such as DNA and protein, to interact with aggregates of cyanine dyes. The specific interaction lays the root for the application of aggregates of cyanine dyes in the recognition of different biomolecules. In the part, we will review their latest application in biology, such as the recognition of aggregates of cyanine dyes to different biomolecules.

Guadruplex-folded telomeric DNA has been found to perturb telomere function and inhibit the activity of telomerase,[57,58] an enzyme over-expressed in above 85% of human cancers, hence opening up a novel avenue for cancer therapy in G-quadruplex stabilizing agents.[59–62] G-quadruplex structure is stabilized by Hoogsteen hydrogen bonding among four guanine bases arranged in a square planar configuration, which exhibits extensive structural polymorphism; the DNA-strand orientation may be either parallel or anti-parallel, even both conformations (termed hybrid) in some cases.[63] Identifying the existence and recognizing particular quadruplex structures in human telomeres both in vitro and in vivo are important in the fields of cell proliferation, cancer research, and drug development.

Cyanine dye supramolecular assembly is sensitive to solution condition and as well as cyanine dye assembly could be used to detect PNA/DNA hybrids[64]. Based on this unique property, Tang YL and colleagues designed[65] a novel cyanine dye supramolecular assembly 3,3′-di(3-sulfopropyl)-4,5,4′,5′-dibenzo-9-ethylthiacarbocyanine triethylammonium salt (ETC, as shown in Chart 3.1) to recognize the unimolecular G-quadruplex in human telomeres in both solution spectroscopy and PAGE, which may offer a new approach for recognition of specific DNA motifs. They discovered M24 (sequence in human telomeres, which is considered as a mixture of several different quadruplex forms, whose exact structure is still in debate[66]) may prefer binding with ETC monomers and twisting their spatial structure, thus leading to disassembly of J-aggregates of ETC. Nevertheless DNAs with other motifs only interact with J-aggregates of ETC without interaction with ETC monomers. Because the different mechanism could be presented in absorption and fluorescence spectra, the significant signature (two independent absorption peaks which are separated by about 80 nm) leads ETC to be a potential probe of M24 from other DNA motifs (Fig. 3.1(b)). The strong enhancement of ETC monomer fluorescence intensity can be considered as another unique signature[67,68] to distinguish mixed G-quadruplex in human telomeres.

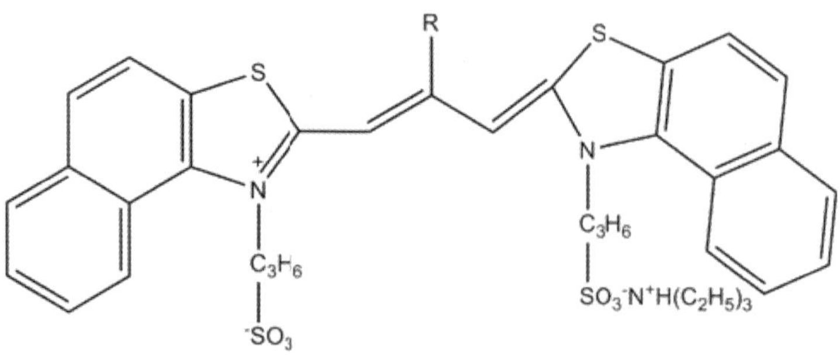

ETC R=$-CH_2CH_3$

Chart 3.1. The molecular formula of ETC.

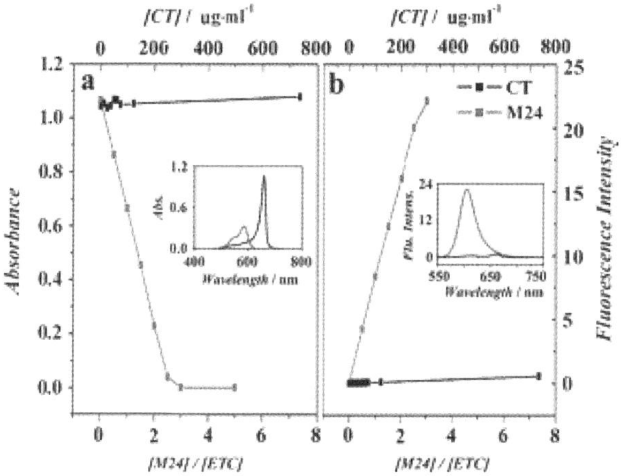

Figure 3.1 The changes of ETC J-aggregates absorbance (a) and ETC monomer fluorescence intensity (b) against the ratio of [M24]/[ETC] and [CT] (μg ml^{-1}), respectively. The inset shows the absorption and fluorescence spectra of 4 μM ETC with 12μM M24 and 4 μM ETC with 736.5 μg ml^{-1} CT, respectively. (Copyright: Royal Society of Chemistry)

Tang YL and coworkers successfully recognized mixed G-quadruplex in human telomeres from other DNA motifs by the supramolecular assembly of a novel cyanine dye salt ETC by the distinct change of absorption, fluorescence spectra and CD, furthermore discussed[69] the binding characterization of ETC to different intramolecular G-quadruplexes, such as hybrid G-quadruplex, parallel G-quadruplex and antiparallel G-quadruplex.[70] They find that end-stacking is the preferring binding of ETC to specific intramolecular G-quadruplexes by the experiments of ^{1}H-NMR and this binding induces disassembly of ETC J-aggregates and strong enhancement of ETC monomer fluorescence intensity. Their study showed that the loop structures nearby also play an important role. The longer lateral loop in G-quadruplex opposite to the propeller one has less steric hindrance for the binding of ETC monomer, which would facilitate ETC binding to hybrid G-quadruplex. If the binding force (including end-stacking and loop interaction) between ETC and specific G-quadruplex is stronger than that among ETC molecules themselves, ETC molecule would be 'snatched' from J-aggregate and bound on the end of G-quadruplex in the form of monomer.[32]

Consequently, ETC J-aggregates are disassembled and the unique spectral signatures appear (Fig. 3.2).

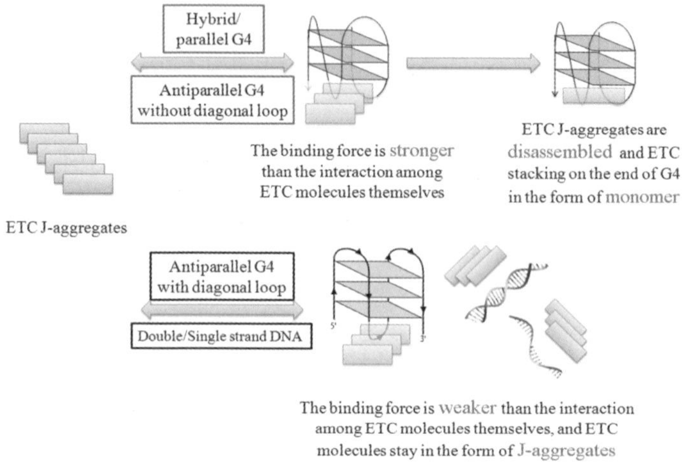

Figure 3.2 The recognition mechanism of specific G-quadruplex by ETC supramolecular assembly compared with other DNA motifs. (Copyright: Oxford University Press)

Figure 3.3 are the plots of the structures of ETC–DNA complex by using the Insight II 2005 software. Clearly, the lateral loops opposite to the propeller ones (pink bases) result in a cavity and 'snatch' part of ETC molecule, and, consequently, facilitate ETC stacking on the end G-quartet.

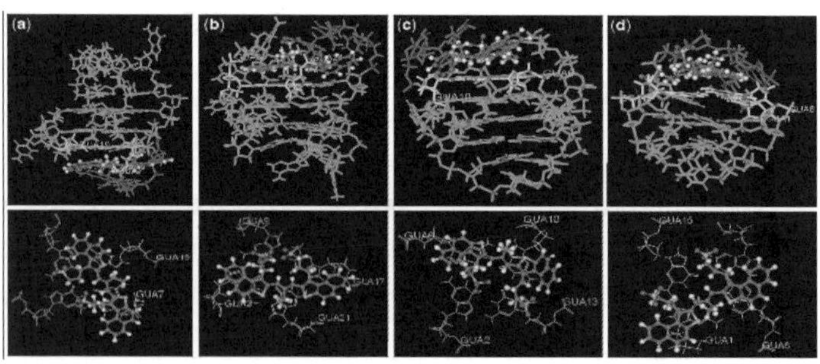

Figure 3.3. The plots of the structure of (a) ETC-bcl-2 2345; (b) ETC-H24; (c) ETC-c-kit1 and (d) ETC-TBA complex from molecular mechanics simulation (on the top), and the top projections of the locations of ETC stacking onto the end G-tetrad (on the button). (Copyright: Oxford University Press)

Beside in the solution system, Tang YL and colleagues[71] also successfully recognized hybrid/mixed G-quadruplex in human telomeres with confocal laser scanning microscopy (CLSM) on interface. Interaction of 3,3′-di(3-sulfopropyl)-4,5,4′,5′-dibenzo-9-methyl-thiacarbocyanine triethylammonium salt (MTC) with hybrid/mixed G-quadruplex results in a set of unique spectrophotometric signatures. Such feature is extended to map the locations of DNAs on interface. Linear duplex and mixed G-quadruplex in human telomeres assembled on Au film and stained by MTC were directly recognized by CLSM. Based on this result, further verifying and mapping hybrid/mixed G-quadruplex structure in vivo may be achieved by MTC supramolecule.

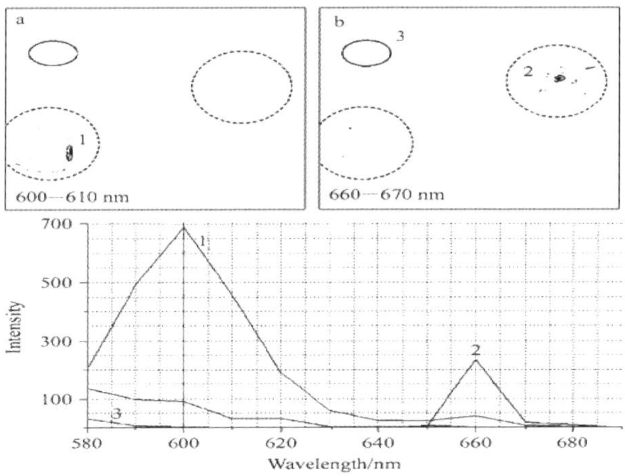

Figure 3.4 The fluorescence microscopy images of MTC-stained DNAs on Au film at different wavelength ranges: (a) 600–610 nm and (b) 660–670 nm. Areas selected by dashed circles are DNA samples assembled on Au film. The signals in the left circle represent MTC-stained M24, while those in the right circle represent stained D24. The emission spectra of the regions selected by solid circle (c) were collected by scanning 5-nm wavelength windows for each image. (Copyright: Wiley)

Tang YL and his coworkers accomplished recognition hybrid/parallel intramolecular G-quadruplexes from linear duplex and single-strand DNAs under physiological conditions by using the H-aggregates of MTC both in solution and at the interface. The selective interaction between MTC H-aggregates and different G-quadruplexes, which result in the transition between H-aggregates and monomers and consequently the change of solution color, is the basis of recognition. The fluorescence intensity of MTC monomer could be enhanced more than 1000 times by interacting with a specific G-quadruplex, which is about 200 times of that enhanced by other DNA motifs. At the interface, owing to the dramatic fluorescence signature, common fluorescence microscopy could easily map the locations of specific G-quadruplex on Au film by using the surface plasmon resonance (SPR) instrument. MTC H-aggregates thus may be applied as a potential excellent probe for DNA structure, both in vitro and in vivo.

Certain cyanine dyes such as merocyanines and ketocyanines have been studied extensively as potential photosensitizers for photodynamic therapy (PDT) and as radiation sensitizers for solid tumor treatment. Attempts have been made to elucidate the mechanism of photosensitizer transport and the factors leading to localization and retention in the tumor. Dhami etc.[72] focused on the comparative photophysical study of cis-disulfonated aluminum phthalocyanine ($AlPcS_2$) in unilamellar vesicles and leukemic K562 cells. Both the fluorescence and triplet quantum yields decreased with increasing sensitizer concentration. Information on the photodynamic transport and localization mechanism had been obtained by drawing correlations between unilamellar vesicles and leukemic K562 cells, indicating active transport of the phthalocyanine into tumor cells involving lysosomal accumulation.

From the similarities observed in the decay time distributions of cis-$AlPcS_2$ in liposomes and leukemia cells it can be concluded that the localization mechanism involves initial accumulation in organelle membranes such as lysosomes and possibly an endocytotic pathway into tumor cells. No aggregation was observed at any of the phthalocyanine concentrations in the microheterogeneous systems indicating that this phthalocyanine derivative may be a potentially useful sensitizer for use in PDT. Most importantly, this study has shown that remarkably good agreement can be found between model and cellular systems (Fig. 3.5).

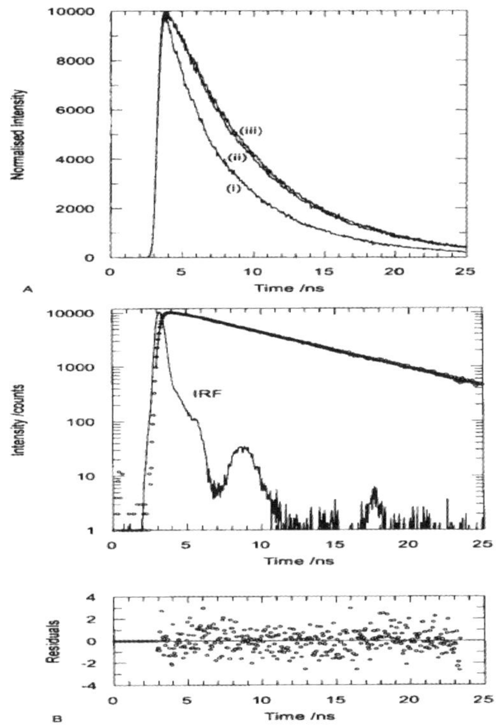

Figure 3.5(a) Fluorescence decay curves of cis-$AlPcS_2$ in leukemic K562 cells after an irradiation exposure time at 633 nm (1Wcm^{-2},) of (i) 1 s, (ii) 30 s and (iii) 90 s. (b) Plot of instrument response function (IRF), fluorescence decay recorded after 90 s irradiation (o) and fitted biexponential function (solid line) and the corresponding plot of weighted residuals. (Copyright: John Wiley and Sons)

The neuropeptides have the potential to disrupt the balance between cell death and cell growth in the tumor. Salido etc.[73] researched the role of bombesin in modulating the mitochondrial membrane potential ($\Delta\psi_m$) in cell death induced by etoposide on PC-3 cells. Cells were cultured and stained with 5,5′,6,6′-tetrachloro-1,1′,3,3′-tetraethylbenzimidazolylcarbocyanine iodide (JC-1). At low membrane potentials, JC-1 produces a green fluorescence, and at high membrane potentials, it forms ‘‘J-aggregates’’ with red fluorescence, which is examined by confocal microscope. The size, number of pixels, and ratios between fluorescence intensity in the red and green channels in each region of interest were calculated. The loss of$\Delta\psi_m$ in etoposide-treated

PC-3 cells was prevented by bombesin. The quantitative analysis of JC-1–stained cells revealed a significant decrease in the red (high$\Delta\psi_m$) to green (low$\Delta\psi_m$) ratio in etoposide-treated cells when compared with control cells, which was restored in the presence of bombesin ($P < 0.00001$). The interaction between treatments and area ($P = 0.0002$) was highly significant, and confirms that PC-3 cells keep their apoptosis machinery, showing an apoptotic volume decrease in response to etoposide. The protection by bombesin occurs by inhibition of apoptosis and maintenance of mitochondrial integrity. New therapeutic protocols and trials need to be developed to test drugs acting through the neutral0ization of antiapoptotic intracellular pathways mediated by neuroendocrine hormones.

4. Prospects

Aggregates of cyanine dye are sensitive to environment, which has been utilized in different fields especially in fields of biology. In this chapter, the interaction between aggregates of cyanine dyes and biomolecules, such as DNA and protein, is presented. Due to different structure of biomolecules, binding mode is different. For example, DNA interacts with aggregates of cyanine dye by groove binding mode, but for proteins, aggregates of cyanine dye can bind to the α-helix or random-coil conformation of proteins, which induces different chiralities of aggregates, for G-quadruplex, end-stacking is the preferring binding of aggregates to specific G-quadruplexes, which has been used in recognizing specific G-quadruplex in human telomeres in solution and at the interface.

Furthermore, because of the successful recognizing specific G-quadruplex in human telomeres in solution and at the interface, it is promising to detect tumors in clinical specimens by the probe of cyanine dye, which opens up a novel avenue for cancer diagnoses. It is reasonable to believe that aggregates of cyanine dyes will play more and more important role in application in biology, diagnoses and therapy of disease.

References

1. R. F. Pasternack, A. Giannetto, P. Pagano and E. J. Gibbs, *Journal of the American Chemical Society* **113**, 7799 (1991).
2. O. K. Kim, J. Je, G. Jernigan, L. Buckley and D. Whitten, *Journal of the American Chemical Society* **128**, 510 (2006).
3. F. Wurthner, S. Yao and U. Beginn, *Angewandte Chemie-International Edition* **42**, 3247 (2003).
4. A. Lohr, M. Lysetska and F. Wurthner, *Angewandte Chemie-International Edition* **44**, 5071 (2005).
5. T. S. Balaban, A. D. Bhise, M. Fischer, M. Linke-Schaetzel, C. Roussel and N. Vanthuyne, *Angewandte Chemie-International Edition* **42**, 2140 (2003).
6. A. R. Pease, J. O. Jeppesen, J. F. Stoddart, Y. Luo, C. P. Collier and J. R. Heath, *Acc Chem Res* **34**, 433 (2001).
7. S. Zahn and J. W. Canary, *Angewandte Chemie-International Edition* **37**, 305 (1998).
8. D. R. Link, G. Natale, R. Shao, J. E. Maclennan, N. A. Clark, E. Korblova and D. M. Walba, *Science* **278**, 1924 (1997).
9. K. S. Min, H. T. Chung, O. K. Cho, G. Y. Kim and K. Y. Park, *Molecular Crystals and Liquid Crystals Science and Technology Section a-Molecular Crystals and Liquid Crystals* **247**, 161 (1994).
10. O. K. Cho and K. Y. Park, *Molecular Crystals and Liquid Crystals Science and Technology Section a-Molecular Crystals and Liquid Crystals* **267**, 393 (1995).
11. W. M. Anderson, B. B. Chambers, J. M. Wood and L. Benninger, *Biochem Pharmacol* **41**, 677 (1991).
12. P. D. Sima and J. R. Kanofsky, *Photochem Photobiol* **71**, 413 (2000).
13. B. Becker, J. Clapper, K. R. Harkins and J. A. Olson, *Anal Biochem* **221**, 78 (1994).
14. T. Ogul'chansky, V. M. Yashchuk, M. Losytskyy, I. O. Kocheshev and S. M. Yarmoluk, *Spectrochim Acta A Mol Biomol Spectrosc* **56**, 805 (2000).
15. Y. Nakamura, H. Yamada, K. Yoshida, T. Manno, Y. Ubukata, M. Suzuki and Y. Yoshimura, *Horm Res* **33 Suppl 2**, 43 (1990).
16. I. D. L. Albert, T. J. Marks and M. A. Ratner, *Journal of Physical Chemistry* **100**, 9714 (1996).
17. F. Wurthner, S. Yao, T. Debaerdemaeker and R. Wortmann, *Journal of the American Chemical Society* **124**, 9431 (2002).
18. D. J. Owen, D. VanDerveer and G. B. Schuster, *Journal of the American Chemical Society* **120**, 1705 (1998).
19. H. S. Rye, S. Yue, D. E. Wemmer, M. A. Quesada, R. P. Haugland, R. A. Mathies and A. N. Glazer, *Nucleic Acids Res* **20**, 2803 (1992).

20. D. A. Higgins, J. Kerimo, D. A. VandenBout and P. F. Barbara, *Journal of the American Chemical Society* **118**, 4049 (1996).
21. N. Li, C. J. Yu and F. Q. Huang, *Nucleic Acids Research* **33**,e37 (2005).
22. S. Kudrevich, N. Brasseur, C. LaMadeleine, S. Gilbert and J. E. vanLier, *Journal of Medicinal Chemistry* **40**, 3897 (1997).
23. V. Nair and C. S. Cooper, *Journal of Organic Chemistry* **46**, 4759 (1981).
24. J. L. Seifert, R. E. Connor, S. A. Kushon, M. Wang and B. A. Armitage, *J. Am. Chem. Soc.* **121**, 2987 (1999).
25. M. M. Wang, G. L. Silva and B. A. Armitage, *J. Am. Chem. Soc.* **122**, 9977 (2000).
26. E. E. Jelly, *Nature* **138**, 1009 (1936).
27. A. Mishra, R. K. Behera, P. K. Behera, B. K. Mishra and G. B. Behera, *Chem Rev* **100**, 1973 (2000).
28. W. J. Harrison, *The journal of physical chemistry* **100**, 2310 (1996).
29. J. E. Maskasky, *Langmuir* **7**, 407 (1991).
30. T. Kobayashi, *J-aggregates*. (World Scientific, Singapore ; River Edge, NJ, (1996).
31. T. D. Slavnova, A. K. Chibisov and H. Gorner, *J Phys Chem A* **109**, 4758-4765 (2005).
32. A. H. Herz, *Advances in Colloid and Interface Science* **8**, 237 (1977).
33. A. S. R. Koti, J. Taneja and N. Periasamy, *Chemical Physics Letters* **375**, 171 (2003).
34. A. K. Chibisov, H. Gorner and T. D. Slavnova, *Chemical Physics Letters* **390**, 240 (2004).
35. J. Xiang, X. Yang, C. Chen, Y. Tang, W. Yan and G. Xu, *J Colloid Interface Sci* **258**, 198 (2003).
36. T. Tani, T. Suzumoto, K. Kemnitz and K. Yoshihara, *Journal of Physical Chemistry* **96**, 2778 (1992).
37. E. Rousseau, M. Van der Auweraer and F. C. De Schryver, *Langmuir* **16**, 8865 (2000).
38. Y. Z. Zhang, J. F. Xiang, Y. L. Tang, G. Z. Xu and W. P. Yang, *Chemistry Letters* **35**, 1316 (2006).
39. C. Guo, J. Xiang, J. Feng, Y. Tang, C. Chen and G. Xu, *J Colloid Interface Sci* **246**, 401 (2002).
40. G. W. Gokel, W. M. Leevy and M. E. Weber, *Chem Rev* **104**, 2723 (2004).
41. Y. Z. Zhang, J. F. Xiang, Y. L. Tang, G. Z. Xu and W. P. Yan, *Dyes and Pigments* **76**, 88 (2008).
42. T. M. Cooper and M. O. Stone, *Langmuir* **14**, 6662 (1998).
43. J. L. Seifert, R. E. Connor, S. A. Kushon, M. Wang and B. A. Armitage, *Journal of the American Chemical Society* **121**, 2987 (1999).
44. R. Lauceri, A. Raudino, L. M. Scolaro, N. Micali and R. Purrello, *J Am Chem Soc* **124**, 894 (2002).
45. E. J. Gibbs, I. Tinoco, Jr., M. F. Maestre, P. A. Ellinas and R. F. Pasternack, *Biochem Biophys Res Commun* **157**, 350 (1988).

46. M. M. Wang, G. L. Silva and B. A. Armitage, *Journal of the American Chemical Society* **122**, 9977 (2000).
47. K. C. Hannah and B. A. Armitage, *Accounts Chem. Res.* **37**, 845 (2004).
48. R. A. Garoff, E. A. Litzinger, R. E. Connor, I. Fishman and B. A. Armitage, *Langmuir* **18**, 6330 (2002).
49. T. Ogul'chansky, M. Losytskyy, V. B. Kovalska, V. M. Yashchuk and S. M. Yarmoluk, *Spectrochim Acta A Mol Biomol Spectrosc* **57**, 2705 (2001).
50. A. Pawlik, S. Kirstein, U. DeRossi and S. Daehne, *Journal of Physical Chemistry B* **101**, 5646 (1997).
51. Q. F. Yang, J. F. Xiang, Q. Li, W. P. Yan, Q. J. Zhou, Y. L. Tang and G. Z. Xu, *Journal of Physical Chemistry B* **112**, 8783 (2008).
52. Y. Z. Zhang, J. F. Xiang, Y. L. Tang, G. Z. Xu and W. P. Yan, *chemphyschem* **8**, 224 (2007).
53. L. Stryer and E. R. Blout, *Journal of the American Chemical Society* **83**, 1411 (1961).
54. N. Berova, K. Nakanishi and R. W. Woody, Circular Dichroism-Principle and Applications, 2nd ed. Wiley, New York, **chap. 12.** (2000).
55. A. S. Tatikolov and S. M. Costa, *Biophys Chem* **107**, 33 (2004).
56. S. M. Andrade and S. M. Costa, *Biophys J* **82**, 1607 (2002).
57. J. R. Williamson, *Annu Rev Biophys Biomol Struct* **23**, 703 (1994).
58. E. H. Blackburn, *Nature* **350**, 569 (1991).
59. Q. Zhou, L. Li, J. F. Xiang, Y. L. Tang, H. Zhang, S. Yang, Q. Li, Q. F. Yang and G. Z. Xu, *Angew Chem Int Edit* **47**, 5590 (2008).
60. B. Fu, J. Huang, L. Ren, X. Weng, Y. Zhou, Y. Du, X. Wu, X. Zhou and G. Yang, *Chem Commun (Camb)*, 3264 (2007).
61. E. Pennisi, *Science* **312**, 1467 (2006).
62. D. Sen and W. Gilbert, *Nature* **334**, 364 (1988).
63. S. Neidle and G. N. Parkinson, *Current Opinion in Structural Biology* **13**, 275 (2003).
64. B. A. Armitage, J. O. Smith and D. A. Olson, *Abstracts of Papers of the American Chemical Society* **217**, U121 (1999).
65. Q. F. Yang, J. F. Xiang, S. Yang, Q. J. Zhou, Q. Li, Y. L. Tang and G. Z. Xu, *Chemical Communications*, 1103 (2009).
66. E. Pennisi, *Science* **312**, 1467 (2006).
67. C. C. Chang, J. F. Chu, F. J. Kao, Y. C. Chiu, P. J. Lou, H. C. Chen and T. C. Chang, *Analytical Chemistry* **78**, 2810 (2006).
68. C. C. Chang, I. C. Kuo, I. F. Ling, C. T. Chen, H. C. Chen, P. J. Lou, J. J. Lin and T. C. Chang, *Anal Chem* **76**, 4490 (2004).
69. Q. F. Yang, J. F. Xiang, S. Yang, Q. Li, Q. J. Zhou, A. J. Guan, X. F. Zhang, H. Zhang, Y. L. Tang and G. Z. Xu, *Nucleic Acids Research* **38**, 1022 (2010).
70. S. Burge, G. N. Parkinson, P. Hazel, A. K. Todd and S. Neidle, *Nucleic Acids Res* **34**, 5402 (2006).

71. Q. F. Yang, J. F. Xiang, S. Yang, Q. J. Zhou, Q. A. Li, A. J. Guan, X. F. Zhang, H. Zhang, Y. L. Tang and G. Z. Xu, *Chinese Journal of Chemistry* **28**, 1126 (2010).
72. S. Dhami, G. Rumbles, A. J. MacRobert and D. Phillips, *Photochemistry and Photobiology* **65**, 85 (1997).
73. M. Salido, J. L. Gonzalez and J. Vilches, *Molecular Cancer Therapeutics* **6**, 1292 (2007).

CHAPTER 6

COUPLING BETWEEN J-AGGREGATES AND INORGANIC EXCITONS

Brian J. Walker and Moungi G. Bawendi
Department of Chemistry, Massachusetts Institute of Technology
Room 6-221, 77 Massachusetts Avenue,
Cambridge, MA USA
E-mail: mgb@mit.edu

J-aggregates, once widespread in film photography,[1] have fostered renewed interest due to their non-linear optical properties, their quantum mechanical coupling at ambient conditions, and their delocalized molecular excitons. Such Frenkel excitons[2] contrast with the Wannier-Mott excitons characteristic of inorganic semiconductors, and J-aggregates have been combined with various inorganic semiconductor materials to study the effects of coupling in these hybrid systems.[3] This chapter reviews the progress to date on J-aggregates coupled with excitonic inorganic materials. After summarizing the motivation and first proposals in the use of these materials, we discuss the experimental realization of semiconductor quantum wells and quantum dots coupled to J-aggregated dyes. An outlook on future directions is then presented.

Keywords: exciton, energy transfer, organic/inorganic interface, electrostatic self-assembly, light harvesting

1. Motivation: why study J-aggregates?

Although systems with correlated optical and structural properties are plentiful, J-aggregates are important model systems for two reasons. First, J-aggregates have distinct, addressable states that are readily studied via spectroscopic and materials characterization techniques and that make them interesting in photonic applications.[4] Studies on the chemical and electronic structures of J-aggregates have thus lead directly to new theories regarding the intermolecular coupling of electronic excited states.[5,6,7,8]

Secondly, and in contrast to light harvesting complexes with similar physics,[9] J-aggregates can form from their constituent monomers using fairly few inexpensive chemical or physical treatments. Thus, J-aggregates are more practical for experiments than many alternatives (e.g. photosynthetic systems).

2. Planar Structures: The First Coupling Between Organic and Inorganic Materials

Organic dyes have been long-established in film photography to enhance the spectral sensitivity of silver halide (AgX) particles, both in their aggregated and un-aggregated states.[10] Understanding the basic physics associated with energy transfer[11,12] and electron transfer[13,14] thus constituted a major motivation for research on J-aggregates.

In 2007, Zhang et al. reported the coupling of J-aggregates with semiconductor nanocrystals, the first example of energy transfer between J-aggregates and an inorganic semiconductor.[15] The photonic structures used for the experiments were composed of films made via layer-by-layer self-assembly[16,17,18] (Fig. 2.1). The CdSe(ZnCdS) nanocrystals in these experiments were coated with thioglycolic acid, which conferred both water solubility and a negative charge to the nanocrystals. J-aggregates in this study were composed of negatively-charged TDBC. To associate the two negatively-charged excitonic materials, a thin (1.7 nm) layer of transparent, positively-charged polyelectrolyte PDDA was deposited in an alternating fashion before either the J-aggregates or the nanocrystals.[14]

Initially, Zhang *et al.* discussed films in which the nanocrystals had a larger band gap than the J-aggregate.[14] These nanocrystal/J-aggregate films demonstrated >95% energy transfer efficiency from the QDs to the J-aggregate, and resulted in a gain of the J-aggregate emission intensity by a factor of 10. Because the band gap of the nanocrystals exhibits a quantum size effect, the photonic device could change entirely by fabricating a different photonic structure that used larger quantum dots. In the new structure, the J-aggregate had a larger band gap energy than the nanocrystal, thus reversing the direction of energy transfer. For the second structure, energy transfer from the J-aggregate donors resulted in a photoluminescence enhancement in the nanocrystal acceptors. The energy transfer from the J-aggregate films to the nanocrystals resulted in a 10-fold increase in the photoluminescence intensity compared to layer-by-layer films of quantum dots alone when illuminated at the J-aggregate absorption maximum.

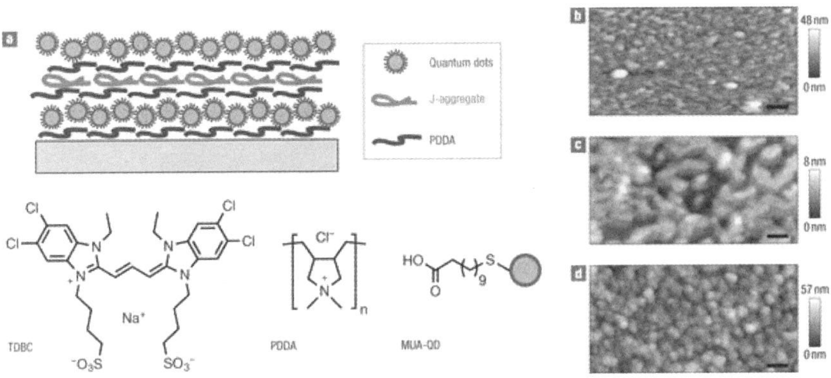

Figure 2.1. (a), Schematic of the hybrid film layer structure: a monolayer J-aggregate of TDBC is sandwiched between two monolayers of CdSe–ZnS nanocrystals, joined by monolayers of PDDA. Molecular structures of TDBC, PDDA and MUA are illustrated in the lower panel. b–d, AFM images of the LBL-grown films, consisting of a monolayer of QDs ($\lambda = 653$ nm) (PDDA/QD) (b) a monolayer of J-aggregate (PDDA/TDBC/PDDA) (c), and the hybrid film II (d). The scale bars in b–d are 100 nm. Reprinted by permission from Macmillan Publishers Ltd: *Nat. Nanotech.* 2007 ref. 15.

The ability to tune the spectral properties of J-aggregates was a major advance. As semiconductor nanocrystals can be excited at any energy greater than their band edge, the use of nanocrystal donors allows a J-aggregate/quantum dot system to be pumped with a broad range of excitation wavelengths, while retaining the narrow, ultrafast emission of the J-aggregate. In the opposite direction, the quantum dot could be pumped in a narrow excitation band defined by the J-aggregate, while preserving the size-tunable emission characteristic of nanocrystals.

Soon afterward, Chanyawadee *et al.* demonstrated that a two-dimensional semiconductor quantum well could be used to transfer energy to J-aggregates of the cyanine dye U3.[19] The photonic structure consisted of GaAs/AlGaAs single quantum wells fabricated by molecular beam epitaxy, and the J-aggregates were deposited via spin-casting directly onto the top-most AlGaAs spacer layer directly from methanol. While spin casting had been used earlier by Tani *et al.* to deposit a range of J-aggregating dyes,[20] the report was the first demonstration of a non-aqueous method for the deposition of J-aggregates into a hybrid organic/inorganic structure.

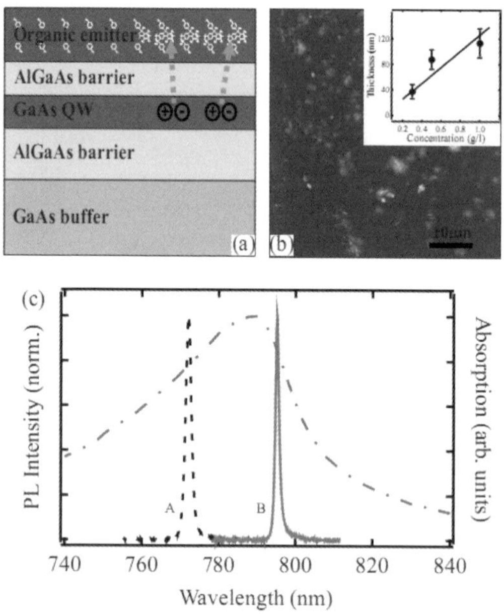

Figure 2.2. Organic/Inorganic hybrid structures. Organic dye is deposited on 4.0 and 6.5 nm thick quantum wells. (b) Atomic force microscopy image of organic dye layer deposited on a glass substrate. Inset: thickness of the organic film vs. concentration of organic dye. (c) Spectral overlaps between the emission of sample A (solid) and of sample B (solid curve, red) at 25 K and the absorption of organic dye dashed-dotted curve, red online. Reproduced with permission from Ref 19.

Because the quantum well photoluminescence peak position depends on the thickness of the GaAs layer, the device architecture offers a convenient method to vary the spectral overlap while largely maintaining the separation between the quantum well donor and the J-aggregate acceptor. On the basis of transient reflectivity measurements and time correlated single photon counting, Chanyawadee *et al.* constructed a kinetic model that fully accounted for the energy transfer from the QW to the J-aggregates.

As noted by Chanyawadee *et al.*, the stacked device architectures have potential in device applications, where both the benefits of J-aggregation and of inorganic semiconductor structures can be gained.[19] Many other device architectures are also possible, depending on the nature of the link that mediates that organic/inorganic conjugation.

3. Nanocrystal/J-aggregate constructs in solution

Until this point, the two reported device structures for energy transfer from inorganic semiconductors to a J-aggregate layer have been arrayed in a parallel planar geometry: either energy transfer from a quantum well or a layer of semiconductor nanocrystals (quantum dots) arranged in a two-dimensional structure. However, the three-dimensional quantum confinement of semiconductor nanocrystals suggested the possibility that discrete quantum dots (QDs) could be coupled to J-aggregates, forming QD/J-aggregate constructs that could be made and studied in solution. Such a solution-phase QD/J-aggregate construct would allow the direct examination of the photophysical interactions at organic/inorganic interfaces, as the QD/J-aggregate constructs in the dilute solution limit are too isolated from one another for long range transport to occur. The relative isolation between QD/J-aggregate constructs is verifiable through dynamic light scattering experiments, which also provide information about the arrangement of QDs and J-aggregates in solution.

QD/J-aggregate constructs in solution also have potential applications in biological labeling and fluorescence multiplexing applications.[21] The maximum spectral density of color channels varies inversely with the emission linewidth, so the narrow J-aggregate emission (about 50 meV full width at half maximum) allows an increased number of different color channels to be used in a given spectral range. Practical multiplexing depends on the use of a single excitation source for all of the different color channels, so it is beneficial to impart a broad absorption spectrum to the J-aggregate in solution. By tuning the nanocrystal (donor) emission to overlap with the J-aggregate (acceptor) absorption, efficient energy transfer from the nanocrystal to the J-aggregate thus results in nanocrystal/J-aggregate constructs that have a broad nanocrystal absorption while retaining the lineshape of J-aggregate emission.

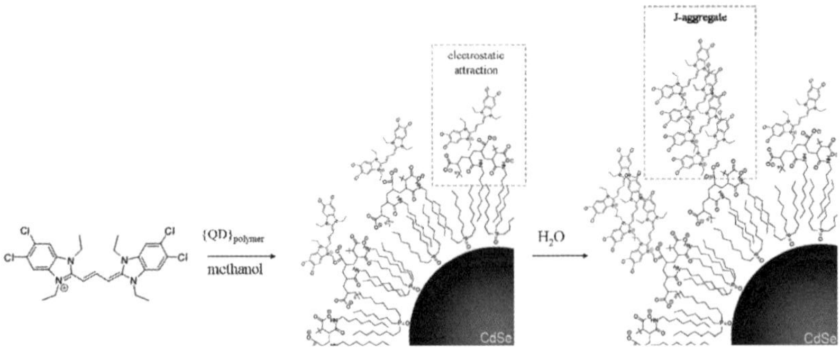

Figure 3.1. Representation of the reaction steps leading first to association of the TTBC monomer with the anionic polymer coating and second to the aggregation of excess TTBC monomers with the anchoring TTBC molecules. Reprinted with permission from Ref 22. Copyright 2009 American Chemical Society.

These QD/J-aggregate constructs in solution were first reported by Halpert et al.[22] Using an amphiphilic polymer synthesized from a poly(acrylic acid) backbone, CdSe(ZnCdS) QDs were rendered water soluble while preserving the native quantum yield of the underlying core(shell) inorganic nanocrystal. The negatively charged carboxylate groups on the polymer were used to associate the QD with the J-aggregates of the positively-charged dye TTBC.

The resulting structures were composed of 100-200 dye molecules per nanocrystal, and the amphiphilic polymer ligand also induced the formation of aggregates. Such an effect differs from the methods of inducing J-aggregate formation using amphiphiles above the critical micelle concentration, as the outward ligand functional groups also biased the formation of J-aggregates. Although suspensions of TTBC J-aggregates in aqueous solutions commonly span several orders of magnitude in physical size, light scattering measurements on the nanocrystal/J-aggregate constructs demonstrated that there was a uniform 70% increase in size compared to the water-soluble nanocrystals alone.

The nanocrystal/J-aggregate constructs also demonstrated efficient energy transfer from the nanocrystals to the J-aggregates (>99%), as observed in J-aggregate luminescence and in QD quenching. The measured energy transfer rate was consistent with an energy transfer mechanism between two coupled transition dipoles. Due to the nature of the distinct, addressable excited states in the QD[23] and the J-aggregates[4] the energy transfer between them constitutes a model process for coupling in hybrid systems.

As in the case of the layer-by-layer films, the roles of donor and acceptor materials can be reversed so that there is resonant energy transfer from a higher energy J-aggregate to a lower-energy-emitting QD. The intense excitation feature of the J-aggregate entails a high molar absorptivity in a narrow band ($\varepsilon > 5 \times 10^5$ L cm^{-1} per mole of cyanine dye at maximum absorbance).[24] When combined with the rapid, efficient diffusion of excitons in a J-aggregate[25–27] the J-aggregate absorbers thus act like optical antennae that can transfer excitation energy to a suitable acceptor. In this context, the QD/J-aggregate constructs thus retain the size-tunable emission properties of the QD while enhancing their absorption in a specific narrow band.

To achieve the energy transfer previously reported for planar architectures, a blue-emitting (negatively-charged) thiacyanine J-aggregate was used with a red-emitting nanocrystal.[24] Energy transfer from organic dyes to nanocrystals is generally inefficient,[28] and thus any efficient process required a redesign of the ligand used to mediate the coupling. In this case, 2-mercaptoethyl-(N,N,N-trimethylammonium) chloride was synthesized and used to impart water solubility and positive charge to the nanocrystals.

The short ligand reduced the donor-acceptor distance between the nanocrystal and the J-aggregate, and also mirrored the longer ligand's templating effect. Although J-aggregates of the thiacyanine dye did not form in deionized water alone, the presence of nanocrystals in an otherwise deionized aqueous solution resulted in the formation of J-aggregates for the same concentration of cyanine dye. The result extends the effects observed previously for incorporating J-aggregates as guests into host-type structures, in which the J-aggregate architecture can be induced and protected using aqueous phase sol-gel processing.[29] In this case, the excitonic nanocrystal not only provides a templating site for J-aggregate nucleation, it also serves as a distance-dependent probe of the proximity of the J-aggregate donor.

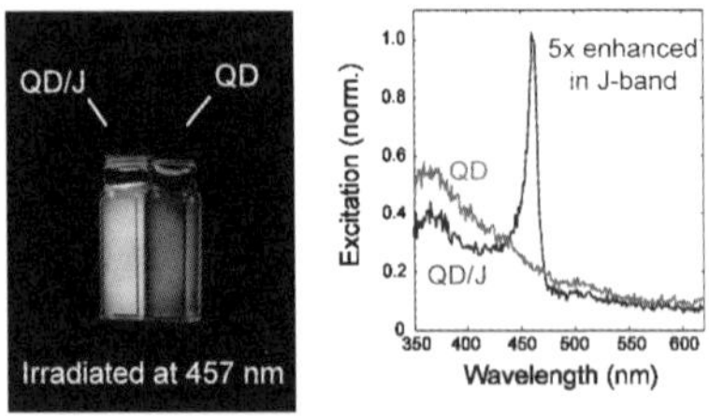

Figure 3.2. Demonstration of narrowband absorption-enhanced QD/J-aggregate conjugates. (a) Photograph of QD/J-aggregate conjugates and QDs alone at identical QD concentration, with 457 nm excitation. (b) Photoluminescence excitation spectra with emission at 635 nm. Energy transfer from the J-aggregates to the QDs results in 5-fold enhancement of QD emission. Reprinted with permission from Ref 24. Copyright 2009 American Chemical Society.

Following the electrostatic association of the nanocrystal and the J-aggregate, the resulting hybrid materials demonstrated energy transfer with near-unity efficiency as measured both through steady state photoluminescence spectroscopy and transient photoluminescence. This energy transfer efficiency was accompanied with a distinct J-aggregate feature in the photoluminescence exciton spectrum (with emission at the nanocrystal emission wavelength), consistent with efficient energy transfer from the J-aggregates to the QDs.

4. J-aggregate Blended Films in Device Applications

With the current prominence of digital photography, photovoltaic research, and LEDs, the excitonic properties of J-aggregates suggest the application of J-aggregating dyes to optoelectronic devices. Such applications might make use of the ultrafast radiative J-aggregate lifetime, the narrow emission, or the optical non-linearity of J-aggregates.

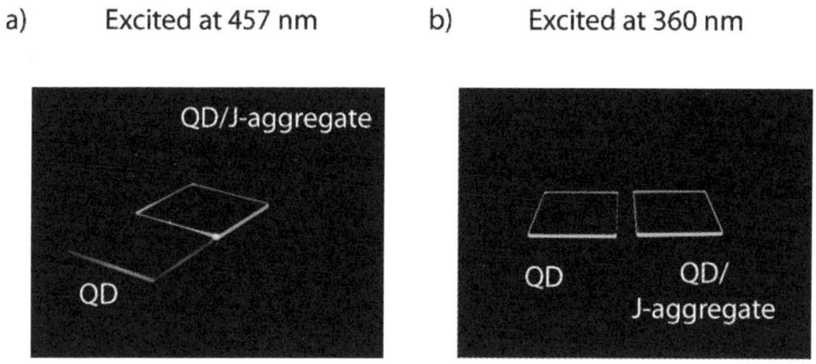

Figure 4.1. Photographs of quantum dot (QD) and QD/J-aggregate thin films with the same quantity of QDs taken with λ=630 nm band pass films to remove excitation light. (a) Films excited using collimated λ=457 nm illumination at the intersection of the two films. The average, whole-film brightness enhancement from QD/J-aggregates is apparent. (b) Films excited using a broad band UV source centered at λ=360 nm. The near- identical brightness is consistent with the spectral specificity of the J-aggregate. Reprinted with permission from Ref 36. Copyright 2010 American Chemical Society.

In practice, two challenges have hindered the application of J-aggregates to optoelectronics. First, the aggregation process depends strongly on the macromolecular structure of constituent chromophores, making the solvents,[30] ions,[31] and other aspects of the chemical environment critical to achieving reproducible J-aggregation. Due to the nature of the intermolecular functional group interactions, much of the J-aggregate literature depends on aqueous processing. These aggregation methods include the formation of J-aggregates via Langmuir-Blodgett films,[32] micelle formation,[33] or layer-by-layer (cited above) or similar aqueous self-assembly.[34]

The second challenge is that J-aggregates are current insulators. Several time of flight studies on evaporated J-aggregate films have demonstrated that both electron and hole mobilities in J-aggregates are low (10^{-6} to 10^{-4} cm^2 $V^{-1}s^{-1}$).[35] However, the incorporation of J-aggregates into other electronic architectures is often incompatible with aqueous processing, and hence there is a demand for non-aqueous methods for the controllable deposition of J-aggregate films.

In particular, it is advantageous to have processing methods for J-aggregates that are compatible with spun cast films of quantum dots, as the blended film structure is a common design theme in hybrid nanocrystal/organic devices. Compared with a planar interface between the J-aggregate and nanocrystal, the blended interface allows a greater surface area of interaction and affords the greatest opportunities for energy transfer between the J-aggregates and the quantum dots.

Blended films of J-aggregate/nanocrystals were prepared by spin casting a mixture of the unaggregated cyanine dye and the nanocrystals from an alcohol solvent.[36] The ligand exchange with short, positively-charged ligands described above was instrumental to the construction of the film for two reasons. First, the length and charge of the ligands promoted the close association of the negatively-charged sulfonate functional groups on the J-aggregating dye, which enabled efficient energy transfer. Second, the change in the solubility property of the nanocrystals allowed the organic and inorganic components of the blended films to be spun without phase segregation. This nanocrystal /organic phase segregation is a reliable means of constructing a layered device,[37] and the blended film structure is now an important alternative.

The resulting blended nanocrystal/J-aggregate films demonstrated efficient energy transfer, largely preserving the effect observed from solution in the solid state thin film.[36] As the J-aggregates likely adopt a different morphology in solution and thin film, the retention of efficient energy transfer is a key feature of the blended architecture. Atomic force micrographs of the films revealed uniform tip-to-surface interactions and a lack of phase segregation in the samples (confirmed by transmission electron microscopy), both of which indicate that the J-aggregates and nanocrystals were well-blended in the solid state film.

5. Conclusions and Outlook

J-aggregate hybrid materials continue to have potential to yield new insights into light/matter interactions, as well as interactions between disparate materials, and there are several promising areas engendered by recent progress. As already observed in the studies above and elsewhere, the extent of the J-aggregation process itself serves as a probe of the dye environment. In an environment that promotes close, ordered packing the J-aggregate spectral features become far more apparent, and the intensity of the J-aggregate resonance thus provides an optical indicator of local order. For example, such a localized probe was used with $AgNO_3$

solution[38] to oxidize the outer wall of a double-walled J-aggregate, while retaining the resonance from the inner wall.[39] This selective transformation demonstrated that the inner and outer walls were electronically isolated from one another, thus helping to resolve the nature of electronic coupling in double-walled J-aggregates.[40]

There has also been encouraging progress toward using J-aggregates in optoelectronic device architectures.[41] For example, Kim *et al.* recently used the helical architecture of carboxymethyl amylose in solution to template the formation of cyanine J-aggregates around an a rod-like donor-acceptor chromophore,[42] demonstrating photoinduced electron transfer in a supramolular excitonic structure. Walker *et al.* demonstrated a new light emitting device architecture, in which surface-bound tubular J-aggregates utilized electrically-induced redox reactions in an aqueous solution to produce spectrally narrow emission from the J-aggregates.[43]

In recent years, new excitonic couplings with J-aggregates have been explored with many inorganic materials. Because the delocalized Frenkel excitons in J-aggregates have notable analogies with the excitations in light harvesting photosynthetic systems,[44] J-aggregate/inorganic structures may serve as synthetic model systems in which rapid energy transfer and charge carrier separation are distributed among different materials. Having already achieved practical use in photographic technology, there is promise for the use of J-aggregates and coupled inorganic excitonic materials in a new generation of optoelectronic devices.[45]

6. Acknowledgments

B. J. W. was supported by a National Science Foundation Graduate Research Fellowship. This work was supported by the DOE Center for Excitonics, an Energy Frontiers Research Center funded by the U.S. Department of Energy, Office of Basic Energy Sciences (DE-SC0001088); the Institute for Soldier Nanotechnologies (W911NF-07-D-0004); the DOE Basic Energy Sciences (DE-FG02-07ER46454); and the Chesonis Family Foundation. Special thanks to colleagues Gleb M. Akselrod, Dylan Arias, Vladimir Bulović, Jonathan E. Halpert, Gautham P. Nair, Keith A. Nelson, Katherine W. Stone, Yaakov R. Tischler,

William A. Tisdale, and Elizabeth R. Young for discussions, collaboration, and feedback.

References

1. T. Kobayashi, ed. *J-aggregates* World Scientific: Singapore (1996).
2. H. Fidder, D. A. Wiersma, Phys. Rev. Lett. **66**, (1991) 1501-1504.
3. S. Blumestengel, S. Sadofev, C. Xu, J. Puls, F. Henneberger, Phys. Rev. Lett **97** (2006) 237401.
4. F. Würthner, T. E. Kaiser, C. R. Saha-Möller. Angew. Chem. Int. Ed. **50**, (2010) 3376-3410.
5. M. van Burgel, D. A. Wiersma, K. Duppen, J. Chem. Phys. 102 (1995) 20-33.
6. S. M. Vlaming, V. A. Malyshev, J. Knoester, Phys. Rev. B **79** (2009) 205121
7. A. Eisfeld, S. M. Vlaming, V. A. Malyshev, J. Knoester, Phys. Rev. Lett. **105** (2010) 137402
8. A. S. Davydov, *Theory of Molecular Excitons*, Plenum Press: New York (1971).
9. H. van Amerongen, L. Valkunas, R. van Grondelle. Photosynthetic Excitons, World Scientific: Singapore (2000)
10. J. W. Mitchell, Rep. Prog. Phys. **20**, (1957) 433-515.
11. T. Förster, Ann. Physik **437** (1948) 55-75.
12. H. Kuhn, J. Chem. Phys. 53 (1970) 101-108.
13. B. Troesken, F. Willig, K. Schwarzburg, A. Ehert, M. Spitler, *J. Phys Chem.* **99**, (1995) 5152-5160.
14. B. Trösken, F. Willig, K. Schwarzburg, A. Ehret, M. Spitler, *Adv. Mater.* **7**, (2004) 448-450.
15. Q. Zhang, T. Atay, J. R. Tischler, M. S. Bradley, V. Bulović, A. V. Nurmikko, Nat. Nanotechnol. **2** (2007) 555-559.
16. H. Fukumoto, Y. Yonezawa, *Thin Solid Films*, **327-329** (1998) 748-751.
17. M. S. Bradley, J. R. Tischler, V. Bulovic, Adv. Mater. **17** (2005) 1881-1886.
18. N. A. Kotov, Mater. Res. Bull. **26**, (2001) 992–997.
19. S. Chanyawadee, P. G. Lagoudakis, R. T. Harley, D. G. Lidzey, M. Henini. Phys. Rev. B **77** (2008) 193402.
20. K. Tani, K. Matsuzaki, Y. Kodama, M. Fukita, T. Kodaira, H. Horiuchi, T. Okutsu, H. Hiratsuka, J. Photochem. Photobiol. A **199** (2008) 150-155.
21. T. Liptay. PhD Thesis (2007): Massachusetts Institute of Technology.
22. J. E. Halpert, J. R. Tischler, G. P. Nair, B. J. Walker, W. Liu, V. Bulović, M. G. Bawendi, J. Phys. Chem. C **113** (2009) 9986-9992.
23. D. J. Norris, M. G. Bawendi, Phys. Rev. B **53**, (1996) 16338-16346.
24. B. J. Walker, G. P. Nair, L. F. Marshall, V. Bulović, M. G. Bawendi. J. Am. Chem. Soc. **131**, (2009) 9624-9625.
25. P. J. Reid, D. A. Higgins, P. F. Barbara, J. Phys. Chem. **100** (1996) 3892-3899.

26. D. A. Higgins, P. J. Reid, P. F. Barbara, J. Phys. Chem. **100** (1996) 1174-1180.
27. D. A. Higgins, P. F. Barbara, J. Phys. Chem. **99** (1995) 3-7.
28. H. Xu, X. Huang, W. Zhang, G. Chen, W. Zhu, X. Zhong, Chem. Phys. Chem. **11**, (2010) 3167–3171.
29. C. A. Steinbeck, M. Ernst, B. H. Meier, B. F. Chmelka, J. Phys. Chem. C **112**, 2565-2573.
30. H. von Berlepsch, S. Kirstein, C. Böttcher, Langmuir **18** (2002) 7699-7705.
31. I. A. Struganova, H. Lim, S. A. Morgan, J. Phys. Chem. B **106** (2002) 1104-11050.
32. S. Watanabe, H. Shibata, S. Hoiuchi, R. Azumi, S. Hideki, M. Abe, M. Matsumoto, J. Colloid. Interface Sci. **343** (2010) 324-329.
33. N. C. Maiti, S. Mazumdar, N. Periasamy, J. Phys. Chem. B **102** (1998) 1528-1538.
34. O.-K. Kim, J. Melinger, S.-J. Chung, M. Pepitone, Org. Lett. **10**, (2008) 1625-1628.
35. A. R. Tameev, A. V. Vannikov, H. F. M. Schoo, Thin Solid Films **451-452** (2004) 109-111.
36. B. J. Walker, V. Bulović, M. G. Bawendi, Nano Lett. **10** (2010) 3995-3999.
37. S. Coe, W.-W. Woo, M. Bawendi, V. Bulović Nature **420** (2002) 800-803.
38. D. M. Eisele, Hans v. Berlepsch, Christoph Böttcher, Keith J. Stevenson, David A. Vanden Bout, Stefan Kirstein, Jürgen P. Rabe, J. Am. Chem. Soc. **132** (2010) 2104-2105.
39. D. M. Eisele. *Optical, Structrual, and Redox Properties of Nanotubular J-aggregates of Amphiphilic Cyanine Dyes.* PhD thesis, Humboldt-Universität zu Berlin (2009).
40. F. Milota, J. Sperling, A. Nemeth, H. F. Kauffmann, Chem. Phys. **357** (2008) 45-53.
41. D. M. Bassani, L. Jonusauskaite, A. Lavie-Cambot, N. D. McClenaghan, J.-L. Pozzo, D. Ray, G. Vives, Coordination Chemistry Reviews **254** (2010) 2429-2445.
42. O.-K. Kim, J. Melinger, S.-J. Chung, M. Pepitone, Org. Lett. **10** (2008) 1625-1628.
43. K. E. Walker, D. A. Vanden Bout, K. J. Stevenson, J. Phys. Chem. C **ASAP** doi: 10.1021/1108015.
44. J. Knoester, Int. J. Photoenergy **61364**, (2006) 1-10.
45. Nat. Nanotechnol. 4, (2009), 607-609.

CHAPTER 7

J-AGGREGATE FORMATION OF PSEUDO-ISOCYANINE AND RELATED DYE MATERIALS UNDER INFLUENCE OF DNA

Sho Kato and Yutaka Kawabe*

Chitose Institute of Science and Technology, Chitose, 066-8655 JAPAN
**E-mail: y-kawabe@photon.chitose.ac.jp*

We employed biopolymer DNA as a template to form J-aggregates of pseudoisocyanine (PIC), and succeeded to optimize the conditions for their formation in solutions and solid films. The optical characteristics of J-aggregates were investigated by absorption, fluorescence and circular dichroic (CD) spectra. Polyvinylalcohol (PVA) introduced as a matrix for the films was proved to play a role to prevent the precipitation or recrystallization of the dyes in solutions and films. We investigated the dependence on concentrations of dye and DNA, and the ratio of these, and effects from counter ions for PIC. We also prepared the samples for more than 20 types of cyanine dyes, finding few dyes showing J-aggregate peak. These results show possibility of application of J-aggregate into novel optical devices requiring optical nonlinearity or superradiant behavior.

Keywords: J-aggregate, pseudo-isocyanine, PIC, cyanine dye, DNA, CD spectrum, circular dichroism

1. Introduction

J-aggregate is considered to be important from both scientific and application point of view. Its remarkable features, such as line narrowing and superradiance originate from the delocalization of molecular excitons.[1] As known from the preceding works for pseudo-isocyanine (PIC), the most well-studied J-aggregate dye, the conditions for J-aggregate formation depend sensitively on many parameters such as types of solvents, dye concentration, pH of the solutions, types of counter ions, additives, temperature, and so on.[2] Therefore, it would be necessary to control the environments for aggregating molecules in order to realize and stabilize the J-aggregates in solid matrices, which would be polymers for practical optical applications.[3–7]

We have used bio-polymer DNA (deoxyribonucleic acid) as a template for aggregate formation of PIC both in solutions and in solid films, and have investigated the conditions for their formation by using mainly absorption and circular dichroic (CD) spectroscopies under the varieties of the dye and DNA concentrations.[5–7] As well known, DNA have a double stranded structure which can include organic molecules through the intercalation between nucleo-base pairs or binding them into major or minor grooves. When the dyes have cationic characteristic, they would be easily attracted to poly-anionic DNA strand by ion exchange processes.[8] Luminescence from several dyes is often strongly enhanced when they are interacted with DNA, and light amplification and laser actions were observed for several dye species.[9–14] DNA not always enhances the dye luminescence, but sometimes suppresses the light emission, and these behaviors depend on the types of dyes and also on the configuration which might be determined by many physical parameters of dyes. Therefore, it is important to survey the DNA effects on dyes for future applications.

In this chapter, we discuss the influence of DNA on the optical characteristics for various dye concentration, molar ratio between DNA and dye, and types of counter anions. We also describe the effects of DNA on more than 20 types of cyanine dyes with different kinds of aromatic rings with various conjugation and alkyl chain length. Then, we give some remarks on the morphology given by the intermolecular interactions among DNA and the dye, through our spectroscopic studies.

2. J-aggregate formation of PIC dissolved with DNA polymer

DNA is a commercially available biological polymer, which can be extracted from sperm of salmon, calf thymus, and so on. It has a complex double helix structure as schematically sketched in Fig. 2.1. It is known that several organic dyes can obtain specific optical characteristics by being taken into DNA via intercalation, groove binding or exchanged binding. Contrary to most of functional organic materials, DNA is insoluble into organic solvents, but invention of DNA-surfactant complex made it solvable and film formable from common organic solvents.[15,16] The materials formed by ion exchange are poly-ionic complexes which can incorporate many organic dyes into their DNA structure. Then, these complexes have been studied as materials for thin film dye lasers, organic EL devices and other optoelectronic systems.[15–19]

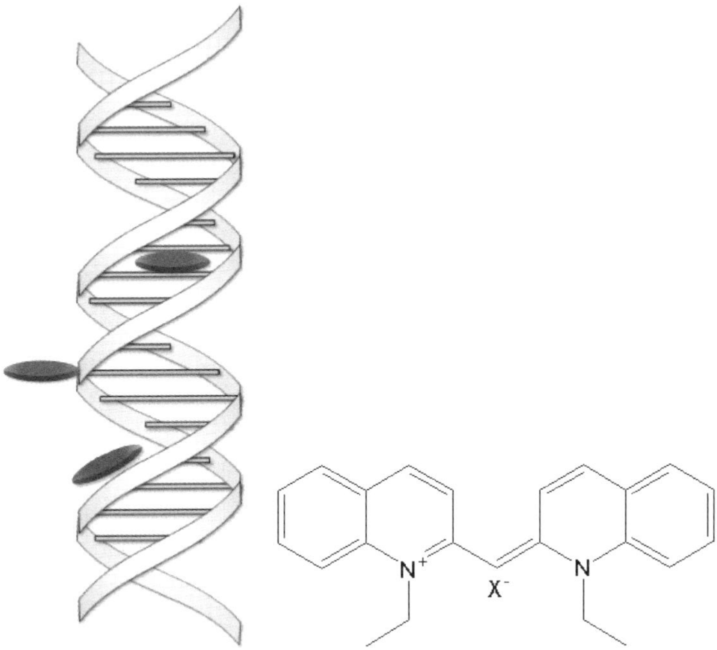

Figure 2.1 Schematic diagram of the coupling between DNA and cationic dyes (left), and the molecular structure of pseudo-isocyanine halide in which X^- is counter ion as iodine, bromine and chlorine (right).

Although most of studies of DNA as a photonic material have employed DNA as the complex with lipids because of its good processability, it is attractive to use bare DNA, since we can utilize its anionic properties more directly to couple DNA and cationic dyes with electrostatic force. Until now, we have investigated the optical properties of some water soluble dyes interacting with DNA in water, finding that their optical absorption were strongly affected by the existence of DNA, resulting in peak shape modulation and peak shift accompanied by fluorescence enhancement or suppression depending on the molar ratio of DNA base pair and dyes.[7,11] In this study, we chose PIC and investigated the conditions for their J-aggregate formation in solutions and films by monitoring their absorption and CD spectra.

2.1 *Sample Preparation*

Structural formula of PIC is shown in Fig. 2.1. PIC is a cationic dye with a counter ion X^- which means iodine, bromine or chlorine. In our case, DNA salt has two sodium ions per base pair, and it behaves as poly anion.

We purchased PIC-I (Sigma Aldrich), PIC-Br and -Cl (Hayashibara), and highly pure DNA extracted from salmon sperm (Ogata Material Science Laboratory), and used them without further purification. We prepared solution samples by the recipe following. It would be preferable to use water as solvent, because DNA is only soluble in water. However, PIC solubility in water depends on the types of counter ions, so we have added methanol for some cases. First, we made the aqueous solution of PIC and polyvinylalcohol (PVA) with concentration of 1.26×10^{-3} mol/l and 48 g/l, respectively. For the case of PIC-I which is less soluble in water, we used the mixture of methanol and water (the ratio 1:4 which is the optimum ratio determined by the trade-off between solubility and aggregate formability). Then we added aliquots of DNA aqueous solution with the concentration in the range of $2 \sim 10 \times 10^{-3}$ mol/l. For the calculation of molar concentration of DNA, we used the average molecular weight of one base pair unit, that is, 661.5. We added PVA which works as film matrix of the dyes because it is difficult to fabricate high quality neat DNA films from aqueous solutions. Also, PVA prevents the precipitation of crystallites both in solutions and solids. Films were formed by spin coating or casting methods from the same solutions used for the optical measurements.

2.2 *Absorption Spectra*

Daltrozzo *et al.* have studied the J-aggregate of PIC dye, finding that the J-aggregate formability as well as the solubility are affected seriously from its types of counter anions.[2] Among the three halide compounds we used, chloride is the most soluble in water, and bromide next followed by iodide. Their absorption spectra in dilute aqueous solutions (concentration less than 1.0 x 10^{-4} mol/l) show the typical shapes common to cyanine dyes with the main peak at 530 nm, and no difference was observed for three types of halogen ions (not shown as figures). The spectral shapes of high concentration solutions showed different behaviors when DNA was added in solutions.

We measured the absorption spectra by using a conventional spectrophotometer with a quartz cell of 0.1mm optical path length due to relatively high concentration of the dyes. Figure 2.2 gives absorption spectra of PIC halides when small amounts of DNA were added. In the case of iodide, we observed the typical J-peak, while a new peak at the longer wavelength side has broader width for bromide and chloride. But the dependences on DNA concentration were similar in all three cases, that is, new peaks grew with DNA when small amounts were added, but they became small and diminished with excess amounts of DNA. The final positions of the original main peaks (which can be assigned to monomers) have slightly shifted from their initial locations. Similar behaviors were also observed for other cyanine dyes as described in our previous study and given in the following section of this chapter.[7,11]

We investigated the detail of the J-aggregate formation and dissociation processes by varying the concentration of DNA additive as shown in Fig. 2.3. In these figures, we plotted the J-peak height as a function of the dye-DNA molar ratio. In the case of iodide, the highest peak was observed when the ratio was about 4, and the peak completely disappeared when the ratio exceeded about unity. But it was 2 for bromide and chloride (data for chloride was not shown because of the similarity to the bromide case). In both cases, a small amount of DNA promotes the J-aggregate formation, but when it exceeded some values, J-aggregate seemed to dissociate. The optimum conditions would relate to the aligning way of the dyes attaching to DNA strands.

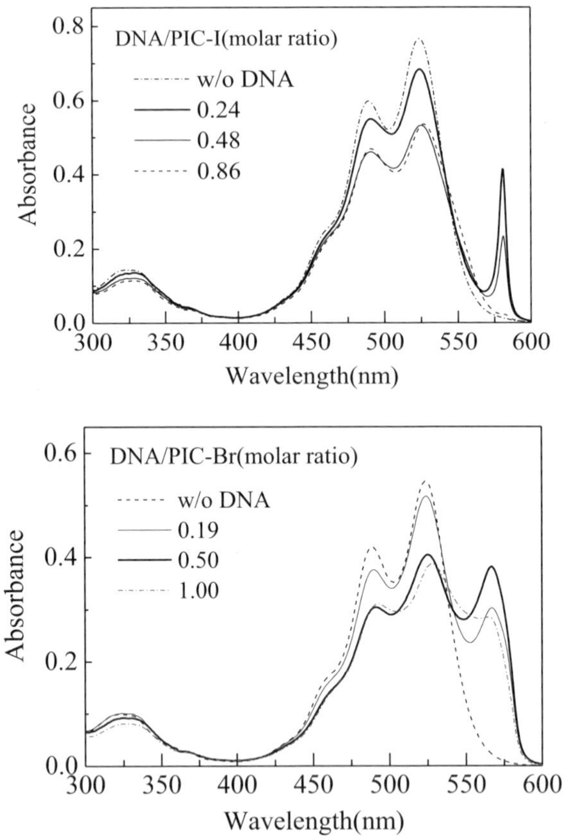

Figure 2.2 Absorption spectra of PIC-I (upper) and PIC-Br (lower) dissolved in water including PVA and different concentrations of DNA.

For comparison, we also studied the absorption spectra when no DNA nor PVA were included, although there were several preceding studies.[2,20-22] For the dye concentration ranges we investigated in this study, no J-peak was observed without DNA, and the existence of PVA did not matter for the J-aggregate formation but it certainly prevented the dye precipitation. As shown by some groups, other ionic additives, such as sodium salts, also promote the formation of J-aggregates.[20,21] In our case, the role of DNA was quite unique and advantageous, because DNA is water soluble ionic polymer which can incorporate J-characteristics into film just by fabricating film from the solution as described later.

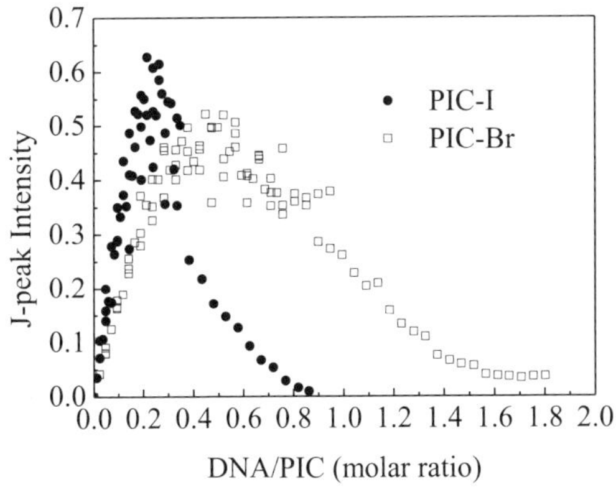

Figure 2.3 Dependence of J-peak intensity as a function of the molar ratio between DNA/PIC for PIC-I and PIC-Br.

2.3. *Circular Dichroism Spectra*

When a dye molecule interacts with materials having a chiral structure such as DNA, chirality induces CD signals in dyes' spectra. DNA itself shows the positive Cotton effect from 200 to 300nm, which indicates a clockwise helical structure.[15,16] We measured the CD spectra of PIC interacting with DNA, showing J-aggregate peak.

PIC-I gave several CD peaks when it was dissolved with DNA, while no peaks were observed when DNA did not exist as shown in Fig. 2.4. With a small amount of DNA, large peaks appeared in the wavelength region of absorption and disappeared by adding excess DNA. A sharp peak at 581 nm certainly originated from J-aggregates because it was observed only when J-peak appeared in the absorption spectra, and its position was corresponding to the absorption peak. There is another peak at 505 nm with negative sign, which also appeared when J-peak did but its position was not corresponding to the absorption peak.

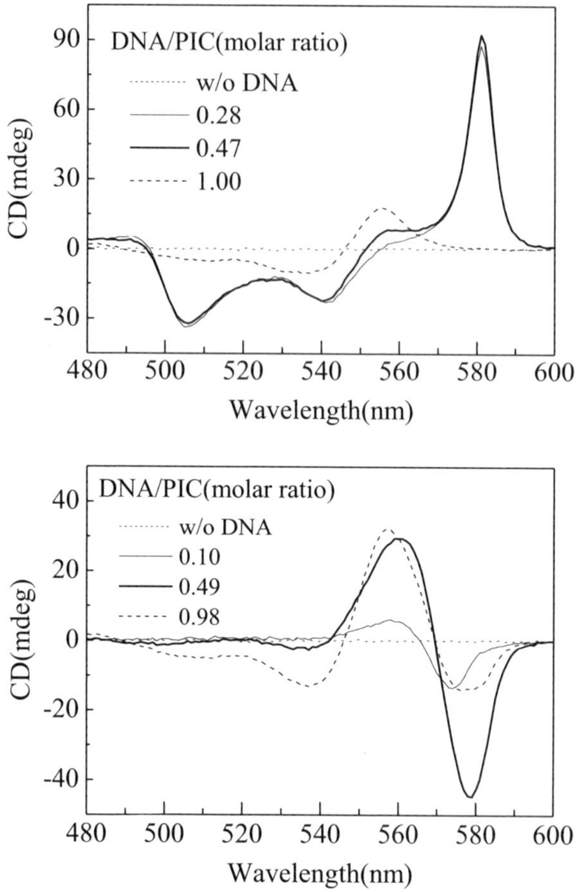

Figure 2.4 CD spectra for PIC-I (upper) and PIC-Br (lower) when mixed with DNA and PVA in aqueous solutions.

Most probable explanation for these results is that the peaks at 580 nm and at 505 nm originate from the same substances, that is, J-aggregate. As well-known from molecular exciton theory, electronic states will split into two branches forming two bands when aggregates are composed by unit cells including a dimer.[23–25] When rod-like molecules align head-to-tail, the lower energy branch will be optically active, appearing as the J-band. The other component (the upper branch) would be dominant in H-aggregates where molecules align side-by-side.

In the case of PIC aggregate, it allegedly forms herring-bone structure, in which the transition to the upper branch will often be partially allowed due to side-by-side aligning component of the structure. Therefore, it is reasonable to assign these two peaks to the same aggregates.

On the other hand, there is a structure with dispersive shape around 550 nm. This signal can be assigned to that from the monomeric molecules interacting with DNA, because it appeared even when no J-peak was observed. But the assignment is not completely clear because it does not correspond to the main peak of the monomer absorption.

The CD spectra for PIC-Br are also shown in Fig. 2.4. In this and also the case for PIC-Cl (not shown because of similarity), a dispersive structure was observed in the region of the red-shifted peak caused by DNA, of which fact indicates that DNA strongly interacts with PIC-Br or -Cl. Indeed, we also observed the CD signals in UV regions for all PIC halide solutions as well as those from DNA itself (not shown). Considering the broader spectral shape, the molecules do not seem to form J-aggregate accompanying exciton delocalization for these cases. However, the spectral shift and the concentration dependence of CD shapes similar to the absorption spectra suggest that DNA stimulated the formation of some kinds of dye aggregates, such as head-to-tail type dimers formed in minor grooves.

2.4 *J-aggregates in Films*

In order to utilize J-aggregates in practical devices, it is necessary to fabricate thin films with polymer matrix. However, most of science oriented studies have been concentrated in solution samples because of easiness of their preparation. Among a few examples for films, Misawa *et al.* succeeded to prepare PVA thin film including high concentration of PIC-Br showing J-aggregate peak, by so-called vertical spin coating method.[1,3] We also have employed PVA as a host, and succeeded to fabricate J-aggregate films by adding small amounts of DNA.[5–7]

For the film formation, we used solutions including PVA, DNA and PIC-I dye, which were the same as those prepared for the solution studies. Films were fabricated by spin coating (1,500 rpm) or casting methods. Typical weight ratio of the composition was PVA:DNA:PIC-I =

100 : 0.4 : 1.2, where the molar ratio of PIC/DNA was corresponding to 4, the optimized value for J-aggregate formation in solution.

Fig. 2.5 shows absorption and CD spectra of the films formed with or without additive DNA for comparison. These results indicate that DNA strongly promotes the forming of J-aggregates.

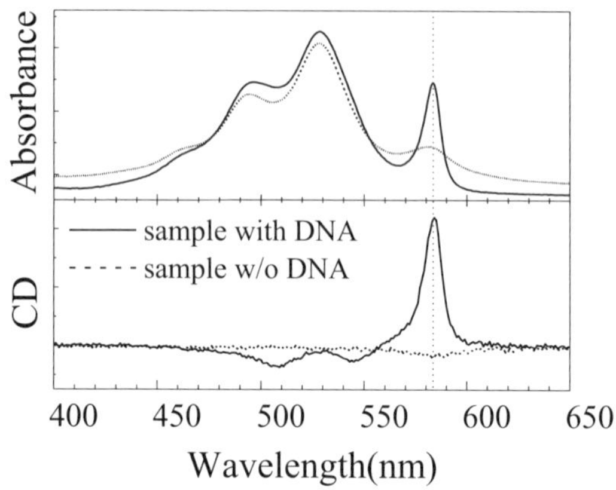

Figure 2.5. Absorption and CD spectra of thin film sample composed of PIC-I and PVA including and not including DNA.

For casting method, sample films were prepared in teflon dishes (5 cm diameter) by drying the solutions of 0.1 ~ 1 ml under heating at about 40°C for one day. The characteristics of the cast films were similar to those made by spin coating, that is, the absorption and CD spectra were almost corresponding to each other except the effects from the thickness.

The films composed of PIC-Br and -Cl did not show significant absorption or CD signals in the wavelength region where new peaks were observed in their solution samples.

In order to confirm the effects of DNA in films, we fabricated films without DNA for comparison. We could form the films showing J-peaks, but we must have increased the concentration of PIC dye in order to observe J-peaks in PVA films without DNA, that is, 2.3% for PIC-I and 12.5% for PIC-Br and -Cl. These ratio were twice or about an order larger than the cases with DNA, and were comparable to those given in literature.[3] Microscopic observation revealed that the sample without

DNA showed the precipitation of ribbon like crystallites, which were never seen in the films including DNA. Even for the samples without DNA, CD signals were observed at the J-peak wavelength, meaning that the aggregates in PVA polymer spontaneously have helical structure without chiral templates as reported in early days.[2,26] However, the sign and the shape of CD signals varied depending on the observation position of the samples, while they were usually invariant for the samples with DNA. Thus, we can conclude that DNA works as an effective template for the formation of J-aggregates of PIC dye both in solutions and films.

2.5 *Discussions*

From the results described above, some speculation for the J-aggregate structure would be possible. The reason why the J-aggregate was formed with only small amount of DNA may be related to the relative strength of intermolecular interaction between dye molecules and that between the dye and DNA strand. In dilute dye solutions, cationic dye would exist as isolated ions resulting in a conventional absorption spectrum. Ionic DNA introduced would modify the circumstances and stimulate the mutual interaction among dyes. When the amount of DNA is relatively small, interaction among dyes is dominant, resulting in the aggregate formation. But when enough amount of DNA is supplied, electrostatic force attracting the dye to DNA becomes dominant, then it dissociates the aggregates to form DNA-dye complex. Attaching of the dye to DNA is the reason of the peak shift from the original position when excess DNA is supplied. Some preceding studies also gave similar speculation proposing the configuration transition from dimer to other kinds of complexes.[27]

In our study, there is very large difference between PIC-I and the other two species, so the role of counter ions should be also important. There were several structure studies for single crystalline PIC dyes having several types of counter anions.[28–30] Although most of the crystalline structures are monoclinic, their space groups and lattice constants depend on the ions and there is also polymorphism. Unfortunately, while the spectral characteristics of these single crystals have not been well studied, it seems that these crystals do not always show sharp J-peaks at room temperature. In our solution cases, the difference appeared in Fig. 2.2 indicates that the J-aggregate formation would be very sensitive to structural difference caused by molecular alignment modulation due to the anion types. These detailed studies will give a key to elucidate the physics of J-aggregate which cannot be explained simply by exciton delocalization predicted even from the most simple molecular exciton theory.[23–25]

3. Aggregate Formation of Related Cyanine Dyes Interacting with DNA

As described in the former section, the conditions for J-aggregate formation of cyanine dyes are very sensitive to the concentrations of dye and DNA, the ratio of them, and existence of PVA, so they might be determined by complex relationships among the relative strength of the dye-DNA interaction and dye-dye interaction and other physical parameters. The details of interaction between DNA and dyes are still under discussion.[31] In the case of PIC dye, even the change of halogen ions affected their structure drastically. There are several possibilities for binding mode such as intercalation into double strand, binding to major or minor grooves, or being attracted by electrostatic force. In order to get insight into details, we employed more than 20 types of water soluble organic dyes and observed the influence of DNA on their optical properties.[32] In this section, we briefly describe some typical results.

We have already investigated the absorption and fluorescence spectra of several cyanine dyes, showing that spectrum peaks broadened and their fluorescence was quenched by the addition of DNA.[7] In this study, we have used molecules resembling PIC with varieties of the conjugated length, types of aromatic rings, or alkyl chain length. The diameter of DNA strand is about 2 nm and the distance between base pair stacks is 0.8 nm, while the length of PIC is 1.4 nm. In order to study the effects of the molecular size, we first compared PIC-I (also indicated as $DiQC_2(1)$) and $DiQC_2(3)$, the structure of the latter shown in Fig. 3.1.

We prepared the solutions of these dyes dissolved with PVA and DNA into water/methanol mixture (the ratio is 4/1), and measured the absorption, fluorescence and CD spectra, while changing the amounts of additive DNA. Figure 3.2 depicts the results which show that little amount of DNA promotes the formation of J-aggregates of $DiQC_2(1)$ as described in the former section (the concentration was about the half of those used for Fig. 2.2) without changing the absorption spectral shap, but no J-aggregate peaks were observed in the $DiQC_2(3)$ spectra. Instead, the absorption spectrum of $DiQC_2(3)$ spread into both longer and shorter wavelength regions. The fact means that the molecules were aggregated by the stimulation from DNA, but their alignment should have been so irregular that both J-type (longer side) and H-type (shorter side) alignments coexisted. Fluorescence seems to reflect the situation, since the intensity was enhanced by DNA for $DiQC_2(1)$, while it was quenched for $DiQC_2(3)$. CD spectra showing signals in the visible wavelength regions give the evidence of the interaction of the both dyes and DNA.

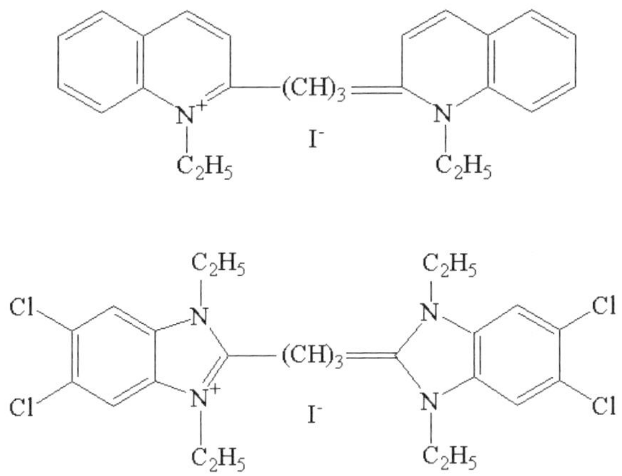

Figure 3.1 Molecule structures of $DiQC_2(3)$ (upper) and NK-1420 (lower).

We also conducted the experiments for the cyanine dyes called $DiXC_n(m)$ (for notation, see Refs. 7 and 30) and found that the behavior under DNA addition depended on the size of dye molecule. When the length of conjugated length given by m was 1, no drastic change was observed in absorption spectrum, and the fluorescence intensity was enhanced by the addition of DNA. On the other hand, when molecular length was comparable to or longer than $DiQC_2(3)$, small amounts of

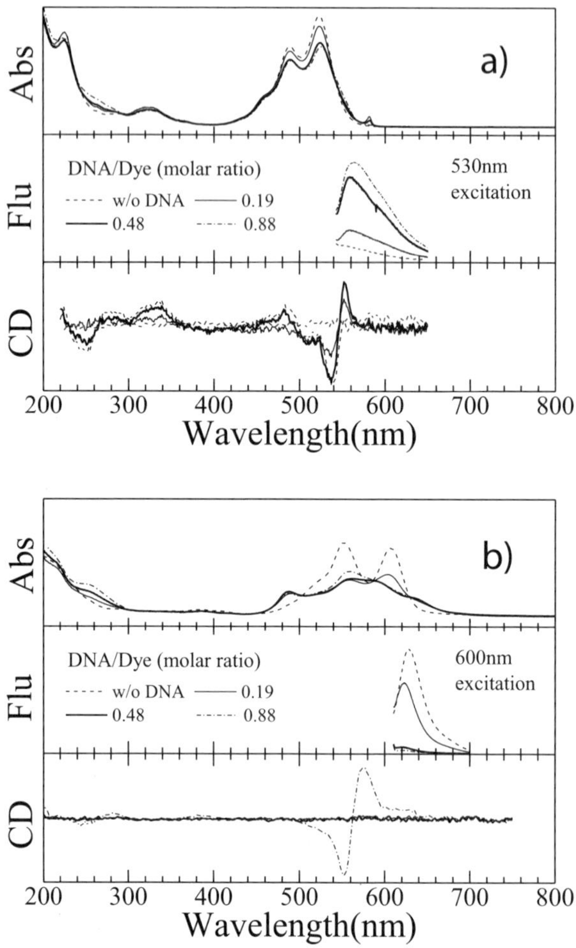

Figure 3.2 Spectra of absorption(upper), fluorescence (middle) and CD (lower) for a) $DiQC_2(1)$ and b) $DiQC_2(3)$ at various DNA/dye molar ratios.

DNA made the spectral width broad and quenched fluorescence but excess amounts recovered them. Although similar suggestion was already addressed in ref. 11, we will emphasize that the molecular size would be quite important in DNA-dye interaction and also strongly influence the formation and structure of the aggregates.

Among these cyanine dyes, unfortunately we could find few species giving J-aggregate in aqueous solution. For example, we show the results from NK-1420 (Hayashibara, molecular structure in Fig. 3.1 in Fig. 3.3. In this case, J-aggregate peak was clearly observed in the luminescence spectrum, however its intensity was reduced by adding DNA. While J-peak in the absorption spectra was not clearly distinguished, CD spectra showed the signal with negative direction at the wavelength corresponding to the J-band. The difference suggests that there are various types of aggregate formation or structure, and DNA might give a method to probe and control these structures, and the fact is important for application to optical devices.

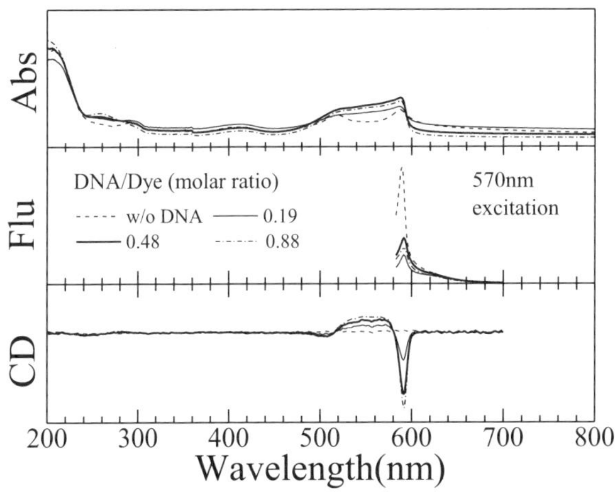

Figure 3.3 Spectra of absorption (upper), fluorescence (middle) and CD (lower) for a) NK-1420 at various DNA/dye molar ratios.

4. Conclusion

We succeeded to control J-aggregate of PIC by adding DNA in solutions and films. While a little amount of DNA promotes the formation of J-aggregates, excess addition dissociates them. In order to control the conditions, we added PVA and optimized the concentration of DNA and dye, and molar ratio of dye/DNA. We succeeded to make thin films showing J-aggregate characteristics with lower dye concentration. We also investigated the effects from counter anion, and discussed their structures. Finally, we investigated the influence of DNA for the aggregation of more than 20 cyanine dyes, finding that few dyes show

J-aggregation and the length of molecules was important parameter for its formation.

References

1. T. Kobayashi ed, *J-Aggregates*, (World Scientific, Singapore, 1996).
2. E. Daltrozzo, G. Scheibe, K. Gschewind and F. Haimerl, Photogr. Sci. Eng., **18**, 441 (1974).
3. K. Misawa, H. Ono, K. Minoshima and T. Kobayashi, Appl. Phys. Lett., 63, 577 (1993).
4. T. Tani, M. Saeki, Y. Yamaguchi, T. Hayashi and M. Oda, J. Lumin., 107, 339 (2004).
5. S. Kato and Y. Kawabe, in *Optical Devices and Systems, Biomedical and Green Technologies*, eds. by S. Kobayashi and H. Sasabe (PWC Pub., Chitose, 2010)
6. S. Kato and Y. Kawabe, Mol. Cryst. Liq. Cryst. 520, 165 (2010).
7. S. Kato, M. Honda, Y. Kawabe and J. Yoshida, Proc. SPIE, 7765, 776503 (2010).
8. B. A. Armitage, in *DNA binders and related subjects*, eds. by M. J. Waring and J. B. Chaires (Springer, Berlin, 2005).
9. Y. Kawabe, L. Wang, S. Horinouchi and N. Ogata, Adv. Mater. 12, 1281-1283 (2000).
10. Y. Kawabe, L. Wang, T. Nakamura and N. Ogata, Appl. Phys. Lett. 81,1372-1374 (2002).
11. M. Honda, N. Nakai, M. Fukuda and Y. Kawabe, Proc. SPIE 5546, 664609 (2007).
12. P. Gupta, P. P. Markowicz, K. Baba, J. O'Reilly, M. Samoc, P. N. Prasad and J. G. Grote, *Appl. Phys. Lett., 88,* 213109 (2006).
13. J. Mysliwiec, L. Sznitko, S. Bartkiewicz, A. Miniewicz, Z. Essaidi, F. Kajzar and B. Sahraoui, *Appl. Phys. Lett.,* 94, 241106 (2009).
14. J. Mysliwiec, L. Sznitko, A. Sobolewska, S. Bartkiewicz and A. Miniewicz, *Appl. Phys. Lett., 96,* 141106 (2010).
15. K. Tanaka and Y. Okahata, J. Am. Chem. Soc. 118, 10679 (1996).
16. L. Wang, J. Yoshida, N. Ogata, S. Sasaki and T. Kajiyama, Chem. Mater. 13, 1273 (2001).
17. J. G. Grote, J. A. Hagan, J. S. Zetts, R. L. Nelson, D. E. Diggs, M. O. Stone, P. P. Yaney, E. Heckman, C. Zhang, W. H. Steier, A. K.-Y. Jen, L. A. Dalton, N. Ogata, M. J. Curley, S, J. Clarson and F. K. Hopkins, J. Phys. Chem. B, 108, 8585 (2004).
18. K. Hirata, T. Oyamada, T. Imai, H. Sasabe, C. Adachi and T. Koyama, Appl. Phys. Lett. 85, 1627 (2004).
19. Y-W. Kwon, C. H. Lee, D-H. Choi and J-I. Jin, J. Mater. Chem. 19, 1353 (2009).
20. K. D. Belfield, M. V. Bondar, F. E. Hernandez, O. V. Przhonska and S. Yao, Chem. Phys. 320, 118 (2006).
21. G. Ya. Guralchuk, I. K. Katrunov, R. S. Grynyov, A. V. Sorokin, S. L. Yefimova, I. A. Borovoy and Y. V. Malyukin, J. Phys. Chem. C 112, 14762 (2008).

22. Y. Tanaka, H. Yoshikawa and H. Masuhara, J. Chem. Phys. B, 110, 17906 (2006).
23. A. S. Davydov, *Theory of Molecular Excitons* (Plenum, New York 1971).
24. M. Pope, C. E. Swenberg, *Electronic Processes in Organic Crystals and Polymers* (Oxford Unv. Press, Oxcford, 1999).
25. V. M. Agranocich *Excitations in Organic Solids* (Oxford Univ. Press, Oxford, 2009).
26. H. J. Norte, Chem. Phys. Lett., 31, 134 (1975).
27. B. Nordén and F. Tjerneld, Biophys. Chem., 6, 31 (1977).
28. H. Yoshioka and K. Nakatsu, Chem. Phys. Lett., 11, 255 (1971).
29. A. P. Marchelli, C. D. Salzberg and E. I. P. Walker, Photogr. Sci. Eng. 20, 107 (1976).
30. K. Nakatsu, H. Yoshioka and H. Morishita, Acta Cryst. B**33**, 2181-2188 (1977).
31. M. J. Waring and J. B. Chaires eds. *DNA binders and related subjects*, (Springer, Berlin, 2005).
32. Y. Kawabe and S. Kato, Proc. SPIE 8103, 81030D, (2011).

CHAPTER 8

GIANT ELECTROOPTIC EFFECT OF PORPHYRIN J-AGGREGATES IN POLYMER FILM AND IN AQUEOUS SOLUTION

Eiji Tokunaga* and Kazuaki Nakata

Department of Physics, Faculty of Science, Tokyo University of Science, 1-3 Kagurazaka, Shinjuku-ku, Tokyo 162-8601, Japan
**E-mail: eiji@rs.kagu.tus.ac.jp*

The effect of electric field on porphyrin J-aggregates both in polymer film and in aqueous solution was studied by electroabsorption spectroscopy with a multichannel lock-in detection technique. A large enhancement of electrooptic (EO) Kerr response was observed due to the formation of J-aggregates from monomer molecules. The enhancement factor of EO Kerr parameters (polarizability difference between the excited and ground states) is as large as 100 for the polymer film and more than 100 000 for the aqueous solution. This is explained by the molecular rearrangement model, where the dipole–dipole interaction energy is modified by field-induced change in the angle between the molecular transition dipole moment and the aggregation axis. The model was supported by various experimental facts. We also found electric-field induced dissociation and association of J-aggregates in aqueous solution. This may lead to the way to control the coherent aggregation number of J-aggregates with a stationary electric field.

Keywords: porphyrin J-aggregates, electrooptic effect, Kerr effect, electroabsorption, molecular rearrangement, coherent aggregates, microaggregate, macroaggregate, electric-field induced dissociation and association, aqueous solution

1. Introduction

In recent years, self-assembled organic and inorganic nanowires and nanotubes receive much interest from the potential applicability to functional nanomaterials. Among them, self-assembled molecular

J-aggregates are prominent candidates for nano-scale opto-electronic devices. This is because J-aggregates have a sharp absorption band and a large transition dipole moment with a large optical nonlinearity due to Frenkel exciton formation[1–13]. Among them, porphyrin J-aggregates, tetraphenyl porphyrin tetrasulfonic acid (TPPS), show extraordinarily large electrooptic effects: electrooptic Kerr signals are enhanced by 100 times in J-aggregates compared with in monomers.[1,2] In this chapter, we present enhanced optical nonlinearity (third-order nonlinearity or second-order electrooptic effect) in porphyrin J-aggregates and its mechanism based on the molecular rearrangement model.[1–3] The model is shown to be supported by various experimental facts. That is, the dependence of the electrooptic signal on electric-field strength, temperature, and orientational distribution of aggregates, are very well explained by the model. For this study, multichannel electroabsorption spectroscopy is introduced[1–4] with the use of a newly developed multichannel lock-in detection system electrooptic signals of J-aggregates not only in polymer film but also in aqueous solution to obtain. Electromodulation spectroscopy of J-aggregates in aqueous solutions[3,4] has never been reported to the best of our knowledge. The unique properties of electrooptic response in J-aggregates are expected to emerge in aqueous solution because they are self-assembled in aqueous solution. In fact, we found the possibility that the coherent aggregation number and the hierarchical structure of J-aggregates could be controlled by a stationary electric field.[4] This is discussed at the last section of the chapter.

2. TPPS J-aggregates

The tetraphenyl porphyrin tetrasulfonic acid (TPPS) is based on a porphyrin ring structure and bonded to phenyl-sulfonic acid with D_{2h} symmetry as shown in Fig. 2.1.

In an aqueous solution of TPPS, there are two forms of monomer. The F-monomer, or free-base, is a tetravalent anion ($H2TPPS^{4-}$) with D_{2h} symmetry, and the D-monomer, i.e., diacid, is a divalent anion (H_4TPPS^{2-}) with D_{4h} symmetry. The D-monomer is formed when two protons are inserted in the central cavity of the porphyrin ring of the F-monomer. This is realized, for example, in the low pH solution. Both monomer molecules show two absorption bands, which are called B, or

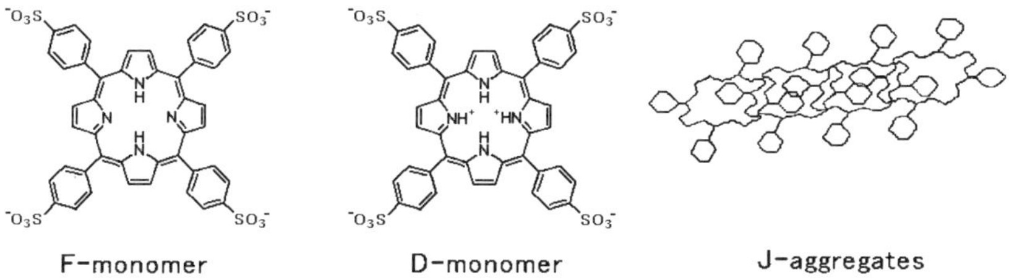

Figure 2.1. Molecule structure of the TPPS in an aqueous solution. The F-monomer (left), the D-monomer (center) and the J-aggregate (right). The J-aggregate is a linear assembly of D-monomers with a slipped face-to-face stacking.[14]

(Soret) and Q bands, in the visible region, as shown in Fig. 2.2. The absorption peaks of the F-monomer are 413 nm (B band), 515, 551, 590, 645 nm (Q band), and those of the D-monomer are 433 nm (B band), and 647 nm (Q band).

The monomer constituting the J-aggregates is considered to be the D monomer, because the positively charged center of one H_4TPPS^{2-} molecule attracts the negatively charged peripheral substituents of the adjacent molecules to form a linear assembly with a slipped face-to-face stacking.[14] Since the monomer has two orthogonal transition dipole moments in the molecular plane, the assembly forms J and H aggregates simultaneously. Among these two forms, the optical properties of the D-monomers are less understood because they tend to be assembled to form the J-aggegates in solution and are hardly observed in polymer matrix. Absorption peaks of the J-aggregates are 423 nm (H band), 490 nm (J band) originating from the monomer B band and 708 nm (J band) from the monomer Q band. The J-aggregates have strong optical anisotropy; the dipole moment of the J band is parallel to the J-axis and that of the H band is perpendicular to the J-axis.

It is known[15–20] that J-aggregates have a hierarchical structure. The spectroscopic feature is due to the micro-(coherent) aggregates, which are a linear assembly of monomer molecules as described above. Macroaggregates are those observed in high-resolution microscopy such as near-field optical microscopy[16–18], electron microscopy,[19] and atomic force microscopy,[20] and are a higher-order structure of J-aggregates formed as a stack of microaggregates. For TPPS J-aggregates, various structures for macroaggregates are proposed.[15,20]

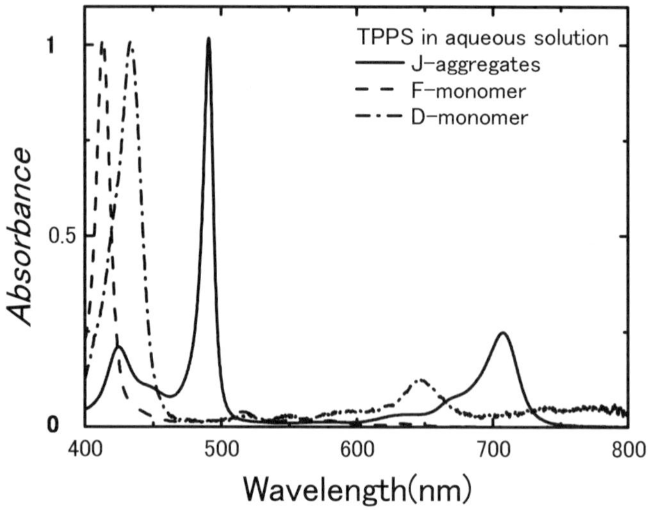

Figure 2.2 Normalized absorption spectra of F-, D-monomers and J-aggregates of TPPS. Solid, dashed and dash-dotted lines are the absorption spectra of J-aggregates, F-monomers and D-monomers, respectively. Both monomer molecules have two absorption bands, i.e., B and Q bands. For J-aggregates, the excitonic J bands appear around 490 nm and 707 nm while the H band appears around 423 nm.

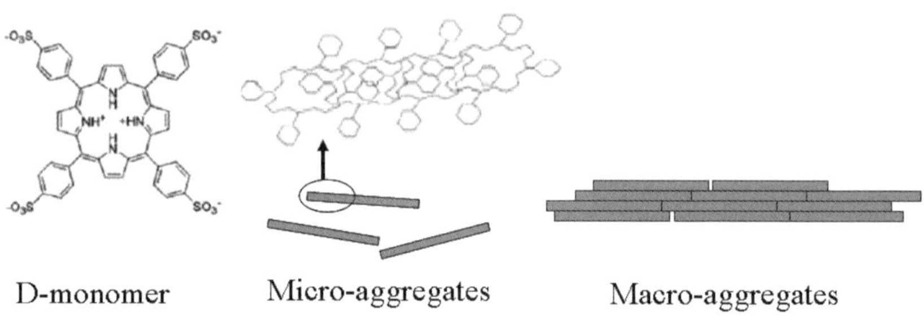

Figure 2-3 A hierarchical structure of J-aggregates.

3. Experimental

3.1 Multichannel Elecromodulation Spectroscopy

The absorbance changes detected at the modulation frequency f and its second harmonic $2f$ are due to the Pockels and Kerr effects, respectively. Here we focused on the Kerr response, for which the $\Delta\alpha$ was deduced from the size of the red shift of the B band (Soret band) of J-aggregates.

The method of data analysis followed Refs. 1 and 21. The absorption change can be written as

$$\Delta A = A_0 \frac{dA}{dE} \Delta\mu \cdot s \cdot F + \frac{1}{2} B_0 \frac{dA}{dE} \Delta\alpha \cdot (s \cdot F)^2 + \frac{1}{2} C_0 \frac{d^2 A}{dE^2} |\Delta\mu|^2 \cdot (s \cdot F)^2 , \quad (3.1)$$

where F is the applied electric field, s is a local field factor (for the Lorentz field, $s=(\varepsilon+2)/3$ with the static dielectric constant ε), and A_0, B_0, and C_0 are fitting parameters relevant to the degree of molecular orientation. Equation 3.1 indicates that difference in the polarizability ($\Delta\alpha$) and difference in the static dipole moment ($\Delta\mu$) between the excited and ground states can be evaluated by fitting the change in the absorbance with the first and second derivatives of the absorption spectrum, respectively. Since there is some uncertainty in the values for such constants as B_0 and s, the electrooptical parameters in the present study contain them as follows:

$$s^2 Tr(\Delta\alpha) B_0 = C_1 \cdot \frac{hc}{F_0^{\ 2}} \cdot \frac{10^{40}}{1.113} \quad (3.2)$$

$$s |\Delta\mu| C_0^{1/2} = C_2^{\frac{1}{2}} \cdot \frac{hc}{F_0} \cdot \frac{10^{30}}{3.34} \quad (3.3)$$

Here h is the Planck constant, c is the speed of light, C_1 and C_2 are fitting constants from the first and second derivatives of the absorption spectrum, respectively.

Figure 3.1a depicts the experimental setup. White light from a xenon-lamp (Hamamatsu, L2273) was collimated after focused through a 200-μ lmpinhole and then loosely focused on the sample to cover the multiple electrode gaps. An ac electric field of $F_{ext} = F_0 \sin(2\pi f t)$ with $F_0 \sim 10^6$ V/m and f = 20–25000 Hz was applied between the electrodes by using a function generator (NF, 1956 multifunction synthesizer) combined with a high-voltage amplifier (Matsusada, HEOPT-5B20) to detect the field-induced absorbance change through a spectrometer with a multichannel lock-in amplifier.[1–4] Since the multichannel lock-in amplifier has 128 detection channels connected to 128 avalanche photodiodes, all wavelength regions of interest were detected simultaneously, reducing a measurement time by two orders of magnitude, or improving the signal-to-noise ratio by one order of magnitude for the same measurement time. The absorbance changes due to the Pockels and Kerr effects were detected at the modulation frequency f and its second harmonic $2f$, respectively.

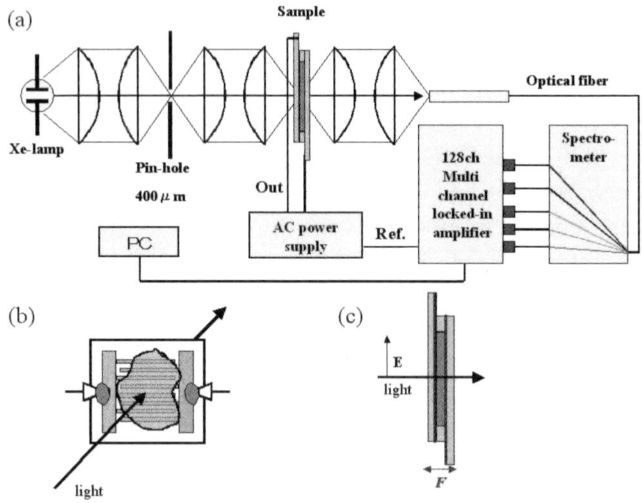

Figure 3.1 Experimental setup. (a) Setup for electroabsorption spectroscopy, where the field-induced absorbance change was detected by a polychromater with a multichannel lock-in amplifier. The multichannel lock-in amplifier has 128 detection channels connected to 128 avalanche photodiodes to detect all the wavelength regions of interest simultaneously. (b) and (c): polymer film and aqueous solution samples.

3.2 Sample preparation

3.2.1 *Film sample*

For film samples[1–3], an array of 8 or 14 interdigitated aluminum electrodes whose gaps were 0.5 or 0.4 mm , respectively, was deposited on the slide glass by vacuum evaporation as shown in Fig. 3.1b, before spin coating of the polymer film described below.

Tetraphenyl porphyrin tetrasulfonic acid (reagent grade, Tokyo Kasei Kogyo), called TPPS, was used without further purification. Polyvinil alcohol (PVA, Kanto Kagaku) was used as a host matrix in a polymer film form. For 2D oriented J-aggregates sample, 8 mg, i.e., 4.3 mmol/l, TPPS and 80 mg PVA were dissolved in water of 2 ml, and the glass substrates, where electrodes were deposited where coated with the solution by a spin coating method with 1000 rpm for 1 minute.

For monomer sample, 1.5 mg, i.e., 0.81 mmol/l, TPPS and 80 mg PVA were dissolved in water of 2 ml, and the glass substrates were coated with the solution by a spin coating method with 500 rpm for 1 minute.

3.2.2 *Solution sample*

For the aqueous solution,[3,4] a pair of indium tin oxside, ITO, glasses whose gap was 0.07-0.1 mm was used as electrodes as shown in Fig.3.1(c). The thickness of the ITO layer, resistivity, and carrier density are 300 nm, $1.3 \times 10^{-4}\ \Omega$ cm, and 1.2×10^{21} cm^{-3}, respectively. A plastic film of 0.07 – 0.1 mm thickness with 12×12 mm^2 hole, where the solution was contained, was sandwiched between the ITO glass plates of $25 \times 25 \times 1$ mm^3.

Two types of solution samples were used. One is a solution without KCl prepared by dissolving specific amounts of TPPS (4 mg, 0.8 mmol/l) in water of 5 ml. Figure 4.2a shows its absorption spectrum. There are two absorption peaks: the J band due to the Frenkel exciton located at 490 nm and the monomer band attributed to the D-monomers showing the broad peak at 433–435 nm. This indicates that the solution contains both aggregates and monomers. The other is a solution with KCl prepared with the appropriate amount of TPPS (3.6 mg, 0.77 mmol/l) and KCl (39 mg, 0.1 mol/l) in water of 5 ml. The two absorption peaks, which are the J-band and the H-band (423 nm), are shown in Fig.4-2c. This indicates that the solution contains only J- and H-aggregates.

4. Giant Electrooptic Effect of Porphyrin J-aggregates in Polymer Film and Aqueous Solution

Remarkable experimental finding is a large enhancement of electrooptic Kerr signal due to the formation of J-aggregates. The enhancement factor amounts to as large as 100 in polymer film and 100 000 in aqueous solution. (The EO signal for monomers in aqueous solution have not been detected. The enhancement factor of 100 000 is compared with the signal for J-aggregates in polymer film.)

Figures 4.1 and 4.2 show absorption and electroabsorption spectra of monomers and J-aggregates in polymer film and J-aggregates in aqueous solution, respectively. Note that a large difference of the applied electric field between the polymer film and the aqueous solution.

Table 4.1 summarizes the results for the polymer films.[1,2] The results for the electric-field strength dependence measured in Sec. 6.1.[2] is also included. Although the experimental values for $\Delta\alpha$ are distributed due to sample dependence both for the monomers and the aggregates, the ratio of $\Delta\alpha_{agg}$ to $\Delta\alpha_{mon}$ (enhancement factor) ranges from 500 (low field) to 15 (high field).

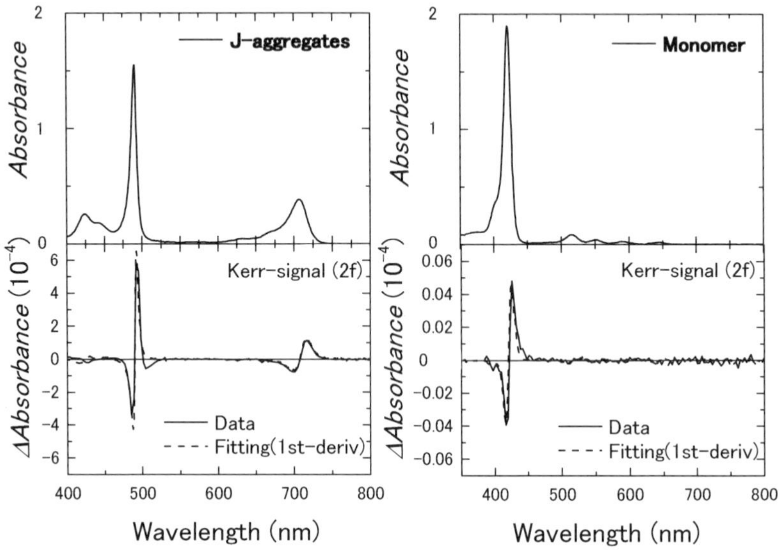

Figure 4.1. Absorption and electroabsorption spectra of the TPPS J-aggregates in 2D orientation. Left: absorption (top) and electroabsorption (bottom) spectra of the TPPS J-aggregates in 2D orientation. Right: Those of the F-monomer. The applied, electric field $\boldsymbol{F}$ was F_0=2.8×10^6 V/m, and the reference frequency was f=235 Hz. Fitting curves (dashed lines) are calculated from the first derivative of the absorption spectra.

Table 4.1. Stark parameters of the TPPS J-aggregate and monomer.

Work	B-band	$B_0\Delta\alpha$[a] High field (>3×10^{6}V/m)	$B_0\Delta\alpha$[a] Low field (<2×10^{6}V/m)
Ref.1	Aggregate Monomer	280(3.1×10^6 V/m) 5(3.1×10^6 V/m)	
Ref.2	Aggregate Monomer	100-300/200~2000 4~7/4~7	200~2000 4~7

[a] $\Delta\alpha$ is a diagonal component of the $\boldsymbol{\Delta\alpha}$ tensor. For aggregate, $\Delta\alpha=\mathrm{Tr}\boldsymbol{\Delta\alpha}$ while for monomer $2\Delta\alpha=\mathrm{Tr}\boldsymbol{\Delta\alpha}$ [1]. B_0 ranges from 7/20 to 1/2 for monomer and from 1/4 to 1/2 for aggregate. The local field factor is taken as $s=(\varepsilon+2)/3=2.63$. The values without this correction are given by $s^2B_0\Delta\alpha$. The modulation frequency is in the range of 100 Hz $< f <$ 1 kHz, where no substantial frequency dependence was observed.

The electroabsorption spectra of TPPS J-aggregates in an aqueous solution[3,4] have a different feature from that in polymer films. Figure 4.2 shows electroabsorption spectra of TPPS J-aggregates in aqueous solutions without and with KCl, respectively. The broadening signal was observed in aqueous solutions, and the fitting curves (thin lines) are the second derivative of the absorption spectra.

Table 4.2 lists the estimated electrooptic parameters, $\Delta\alpha$, $\Delta\mu$ and the third-order nonlinear optical susceptibility $\chi^{(3)}$ for isotropic distribution of the TPPS J-aggregates in aqueous solution and for the two-dimensionally distributed film. $\chi^{(3)}$ for the solution is 100 000 times larger than that for the polymer film. Note that the magnitude of $\Delta\alpha$ is larger with than without KCl. The broadening signal typically reflects the difference in the static dipole moment $\Delta\mu$; however, in the rearrangement model, it is explained by the change in the polarizability difference, $\Delta\alpha$.

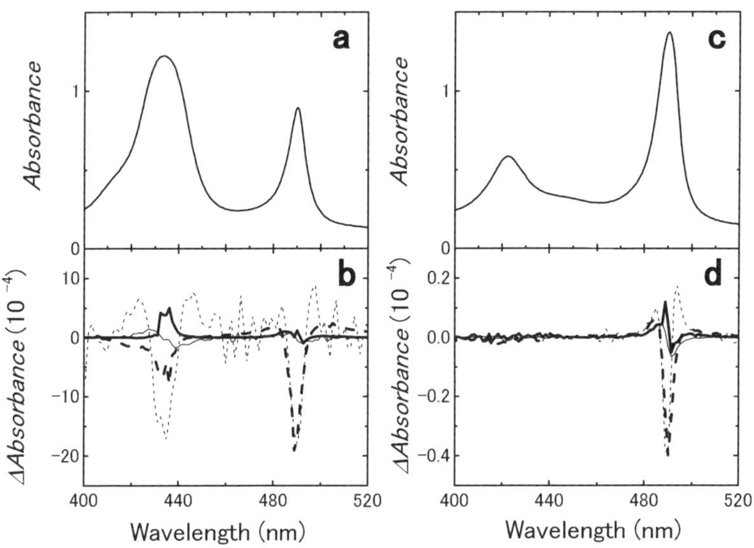

Figure 4.2 Absorption and electroabsorption spectra of TPPS J-aggregates in aqueous solution without (a, b) and with (c, d) KCl. Bold solid, bold dashed, thin solid, and thin dashed lines are X-phase spectra, Y-phase spectra, fitting curves to Y, and fitting curves to X for electroabsorption spectra, respectively. The applied electric fields were $F_0 = 8.6 \times 10^3$ V/m (b, without KCl) and $= 7.1 \times 10^2$ V/m (d, with KCl). The reference frequency was 221 Hz. "Copyright: Royal Society of Chemistry"

Table 4.3 lists the same data with explicit consideration for the Lorentz local-field factor s, which is expressed by $(\varepsilon+2)/3$, where ε is the dielectric constant with $\varepsilon_{H2O} = 80$ and $\varepsilon_{PVA} = 5.9$. After this correction, $\chi^{(3)}$ for the solution is 10^2 times larger than that for the polymer film.

Table 4.2. Electrooptic parameters of TPPS J-aggregates for isotropic distribution in aqueous solutions and for the two-dimensionally distributed film.

	$s^2\mathrm{Tr}\Delta\alpha B_0$ (Å^3)	$s\Delta\mu C_0^{1/2}$ (D)	$\lvert\chi^{(3)}\rvert$ (esu)
With KCl N>10	-4×10^8	50	5.3×10^{-6} 5.4×10^{-6}
With KCl $N\approx10$	1.4×10^9	35	6.1×10^{-6} 7.2×10^{-6}
With KCl N<10	6×10^9		1.8×10^{-6}
Without KCl	$1.2\sim1.9\times10^8$	600~1000	$0.3\sim1.1\times10^{-6}$ $5.0\sim7.5\times10^{-6}$
PVA film	1900	0.1	1.7×10^{-11}

Table 4.3. Electrooptic parameters with the local-electric factor s equal to $(\varepsilon+2)/3$ in the Lorentz field, where ε is the dielectric constant; $\varepsilon_{H2O} = 80$ and $\varepsilon_{PVA} = 5.9$.

	$\mathrm{Tr}\Delta\alpha B_0$ (Å^3)	$\Delta\mu C_0^{1/2}$ (D)	$\chi^{(3)}$(esu)
With KCl N>10	-5.5×10^5	1.9	7.3×10^{-9} 1.9×10^{-8}
With KCl $N\approx10$	1.9×10^6	1.3	8.4×10^{-9} 9.9×10^{-9}
With KCl N<10	8.2×10^6		2.4×10^{-9}
Without KCl	$1.6\sim2.6\times10^5$	22~37	$0.4\sim1.5\times10^{-9}$ $0.7\sim10\times10^{-8}$
PVA film	280		2.5×10^{-12}

5. Molecular Rearrangement Model

There is a model to explain that $\Delta\alpha$ scales with the coherent aggregation number N_c, the size of a micro-aggregate.[5–7] From this model, it is concluded that TPPS J-aggregates have the coherent size as large as 70 molecules.[1] However, other experimental facts indicate a smaller number (from 5 to 20) of size.[8–12] In addition, this model takes into account only two levels (the ground and single exciton states), where $\Delta\alpha$ is always negative, in contradiction with the positive values for $\Delta\alpha$ in most of experiments. In order to explain these positive values, one should take into account higher excited states, i.e., more than the two levels. In such a realistic model, it is discussed that there is no enhancement in $\Delta\alpha$ due to N_c if the two-exciton state is also taken into account.[13] As a mechanism for electrooptic response in molecular crystals, it is also proposed that the coupling between charge-transfer (CT) and the Frenkel states is crucial for the enhancement of $\Delta\alpha$, and this model has accounted for the experimental electroabsorption spectra in polyacene crystals.[22–24] However, both models can not account for all the experimental results presented in the later sections.

We proposed a new model for explaining the large enhancement for $\Delta\alpha$ by formation of J-aggregates.[2] It is the molecular rearrangement model, where an electric-field-induced change in the angle between the molecular transition dipole moment and the bonding axis of the micro-aggregate is responsible for the giant electrooptic response.

For J-aggregates, the excitation energy depends on the aggregation number and molecular arrangement of the constituent monomers as [2,25]

$$E_k = E_0 + 2J\cos\left(\frac{k\pi}{N+1}\right) = E_0 + 2\frac{M^2}{4\pi\varepsilon_B r^3}\left(1-3\cos^2\theta\right)\cdot\cos\left(\frac{k\pi}{N+1}\right), \quad (5.1)$$

where E_0 is the excitation energy of monomer molecules, N is the aggregation number forming the coherent (micro-) aggregate, J is the transition-dipole transition-dipole interaction emergy expressed by M, r, ε_B, θ. Here, M is the transition dipole moment between the relevant electronic states in the molecule, r is the intermolecular distance, ε_B is the background permittivity, and θ is the angle between the molecular transition dipole moment and the intermolecular bonding axis connecting the centers of the transition dipoles, i.e., ***J***-axis.

The rearrangement model, which was introduced in Ref. 2, can explain the energy shift of TPPS J-aggregates by the change in J, i.e., the interaction energy between monomer transition dipoles, induced by the applied field. In the model, the relative direction of the ***J***-axis with respect to the electric field vector ***F*** influence the change in the excitation energy of J-aggregates.

For TPPS J-aggregates, $2J = -0.4$eV and the field-induced red-shift energy is 1.4×10^{-5} eV. If we take, for example, r=1 nm and $\theta = 20^\circ$, then the energy shift observed in the film samples[1,2] is explained by the change in θ as small as $\Delta\theta = 0.002^\circ$, or the change in r as small as $\Delta r = 0.0001$ Å. The change in the angle seems to be more probable than that in the distance for the following reason: The dipole moment is induced in the constituent monomers by the field, and it rotates due to the torque exerted by the field. In this case, no change occurs in the center of mass for each molecule. When the mutual distance changes, however, the monomers located on both ends of the aggregates should change its position by a much larger distance. This is unlikely to occur.

Figure 5.1 depicts the molecular rearrangement model to explain the experimental finding of the electric-field strength dependence presented in section 6.1. Initially, i.e., before the electric field is applied, the constituent molecules in the J-aggregates are located at certain stable positions, making a finite angle between the molecular transition dipole moment and the intermolecular bonding axis. The positions are determined by the intermolecular interaction and the polymer potential. When an alternating electric field is applied, the electric-field-induced dipole moments in the molecules tend to be aligned along the electric-field direction. As a result, the angle decreases and the transition dipole-transition dipole interaction J increases, resulting in the red shift of the J band.

The upper part of Fig.5.1 shows the simplest model for a rod-like molecule, where the induced and transition dipole moments are both parallel to the molecular axis. The lower part of Fig.5.1 shows a more realistic model for a disk-like TPPS molecule. Since the monomer has two orthogonal transition dipole moments in the molecular plane, the assembly forms J- and H-aggregates simultaneously. The lower part shows the top view of the assembly, while the upper part is regarded as the side view in this respect. Fig.5.1 also shows that as the applied electric field is increased, the signal intensity tends to be saturated in this model.

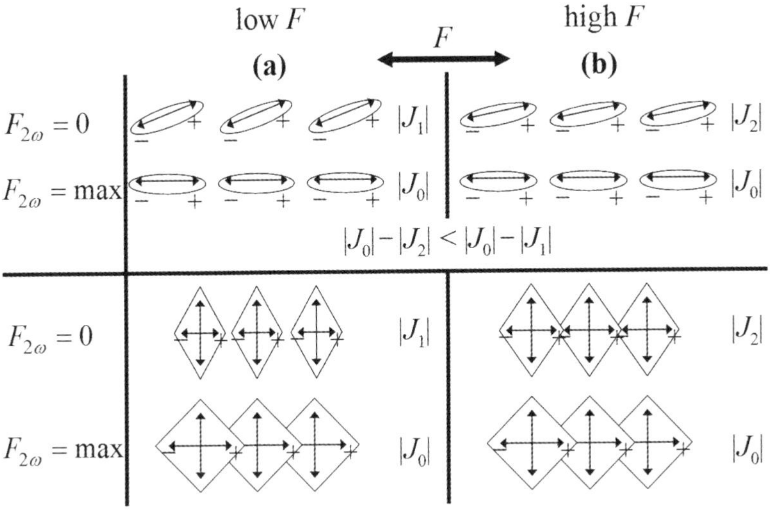

Figure 5.1 Schematic of the rearrangement model. Above: (a) With (***F***≠0) and without (***F***=0) electric field, each molecule in J-aggregates changes its direction such that the molecular axes are more aligned due to the electric-field induced dipole moment, which is expressed by + and −, in the ground state. It does not matter whether it is in the ground or excited state because there is no substantial difference in the polarizability between the two states, as measured for the monomer. Since $|J_1|<|J_0|$, the B band is red-shifted by $2(|J_0|—|J_1|)$. The transition dipole moment is denoted by the thin double-sided arrows. The direction of the applied alternating electric field is shown by the thick double-sided arrow. (b) If the electric field is large, the molecules are rearranged into the more aligned configurations with $J=J_2$ by the $F_\omega=0(\propto F^2)$ component to give a smaller $F_{2\omega}(\propto F^2)$-induced red shift by $2(|J_0|-|J_2|)$.

Below: more realistic model for the molecular arrangement with a slipped face-to-face stacking from top view, which takes account of the molecular structure for the diacid H_4TPPS^2 monomer. There are two orthogonal transition dipole moments in the molecular plane. The side view is similar to the arrangement shown in the above figure.*

6. Experimental Evidence for the Model

6.1. Electric-field strength dependence of EA signal

Figure 6.1 shows the electric-field strength dependence of the signal intensity, i.e., the change in the peak absorption of the B band due to the red shift. For the J-aggregates, the signal intensity deviates from the quadratic dependence on F_{ext} for higher F_{ext}, while that for the monomers obeys the quadratic dependence. The electric-field strength dependence for the J-aggregates can be fit by the function $\boldsymbol{F}_{ext}^2/(a + \boldsymbol{F}_{ext}^2)$. This dependence on the external electric field $\boldsymbol{F}_{ext}$ is reflected also on Table 4.1 and can be deduced in the rearrangement model as follows.

The rearrangement model can be formulated by a simple mechanical model.[2] As shown in Fig. 6.2, monomer molecules are positioned in the mechanical equilibrium such that its molecular axis is tilted by a small angle θ_0 (<<1) with respect to the intermolecular bonding axis. When the alternating electric field $\boldsymbol{F}$ is applied in the direction of the bonding axis, a dipole moment is induced in the molecular axis of each monomer and is made aligned in the field direction. We assume the Lorentz model with displacement x for the induced dipole moment qx in the monomer with q being the electronic charge. The monomer can rotate as a rigid body around its center of mass by the angle of θ with respect to θ_0 against the restoring force. Then, the equations of motion are represented by

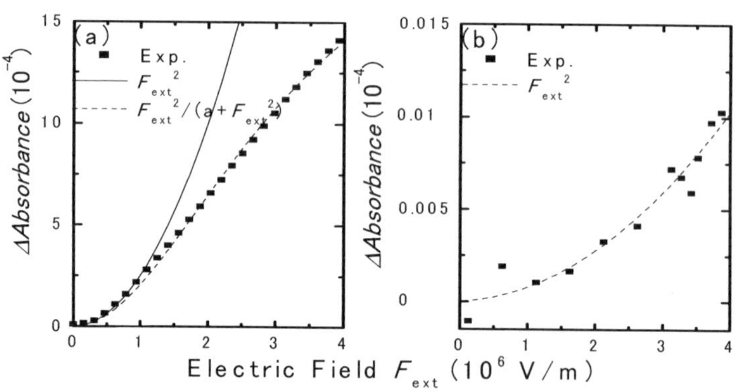

Figure 6.1 The signal intensity as a function of the external electric field for J-aggregates and for monomers. The peak absorption change in the B band due to the red shift at $2f$ in the Stark spectra for J-aggregates (a) and for monomer (b) at f=235 Hz. For J-aggregates, the fitting curves are proportional to $\boldsymbol{F}_{ext}^2$ (black line) and $\boldsymbol{F}_{ext}^2/(a + \boldsymbol{F}_{ext}^2)$ (dashed line). In the unit of 10^6 V/m for $\boldsymbol{F}_{ext}$, a = 10.3. For monomers, the fitting curve is proportional to F_{ext}^2 (dashed line).

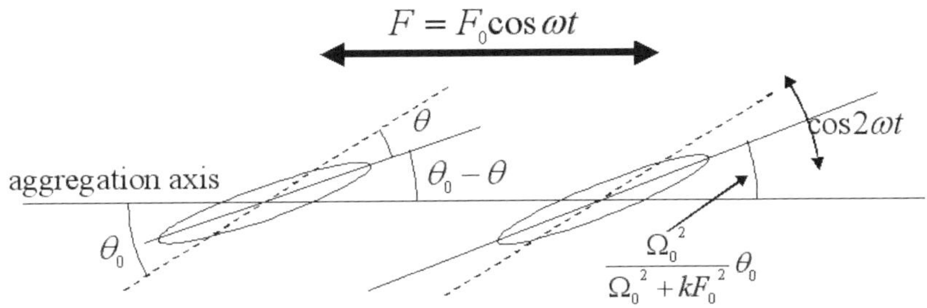

Figure 6.2 A mechanical model for the molecular rearrangement model. Left: definition of the angles θ and θ_0 in the mechanical model for the molecular rearrangement model. The magnitude of the angles is exaggerated. Right: with $F = F_0 \cos\omega t$ applied, the axis of the monomer molecule oscillates with 2ω about the angle of $\Omega_0^2\theta_0/(\Omega_0^2 + kF_0^2)$ (Eq. 6.5), which is decreased for larger F_0. "Copyright: Elsevier B.V."

$$m\left(\frac{d^2x}{dt^2} + \omega_0^2 x\right) = qF\cos(\theta_0 - \theta) \approx qF, \tag{6.1}$$

$$I\left(\frac{d^2\theta}{dt^2} + \Omega_0^2\theta\right) = qxF\sin(\theta_0 - \theta) \approx qxF(\theta_0 - \theta) \tag{6.2}$$

where m is the mass of the charge, I is the moment of inertia for the monomer, ω_0 and Ω_0 are the resonance frequencies due to the restoring force, and $\theta_0 - \theta$ is the angle from the intermolecular bonding axis. Substituting $F=F_0 \cos\omega t$ and $x=x_0 \cos\omega t$ with $\omega=2\pi f$ (ω_0, $\Omega_0 >> \omega$), we obtain

$$x_0 = \frac{q}{m}\cdot\frac{F_0}{\omega_0^2 - \omega^2} \tag{6.3}$$

$$xF = \frac{q}{2m}\cdot\frac{F_0^2}{\omega_0^2 - \omega^2}(1 + \cos 2\omega t) \equiv F_{\omega=0} + F_{2\omega}\cos 2\omega t \tag{6.4}$$

The first term in Eq. (6.4) rearranges the molecular axes in a more aligned position, as shown in Fig. 5.1(b), and the second term modulates them at $2f$. Neglecting the higher-order terms than F_0^2, one can obtain

$$\theta_0 - \theta \approx \frac{\Omega_0^2}{\Omega_0^2 + A}\theta_0 - \phi_0 \cos 2\omega t \approx \frac{\Omega_0^2}{\Omega_0^2 + A}\theta_0\left(1 - \frac{A}{\Omega_0^2 + A}\cos 2\omega t\right) \tag{6.5}$$

with

$$A = \frac{q^2}{2mI} \cdot \frac{F_0^{\,2}}{\omega_0^2 - \omega^2} \approx \frac{q^2}{2mI} \cdot \frac{F_0^{\,2}}{\omega_0^2} \equiv kF_0^2 \tag{6.6}$$

$$\varphi_0 = \frac{A\Omega_0^2}{\left(\Omega_0^2 + A - 4\omega^2\right)\left(\Omega_0^2 + A\right)}\theta_0 \approx \frac{A\Omega_0^2}{\left(\Omega_0^2 + A\right)^2}\theta_0 \tag{6.7}$$

Eq. (6.5) explains the electric-field dependence of the signal magnitude depicted in Fig. 5.1, as illustrated in Fig. 6.2. To be more precise, the F_0 dependence of the modulation of J is estimated with Eq. (5.1) as

$$J = \frac{M^2}{4\pi\varepsilon_B r^3}\left(1 - 3\cos^2(\theta_0 - \theta)\right) \approx \frac{M^2}{4\pi\varepsilon_B r^3}\left(-2 + 3(\theta_0 - \theta)^2\right) \tag{6.8}$$

with

$$\begin{aligned}(\theta_0 - \theta)^2 &= -2\frac{\Omega_0^2}{\Omega_0^2 + A}\varphi_0\theta_0 \cos 2\omega t + C \\ &\approx -2\frac{k\Omega_0^4 F_0^2}{\left(\Omega_0^2 + kF_0^2\right)^3}\theta_0^2 \cos 2\omega t + C \\ &\approx -2\frac{kF_0^2}{\Omega_0^2 + 3kF_0^2}\theta_0^2 \cos 2\omega t + C\end{aligned} \tag{6.9}$$

where C represents terms other than that oscillating with 2ω and the minus sign indicates the red shift in the J band. For order estimation, A is roughly estimated to be 10^{19} s^{-2} when ω_0 is of the electronic origin such that $\omega_0 \sim 10^{15}$ Hz, q is taken as the charge of the electron and m as its mass, I is taken as 10×(proton mass) ×(1 Å)^2, and $F_0 \sim 10^6$ V/m. If Ω_0 is in the range of 10 GHz, as is reasonable for the rotational frequency, $A(=kF_0^{\,2})$ and $\Omega_0^{\,2}$ are comparable in Eq.(6.9), resulting in a deviation from the $F_0^{\,2}$ dependence of the 2f signal intensity (red shift in the J band), as observed in Fig. 6.1.

As for the monomer, by contrast, the electric-field dependent energy is expressed simply by

$$E = -qxF\cos(\theta_0 - \theta) \approx -qxF \approx -\frac{q^2 F_0^2}{2m\omega_0^2}(1 + \cos 2\omega t) \tag{6.10}$$

This is purely the electronic response without a saturation effect. Although there is a contribution from the molecular rotation to the polarizability as well, it is readily deduced that the rotational contribution is of higher order than $F_0^{\,2}$ such that the electronic contribution dominates for the 2f response.

6.2. Orientational Distribution Dependence of EA Signal

For the film samples, we used the spin-coated PVA film, under which the electrodes were deposited. Therefore, the axes of the aggregates were orientationaly distributed in the film plane, to which the in-plane electric field was applied. This is exactly the case depicted in Fig. 6.3, where the red shift dominates over the blue shift in agreement with the experiment in Fig.4.1.

If the TPPS aggregates are isotropically distributed in the matrix, the electric-field induced shift occurs evenly in both (red and blue) directions. In the aqueous solution, the signal broadening was observed as shown in Figs. 4.2(b) and 4.2(d). A broadening signal usually reflects the orientation distribution of the static dipole moment. By contrast, the origin of this signal can be attributed to the orientation distribution of the J-axis with respect to the applied electric field ***F***, as explained below. In this case, as depicted in Fig. 6.4(a), the model predicts that red, for ***J*** parallel to ***F***, and blue, for ***J*** perpendicular to ***F***, shifts evenly occur, because there are equivalent contributions of these shifts.

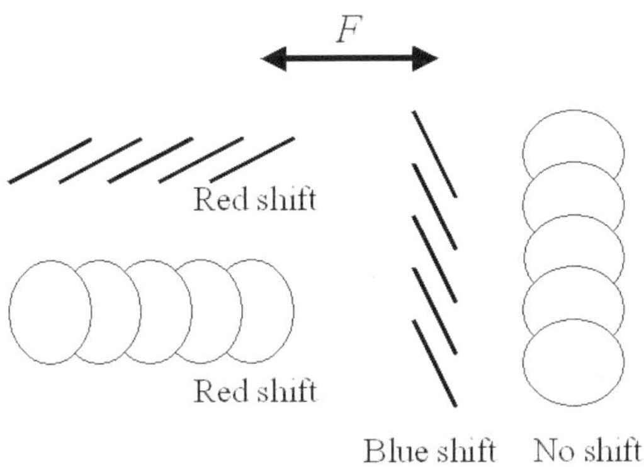

Figure 6.3 A model of TPPS aggregates in the spin-coated PVA film. Four possible directional arrangements of TPPS aggregates in the spin-coated PVA film. Any arrangement of the aggregate is expressed by the combination of these four, all of which occur at the same probability. Thus, the red shift is dominated over the blue shift when the electric field is applied in the film plane. "Copyright: Elsevier B.V."

Even with considering the fact that the applied electric field ***F*** is perpendicular to the optical electric field ***E***, as shown in Fig. 6.4(b), the result is the same.[3,4]

Signal broadening usually reflects the squared absolute value of difference in the static dipole moment, $|\Delta\mu|^2$, in the third term in Eq. (3.1). More precisely, it is expressed by

$$|\Delta\mu|^2 = \sum_{i=1}^{n} |\Delta\mu_i|^2 / n. \tag{6.11}$$

Here n is the number of J-aggregates being in the probed region of the sample. This signal could appear when each J-aggregate, i.e., macro-aggregate, has a static dipole moment with random orientation. Without symmetry-breaking perturbation, the TPPS J-aggregate has no static dipole moment because of its centrosymmetry originating from the symmetry of the monomer molecule. External perturbation breaks the centrosymmetry, resulting in a finite static dipole moment in both excited and ground states. This effect induces broadening of the signal, but perturbation-induced $\Delta\mu$ usually causes much smaller electrooptic signals than intrinsic $\Delta\alpha$, as observed for film samples in Refs. [1 and 2]. By contrast, ΔA of TPPS in the aqueous solution is unusually large.* The signal size in the is about 100 000 times larger in $\Delta A/F^2$ than that in a polymer film. This is probably because constituent monomers in J-aggregates are more readily realigned in liquid water than that in a solid polymer. For J-aggregates, the signal strength is given by Eq. (6.9) as

$$J \propto -2\frac{kF^2}{(\Omega^2 + 3kF^2)}\theta_0^2 \cos 2\omega t. \tag{6.12}$$

Here Ω is the resonance frequency of rotation of the monomer molecule depending on the surroundings, k is the constant, θ_0 is the equilibrium angle between the molecular transition dipole moment and the aggregation axis, and ω is the modulation frequency of the applied electric field. A much higher signal intensity in the aqueous solution is considered to be due to a smaller Ω than that in the polymer film because J-aggregates in the solution are more loosely bound in liquid water than those in the polymer matrix. The observed magnitude of $\Delta\alpha$ is explained if $\Omega = 2\pi\nu \fallingdotseq 5\times10^{10}$ Hz, or $\nu \fallingdotseq 10$ GHz in the polymer film and $\Omega = 2\pi\nu \fallingdotseq 5\times10^{8}$ Hz, or $\nu \fallingdotseq 3\times10$ MHz in the solution.

* Thus, the broadening signal is not caused by $\Delta\mu$, but by $\Delta\alpha$, as explained by Fig. 6.4. (For comparing $\chi^{(3)}$ between the solution and the polymer film in Tables 4.2 and 4.3, $\chi^{(3)}$ for the solution is formally evaluated from $\Delta\mu$ but intrinsically originated from $\Delta\alpha$.)

The local field factor, represented by the parameter s in Eq. (3.2), is another possible origin for the larger electrooptic signal in the aqueous solution than in polymer film. Since the static dielectric constant for liquid water is 80 while that for PVA is 5.9, the Lorentz local field for the aqueous solution sample is 9 times as high as that for the polymer sample. In Eq. (3.2), s^2 is 81 times larger for the solution than for the film, thus yielding about a 100-fold difference. There still remains, however, a 1000-fold difference, requiring consideration on the rearrangement model as stated above. The local field effect could explain partly the larger difference in signal size, but there are questions whether the Lorentz field is most appropriate and whether the local field consideration is valid for relatively large macro-aggregates. To answer a part of these questions, the signal magnitude for the monomer in aqueous solution should be experimentally evaluated. At present, however, it is not clear because the signal magnitude for the monomer is much smaller than that for aggregates. To make matters worse, the monomer signal appears in the close vicinity of the H-band such that the signal is mixed with that due to coexisting H-aggregates.

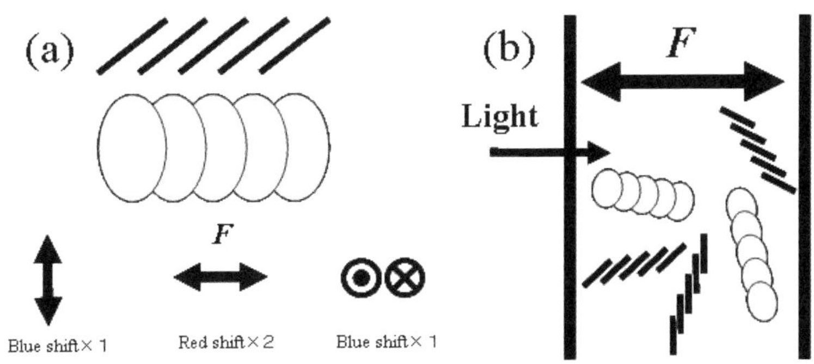

Figure 6.4 A model of TPPS for isotropic distribution in an aqueous solution. (a) Six possible directional arrangements of TPPS J-aggregates in an aqueous solution. Any arrangement of the aggregate is expressed by the combination of these six. The broadening effect is dominant when the electric field is applied because red and blue shifts take place evenly. (b) Isotropic orientation of TPPS J-aggregates between ITO glasses.

There is also a possiblility of the contribution of the electric double layer to the large electroabsorption signal in aqueous solution.[4] The electrooptic Kerr signal S is related to the applied electric field F such that

$$S = KF^2, \tag{6.13}$$

where K is a Kerr constant of the sample. If the electric field is evenly distributed between the electrodes (without EDL),

$$S = kNF^2 = knL(V / L)^2 = knV^2 / L\,, \tag{6.14}$$

where k is the Kerr constant per molecule, N is the number of molecules, L is the sample thickness, V is the applied voltage, and n is the number density of molecules per thickness (N=nL). From Eqs. (6.13) and (6.14),

$$KF^2 = K(V / L)^2 = knV^2 / L \tag{6.15}$$

to obtain

$$K = knL = kN\,. \tag{6.16}$$

This is the result as it should be. On the other hand, if the EDL is formed such that the electric field is concentrated on a shorter length of $xL = D$ ($0<x<1$, $D=2d$, d: thickness of EDL) and only the molecules in the EDL respond to the field, then the signal is given by

$$S = k(nxL)(V / xL)^2 = knV^2 / (xL)\,. \tag{6.17}$$

In order to calculate the Kerr constant of the sample, one assumes that the electric field is uniformly distributed over the sample volume to be given by V/L, so that

$$S = KF^2 = K(V / L)^2 = knV^2 / (xL)\,. \tag{6.18}$$

As a result, for the apparent Kerr constant of the sample is given by

$$K = knL / x = kN / x \tag{6.19}$$

For the apparent Kerr constant per molecule K/N, which is proportional to $\chi^{(3)}$ per molecule, is given by

$$K / N = k / x = k(L / D)\,. \tag{6.20}$$

Both K and kn are enhanced by a factor of $1/x$ (=L/D). This might account for the large enhancement of the EO signal and $\chi^{(3)}$ for TPPS in the aqueous solution compared with those in the polymer film.

Generally, thickness of EDL (d) is expressed by the Debye length

$$d = 0.304[\mathrm{Ion}]^{-0.5}. \tag{6.21}$$

In the experiment, on the basis of this formula, thickness of the EDL and the number ratio of the molecules present in the EDL vs all the molecules between the electrodes are calculated to be d=10nm for $[H^+]$=0.001mol/l and the ratio=10nm×2/10^5nm=2×10^{-4} without KCl, and d=1nm for $[KCl]$=0.1mol/l and the ratio=1nm×2/10^5nm=2×10^{-5} with KCl. However, in order for the signal magnitude of $\Delta A=2\times10^{-3}$ in Fig.4.2b to be explained with the EDL, the EDL thickness has to be thicker than 50nm.[4] This is reasonable if we consider a finite formation time of the EDL with the AC electric field.

6.3 Temperature Dependence of EA Signal

Finally, we present more solid evidence for the model with the temperature dependence of the electro-optic signal[3] in Fig. 6.7. A more than 10-fold decrease in the signal magnitude was observed as the temperature was decreased from 293 to 77 K, for the two-dimensionally oriented polymer film. This result cannot be explained by the purely electronic origin. It requires the molecular orientational contribution as the origin to explain the result. Temperature dependence of electroabsorption spectra was studied, for example, using photosynthetic proteins and quinoline derivatives embedded in polymer films to obtain useful information.[26,27] In Ref. 26, quadrature phase components in electroabsorption signals are shown to be reduced with decreasing temperature. In Ref. 27, it is shown that molecules are immobilized below 100K because molecular orientational motion is strongly restricted in a polymer environment. Similarly, our observation is also explained by the restriction of the orientational motion of the constituent monomer molecules at this low temperature, supporting the validity of the molecular rearrangement model.

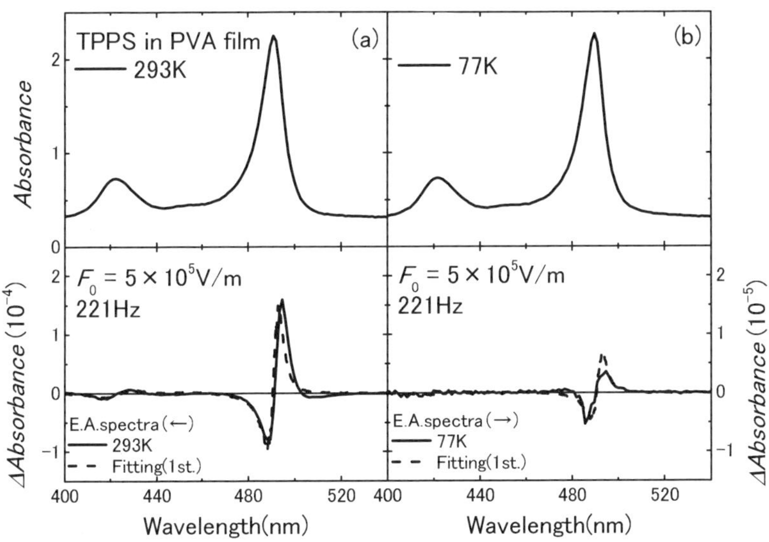

Figure 6.7 Temperature dependence at 293 K (a) and 77 K (b) of the electro-optic signal for the two-dimensionally oriented film with F_0=5×10^5 V/m at f=221 Hz, which is 1/6 of F_0 in Figs. 4.1 Fitting curves (dashed line) calculated from the first derivative of the absorption spectra. "Copyright: Elsevier B.V."

7. Control of the Aggregation State by Electric Field

There is experimental evidence[13–18] that J-aggregates have a hierarchical structure. Those observed as rod-like shapes by near-field optical microscopy,[16–18] electron microscopy[19] and atomic force microscopy (AFM)[20] are macroaggregates, which are composed of an incoherent ensemble of the microaggregates. The microaggregate is a one-dimensional assembly of monomer molecules, whose arrangement determines the interaction energy *J*, between the transition dipole and neighboring transition dipole. An image of microaggregates has never been obtained in microscopic experiments. Coherent extension of the Frenkel exciton defines the coherent aggregates, the size of which is the same as that of the microaggregates. Spectroscopic investigation is important to provide information on the coherent aggregates; however, presently observed spectra are inhomogeneously broadened due to the formation of macroaggregates. Therefore, isolation of microaggregates from macroaggregates and spectroscopic measurement of a single microaggregate have been one of the most challenging subjects.

During the measurement of electroabsorption spectra of TPPS-J aggregates in aqueous solution, we found that the absorption spectra exhibited a significant spectral shift before and after the measurement. Therefore we conducted a systematic measurement of the electric-field induced effects in the absorption spectra of J-aggregates in aqueous solution under the stationary DC or AC electric field.[4] It was found that the resulting absorption change was mostly irreversible and dependent on KCl concentration in aqueous solution. The analysis of the results showed evidence that we obtained microaggregates with the reduced coherent aggregation number.

7.1 *Electric-Field Induced Change in Absorption and EA Spectra*

An irreversible change in the absorption spectra was induced by a stationary electric field for a period of about 1 h. This is referred to as the long-time response of the absorption spectra to a constant electric field. An electric field with $F_0 = 2.9 \times 10^4$ V/m (DC) or 3.4×10^4 V/m (AC) was applied between the electrodes using the function generator. Spectra were obtained every minute by scanning the wavelength repetitively with a spectrophotometer (Shimadzu, UV-3150). In addition, during the time course of the irreversible change under the AC electric field, EA spectra or electromodulation (EM) spectra were successively measured every

7 min. The prompt acquisition of EA spectra was facilitated with the use of the multichannel lock-in detection method. In order to discriminate the EA spectra from the electric-field induced irreversible change in the absorption spectra, EM spectra are often used instead of EA spectra below.

Figure 7.1 shows the long-time response of the absorption spectra with DC (Fig. 7.1(a) and 7.1(b) and AC (Fig. 7.1(c)) electric fields over 60 and 120 min, respectively. In the case of the solution without KCl (Fig. 7.1(a)), the absorption strength of the J band increased, while that of the D-monomer band decreased with time. In contrast, for the solution with KCl (Figs. 7.1(b) and 7.1(c)) the peak energy, intensity, and spectral bandwidth of the J-band blue-shifted, decreased, and increased, respectively. At the same time, the F-monomer band peak at 412–413 nm increased on the shorter wavelength side of the H band, while that of the

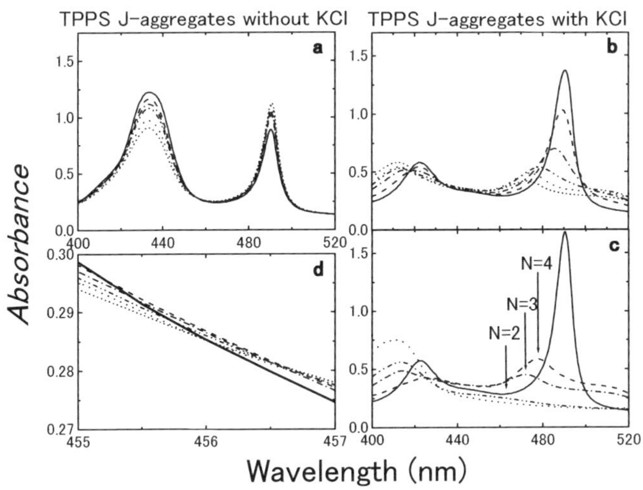

Figure 7.1 The long-time response of absorption spectra measured with DC (a, b) and AC (c) electric fields for one hour and for two hours, respectively. Without KCl, increase and decrease in the height of the J-band and the D-monomer band was observed, respectively (a). The spectra from 455 to 457 nm are magnified in (d) to show an isosbestic point. With KCl, dissociation of J-aggregates was observed under the electric field (b, c). For a and b, solid, dashed, dash-dotted, dash-dot-dotted, dotted, and short dashed lines are the absorption spectra at 0, 10, 15, 20, 40, and 60 minutes after the time when the DC electric field was applied, respectively, with $F = 2.9\times10^4$ V/m. In c, absorption spectra of TPPS J-aggregates (N>>10), TPPS tetramers (N=4), trimers (N=3), dimers (N=2), and F-monomers are shown by solid, dashed, dash-dotted, dash-dot-dotted, and dotted lines, respectively, with $F_0 = 1.7\times10^4$ V/m. The reference frequency was 221 Hz. "Copyright: Royal Society of Chemistry"

H band decreased, as shown in Fig. 7.1(b). The spectra in Fig. 7.1(c) under an AC electric field showed a gradual blue-shift of the J-band accompanied by an increase of the F-monomer peak. The peak energy represented by N, which corresponds to N-mers, continuously varied from $N > 16$ to $N = 4$ (tetramers), $N = 3$ (trimers), and $N = 2$ (dimers) as intermediate states, and finally to that of the F-monomer. The coherent aggregation number N was estimated using Eq. (5.1). When the dissociation made substantial progress, the J band no longer recovered within 2 h after the electric field was turned off, which implies that dissociation occurred irreversibly. In addition, the blue shift of the F-monomer band was observed when the electric field was applied for more than 2 h.

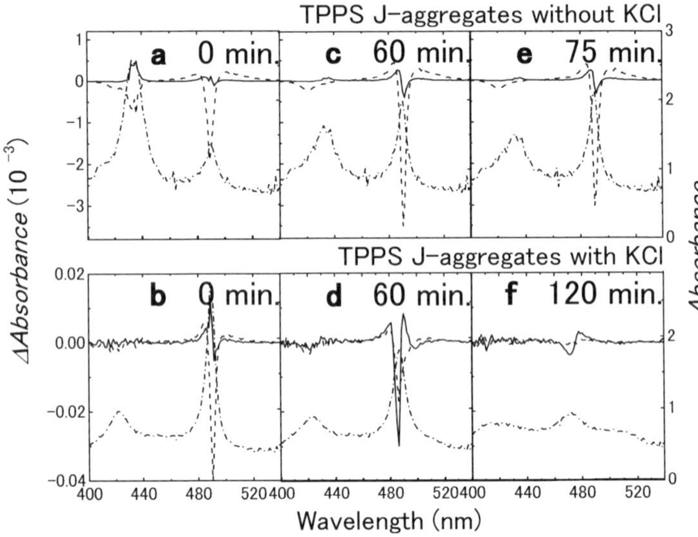

Figure 7.2 Long-time response of EA and absorption spectra of TPPS J-aggregates without (a, c, e) and with KCl (b, d, f) measured under the electric field for 2 hours. For samples with KCl, the aggregation number of the coherent TPPS J-aggregates N in (b, d and f) are $\gg 10$, ≈ 10 and <10, respectively. Solid, dashed and dash-dotted lines are EA spectra of X-phase, those of Y-phase and absorption spectra, respectively. The spectra measured at 0, 60, 75, and 120 minutes after the time when the electric field was applied are shown in a-b, c-d, e and f, respectively. The electric fields were $F_0 = 8.6\times10^3$ V/m (a, c, e) and $=7.1\times10^2$ V/m (b, d, f), respectively.

Figure 7.2 shows the successive change in the EA spectra during the time course of the change in the absorption spectra under constant electric field for solutions without (Figs. 7.2(a), 7.2(c) and 7.2(e)) and with KCl (Figs. 7.2(b), 7.2(d) and 7.2(f)). The absorption and EA spectra are shown at t=0 (Figs. 7.2(a) and 7.2(b)), 60 (Figs. 7.2(c) and 7.2(d)), 75 (Fig. 7.2(e), and 120 Fig. 7-2(f)) min. after the onset of the electric field. The absorption spectra were calculated from the multi-lock-in signals using $A = -\log(T_t/T_0)$, where T_0 is the intensity of the probe light without the sample and T_t is that with the sample at the corresponding time. There was no substantial change in the EA signal for the solution without KCl, while a characteristic change was observed for the solution with KCl; the initial broadening signal changed to a red-shifted signal as the absorption peak blue-shifted with time.

7.2 *Possible Mechanism of Field Induced Dissociation and Association*

The long-time spectral response under an electric field is of significant interest. Without KCl, the respective increase and decrease in the intensities of the J- and D-monomer bands are simultaneously observed, as shown in Fig. 7.1(a). These changes are considered to correspond to the progress in aggregation associated with the applied electric field. One possible mechanism for the enhancement of aggregation is the electric-field induced alignment of monomer molecules. The diacid TPPS molecule is a disk-like molecule with D_{4h} symmetry, so that the dipole moment is induced in the molecular plane by the electric field and it is oriented in the direction of the electric field. This results in alignment of the molecular planes of the D-monomer molecules parallel to the electric field. When such monomers encounter each other, aggregates are formed with higher probability than monomers of random orientation.

In contrast, Fig. 7.1(b) shows dissociation of J-aggregates in the solution with KCl under an electric field. From Eq. (5.1), the blue (red)-shift of the absorption peak of J-(H-)aggregates indicates reduction in the coherent aggregation number N. The increase in the height of the F-monomer band indicates an increase in the number of F-monomer molecules. The peak energies of the coherent aggregates for dimers ($N = 2$), trimers ($N = 3$) and tetramers ($N = 4$) are estimated to be 460, 472 and 478 nm, respectively, according to Eq. (5.1) with $k = 1$, which corresponds to the lowest-energy dipole-allowed exciton state in the J-aggregate, such that the red-shift from the D-monomer peak is the largest. The spectra in Fig. 7.1(c) strongly suggest that coherent aggregates of such a small number of N were formed in the course of dissociation, although it is not clear whether *N*-mers are isolated from one another or constitute macroaggregates.

One of the possible dissociation mechanisms in J-aggregates is the involvement of K^+ ions, which neutralize excess charges in D-monomers to promote the formation of J-aggregates. If the electric field causes detachment of K^+ ions from the constituent D-monomers, the J-aggregates are split into fragments under the electric field. For this mechanism to take place requires that K^+ ions are more loosely bound in the D-monomer than H^+. This is because aggregation in the solution without KCl was found to be promoted by the electric field, where H^+ plays a major role in neutralization instead of K^+.

It is noteworthy that the D-monomer band did not appear during this dissociation process in Figs. 7.1(b) and 7.1(c). Considering that J-aggregates are composed of D-monomers, it is expected that the D-monomer band should appear before the F-monomer band is substantially increased by dissociation.

It appears that there was an isosbestic point at 456 nm in the long-time response for the TPPS solution without KCl in Fig. 7.1(a). If this point exists, then the absorbance changes are not induced by the change in the coherent size of the J-aggregates, but by the change in the concentration ratio between the two species, the F-monomer and the J-aggregate of certain size. However, when the magnified spectra are closely examined, as shown in Fig. 7.1(d), the initial spectrum (soid line) intersects the time elapsed spectra (dashed or dotted lines) and the other spectra have an isosbestic point at 456.7 nm. This indicates that the coherent size was initially increased, but was immediately saturated (from the initial to time elapsed spectra). The concentration of J-aggregates of a certain saturated size (sufficiently large *N*) then increased against that of the F-monomer. In contrast, no isosbestic point was observed for the solution with KCl, which demonstrates that the coherent size of the J-aggregates was gradually varied by dissociation into a smaller size.

There have been reports that the coherent length of aggregates is controlled by a change in the experimental parameters[28–31], where the aggregation number *N* is evaluated from a small energy shift or change in the bandwidth of the absorption spectra. There is no example where such a large absorption change is observed, as shown in Fig. 7.1, by simply applying an electric field.

7.3 Equilibrium condition of J-aggregate formation with and without KCl [4]

There is a large difference in the long-time response between the two samples with and without KCl. This should reflect the role of KCl in the aggregation mechanism. A sufficient concentration of H^+ (proton), i.e., sufficient TTPS concentration, because it ionizes into $F^{4-} + 4H^+$, results in di-protonation of the F-monomer (H_2TPPS^{4-}) into the D-monomer (H_4TPPS^{2-}). For the J-aggregate to be formed, two more cations are required for neutralization. With a sufficiently high concentration of K^+, the J-aggregates are probably comprised of D-monomers and K^+ ions donated by KCl, instead of protons.

The mechanism for the formation of TPPS J-aggregates without KCl is described in terms of the equilibrium among protons, F-monomers, D-monomers and J-aggregates. The supplied concentration of TPPS molecules $[TPPS]$ and that of protons $[H_{\mathrm{all}}]$, which are ionized from TPPS molecules, satisfy the following equations:[4]

$$[TPPS] = [F] + [D] + 2[J_2] + 3[J_3] + \cdots + n[J_n]\,, \qquad (7.1)$$

$$[H_{all}] = [H] + 2[D] + 8[J_2] + 12[J_3] + \cdots + 4n[J_n]\,, \qquad (7.2)$$

where $[F]$, $[D]$, $[J_{\mathrm{i}}]$ and $[H]$ are the concentrations of the F-monomer, D-monomer, J-aggregates with aggregation number i, and protons, respectively. When TPPS molecules are entirely ionized, i.e., TPPS→F^{4-} $+4H^+$, the equilibrium between $[H]$, $[F]$, $[D]$, $[J_2]$(dimer), $[J_{\mathrm{i-1}}]$, and $[J_{\mathrm{i}}]$ is given by the following relations:[32]

$$\begin{aligned} &[F][H]^2 = k_d[D] \\ &[D]^2[H]^4 = k_2[J_2] \\ &[J_{i-1}][D][H]^2 = k_i[J_i] \end{aligned}\,, \qquad (7.3)$$

where k_d and k_i are equilibrium constants for the relevant reaction.

The reaction between J-aggregates and monomer molecules is usually assumed to be $N[M] = [J]$, as previously reported[33–35]. That is, J-aggregates are considered to be a single species with a fixed aggregation number N. In the present experiment, however, the aggregation number N takes various values, as shown in Fig. 7-1, so that Eq. (7.3) is a more appropriate representation than an expression such as $N[M] = [J]$.

By assuming $k_2=k_3=\cdots=k_n=k$, the concentrations of F-, D-monomers and J-aggregates estimated from Eqs. (7.1)–(7.3) with $k_{\mathrm{d}}=2\times10^{-6}$ and $k=1.5\times10^{-9}$ are listed in Table 7.1 with $n=20$ as the largest aggregation number for a coherent aggregate. k_d was estimated from Eq. (7.3) with

$[TPPS]=3.5\times10^{-5}$ mol/L and $[D]<<[F]$. Table 7.1 explains why only those J-aggregates that are substantially red-shifted are observed for any concentration. Well-grown aggregates with $n>15$ occupy most of the concentration of the aggregates if $[TPPS]$ > 0.0012 mol/L. When the concentration ratio of monomers ($F+D$) to aggregates ($J_{2\text{-}20}$) is 99.9:0.1, J-aggregates with $N\geq16$ are already dominant in concentration over those with $N<15$. The calculated results given in Table 7.1 reproduce the features of the absorption spectra of the TPPS aqueous solution very well.

With KCl, K^+ ions in addition to H^+ serve to neutralize excess charge in the D-monomer for the formation of J-aggregates. Figure 7.3 shows the absorption spectra of TPPS as a function of the KCl concentration. For TPPS of 3×10^{-4} mol/L with a low concentration of KCl, the absorption spectrum shows two peaks characteristic of the F and D monomers at 412 and 433 nm, respectively. With increasing KCl, the J-bands at 490 and 707 nm increase while the monomer bands decrease. Finally the D-monomer band is completely quenched while the F-monomer band has significant magnitude.

Table 7.1. Calculated concentrations of F-monomer, D-monomer, and J-aggregates in the solution without KCl, from Eqs. (7.4) and (7.6).

$[TPPS]$	$[F]$(%)	$[D]$(%)	$[J(16\leq N\leq20)]$(%)
1×10^{-5}	99.9	0.079	0
2.5×10^{-5}	99.5	0.49	0
1×10^{-4}	92	7.4	0
3×10^{-4}	67	33	0
8.1×10^{-4}	31	62	6
0.0011	24	50	23
0.012	2.2	5.0	89
0.12	0.22	0.53	97

In this situation, Eq. (7.3) can be rewritten with $[K]$ as Eq. (7.4).[4]

$$\begin{aligned}&[F][H]^2 = k_d[D]\\&[D]^2([H]+c[K])^4 = k_2[J_2]\\&[J_{i-1}][D]([H]+c[K])^2 = k_i[J_i],\end{aligned} \qquad (7.4)$$

where c is a parameter to introduce the difference in the equilibrium constants between K^+ and H^+. The concentrations of the F-, D-monomers and J-aggregates can then be estimated using Eqs. (7.1), (7.2) and (7.4), if $[H]$ is given, as listed in Table 7.2. The experimental spectral features of TPPS in aqueous solution with KCl shown in Fig. 7.3, where the D-monomer peak is quenched more rapidly than the F-monomer peak as the concentration of K^+ is increased, are reproduced fairly well by the calculation with c=0.12. The result with c<1 indicates that the K^+ ions play a less important role in the formation of J-aggregates than H^+. In other words, K^+ ions are more loosely bound in the D-monomer than H^+, which is consistent with the dissociation mechanism under an electric field as considered in Figs. 7.1(b) and (c).

In the present analyses, the complexity of the equilibrium between macroaggregates and microaggregates is omitted, because the absorption spectra of J-aggregates are mainly characterized by microaggregates rather than macroaggregates. The present analyses are appropriate as long as microaggregates are not spectrally discriminated from macroaggregates. In the next section, a method to distinguish them from EM spectra of aggregates is investigated. In this case, more parameters are required to explain the equilibrium between micro and macroaggregates.[32,33]

Table 7.2. Calculated concentrations of F-monomer, D-monomer, and J-aggregates for the solution with KCl, from Eqs. (7.1), (7.2) and (7.4).

$[K_{all}]$	$[TPPS]$	$[F]$(%)	$[D]$(%)	$[J(16\leq N\leq 20)]$(%)
0.00045	0.0003	67	33	0
0.03	0.0003	65	30	4
0.06	0.0003	37	11	46
0.12	0.0003	15	3	74

7.4 *Electromodulation Spectra of the Coherent Aggregates*

The absorbance changes induced by application of an electric field for 2 h were accompanied by changes in the EM spectra. The electrooptic parameters for the samples without KCl showed little change over 75 min, because the peak energy of the J band was invariant, as shown in Figs. 7.2(a), 7.2(c) and 7.2(e). In contrast, the samples with KCl had significant changes in the EM spectra, as shown in Figs. 7.2(b), 7.2(d) and 7.2(f).

The coherent aggregation number N can be estimated from the J-band peak energy. For N>>10, the broadening component dominates over the

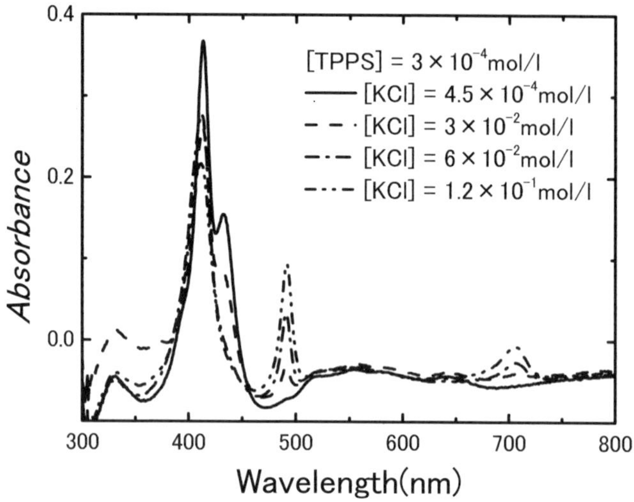

Figure 7.3 Absorption spectra of TPPS in aqueous solutions with different concentrations of KCl. solid, dashed, dash-dotted and dash-dot-dotted lines show absorption spectra with the KCl concentration of 4.5×10^{-4}, 3×10^{-2}, 6×10^{-2} and 1.2×10^{-1} mol/L, respectively. "Copyright: Royal Society of Chemistry"

shift component, similar to the signal observed for the solution without KCl. However, for N=10, the shift component becomes as large as the broadening one. For N<10, the shift component dominates over the broadening component, as shown in Fig. 7.2(f).

For isotropic distribution of macro-J-aggregates in aqueous solution, the broadening signal is expected according to the rearrangement model. However, the situation should be different for microaggregates. F-monomers were finally formed by application of the stationary electric field with KCl, as shown in Figs. 7.2(b) and 7.2(c); therefore, it is reasonable to assume that the spectrally observed N-mers do not constitute macroaggregates, but are microaggregates dissociated from macroaggregates. If this is the case, then the coherent aggregates would be readily oriented along the direction of the electric-field, since the size of the coherent aggregates is much smaller than that of the macroaggregates considering the following reason.

The torque τ of dipole moments induced by the applied electric field for the J-aggregates is given by

$$\tau = (Force)\times a = I\frac{d\omega}{dt}, \tag{7.12}$$

where a is the length of the J-aggregates, I is the inertial moment of the macro or microaggregates, and ω is the angular velocity of the induced dipole moments. For a rod-like structure, the inertial moment I is proportional to a^2. The angular acceleration can then expressed by

$$\frac{d\omega}{dt} \propto \frac{1}{a}. \quad (7.13)$$

Equation (7.13) shows that the angular acceleration is larger for microaggregates than for macroaggregates, i.e., microaggregates can be readily aligned by the electric field. In the case of quasi-one-dimensional orientation of J-aggregates along the direction of the applied electric field, the rearrangement model predicts a red-shift of the J band, as depicted by Fig. 6.3.

Therefore, we conclude that the changes in the EA spectra with decreasing aggregation number are explained by the rearrangement model and by the difference between macro and microaggregates, as shown in Fig. 7.4. When N is sufficiently large (N>>10), the broadening signal predominates in the EM spectra. This is explained by immobile macroaggregates with N>>10. When N is sufficiently small (N<10), then the red-shift signal predominates in the EM spectra. This is explained by mobile, rotatable microaggregates with N<10. We conclude that macroaggregates break into microaggregates with diminution of N under a stationary electric field

8. Summary

We found that the large enhancement for electrooptic Kerr signal by formation of J-aggregates for TPPS molecules. The enhancement factor is about 100 for J-aggregates in the polymer film. Further, we also studied the electrooptic Kerr response of TPPS J-aggregates in aqueous solution to find that the EO signal for the aqueous solutin is about 100 000 times larger in $\Delta A/F^2$ than that for the polymer film. These phenomena can be explained by the molecular rearrangement model, where the dipole–dipole interaction energy is modified by field-induced change in the angle between the molecular transition dipole moment and the aggregation axis. Three pieces of evidence are presented that indicate the molecular rearrangement model is an appropriate model for explaining the large enhancement of $\Delta\alpha$ due to the formation of TPPS J-aggregates. Firstly, the EO Kerr signal of J-aggregates in the polymer film was saturated with the electric-field strength to show a deviation from the

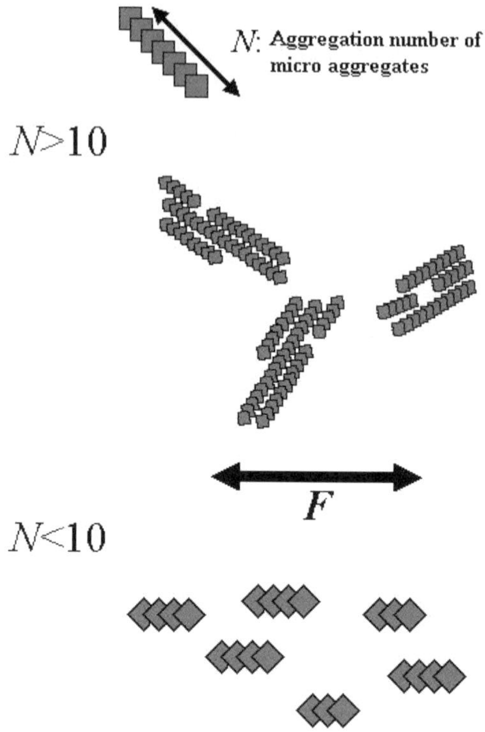

Figure 7.4 Response of the TPPS macro and microaggregates to the applied electric field ***F***. The TPPS macro-J-aggregates are made up of a lot of microaggregates, so that they are hard to be rotated by the electric-field-induced torque because of the large moment of inertia. Microaggregates, on the other hand, are readily rotated by the electric field to be aligned in the direction of the electric field, resulting in a quasi-one-dimensional distribution. "Copyright: Royal Society of Chemistry"

quadratic dependence on the electric field. Secondly, for the J-band in the absorption spectra, the EO Kerr signal exhibits a redshift for the polymer film while a broadening for the aqueous solution because of the difference in the orientational distribution of J-aggregates. Third, the signal intensity was substantially decreased with decreasing temperature for the polymer film. It should be noticed that a large broadening signal was measured in the aqueous solution, and the $\chi^{(3)}$ in the aqueous solution was 100 000 times larger than that in the polymer film. These results are reasonably explained by the rearrangement model. The much larger $\chi^{(3)}$ in the aqueous solution than in the polymer film could also be explained by considering the contribution of the electric double layer formed at the interface between the electrode and the solution.

The dissociation and aggregation processes of J-aggregates in solutions with and without KCl, respectively, were observed as the long-term change in the absorption spectra with application of an electric field over 2 h. Without KCl, the concentration of well-grown aggregates with a large coherent size increased with time. With KCl, the absorption and EM spectra strongly suggest that macroaggregates composed of microaggregates with a large coherent size are dissociated into microaggregates with small coherent size over time. The large difference in the long-time response between the samples with and without KCl reflects the role of K^+ ions in the aggregation mechanism. The equilibrium constants for aggregate formation from monomers were derived and were shown to reproduce the concentration dependence of the absorption spectra. K^+ was found to be more loosely bound to the constituent monomers in J-aggregates than H^+, which is in agreement with the observed dissociation behavior under an electric field. These results provide clues to establish a method for the fabrication of microaggregates with a desired coherent aggregation number *N*. If this is realized, a spectroscopic study of single coherent aggregates would reveal the detailed structure of optically excited states in J-aggregates beyond the simplest Frenkel exciton model.[36]

Acknowledgments

The authors thank T. Ogawa and T. Katsumata for their experimental collaboration in this study and T. Kobayashi for valuable discussion. Especially, T. Ogawa is greatly acknowledged for his finding the molecular rearrangement mechanism. This work was partly supported by the Research Center for Green and Safety Sciences, Tokyo University of Science.

References

1. T. Ogawa, E. Tokunaga, T. Kobayashi, *Chem. Phys. Lett.* **408**, 186 (2005).
2. T. Katsumata, K. Nakata, T. Ogawa, K. Koike, T. Kobayashi, E. Tokunaga, *Chem. Phys. Lett.* **477**, 150 (2009).
3. K. Nakata, T. Kobayashi, E. Tokunaga, *Opt. Rev.* **17,** 346 (2010).
4. K. Nakata, T. Kobayashi, E. Tokunaga, *Phys. Chem. Chem. Phys.* 13, 17756 (2011).

5. K. Misawa, T. Kobayashi, *Nonlinear Optics* **14**, 103 (1995).
6. H. Wendt, J. Friedrich, *Chem. Phys.* **210**, 101 (1996).
7. T. Kobayashi, *Mol. Cryst. Liq. Cryst.* **314**, 1 (1998).
8. O. Ohno, Y. Kaizu, H. Kobayashi, *J. Chem. Phys.* **99**, 4218 (1993).-
9. N.C. Maiti, S. Mazumder, N. Periasamy, *J. Phys. Chem.* **99**, 10708 (1995).
10. N.C. Maiti, S. Mazumder, N. Periasamy, *J. Phys. Chem.* **102**, 1528 (1998).
11. T. Kobayashi, T. Saito, C. Hikage, K. Misawa, *Meeting Abstracts Phys. Soc. Japan* **53,** 221 (1998).
12. K. Misawa, T. Kobayashi, *Technical Digest of IQEC'98*, 1998, p. 99.
13. A. Eilmes, *Chem. Phys. Lett.* **347**, 205 (2001).
14. D. M. Chen, T. He, D.F. Cong, Y.H. Zhang, F.C. Liu, *J. Phys. Chem.* **A 105**, 3981 (2001).
15. Y. Kitahama, Y. Kimura, K. Takazawa, *Langmuir* **22**, 7600 (2006).
16. D. A. Higgins, Paul F. Barbara, *J. Phys. Chem.* **99**, 3 (1995).
17. P. J. Reid, D. A. Higgins, P. F. Barbara, *J. Phys. Chem.* **100**, 3892 (1996).
18. D. A. Higgins, P. J. Reid, P. F. Barbara, *J. Phys. Chem.* **100**, 1174 (1996).
19. H. von Berlepsch, C. Bottcher, and L. Dahne, *J. Phys. Chem.* **B 104**, 8792 (2000).
20. R. Rotomskis, R. Augulis, V. Snitka, R. Valiokas, B. Liedberg, *J. Phys. Chem.* **B 108**, 2833 (2004).
21. W. Liptay, in; E.C. Lim (Ed.), *Excited States,* Academic Press, New York, (1974) p.129.
22. P. Petelenz, *Organ. Electron.* **5**, 115 (2004).
23. M. Slawik, P. Petelenz, *J. Chem. Phys.* **111**,7576 (1999).
24. P. Petelenz, M. Slawik, *Chem. Phys.* **157**,169 (1991).
25. T. Kobayashi, *J-Aggregates*, World Scientific, Singapore, (1996).
26. K. Yanagi, H. Hashimoto, A. T. Gardiner, and R. J. Cogdell, *J. Phys. Chem.* **B 108**, 10334 (2004).
27. M. S. Mehata, T. Iimori, T. Yoshizawa, and N. Ohta, *J. Phys. Chem.* **A 110**, 10985 (2006).
28. A.S.R. Koti, J. Taneja, N. Periasamy, *Chem. Phys. Lett.* **375**, 171 (2003).
29. U. De Rossi. S. Daehne, *Langmuir* **12**, 1159 (1996).
30. A. Chowdhury, L. Yu, I. Raheem, L. Peteanu, L. A. Liu, and D. J. Yaron, *J. Phys. Chem.* **A 107**, 3351(2003).
31. O. Karthaus, Y. Kawatani, *Jpn. J. Appl. Phys.* **42,** 127 (2003).
32. F. Oosawa, M. Kasai, *J. Mol. Biol.* **4**, 10 (1962).
33. R. F. Pasternack, J. I. Goldsmith, S. Szep, E. J. Gibbs, *Biophysical Journal* **75**, 1024 (1998).
34. H. Nagatani, H. Watarai, *Anal. Chem.* **70**, 2860 (1998).
35. A. S. Tatikolov, S. M. B. Costa, *Chem. Phys. Lett.* **346**, 233 (2001).
36. R. Schuster , M. Knupfer, H. Berger, *Phys. Rev. Lett.* **98**, 037402 (2007).

CHAPTER 9

TEMPERATURE DEPENDENCE OF EXCITON TRANSPORT IN J-AGGREGATES

Ivan G. Scheblykin

Chemical Physics
Lund University, Box 124
22100 Lund, Sweden
E-mail: ivan.scheblykin@chemphys.lu.se

Excitons in J-aggregates possess mobility probably superior to that in any other organic system. Recently exciton migration over millions of dyes molecules was demonstrated at cryogenic temperatures for several types of J-aggregates. It is clear that instead of commonly accepted incoherent exciton hopping one has to think in terms of coherent wave-like exciton motion in order to account for the experimental results. However, there is still no consistent theoretical explanation of such excellent exciton mobility. Moreover, efficient exciton transport at low temperatures seems to contradict with the well-accepted theoretical model of the non-monotonous temperature dependence of the Stokes shift in J-aggregates. The latter effect is explained by the idea that excitons at low temperatures have very small mobility and therefore cannot sample exciton states localized on spatially different segments of the chain.

Keywords: Exciton, energy migration, low temperature, thermalization, coherent transport, incoherent hopping, disorder, Stokes shift

1. Exciton Migration

1.1 *Experimental Observation of Exciton Migration in Organic Materials*

Transport of excitons in molecular materials and inorganic semiconductors has been studied theoretically and experimentally for many decays.[1–5] One of the ways to probe exciton migration is to study concentration dependence of trap fluorescence. It has been applied to molecular crystals,[1,4] conjugated polymers[6–8] and J-aggregates.[9–11] Another approach is to study self-quenching of excitons, so-called exciton-exciton annihilation (EEA) process. It can be done by measuring the fluorescence quantum yield or fluorescence decay as a function of the excitation power density. EEA can be seen as energy transfer from one exciton (donor) to another one (acceptor). In this case the first exciton disappears while the second one excites to an upper exciton band. Then rapid relaxation brings the excitation down to the single-excited exciton state. So, only maximum one exciton survives. EEA can be used to probe exciton migration because quite a short distance between two excitons (Förster radius or less) is needed for them to annihilate.

The exciton migration distance (or exciton diffusion radius) varies from system to system. At room temperature values of the order of 20 – 100 nm were reported for the migration distance of singlet excitons in organic molecular crystals (anthracene-type crystals, see review in Ref. 12) and disordered systems like conjugated polymers.[13] This value is not very impressive simply because the Förster radius of energy transfer (single step process) can be as large as 10 nm. The quite short migration distances of singlet excitons in molecular crystals are probably related to small values of the resonance intermolecular interaction (about 100 cm^{-1}), while in conjugated polymers the lack of organization (very large disorder) is the limiting factor. As the reader will see below, J-aggregates possess exciton migration distances orders of magnitude larger.

1.2 *Coherent and Incoherent Exciton Migration*

There are many theoretical approaches to treat exciton migrations in solids. Generally it can be seen as coherent (or band type) or incoherent (or hopping type) process. An extensive review on the existing theories, their limitations and problems can be found in Ref. 14. Strictly speaking it is not possible to predict theoretically the type of transport in a particular system because all existing theories require the knowledge of parameters, which cannot all be measured experimentally.

Coherent transport means the wave-like motion of an exciton with some scattering on defects and phonons. Between the scattering events the exciton moves freely (with a large free-path) keeping a certain direction, every type of scattering leads to diffusion behavior at longer time scales. Incoherent hopping means random jumping of an exciton from one localized position to another. So, the difference between the incoherent and coherent diffusion is the length of the free path, which is very small for incoherent and large for coherent migration. This difference in a particular system results in a much larger diffusion coefficient for the coherent regime in comparison to that for the hopping. Note again, that whatever the particular mechanism is, on long time scales (longer than the exciton free path time) exciton motion looks as random walk.

In the case of a very strong exciton-photon coupling, small resonance intermolecular interaction (J) or large disorder exciton coherence can be completely lost during the first scattering event (within tens of femtoseconds). In this case the originally extended exciton becomes localized and never moves coherently again. The diffusion coefficient in this case is strongly time dependent.

A way to shed light on the type of transport is to measure temperature dependence of the exciton migration efficiency for a particular system over a large temperature range.

From the theory of molecular excitons[2,3,14] developed for molecular crystals it is known that the temperature dependence of the exciton diffusion constant D is completely different for coherent and incoherent transport.[2,4] For incoherent hopping (e.g. from one trap to another) D increases with increasing temperature and scales as exp(-Ea/T) where Ea is an activation energy. In the case of coherent motion the theory predicts decreasing of D with increasing temperature as $1/T^m$ where m ranges from 0 to 3/2 depending on the particular parameters of the exciton-phonon system. Qualitatively similar temperature dependences were also obtained in the framework of the generalized master equation theory[14–16]

Therefore, the characteristic signature of coherent exciton diffusion is an increase of the diffusion coefficient upon cooling the system down due to an increased free path between scattering events, while in the case of incoherent hopping the temperature dependence is the opposite because hopping is a temperature-activated process. Basically hopping is impossible at low temperatures.

In the majority of systems at room temperature the exciton migration was always treated as hopping, no coherent effects were expected. Experimentally at room temperature the diffusion coefficient usually increases with increasing temperature indicating the temperature-activated hopping as the migration mechanism.

In order to expect coherent exciton motion one needs to find a system where the exciton bandwidth (which is proportional to the resonance intermolecular interaction or the exciton coupling J) is much larger than kT and any energetic imperfection. Anthracence-like crystals although very good in terms of having low disorder possess quite narrow exciton band, which limits the coherent transport to very low temperatures ($T < 60K$) only.[2,3,14]

Molecular J-aggregates usually possess approximately one order of magnitude higher exciton coupling ($J = 500$–$1000\ cm^{-1}$, corresponding to the total exciton bandwidth of 2000–4000 cm^{-1}) than that in the crystals. However, the problem for the effective exciton migration might be temperature independent static disorder. Even for PIC aggregates (the system having the narrowest absorption line at cryogenic temperatures reported so far) the calculations give disorder as large as 0.13J (80 cm^{-1}),[17] while for all other known J-aggregates it is even larger.

1.3 *Exciton Migration in J-aggregates*

In comparison to other organic materials J-aggregates possess an extraordinary efficient energy migration, probably the best known for organics so far. Already the very early work of Scheibe[9] showed that photoexcitations in J-aggregates at room temperature could be quenched by adding an extremely low concentration of quenching molecules. It was interpreted as an indication of a very efficient transport of excitons in J-aggregates over as many as 10^6 monomer dyes. As it was pointed out in Ref. 11, it is indeed a surprise that other researchers did not follow this important finding.

Later careful quenching experiments in monolayers[10,11] demonstrated exciton migration over 10^4 monomers at room temperature. In this case the migration volume is a two-dimensional (2D) layer. It allows to calculate the migration radius in 2D as $(10^4)^{0.5}$ = 100 monomers which is about 100 nm. The authors also estimated that the excitons should be able to move with speed of about 2 km/s in order to account for the observed quenching effect.[10] The exciton migration was decreasing with decreasing the temperature.

In 1980-s the development of pulsed lasers and time-resolved spectroscopy techniques allowed one to follow exciton migration by observation of exciton-exciton annihilation.[18] In PIC J-aggregates Sundström et al reported exciton annihilation at exciton concentrations as low as one per 10^4 dye monomers at room temperature. Later on these result have been confirmed in other systems. The suggested mechanism of the exciton migration was incoherent hopping of the excitation from one segment of the molecular assembly to another.

Extensive studies of optical and exciton transport properties have been carried out on J-aggregates of triethylammonium salt of 3,3′-bis-[3-sulfopropyl]-5,5′-dichloro-9-ethylthia- carbocyanine (THIATS) (Fig. 1.1) by Van der Auweraer, Vitukhnovsky and co-workers.[19–26]

It was showed[24] that exciton annihilation in THIATS J-aggregates happens even at cryogenic temperatures indicating an exciton migration volume as large as 6×10^6 monomers at 77K. The important finding is that, contrary to the results of Kuhn and co-workers,[10] the EEA (and exciton migration accordingly) is found to be much more efficient in the temperature range from 20 to 70K than that at room temperature.[21,24,25] This result was quite surprising because in the case of incoherent hopping,[3,4] which was accepted as the migration mechanism by the most of groups working with self-assembled organic structures the diffusion coefficient should decrease with decreasing temperature. Later EEA at cryogenic temperatures was reported for another J-aggregate system too demonstrating even orders of magnitude larger migration volume (10^8–10^9 monomers).[27]

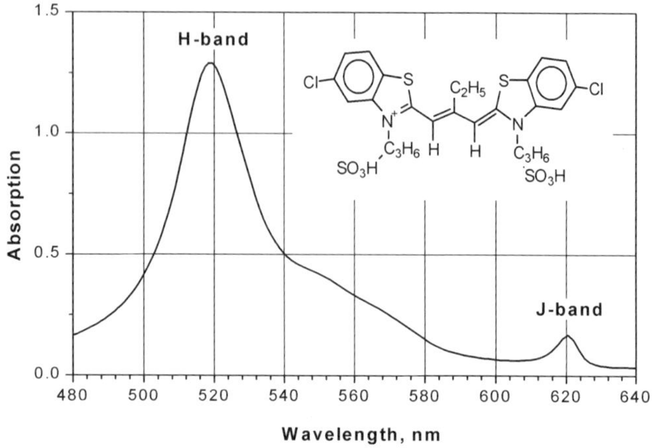

Figure 1.1 Absorption spectrum of J-aggregates of THIATS in water/ ethylene glycol (3/2) at room temperature and chemical structure of the dye. Reproduced from Ref. 25, copyright (2001) American Chemical Society.

1.4 *Random Walk Model. Simple Estimations*

Certainly, there is no reason to think that the exciton migration in J-aggregates is strictly one-dimensional (1D), moreover, as the reader will see below it is simply physically impossible to reach in 1D the migration volumes observed experimentally.

Actually the brickwork packing[28] often proposed for J-aggregates will lead to a 2D-aggregates in solution[29] or crystals.[30,31] Those 2D lamellae can even associate in larger three-dimensional (3D)-aggregates.[32] The existence of large macroaggregates (Fig.1.2) of submicrometer or even larger sizes in three-dimensional (3D) proposed long time ago has been confirmed by SNOM, fluorescence microscopy experiments[33,34] and by X-ray diffraction.[29] A 3D aggregate of 10^6 molecules has a characteristic radius of $(10^6)^{1/3}$ = 100 monomers or about 100 nm. The same estimation for 10^9 molecules (the migration volume measured by of Spitz and Dahne[27]) gives 1000 nm as the characteristic dimension.

In the case of the random walk model in order to be quenched by a single quencher placed in a lattice of 10^6 sites the excitation must visit approximately 2×10^6 sites in a 3D lattice, almost 10^7 sites if the lattice is 2D and 10^{11} sites if the lattice is 1D (see the equations in Ref. 35 at page 121). Let us assume the exciton lifetime is 100 ps = 10^{-10} s. Then a pure

1D situation requires the exciton to spend about 10^{-10}s/10^{11} = 10^{-21} s on each molecule only. Taking 1 nm as a characteristic size of a dye molecule we obtain exciton velocity 10^{-9}m/10^{-21} s = 10^{12} m/s which is almost 4 orders of magnitude larger than the speed of light (3×10^8 m/s). Even in the 3D situation, the time per molecule is about 10^{-16}s (corresponding to the velocity of about 10^7 m/s), which is still too short to be reasonable. In order to solve the situation one obviously must take into account that the exciton has quite large size (about ten molecules at least), the single jump is larger than one molecule and that a long-range interaction between the traps and excitons exists.

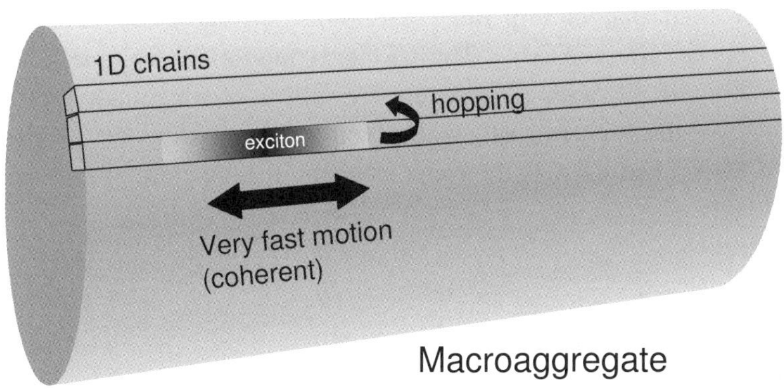

Figure 1.2 Hypothetical organization of a macroaggregate. Due to strong intermolecular coupling excitons move very fast (wave-like coherent motion) along the 1D chains. However, hopping between the chains is probably also quite efficient making the exciton migration essentially 3D.

The only purpose of this estimation is to show that the migration volumes obtained experimentally are very large and indeed close to or above the limits given by the fundamental physical principles, especially for 1D aggregates. Moreover, it is difficult to imagine so fast motion in the framework of incoherent hopping, especially if the results of Spitz and Dahne[27] are concerned. Note that so enormously large number of jumps (especially in 1D case) comes from the diffusive nature of the motion. Basically the exciton goes forth and back all the time instead of going directly to the "target". If an exciton would be able to keep the direction of its motion over large distances (say hundreds of nanometers), it would dramatically decrease the required velocity to probe the migration volume and probably would allow accounting for the experimental results.

1.5 *Temperature Dependence of Exciton Mobility in J-aggregates*

Careful measurements of the temperature dependence of the exciton migration rate (via measurements of EEA) in the temperature interval from 4 to 300K have been reported for THIATS aggregates.[24,25] It turned out that the EEA rate (proportional to the exciton migration volume) possesses a very dramatic temperature dependence, which is much stronger than that of any other excitonic characteristics like spectra, fluorescence lifetime etc. The exciton migration volume changes nonmonotonically with temperature by orders of magnitude. A strong decrease of the EEA process is observed at temperatures below 10 K or above 80 K. The EEA is the most efficient at temperatures between 20 and 70 K.

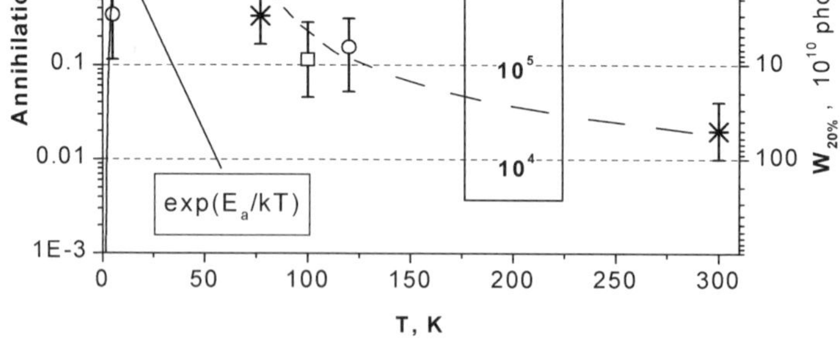

Figure 1.3 Exciton-exciton annihilation rate γ at different temperatures for THIATS aggregates. Solid line, the best-fit Arrhenius function with the activation energy of 15K; dashed line, guide for eyes. The excitation photon flux (W) in the units of number of molecules per created exciton is shown within the bar. Reproduced from Ref. 25, copyright (2001) American Chemical Society.

The efficiency of EEA can be assessed by the excitation photon flux value that is needed to produce a certain quenching effect. In the temperature range from 20 to 70K the exciton emission can be quenched already by 20% under an excitation photon flux as low as 10^9 photons/(cm^2pulse). This intensity corresponds to approximately one absorbed photon per 10^7 dye molecules. If ones assumes a 3D organization of the dye molecules in macroaggregates and the dye volume of $2 \times 1 \times 0.5 = 1$ nm^3 the expected exciton migration volume should be 10^7 nm^3 or about $(200\ nm)^3$ at low temperatures. Note that due to difficulties of quantitative measurements at cryogenic conditions (the main problem is to know the exact value of the excitation photon flux) an error by factor of 3–5 for the migration volume cannot be excluded in this estimation.

To explain such large migration volume at low temperatures it was natural to propose[24] that the exciton migration in THIATS aggregates occurs via a coherent mechanism rather than hopping between different coherent segments. In the framework of this model exciton moves as a coherent wave packet of several pure exciton states with its group velocity until it scatters on lattice defects (disorder) or phonons. Several scattering events lead to the loss of the phase memory in the packet and changing the wave packet composition, motion direction and speed. Usually such "loss of memory" is believed to happen on time scales of tens of femtoseconds at room temperatures. However the phase memory survives much longer at low temperature.

In fact during last years coherent effects in energy transfer in biological light-harvesting antennas at ambient conditions have become a very hot topic. The short coherence lifetime at room temperature has been recently questioned in several experimental and theoretical studies, see e.g. Refs. 36 and 37 and references there in. It is worth to note, however, that the energy transfer that has surprised the researchers recently so much occurs over several tens of chlorophyll molecules, while it is already known for 70 years that excitons in J-aggregates move over hundreds of thousands of dyes molecules at the same conditions.[9]

One may attribute the "freezing" of the coherent exciton motion at $T<10$ K in THIATS aggregates to the exciton blocking by energetic barriers in disordered chains. In this case kT is much smaller than the energetic disorder σ, therefore, the exciton, delocalized over its coherence length of 20–100 molecules, can be blocked between the energetic barriers. However, one has to understand that such "hand-waving" explanation is no more than a speculation. In Figure 3 the experimental data for T < 20 K were fitted to the Arrhenius equation with activation energy E_a =15 K = 10.5 cm^{-1}, which may be related to the energetic disorder σ. However, even at 5 K the EEA is still very pronounced corresponding to an exciton migration over ca. 10^4 dye molecules that is similar to the situation at room temperature.

In the temperature range between 70 and 110 K, the EEA efficiency starts to be strongly temperature dependent again. The exciton migration volume decreases by almost two orders of magnitude when the temperature is increased by 50% only. Such strong dependence cannot be explained in the framework of exciton scattering on acoustic phonons or optical phonons with small frequencies ($\omega \ll kT$) because it could only lead to a rather slow decrease of the diffusion. The much stronger dependence observed experimentally can be attributed to the transition between coherent and incoherent migration regimes. Most probably coherent exciton motion is possible only along 1D chains or within 2D planes, while the migration between the weakly interacting chains or planes forming macroaggregates occurs via a hopping mechanism at all temperatures (Fig. 1.2).

The very high efficiency of EEA at low temperatures (4K) was later confirmed in tubular J-aggregates by Spitz and Daehne.[27] For this system the benchmark of 20% fluorescence quenching is hit at 5 × 10^6 photons/pulse × cm^2 which is more than two orders of magnitude lower intensity as found in THIATS aggregates in the maximum of the annihilation rate around 50K. If possible differences in the optical density at the excitation wavelengths in these two experiments are neglected, this excitation intensity corresponds to the exciton migration volume of approximately 100 times lager than in THIATS aggregates reaching 10^9 monomers. Assuming again a 3D structure for the macroaggregates one obtains an exciton migration volume of $(900\ nm)^3$ probed by an exciton within 85 ps (the exciton lifetime). So, even in 3D case the migration along one dimension almost reaches one micrometer.

Note again that there is no reason to believe that tubular aggregates studied in the discussed work are not forming large 3D macroaggregates.

Although the estimations of the exciton migration distances are every sensitive to the dimension, there is no doubt that J-aggregates possess excellent exciton mobility (far beyond expectations) at cryogenic conditions that cannot be explained in the framework of incoherent hopping concept. It seems that the only way to comprehend the experiments is to consider the exciton migration as a coherent process.

1.6 *Exciton Band Splitting and Exciton Transition Dipole Moment*

The tubular aggregates studied in Ref. 27 and THIATS aggregates[24,25] have a quite similar absorption line width (about 70 cm^{-1}) at low temperatures. Therefore the static disorder level is probably similar in these two systems. Then why is the migration efficiency so different?

THIATS aggregates prepared in water/ethylene glycol 3/2 mixtures possess very pronounced Davydov splitting indicating a periodic structure with two molecules per unit cell.[38] The spectrum contains a very strong and broad H-band and a small narrow J-band (Fig. 1.1).[19] It is important to note that the both bands indeed belong to the same aggregate, which has been confirmed by several spectroscopic studies.[19,22] When excited via the H-band, a very rapid relaxation (tens of femtoseconds as estimated from the H-band width[25]) to the bottom of the exciton band occurs. Due to Davydov splitting and the geometry of the packing the majority of the total excitonic oscillator strength is concentrated in the top of the exciton band (H-states) and only a small part of it resides in the J-states (Fig. 1.4).

The radiative lifetime of an aggregate τ^0_J having several exciton bands in the absorption spectrum is given by:

$$\tau_J^0 = \frac{\tau_m^0}{N_C}(1+Q) \tag{1.1}$$

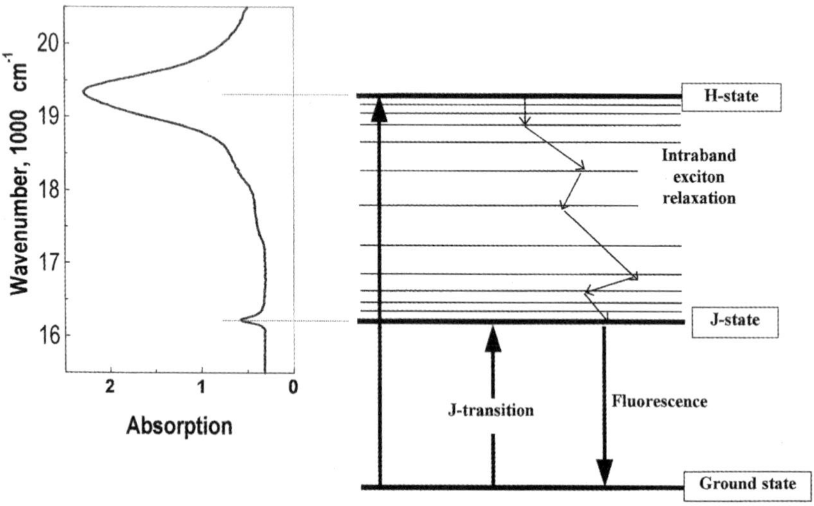

Figure 1.4 Absorption spectra of THIATS J-aggregates at 77 K and their energy diagram illustrating the Davydov splitting of the exciton band. Reproduced from Ref. 23, copyright (2000) Elsevier Science B.V.

where τ^0_m — radiative lifetime of the monomer, N_C — exciton coherence length. The factor Q takes into account the presence of the Davydov (or any other) splitting:[22,23]

$$Q = \frac{\int_{Whole\ spectrum} A(\nu)\nu d\nu}{\int_{J-band} A(\nu)\nu d\nu} - 1 \tag{1.2}$$

where A(ν) — absorption spectrum. The factor Q+1 can be seen as the ratio of the sum of the oscillator strength of all exciton transitions to that of those forming the J-band. In the case of a simple spectrum containing H- and J-bands only the equation simplifies to:

$$Q \approx \frac{\nu_H}{\nu_J} \cdot \frac{\int_{H-band} A(\nu) d\nu}{\int_{J-band} A(\nu) d\nu} \tag{1.3}$$

Aggregates of the pure J-type are characterized by Q=0, of pure H-type by $Q = \infty$. Note that the simplified version of Eq. (1.1) with Q=0 is widely used for estimation of N_C from lifetime measurements even though the exact exciton band structure is usually unknown. Not only the Davydov splitting but also band splitting due to cylindrical geometry[39-41] in tubular aggregates leads to $Q \neq 0$.

THIATS aggregates have Q factor of about 30. As a consequence even for the coherence length of 30 monomers at low temperature the radiative lifetime of THIATS aggregates stays as long as 5 ns at 4K instead of showing extensive superradiance.[23]

The tubular aggregates[27] also possess a complicated absorption spectrum consisting of several bands.[39] By analyzing the spectrum[39] the Q factor can be estimated to be close to 4 for these aggregates at low temperature conditions. Then, taking the coherence length as 30 for THIATS[23] and 95 monomers for the tubular aggregates[42] the ratio of the J-band transition dipole moments squared can be estimated as $(\mu_{tubular}/\mu_{THIATS})^2 \approx (Q_{THIATS}/Q_{tubular})(N_C^{tubular} / N_C^{THIATS}) \approx 40$.

Long-range interactions between exciton states scale as the transition dipole moments squared in a point dipole approximation. Therefore, these interactions, which are responsible for the both exciton transport and exciton annihilation, should be about 40 times stronger in the tubular aggregates than in THIATS aggregates. This can be one of the reasons why EEA is more efficient in tubular aggregates. However when comparing EEA in THIATS and the tubular aggregates, one should consider that the stronger dipole-dipole interactions and the resulting higher exciton mobility in the tubular aggregates is to some extent compensated by the larger radiative lifetime of the THIATS aggregates, which also scales as the transition dipole squared, giving for the excitons more time to migrate. This will especially be the case at low temperature where the decay of the excited states of the J-aggregates is nearly exclusively radiative. Admittedly, a proper theoretical explanation is missing here.

2. Temperature Dependence of the Stokes Shift in J-aggregates

2.1 *Disorder and Density of States*

As it has been recently reviewed[43] there are several fundamentally different theoretical approaches to explain absorption and fluorescence spectra of J-aggregates.

The most popular way is to consider J-aggregates as disordered low-dimensional molecular crystals (for example 1D or 2D) with the excited states described as Frenkel excitons. The disorder effects are usually modeled by introducing an energetic disorder σ, which is the characteristic amplitude of the monomer transition energy fluctuation.[17,44,45] As a result the exciton wavefunctions of the lowest states that carry transition dipole moment become localized on certain parts of the molecular chain and form so-called hidden structure of the DOS.[46,47] Figure 2.1 shows the wavefunctions of the exciton states lying at the low-energy edge of the density of states (DOS) numerically calculated for a particular realization of the Gaussian disorder.[46] The localization length (N_C) measured in monomers is called the coherence length. Numerical calculations for the case of Gaussian diagonal disorder have been very successful to model the shape of J-band in absorption and fluorescence.[17] In order to model the exciton dynamics one has to introduce exciton-phonon interaction, which leads to exciton scattering (transfer) between different states depicted in Fig. 2.1.[46,48] The static disorder can also be temperature dependent, at least it must be different at room (liquid solution) and at low temperature (solid matrix).

In the following sections it will be shown that this approach explains satisfactory the temperature dependence of absorption and emission spectra and the Stokes shift in J-aggregates but completely fails to account for the excellent exciton migration at low temperatures.

2.2 *Temperature Dependent Spectral Properties of THIATS Aggregates*

Although numerous studies of J-aggregates have been done at cryogenic temperatures, probably the first systematic account for temperature

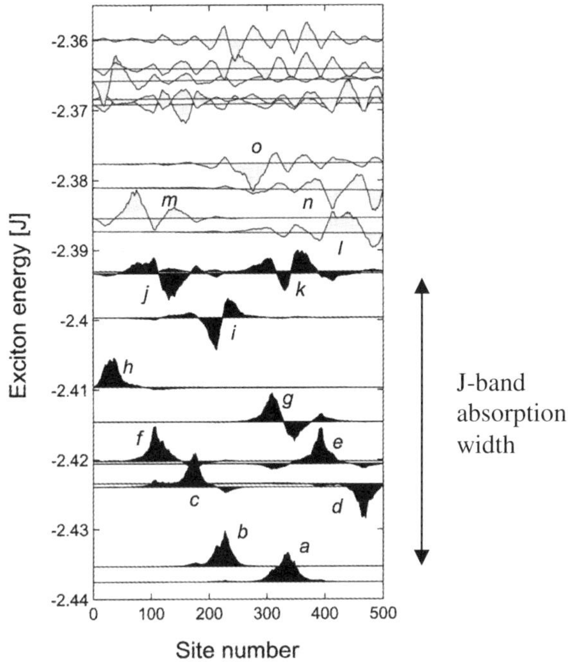

Figure 2.1 Exciton wave functions and energy levels in the vicinity of the bottom of the exciton band of a 1D aggregate with 1 molecule per unit cell for a particular realization of the disorder at $\sigma = 0.1J$. The states are obtained by numerically diagonalizing the exciton Hamiltonian for a chain of 500 sites. The baseline of each state represents its energy in units of J. The wave function amplitudes are in arbitrary units. It is seen that the lower states (filled in black) are localized on segments of the chain with a typical size small compared to the chain length. Some of these localized states can be grouped into local manifolds of two or sometimes three states that overlap well with each other and overlap much weaker with the states of other manifolds [see the doublets of states (b, i) and (f, j), and the triplet (a,g,k)]. The higher states (filled in gray) are more extended and cover several segments at which the lower states are localized. (Reproduced from Ref. 46, copyright (2004) American Institute of Physics).

dependences of exciton spectra and lifetime was done by Van der Auweraer, Vitukhnovsky and co-workers for THIATS aggregates in the temperature range from 4 to 300K.[23,25] The temperature dependences of the J-band fluorescence and absorption width and Stokes shift are presented in Fig. 2.2. In this chapter we will discuss the Stokes shift only. It depends non-monotonically on temperature reaching its maximum at 50K.[25] When temperature goes down or up from this temperature region the Stokes shift decreases. Later a similar non-monotonous behavior was reported also for tubular aggregates.[49]

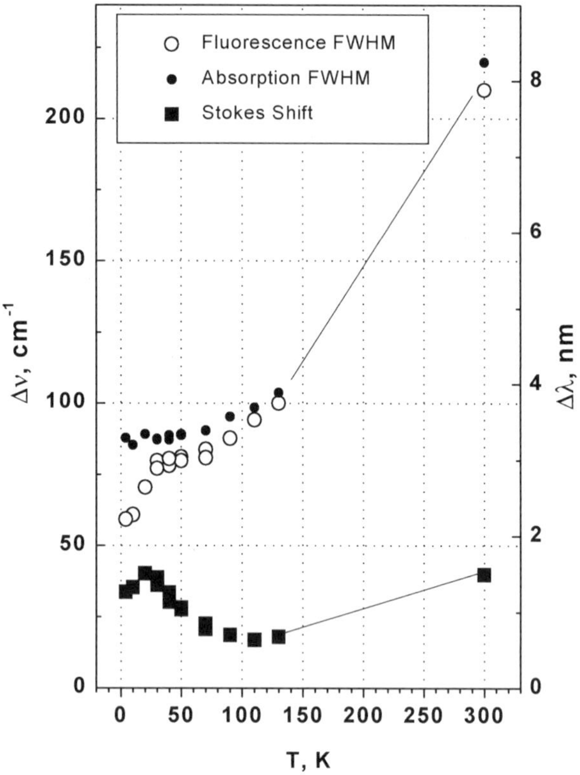

Figure 2.2 Temperature dependence of fluorescence fwhm (full width at half maximum, open circles), absorption fwhm (solid circles) and the Stokes shift (squares). Excitation wavelength was 565 nm. (Reproduced from Ref. 25, copyright (2001) American Chemical Society).

A qualitative explanation of this phenomenon based on the idea that excitons are not able to sample all the states of the DOS was proposed by Scheblykin et al.[25] The idea is that at low temperatures excitons have very limited ability to travel between the states close to the bottom of the DOS (Fig. 2.1). Just after 1–2 jumps an exciton is trapped at a local low-energy state and, therefore, it is unable to reach the global energy minimum. For example, an exciton created at the state ***d*** in Fig. 2.1

cannot reach the low-energy states ***a*** and ***b*** due to poor wavefunction overlapping. This obviously leads to non-thermal population of the DOS at very low temperatures. It is important to see that the probing of different states and the exciton motion are directly connected in this model. As the reader can see already now, it sounds like a contradiction to the excellent exciton migration over millions of molecules at low temperature discussed earlier. We will come back to this problem in Sec. 2.5.

So, at very low temperatures excitons cannot reach the thermal equilibrium in the DOS. With increasing temperature due to the temperature-induced scattering between the localized states depicted in Fig. 2.1 more and more low-energy states of the DOS become available for excitons. Thus, contrary to the common logic, the Stokes shift increases ("antithermalization" behavior as it was called in Ref. 50) with increasing temperature. In the same temperature range an increase of the exciton diffusion by nearly two orders of magnitude is observed upon increasing the temperature (Fig. 1.3). For THIATS the Stokes shift reaches its maximum at T = 20K and then starts to decrease upon further heating simply due to thermally equilibrated population of different levels close to the bottom of the DOS according to the Boltzmann distribution.[25]

It should be noted that a similar non-monotonous temperature dependence of the Stokes shift had been already observed in 1980-s in disordered semiconductor quantum wells and the same qualitative explanation had been given.[51,52] The authors of Ref. 25 were not aware of this at the time the paper was published. Baranovskii and co-workers already in 1998 applied Monte Carlo simulation and modeled successfully the temperature dependences of spectral parameters of the quantum wells including the non-monotonous dependence of the Stokes shift similar to that shown in Fig. 2.2.[50]

Theoretical calculations of the temperature dependences specifically for J-aggregates were done by Knoester and co-workers[46,53] using the same general ideas. The authors were able to model satisfactory the experimental temperature dependences of the Stokes shift and the fluorescence lifetime of THIATS aggregates (Fig. 2.3). The starting point for the model was fitting the absorption and fluorescence spectra at 10K.

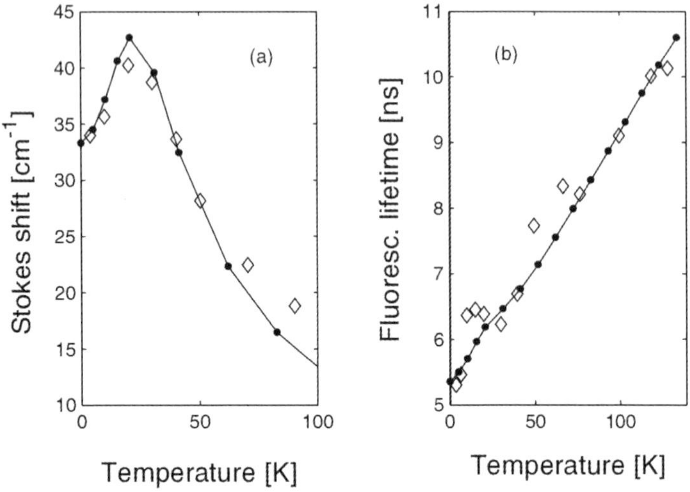

Figure 2.3 (a) Calculated Stokes shift of the fluorescence spectrum as a function of temperature (solid circles). The solid line is a guide to the eye. Diamonds mark the measured Stokes shift for THIATS aggregates reported in Ref. 25. (b) As in (a), for the fluorescence radiative lifetime instead of the Stokes shift. Diamonds are the experimental data from Ref. 23. (Adapted from Ref. 53, copyright (2003) The American Physical Society).

2.3 *Relative Stokes shift. Comparison of THIATS and PBI aggregates. Scaling Phenomena*

Scheblykin et al proposed to plot the Stokes shift in the units of the fluorescence FWHM in a semi-logarithmic scale.[25] The temperature dependence of such relative Stokes shift (SS_{rel}) shows very nicely three distinct temperature regions (Fig. 2.4). It is constant at $T<T_1$, decreases linearly at $T_1 < T < T_2$ (between 20 and 90K for THIATS) and becomes constant again at higher temperatures. Note that the very similar critical temperatures were observed for the exciton migration efficiency (Fig. 1.3). The experimental temperature dependence of the relative Stokes shift was compared with a simple model of the totally accessible DOS.[25,54] The exponential decrease of the relative Stokes shift in the intermediate temperature region ($T_1 < T < T_2$) seems to be explained by this model quite well.[25,54] Basically the Stokes shift changes just because the exciton population of the DOS is determined by the Boltzmann equation. At temperatures higher than T_2 (90K for THIATS) another mechanism must be considered because the experimental Stokes shift becomes constant instead of going further down. Note that this is the same temperature region where the exciton migration efficiency drops orders of magnitude, compare with Fig. 2.3.

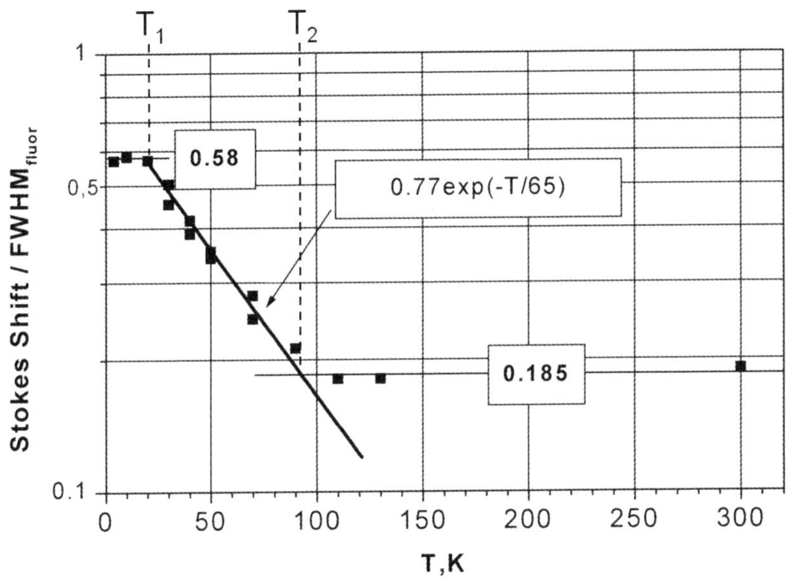

Figure 2.4 Temperature dependence of the Stokes shift relative to the fwhm of the fluorescence spectrum in the semilogarithmic scale. The functions plotted by the solid lines are indicated in the figure. (Adapted from Ref. 25, copyright (2001 American Chemical Society).

Recently temperature dependence of the spectral parameters was measured in a completely different type of J-aggregates, aggregates of PBI 5a dyes.[54] The properties of these aggregates are described in the Chapter 11 of this book. In spite of the great differences in the exciton line width and the absolute values of the Stokes shift (Fig. 2.5) indicating a much larger disorder level in PBI aggregates in comparison to THIATS aggregates qualitatively the same temperature dependence of the relative Stokes shift was observed (Fig. 2.6).

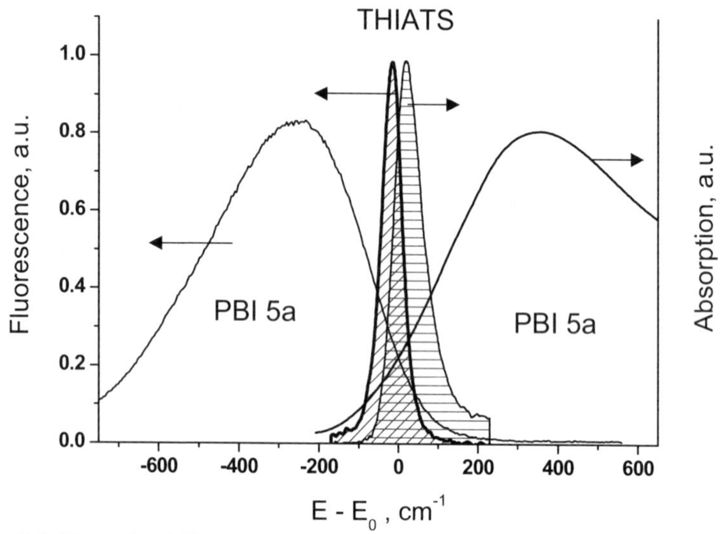

Figure 2.5 Normalized fluorescence and absorption spectra (J-band) of PBI 5a (see Chapter 11) and THIATS J-aggregates at 5 K and 10 K, respectively. The energy is measured relative to the absorption and fluorescence spectra crossing point. Adapted from Ref. 54, copyright (2009) American Chemical Society.

Moreover, by introducing just a single scaling parameter for the all characteristic temperatures and energies the dependence of the relative Stokes shift for THIATS aggregates can be transformed to that for PBI 5a aggregates.

The function shown by the thick gray line in Fig. 2.6 is the dependence of SS_{rel} for THIATS (black like) "multiplied" by a factor of 2.8. The exact equations are shown in the plot. Basically all the energy related parameters, the activation energy E_a and the temperatures T_1 and T_2, are scaled up by the factor of 2.8. Because of the nature of the exponential function, such scaling automatically gives the same scaling factor for the constant SS_{rel} value at $T < T_1$ and SS_{rel} at $T > T_2$. As one can see in Fig. 2.6 it fits the data for PBI 5a quite nicely. This implies that Stokes shift / $FWHM_{fl}$ is a very interesting parameter to characterize thermally induced spectral shifts and broadening in low-dimensional molecular systems, and thus a highly remarkable finding.

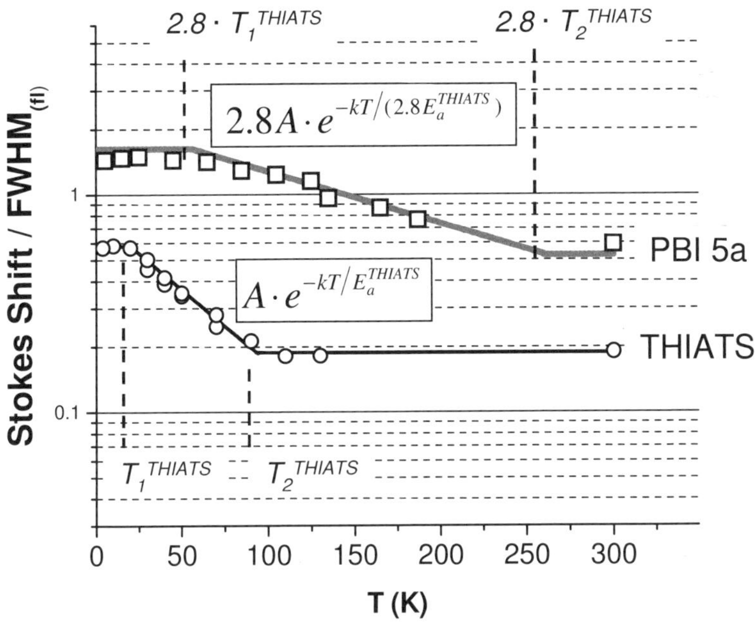

Figure 2.6 Comparison of the relative Stokes shift of PBI 5a aggregates (see Chapter 11, Fig. 2.3) with that of THIATS. The black lines, which fit the experimental data of THIATS, are scaled up by a factor of 2.8 to give the thick gray lines, which fit very well with the experimental data of PBI 5a. The exact equations are shown in the figure. T_1 and T_2 are the characteristic temperatures, and E_a denotes the activation energy, which has for the example of THIATS a value of $E_a^{THIATS} = 45$ cm^{-1}. The activation energy for PBI 5a (E_a^{PBI}) is $2.8 \times E_a^{THIATS} = 130$ cm^{-1}. Adapted from Ref. 54, copyright (2009) American Chemical Society.

The fact that two systems with very different disorder levels demonstrate such a nice scaling behavior should be an interesting aspect for theoreticians to explore. So far there is no reasonable explanation for the origin of the scaling factor 2.8 between THIATS and PBI 5a aggregates. An interesting observation is, however, that the degree of disorder (σ/J) estimated from the absorption spectral width is 2.5 times higher for PBI 5a than that of THIATS aggregates.[54] Possibly, this is just a coincidence, but these two factors are in close proximity.

2.5 *Inaccessibility of the Density of States and Excellent Exciton Migration. Contradiction?*

The reader may have already noticed that the inaccessibility of the complete DOS at low temperatures and the extremely efficient exciton migration in space seems to contradict to each other. Indeed, in terms of the classical model of J-aggregates as molecular chains with static disorder the motion of excitons in space means jumping from one localized state (coherent segment) to another.[46] In this case due to the fast intra and intersegment relaxation all states of the DOS should become available for an exciton, which probes 10^4 monomer molecules at 4K and 10^6 monomers at 50K (data for THIATS aggregates).

Theory of exciton migration in 1D J-aggregates developed in Ref. 55 and 56 states that an exciton can migrate over ca. 1000 monomers only if the temperature is comparable with the J-band spectral width (ca. 100K for THIATS) and therefore cannot explain the extremely efficient exciton diff at lower temperatures.

This obviously conflicts with the explanation of the non-monotonous temperature dependence of the Stokes shift given in the previous sections. There is no explanation for this contradiction so far. What can be said for sure is that the exciton migration theory based on disordered chains[46,53,57,58] and localized excitations completely fails to explain the exciton migration (and exciton-exciton annihilation) at low temperatures, while it accounts nicely for temperature dependence of the exciton spectral features.

It seems that the only way to rationalize the situation at low temperatures is to assume that while migrating the excitons indeed do not probe those deeply localized states of the DOS. It can be imagined that an exciton as a coherent wavepacket does not "sink down" to those low energy traps (e.g. states a,b in Fig. 2.1). It moves "over" them instead without any loss of energy. It seems actually how the coherent exciton migration is supposed to look like. Also the fact that the macro-aggregates where this efficient exciton diffusion was observed are not 1D perhaps also plays a role. When the temperature goes up, due to exciton-phonon interaction the coherent migration slows down giving more chance for excitons to find low energy traps making excitons more and more thermally equilibrated within the DOS. However, such a naïve picture is nothing more than a speculation. Obviously further development of the existing theories is needed to explain co-existence of the temperature dependent Stokes shift and extraordinary exciton mobility at low temperatures.

References

1. M. Pope and C. E. Swenberg, *Electronic Processes in Organic Crystals and Polymers* (Oxford University Press, New York, 1999) 2 ed..
2. V. M. Agranovich and M. D. Galanin, *Electronic Excitation Energy Transfer in Condensed Matter* ,Vol. 3 (North Holland Pub. Co., Amsterdam, 1982).
3. V. M. Agranovich, *Excitations in organic solids* (Oxford University Press, 2009).
4. Silinsh E.A. and Capek V, *Organic molecular crystals: interaction, localization and transport phenomena* (American Institute of Physics, New York, 1994).
5. M. D. Galanin, *Luminescence of Molecules and Crystals* (Cambridge International Science Publishing, 1996).
6. E. J. W. List, C. Creely, G. Leising, N. Schulte, A. D. Schluter, U. Scherf, K. Mullen, and W. Graupner, *Chem. Phys. Lett.* **325**, 132 (2000).
7. H. Wiesenhofer, E. Zojer, E. J. W. List, U. Scherf, J. L. Bredas, and D. Beljonne, *Adv. Mater.* **18**, 310 (2006).
8. I. G. Scheblykin, L. S. Lepnev, A. G. Vitukhnovsky, and M. Van der Auweraer, *J. Lumin.* **94**, 461 (2001).
9. G. Scheibe, A. Schontag, and F. Katheder, *Naturwissenschaften* **27**, 499 (1939)
10. D. Mobius and H. Kuhn, *J. Appl. Phys.* **64**, 5138 (1988).
11. H. Kuhn and C. Kuhn, in *J-aggregates*, Ed. T. Kobayashi (World Scientific, 1996) Ch. 1, pp. 1-40.
12. Silinsh E.A. and Capek V, in *Organic molecular crystals: interaction, localization and transport phenomena* (American Institute of Physics, New York, 1994) Ch. 2, pp. 60-82.
13. I. G. Scheblykin, A. Yartsev, T. Pullerits, V. Gulbinas, and V. Sundstrom, *J. Phys. Chem.* **B111**, 6303 (2007).
14. Silinsh E.A. and Capek V, in *Organic molecular crystals: interaction, localization and transport phenomena* (American Institute of Physics, New York, 1994) Ch. 5, pp. 199-253.
15. V. Capek and R. W. Munn, *Phys. Status Solidi* **B108**, 521 (1981).
16. V. Capek and R. W. Munn, *Phys. Status Solidi* **B109**, 245 (1982).
17. J. Knoester, *Int. J. Photoenergy* **2006**, 61364 (2006).
18. V. Sundstrom, T. Gillbro, R. A. Gadonas, and A. Piskarskas, *J. Chem. Phys* **89**, 2754 (1988).
19. I. G. Scheblykin, M. A. Drobizhev, O. P. Varnavsky, M. VanderAuweraer, and A. G. Vitukhnovsky, *Chem. Phys. Lett.* **261**, 181 (1996).
20. M. A. Drobizhev, M. N. Sapozhnikov, I. G. Scheblykin, O. P. Varnavsky, M. VanderAuweraer, and A. G. Vitukhnovsky, *Chem. Phys.* **211**, 455 (1996).

21. I. G. Scheblykin, O. P. Varnavsky, M. M. Bataiev, O. Sliusarenko, M. Van der Auweraer, and A. G. Vitukhnovsky, *Chem. Phys. Lett.* **298**, 341 (1998).
22. I. G. Scheblykin, O. P. Varnavsky, W. Verbouwe, S. De Backer, M. Van der Auweraer, and A. G. Vitukhnovsky, *Chem. Phys. Lett.* **282**, 250 (1998).
23. I. G. Scheblykin, M. M. Bataiev, M. Van der Auweraer, and A. G. Vitukhnovsky, *Chem. Phys. Lett.* **316**, 37 (2000).
24. I. G. Scheblykin, O. Y. Sliusarenko, L. S. Lepnev, A. G. Vitukhnovsky, and M. Van der Auweraer, *J. Phys. Chem.* **B104**, 10949 (2000).
25. I. G. Scheblykin, O. Y. Sliusarenko, L. S. Lepnev, A. G. Vitukhnovsky, and M. Van der Auweraer, *J. Phys. Chem.* **B105**, 4636 (2001).
26. M. Van der Auweraer and I. Scheblykin, *Chem. Phys.* **275**, 285 (2002).
27. C. Spitz and S. Daehne, *Int. J. Photoenergy* **2006**, 84950 (2006).
28. V. Czikklely, H. D. Forsterling, and H. Kuhn, *Chem. Phys. Lett.* **6**, 207 (1970).
29. G. J. T. Tiddy, D. L. Mateer, A. P. Ormerod, W. J. Harrison, and D. J. Edwards, *Langmuir* **11**, 390 (1995).
30. H. Stoeckli, *Helv. Chim. Acta* **57**, 1 (1974).
31. K. Nakao, K. Yakeno, H. Yoshioka, and K. Nakatsu, *Acta Crystallogr. Section* **B35**, 415 (1979).
32. D. A. VandenBout, J. Kerimo, D. A. Higgins, and P. F. Barbara, *J. Phy. Chem.* **100**, 11843 (1996).
33. K. Misawa and T. Kobayashi, in *J-aggregates*, Ed. T. Kobayashi (World Scientific, 1996) Ch. 2, pp. 41-66.
34. D. A. Higgins, J. Kerimo, D. A. VandenBout, and P. F. Barbara, *J. Am. Chem. Soc.* **118**, 4049 (1996).
35. M. Pope and C. E. Swenberg, in *Electronic Processes in Organic Crystals and Polymers* (Oxford University Press, New York, 2 edn., 1999) Ch. 1F, pp. 96-134.
36. G. D. Scholes, *J. Phys. Chem. Lett.* **1**, 2 (2010).
37. Y. C. Cheng and G. R. Fleming, *Annu. Rev. Phys. Chem.* **60**, 241 (2009).
38. A. S. Davydov, *Theory of Molecular Excitons* (McGraw-Hill, New York: 1962).
39. C. Spitz, J. Knoester, A. Ouart, and S. Daehne, *Chem. Phys.* **275**, 271 (2002).
40. M. Bednarz and J. Knoester, *J. Phys. Chem.* **B105**, 12913 (2001).
41. V. I. Prokhorenko, D. B. Steensgaard, and A. F. Holzwarth, *Biophys. J.* **79**, 2105 (2000).
42. S. S. Lampoura, C. Spitz, S. Dahne, J. Knoester, and K. Duppen, *J. Phys. Chem.* **B106**, 3103 (2002).
43. V. V. Egorov. *Theory of the J-band: from the Frenkel exciton to charge transfer*, International conference on luminescence and optical spectroscopy of condensed matter (2008), Physics Procedia 2, 223 (2009).
44. H. Fidder, J. Terpstra, and D. A. Wiersma, *J. Chem. Phys* **94**, 6895 (1991).
45. M. Schreiber and Y. Toyozawa, *J. Phys. Soc. Jpn* **51**, 1528 (1982).
46. M. Bednarz, V. A. Malyshev, and J. Knoester, *J. Chem. Phys* **120**, 3827 (2004).
47. V. Malyshev and P. Moreno, *Phys. Rev.* **B51**, 14587 (1995).
48. D. J. Heijs, V. A. Malyshev, and J. Knoester, *Phys. Rev. Lett.* **95**, 177402 (2005).
49. A. Pugzlys, R. Augulis, P. H. M. van Loosdrecht, C. Didraga, V. A. Malyshev, and J. Knoester, *J. Phys. Chem.* **B110**, 20268 (2006).

50. S. D. Baranovskii, R. Eichmann, and P. Thomas, *Phys. Rev.* **B58**, 13081 (1998)
51. M. S. Skolnick, P. R. Tapster, S. J. Bass, A. D. Pitt, N. Apsley, and S. P. Aldred, *Semicond. Sci. Tech.* **1**, 29 (1986).
52. S. T. Davey, E. G. Scott, B. Wakefield, and G. J. Davies, *Semicond. Sci. Tech.* **3**, 365 (1988).
53. M. Bednarz, V. A. Malyshev, and J. Knoester, *Phys. Rev. Lett.* **91**, 217401 (2003).
54. T. E. Kaiser, I. G. Scheblykin, D. Thomsson, and F. Wurthner, *J. Phys. Chem.* **B113**, 15836 (2009).
55. A. V. Malyshev, V. A. Malyshev, and F. Dominguez-Adame, *J. Phys. Chem.* **B107**, 4418 (2003).
56. A. V. Malyshev, V. A. Malyshev, and F. Dominguez-Adame, *Chem. Phys. Lett.* **371**, 417 (2003).
57. V. A. Malyshev, G. G. Kozlov, H. Glaeske, and K. H. Feller, *Chem. Phys.* **254**, 31 (2000).
58. I. V. Ryzhov, G. G. Kozlov, V. A. Malyshev, and J. Knoester, *J. Chem. Phys* **114**, 5322 (2001).

Chapter 10

QUANTUM INFORMATION PROCESSING ATTRIBUTES OF J-AGGREGATES

A. Thilagam

Minerals and Materials Science and Technology,
Information Technology, Engineering and the Environment,
Mawson Institute,
University of South Australia,
Australia 5095
E-mail: thilaphys@gmail.com

We examine the unique spectroscopic features which give rise to quantum information processing attributes of J-aggregate systems via several entanglement entities such the von Neumann entropy, Wootters concurrence, and the Wei-Goldbart geometric measure. We also discuss two other measures, the quantum discord and its geometric measure which are intrinsically distinct from other known quantum correlation measures. The role of dispersion and resonance terms in the exciton-phonon interaction are analyzed using Green's function formalism. We show that the scaling properties of the third order optical response $\chi^{(3)}$ with exciton delocalization size is a notable feature that can be harnessed as an experimentally demonstrable measure of quantifying multipartite entanglement in J-aggregate systems. Results of the quantum discord for a linear monomer chain highlights the potential of J-aggregates as sources of robust non-classical correlations which can be utilised for their superior optical properties.

Keywords: Quantum information; J-aggregates; Entropy; Concurrence; non-classical correlation; exciton; phonon; multipartite entanglement; quantum discord

1. Introduction

The tendency of organic dye dye molecules to aggregate when present at higher concentrations in solutions has been well examined in several works [1, 2]. In some cases such aggregation results in the appearance of an unusually sharp and intense red-shifted optical absorption band. These aggregates, first discovered by Scheibe [3, 4] and Jelly [3], and also known as quantum J-aggregates, are low dimensional molecular systems which display enhanced coherent interactions with light. For instance, a well known dye molecule pseudoisocyanine (PIC) forms a quasi-one dimensional arrangement of molecules under suitable conditions, displaying the characteristic feature of a sharp absorption spectrum red shifted from the broader monomer transition band. The J-band feature has been observed for select class of molecules, for instance in dye systems which possess the extended π-electron configuration as well as molecular systems which possess a specific structural framework. It is known from fluorescence quenching experiments [4] that excitation in J-aggregate systems progapate in very efficient times of just a few picoseconds over a large number ($\approx 10^3$) of molecules. Many spectroscopic features of J-aggregates can be interpreted via a model of one-dimensional Frenkel exciton system [5], with the delocalized exciton considered the main reason behind the coherent properties of such systems. There have been suggestions that the total number of molecules which form such collective states varies from ten to several hundreds [6].

The spectroscopic feature of the sharp and intense red-shifted optical absorption band be viewed as a spectacular example of nonlocal effects due to collective properties of the Frenkel exciton [3, 4, 7–9]. Eisfeld et. al. [8] first suggested that the delocalized excitation of the aggregate constituting a system of monomers is a manifestation of the entangled state. The molecular aggregation can be seen to trigger an emergent quantum behaviour, which results in the enhanced rate of spontaneous emission in aggregates compared to that of the monomer. This effect which arises from the collective nature of the excited aggregate states is known as superradiance. The narrow isolated band of J-aggregates usually formed in liquid or solid organic matrix, generally appears adjacent to a monomer based broad absorption band incorporating vibrational features. The absence of a similar vibrational structure in the polymer J-band is a striking phenomenon, almost equivalent to a decoherence-free subsystem, widely discussed in quantum information studies [10].

It is seen that subtle features of quantum J-aggregates cannot be adequately explained using conventional theory of exciton dynamics [5, 11, 12]. Recent technical advances in spectroscopic techniques such as two-dimensional (2D) coherent electronic spectroscopy with fast femtosecond time resolution [13] has made it possible to observe intricate dephasing processes of delocalized exciton states in the single J-band aggregate system [14]. Such detailed experimental probes of the entangled properties highlight the potential in utilizing the opto-electronics properties of J-aggregate systems for quantum information processing. Optical-based applications have already been demonstrated in organic semiconductor microcavities [17]. The unusual characteristics of J-aggregate systems allows their use as solar efficient dye materials, as well as optical sensitizers and as polarizable material systems. The interactions in cyanine dye chromophores [18] is known to self-organize into ordered aggregates with superior optical properties of J-aggregates, and with important applications in artificial light harvesting systems. Moreover photoinduced electron transfer (PET) reactions from the dye aggregates to electron acceptor molecules may bring with it potential applications linked to fabrications of nanostructures as well. Overall, the potential innovative applications based on new understanding of the puzzling effects of quantum sized structures provide much impetus for further examination of J-aggregates, particularly from a quantum information perspective. This approach is expected to assist in the better understanding of the unusually efficient mechanism of energy transfer in J-aggregate systems.

Theoretical studies of J-bands are based on models of identical monomers interacting with vibrations [19, 20] or those based on rigid monomers of varying electronic site energies [6]. The SP parameter, given by the ratio of inter-monomer interaction energy to the vibrational width of the monomer spectrum, was used by Simpson *et al.* [19] to provide a measure of narrowness of the J-band. At high SP values, excitation transfer between monomers occur at a much faster than relaxation processes due to lattice vibrations. In Knapp's Frenkel exciton model [6], specific details of the background medium was discarded with attention focused on the the J-band. The time of propagation of the Frenkel exciton was considered faster than time variations in electronic site energies. The term "exchange narrowing" [6] was used to describe the decrease in linewidth by $1/\sqrt{N}$ in the absence of intersite correlation, N being the number of monomers. The "exchange narrowing" phenomenon is therefore linked with the participation of N monomers as a multipartite system, each monomer thus experiencing an effective disorder of magnitude $\frac{D}{\sqrt{N}}$ instead of D, the full disorder experienced by an isolated monomer. In an alternative approach, the characteristic nature of the aggregate system is given importance, and the charge is considered transported via energy exchanges between a continually changing form in nuclei and the environment [21].

Quantum non-locality is a collective property of a quantum system consituting two or more subsystems, in which the dynamics of distinct constituents depend on spatially separated adjacent subsystems [10]. These subsystems manifest themselves in the form of monomers in J-aggregates. The Pauli exclusion principle excludes the possibility of two excitons occupying the same molecular site, thus exciton-exciton scattering and excitation/de-excitation processes give rise to entanglement effects in excitonic spin chains. It is known that the time taken for individual monomers to interact and form the collective J-band is far less than interaction time with lattice vibration as revealed experimentally [13, 14]. Here we examine the dynamics of J-aggregates using a quantum information approach adopted in earlier works [15, 16], and in which the origin of its sharp J-band is attributed to entanglement and non-local effects associated with a system of excitonic monomers. The decoherence effects [22–25] associated with a noisy environment may disrupt the delicate delocalized excitation which is responsible for the narrowness of J-aggregate spectrum. The extreme limit of an infinite chain with infinitely strong coupling can be considered as a system in which the nonlocal effects is maximal, with the distinct characteristics of an ideal J-band forming an infinitely narrow spike.

However the general problem [26–33] of quantifying the level of entanglement in any arbitrary multipartite system should be emphasized. It seems that the act of partitioning individual subsystems affects multipartite entanglement and introduces an element of subjectivity when using current classification of entanglement such as k-separability. For a selected partition, one can obtain varying degrees of multipartite entanglement, changing from a fully separable to fully inseparable states with each member of the group sharing genuine multipartite entanglement. The complicated nature of studies of quantum states is further highlighted by the fact that there exists a degree of quantum correlations other than entanglement in the form of quantum non-locality without entanglement [34]. One such type of quantum correlation known as quantum discord [35–37] has gain increased attention in recent works involving entangled systems. This quantity which yields the difference between quantum mutual information and classical correlation in a bipartite system, and is different from entanglement measures such as the von Neumann entropy [28, 38], Wootters concurrence [39] and the class of entanglement quantity known as the Wei-Goldbart geometric measure of entanglement [40]. While the Wootters concurrence may be zero for separable mixed states with no entanglement, such states possess non-zero values of the quantum discord.

A very recent work [43] has conclusively shown that the quantum discord is more robust than Wooters concurrence. The significance of studies related to quantum discord is highlighted by a recent work [44] that a vast number of quantum states have non-zero quantum discord. However due to the difficulty in obtaining an explicit expression due to maximization procedures, the quantum discord for two-quantum states has been studied in detail only for a restricted set of two-qubit states [45]. The complicated dynamics of the quantum discord measure is emphasised by the fact that even for the basic configuration of two entangled qubits, there exists no explicit relation between the quantum discord, entanglement, and classical correlation measures. The unique features of quantum discord presents important applications in quantum information processing and it is worthwhile to explore such possibilities within the context of J-aggregate systems in future investigations. In this work, we provide brief details of the quantum discord and a related measure known as the geometric measure of quantum discord which may be of interest to J-aggregate systems.

In view of the complicated nature of categorizing and examining the structure of quantum states as mentioned in the earlier paragraphs, investigations of the quantum correlation aspects of fragile open systems such as excitonic chains is expected to remain unresolved for some time. In this regard, we point to the possibility that the scaling of the third order optical response $\chi^{(3)}$ with exciton delocalization size may provide an experimentally demonstrable means of categorizing genuine multipartite entanglement, at least in the case of J-aggregates. We discuss this possibility in a later section.

This chapter is organised as follows. In Sec. 2., we summarize salient features of the Frenkel exciton in molecular systems, including a brief description of the site basis representation appropriate to excitonic systems. We also include a brief discussion related to the coding of excitonic qubits for a simple two lattice site system. In Sec. 3., we provide a brief description of the von Neumann entropy [28, 38], Wooters concurrence, the Meyer-Wallach measure and the quantum discord. In Sec. 4. we provide mathematical details of derivations based on Green's functions formalism [15] and present results which show the usefulness of the entropy in capturing the transition from coherent to incoherent energy transfer in a finite one-dimensional J-aggregate system. In Sec. 5., we discuss molecular cooperativity as a paradigm of collective entanglement and utilize the well-known Wootters concurrence [39] to interpret superradiance properties of one-dimensional excitonic systems. The theory of two-exciton states and analysis of the intricate link between the third order optical response $\chi^{(3)}$ and the class of entanglement quantity known as the Wei-Goldbart geometric measure of entanglement [40] is presented in Sec. 6. Finally some quantum information attributes of J-aggregates and other decoherent mechanisms are presented in Sec. 7.

2. J-aggregates and Frenkel Excitons

The optical properties of J-aggregates are invariably linked to the geometrical attributes of its molecular arrangement. The molecular structure of the chemical constituent is expected to determine the morphological characteristics of the aggregate system which may assume the shape of tubes or bundles of tubes, and variations of these basic shapes can be obtained by the use of crystal growth modifiers. Knoester et al. [56] used a cylindrical model, in which the molecules are located on rings which are stacked on top of each other, to study the optical properties of aggregates. In general, the aggregate system is generally modelled as a one-dimensional chain, and in some instance as a two-dimensional system. The nearest-neighbor model can be utilized within the one-dimensional structure to introduce numerical simplicity. This simple model can be extended using the long-range interacting cycle (LRIC) graph lattice in which all node pairs at a specific distance are connected, a key feature that ensures that all lattice sites have the same connectivity of 4. These newly added edges serve as shortcuts in the cycle graph, and act to increase the cross paths and hence bandwidth B of the lattice system.

We restrict our study to Frenkel excitons in a quasi one-dimensional system of finite N molecular sites, and which we consider an appropriate system to model the one-dimensional aggregate system. Excitons in molecular crystals are generally treated using the tight-binding model [5] due to the weak overlap of intermolecular wavefunctions. The exact form of these functions are not needed to derive the salient properties of Frenkel excitons, however every molecular site has an equal probability of being excited by an incident photon due translation symmetry, a salient feature which has important implications for quantum parallelism. Various mechanisms result in the loss of excitonic coherence, in this work we consider the effects due to coupling with lattice vibrations as well as the influence of finite temperature.

The absorption of a photon triggers an initialization process with the creation of a delocalized exciton. The exciton represents an equal weighted superposed form of occupation probabilities of an electron-pair at all lattice sites. While ideal crystal in vacuum state is equivalent to the null state $|0\rangle^{\otimes N}$ for N sites, photon absorption process initiates a Hadamard-like transformation [10] at each site. In the absence of lattice vibrations, the exciton eigenvector can thus be written as

$$|K\rangle = N^{-1/2}\sum_l e^{iK.l}B_l^\dagger|\mathbf{0}\rangle \tag{2.1}$$

where the reciprocal lattice vector $K = \frac{2\pi k}{N}$ within the Brillouin zone $k \in [-\frac{N}{2}, \frac{N}{2}]$, $|\mathbf{0}\rangle$ is the vacuum state with all molecules in ground state and $B_l^\dagger$ is the creation operator of exciton at position coordinate l. N is the number of unit cells in the crystal that increases with the number of molecules for each unit cell. In Eq. (2.1), we have dropped notations associated with the exciton spin for convenience, and include notations in scalar form as we restrict the model considered here to a finite one-dimensional linear chain. As mentioned earlier, we consider that all molecular sites have an equal chance of absorbing a photon to raise it to an excited state $|1\rangle$ from a ground state of $|0\rangle$. The transformation process at each site can be represented by the Hadamard-like transformation: $|0\rangle \to |0\rangle + |1\rangle$. It is important to note the vital difference between a photon absorption process and a true Hadamard gate conversion which can reset a state to its original form. For instance, other possible transformations inherent in a true Hadamard gate such as $|0\rangle + |1\rangle \to |0\rangle$ and $|1\rangle \to |0\rangle - |1\rangle \to |1\rangle$ are not present upon successive absorption of two or more photons at a molecular site.

The creation operator of a Frenkel exciton with wavevector K can be easily obtained using the Fourier transform of Eq. (2.1)

$$B_K^\dagger = N^{-1/2}\sum_l e^{iK.l}B_l^\dagger \tag{2.2}$$

The exciton creation operator $B_K^\dagger$ localized in K-space is delocalized in real space [5]. The motion of the exciton in molecular systems is governed by an Hamiltonian derived using a tight-binding model [5, 11, 12]

$$\hat{H}_{ex} = \sum_l \left(\Delta E + \sum_{m\neq l} D_{l,m}\right) B_l^\dagger B_l + \sum_{m\neq l} M_{l,m}B_l^\dagger B_m \tag{2.3}$$

where ΔE, the on-site (intra-site) excitation energy at equilibrium is the same at all sites due to translational symmetry. $D_{l,m}$ is the dispersive interaction matrix element which determines the energy difference between a pair of excited electron and hole at a molecular site and ground state electrons at neighboring sites [5, 11]. $M_{l,m}$ is the electron transfer matrix element between molecular sites at l and m. Using Eq. (2.2), we obtain a Hamiltonian diagonal in K space

$$\hat{H}_{ex} = \sum_{k} E_0(k) B_k^{\dagger} B_k \tag{2.4}$$

$$E_0(k) = \Delta E + \sum_{m \neq 0} D_{0,m} + M_{0,m} \exp(ik.m)$$

where $E(k)$ is energy of the exciton of wavevector k and subscript 0 denotes the absence of lattice site fluctuations. For a simple one-dimensional lattice system, we approximate $M_{l,m} \sim V_{i,j}$ where $V_{i,j}$, the intersite transfer energy is given by

$$V_{i,j} \propto \frac{\mu_i \mu_j}{d^3} \tag{2.5}$$

μ_i the molecular transition moment at site i and d is the separation between the two sites involved during the energy transfer.

2.1 *Site Basis Representation*

It is possible to code excitonic states as qubit states in more than one ways. For instance, Eq. (2.1) may be considered as the result of a single step of a naturally occurring process involving photon absorption. This requires the equivalent of $\log_2(N) + 1$ qubits to store the superposed state. Another approach may involve representing each molecular site by the presence or absence of an exciton, as is the case in a two-level system. Eq. (2.3) can be formulated in terms of the terms of the exciton density matrix as:

$$\rho(t) = \sum_{n=1}^{N} \rho_{nn}(t)|n\rangle\langle n| + \sum_{n<m}^{N} \rho_{nm}(t)|n\rangle\langle m| + \rho_{nm}^{*}(t)|m\rangle\langle n| \tag{2.6}$$

where $|n\rangle$ represents a site basis state where the exciton is present at only the nth molecular site with all other sites unoccupied. The basis state $|n\rangle$ denotes $|0\rangle^{\otimes(n-1)} \otimes |1\rangle \otimes |0\rangle^{\otimes(N-n)}$, where $N \times N$ is the dimension of the global density matrix. $|0\rangle(|1\rangle)$ denotes the absence (presence) of an excitation at a given molecular site. The N-dimensional subspace spanned by vectors with just one qubit in a state that is different from the rest is the simplest subspace of the 2^N-dimensional Hilbert space of N qubits. The hopping exciton can be considered as equivalent to the "single-particle states" configuration in the Hilbert space, and this one-to-one correspondence can be used to evaluate entanglement of a system of N qubits.

A qubit state Ψ associated with $N > 2$ subsystems can be written as generalized Schmidt Decomposition belonging to the Hilbert space of $\hat{H}_{ex}$ in Eq. (2.3)

$$|\Psi\rangle = \sum_{n=1}^{N} c_n \left(|0\rangle^{\otimes(n-1)} \otimes |1\rangle \otimes |0\rangle^{\otimes(N-n)}\right) \tag{2.7}$$

with Schmidt coefficients c_n. The Schmidt Decomposition above is well known for the separability characterization of pure states. For the simpler case of just two molecular sites $N = 2$, the bipartite is separable only if there is one non-zero Schmidt coefficient. The state can be considered maximally entangled if all the Schmidt coefficients are non-zero and equal. The well known Dicke state $|W\rangle = \frac{1}{\sqrt{N}}(|100...0\rangle + |01...0\rangle + ... + |0...01\rangle)$ is symmetric under permutations of each basis state $|n\rangle$ and is thus an equal weighted superposed form of basis states. The absorption of a single photon will give rise to a superposition of all possible states as shown in Eq. (2.1) and also include the special subset of Dicke states.

2.2 *Coding Excitonic Qubits for a Simple Two Lattice Site System*

To illustrate the case of a simple qubit system, we consider two excitons in their ground states at adjacent lattice sites, $R_{\rm a}$ and $R_{\rm b}$. We label the localized excitonic states as $|R_{\rm a}\rangle$ and $|R_{\rm b}\rangle$ and code the excitonic qubits states using the relative position of the exciton via the basis set, $(|L\rangle, |R\rangle)$

$$\begin{aligned} |L\rangle &= |R_{\rm a}\rangle \otimes |0\rangle_b \\ |R\rangle &= |0\rangle_a \otimes |R_{\rm b}\rangle, \end{aligned} \tag{2.8}$$

The states, $|0\rangle_a$ and $|0\rangle_b$, which correspond to the absence of excitons denote the respective ground states of the adjecent lattice sites. We further simplify the approach by considering a two-level system involving only the states $|L\rangle$ and $|R\rangle$. The direct Coulomb interaction which causes the formation of the biexciton state $|R_{\rm a}\rangle|R_{\rm b}\rangle$ is neglected. We also exclude the possibility of entangled states involving the vacuum state, $|0\rangle_a|0\rangle_b$. The two level excitonic qubit Hamiltonian is written as

$$\hat{H}_{\rm ex-qb} = -\hbar(\frac{\Delta\Omega}{2}\,\sigma_z + V\,\sigma_x), \tag{2.9}$$

where the Pauli matrices $\sigma_x = |L\rangle\langle R| + |R\rangle\langle L|$ and $\sigma_z = |L\rangle\langle L| - |R\rangle\langle R|$. $\Delta\Omega = \Omega_{\rm a} - \Omega_{\rm b}$ is the difference in exciton creation energy between the lattice site at $R_{\rm a}$ and that at $R_{\rm b}$. V denotes the intersite interaction amplitude responsible for the transfer of an exciton from one lattice site to the other without involving a tunnelling process.

The eigenstates of the interacting qubit system in Eq. 2.9 appear as

$$|\chi_{\rm s}\rangle = \cos(\beta/2)\,|L\rangle \; + \sin(\beta/2)\,|R\rangle \qquad (2.10)$$
$$|\chi_{\rm as}\rangle = \sin(\beta/2)\,|L\rangle \; - \cos(\beta/2)\,|R\rangle\,,$$

with eigenenergies

$$E_{\rm as(s)} = \Omega_0 + \Omega_1 - \frac{\Delta\Omega}{2} \mp \sqrt{(\frac{\Delta\Omega}{2})^2 + F^2} \qquad (2.11)$$

Ω_0 denotes the ground state energy of the system in which both the lattice sites are not occupied by excitons while Ω_1 denotes a higher energy level of the excitonic qubit system. The energy difference between the eigenstates is $\sqrt{\Delta\Omega^2 + (2V)^2}$ and can be studied as a function of time. In the absence of any decoherence process, the excitonic qubit oscillates coherently between the two lattice sites with the Rabi frequency $E_{\rm s} - E_{\rm as}$. The polar angle β in the Bloch sphere representation of a qubit is related to $\Delta\Omega$ and V via

$$\tan\beta = \frac{2V}{\Delta\Omega} \qquad (2.12)$$

3. The von Neumann Entropy

The local density matrix $\rho_n = u_n|1\rangle_{nn}\langle 1| + (1 - u_n)|0\rangle\langle 0|$ can be used to determine entanglement dynamics of the nth qubit associated with a basis which spans the two dimensional space ($|0\rangle$ and $|1\rangle$). The occupation probability at the nth molecular site is given by [15] $u_n = \langle K|B_n^\dagger B_n|K\rangle$ where B_n is the exciton creation operator localized at site n and $|K\rangle$ is the exciton eigenvector in K-space (see Eq. (2.1)). The reduced density matrix has a simple form

$$\rho_n = \begin{pmatrix} 1 - u_n & 0 \\ 0 & u_n \end{pmatrix}. \qquad (3.1)$$

The well known von Neumann entropy measure [28, 38] appear as

$$S_n = -{\rm Tr}_n \rho_n \ln \rho_n \qquad (3.2)$$

where S_n which is zero for pure states, is obtained by tracing over the degrees of freedom of the remaining $N - 1$ molecular sites. The average von Neumann entropy measure is then given by

$$S = \frac{1}{N}\sum_{n=1}^{N} -{\rm Tr}_n \rho_n \ln \rho_n \qquad (3.3)$$

For the ideal delocalized state in Eq. (2.1), we obtain $S = \frac{1}{N} \ln N$ at very large N. This means that larger the entropy, greater is the entanglement between excitation at a given site with all other remaining sites. On the other hand, $S = 0$ for the localized state and hence,s is a useful measure in characterizing quantum phase transitions at the localization-delocalization borders. For the case of an extended state in a finite system of size N, we obtain $S = \frac{1}{N} \ln N - (1 - \frac{1}{N}) \ln(1 - \frac{1}{N})$.

3.1 *Wootters Concurrence*

Here we briefly examine the Wootters [39] concurrence which is a well known measure of entanglement between two qubit systems. We consider a composite state of the exciton at a lattice site interacting with a reservoir of phonons in the initial vacuum state as follows

$$|1\rangle_{\rm ex} \otimes |0\rangle_{\rm p} \longrightarrow u(t)\ |1\rangle_{\rm ex} \otimes |0\rangle_{\rm p} + v(t)\ |0\rangle_{\rm ex} \otimes |1\rangle_{\rm p}\,, \tag{3.4}$$

where $|0\rangle_{\rm p}$ denotes the reservoir state with zero phonon occupation at allowed modes and $|1\rangle_{\rm p}$ denotes the presence of a single photon at any allowed mode. The functions $u(t)$ and $v(t)$ are assumed to satisfy the relation $u(t)^2 + v(t)^2 = 1$.

The joint evolution of a pair of excitons at neighbouring sites interacting with two uncorrelated reservoir of phonons has been examined [46] in an earlier work via the reduced density matrix of the composite bipartite system $e_1 e_2$ within the basis $(|0\ 0\rangle\,, |0\ 1\rangle\ |1\ 0\rangle\ |1\ 1\rangle)$

$$\rho_{\rm e1,e2}(t) = \begin{pmatrix} f_1(t) & f_5(t) & f_6(t) & f_7(t) \\ f_5(t) & f_2(t) & f_8(t) & f_9(t) \\ f_6(t) & f_8(t) & f_3(t) & f_{10}(t) \\ f_7(t) & f_9(t) & f_{10}(t) & f_4(t) \end{pmatrix}. \tag{3.5}$$

where e_i (i=1,2) correspond to the exciton pair at separate lattice sites. It can be shown that for $t \geq 0$, the matrix elements evolve as

$$\begin{aligned}
&f_1(t) = a^2 + b^2\ v_1(t)^2\ v_2(t)^2 + c^2\ v_2(t)^2 + d^2\ v_1(t)^2, \\
&f_2(t) = b^2\ v_1(t)^2\ u_2(t)^2 + c^2\ u_2(t)^2, \\
&f_3(t) = b^2\ u_1(t)^2\ v_2(t)^2 + d^2\ u_1(t)^2, \\
&f_4(t) = b^2\ u_1(t)^2\ u_2(t)^2, \quad f_5(t) = a\ c\ u_2(t), \\
&f_6(t) = a\ d\ u_1(t) + b\ c\ u_1(t)\ v_2(t)^2, \\
&f_7(t) = a\ b\ u_1(t)\ u_2(t), \quad f_8(t) = c\ d\ u_1(t)\ u_2(t), \\
&f_9(t) = b\ c\ u_1(t)\ u_2(t)^2, \quad f_{10}(t) = b\ d\ u_1(t)^2\ u_2(t).
\end{aligned}$$

where u_1, v_1 are conversion functions associated with the first exciton and u_2, v_2 are associated with the second exciton. The coefficients a, b, c, d are associated respectively, with the basis states, $(|0\,0\rangle, |0\,1\rangle\,|1\,0\rangle\,|1\,1\rangle)$. For $c = d = 0$, reduced bipartite density matrices $\rho_{e1,e2}$ assume a simple form with the well-known X-state structure which preserve its X-form during evolution.

The Wootters concurrence [39] for the appropriate density matrix can be evaluated using $\mathcal{C}(t) = \max\{0, \sqrt{\lambda_1} - \sqrt{\lambda_2} - \sqrt{\lambda_3} - \sqrt{\lambda_4}\}$ where λ_i are eigenvalues in decreasing order of of the Hermitian matrix $\widetilde{\rho} = \rho\left(\sigma_y^1 \otimes \sigma_y^2\right)\rho^*\left(\sigma_y^1 \otimes \sigma_y^2\right)$ where σ_y belongs to the set of Pauli matrices. ρ^* denotes the complex conjugation of ρ in the standard basis Eq. (3.5). For density matrix given in Eq. (3.5) we obtain for $c = d = 0$, $\mathcal{C} = \max\{0, 2u_1u_2(|ab| - b^2v_1v_2\}$. Detailed analysis of the evolution of $\mathcal{C}$ under various conditions has been discussed in an earlier work on the concurrence measure in entangled systems [46].

In the single particle-state picture, juts one particle is ocnsidered to propagate along the N sites of a one-dimensional chain. The Wootters concurrence in this case is obtained by calculating the entanglement among the N sites. The state in which the pairwise concurrence is at its minimum (i.e. $\mathcal{C}_{\min} = \frac{2}{N}$) is at the point of highest entanglement. Hence a completely delocalized exciton state of the J-aggregate system is a highly entangled state, which may account for many appealing properties of this system.

3.2 *The Meyer– Wallach Measure*

A measure that is closely related to the von Neumann entropy is the Meyer-Wallach measure. This entity extends the analysis of quantum correlations beyond the standard pairs of qubit systems to a general N pair. This monotone measure was defined by Meyer and Wallach [32] as a single scalar measure of pure state entanglement for the three and four-qubit cases

$$Q = \frac{1}{n}\sum_{k=1}^{n} 2(1 - \mathrm{Tr}[\rho_k^2]), \tag{3.6}$$

where ρ_k is the reduced density matrix of the kth qubit obtained after tracing out all the remaining qubits. However the Meyer-Wallach measure has inherent deficiencies in that it is unable to distinguish states which are fully inseparable from states which are separable into states of some set of subsystems. Nevertheless it is a useful tool to utilize in the investigation of qualitative features in entangled systems.

For density matrix given in Eq. (3.5) we obtain for $c = d = 0$, an explicit expression of the Meyer–Wallach measure of N pairs of qubit-reservoir systems

$$Q = 2a^2b^2 + 4b^2u^2v^2 \tag{3.7}$$

where we have assumed that the conversion function $u_i = u$ for all the N pairs. The maximum $Q = 1$ is obtained for $a = 0, b = 1$ with $u_1 = u_2 = \frac{1}{\sqrt{2}}$. For the case of N pairs of equivalent qubit-reservoir systems for which $u_i = u_1$ and S pairs of equivalent qubit-reservoir systems for which $u_i = u_2$, $(u_1 \neq u_2)$ we obtain

$$Q = \frac{2N}{N+S}(a^2b^2 + 2b^2{u_1}^2{v_1}^2) + \frac{2S}{N+S}(a^2b^2 + 2b^2{u_2}^2{v_2}^2) \tag{3.8}$$

For the case of two excitons at two lattice sites interacting with uncorrelated reservoirs of phonons, we obtain the maximum $Q = 1$ for $a = 0, b = 1$ with $u_1 = u_2 = \frac{1}{\sqrt{2}}$ for the following state

$$|\Phi\rangle_m = \frac{1}{2}\left[|11\rangle_{\rm ex}|00\rangle_{\rm r} + |00\rangle_{\rm ex}|11\rangle_{\rm r} + |01\rangle_{\rm ex}|10\rangle_{\rm r} + |10\rangle_{\rm ex}|01\rangle_{\rm r}\right] \tag{3.9}$$

This state involves all possible combinations of the quantum state where either the qubit or its corresponding reservoir is occupied but not both at the same time. For $a = b = \frac{1}{\sqrt{2}}$, we obtain $Q = \frac{1}{2}(1+u_1^2-u_1^4+u_2^2-u_2^4)$, and a constant value of $Q = 0.5$ is obtained for $(u_1 = 1, u_2 = 0)$, $(u_1 = 0, u_2 = 1)$, $(u_1 = 1, u_2 = 1)$ or $(u_1 = 0, u_2 = 0)$ partitions.

3.3 *Quantum Discord and Geometric Measure of Quantum Discord*

Unlike the entanglement measures described earlier, the quantum discord have several unique and puzzling features, and incorporates more generalized correlations not present in the entanglement measures such as concurrence. For instance non-zero positive discord states can be mixed to produce zero-discord states, and two zero-discord classical states can be combined to form a non-zero discord state.The monogamy rule which is an important feature of the concurrence measure during entanglement sharing is not a requirement in states with non-zero quantum discord. Hence the quantum discord is seen to offer great advantages in its applicability for a wide range of systems such as in spin array and quantum communication systems. The quantum discord can be understood by using an interpretation based on classical analogies.

The correlation between two random classical variables A and B can be defined via the concept of mutual information, $\mathcal{I}(A:B)$:

$$\mathcal{I}(A:B) = H(A) + H(B) - H(A,B) \tag{3.10}$$

where $H(A)$ is Shannon entropy and $H(A,B)$ is the joint Shannon entropy which is defined by $H(A,B) = -\sum_{a,b} p_{a,b} \log_2 p_{a,b}$. $H(A|B)$ is the conditional entropy which quantifies the ignorance of A given known values of B.

The quantum analogy to Eq. (3.10) can be obtained by making use of density matrices and the von Neumann entropy defined in earlier sections. The total classical and quantum correlations of the bipartite density matrix in Eq. (3.5) is quantified by a measure known as quantum mutual information defined as

$$\mathcal{I}(\rho_{\mathrm{e1,e2}}) = S(\rho_{\mathrm{e1}}) + S(\rho_{\mathrm{e2}}) - S(\rho_{\mathrm{e1,e2}}) \tag{3.11}$$

where $S(\rho_{\mathrm{x}})$ is the von Neumann entropy as defined in Eq. (3.2). Quantum mutual information may be written as [35–37]

$$\mathcal{I}(\rho_{\mathrm{e1,e2}}) = \mathcal{C}(\rho_{\mathrm{e1,e2}}) + \mathcal{Q}(\rho_{\mathrm{e1,e2}}) \tag{3.12}$$

where $\mathcal{C}$ is the classical correlation and $\mathcal{Q}$ is the quantum discord [35]. In Eq. (3.12), the classical correlation term $\mathcal{C}$ can be translated as the maximum information that can be obtained about one subsystem e_1 by performing measurements on the other subsystem e_2. The quantum discord therefore vanishes for states with only classical correlations. By virtue of the different outcomes of measurement procedures on distinctly separate systems, $\mathcal{Q}$ is not a symmetric quantity. The value that can be obtained for $\mathcal{Q}$ therefore depends on percularities of the measurement procedure on each of the two subsystems. The mathematical expression for $\mathcal{Q}$ for the bipartite state $\rho_{\mathrm{e1,e2}}$ in Eq. (3.5) is given by [47]

$$\mathcal{Q}(\rho_{\mathrm{e1,e2}}) := \min_{\prod^{e1}}\{\mathcal{I}(\rho_{\mathrm{e1,e2}}) - \mathcal{I}[\prod{}^{e1}(\rho_{\mathrm{e1,e2}})]\} \tag{3.13}$$

The minimum in the above expression is taken over von Neumann measurements on subsystem e_1 where $\prod^{e1}(\rho_{\mathrm{e1,e2}})$ is the state produced after the measurement is performed. The measurements proceed via a process denoted by $\prod^{e1} = \{\prod_k^{e1}\}$ involving orthogonal projectors. Thus the quantum discord is interpreted [47] as the minimum loss of quantum mutual information due to measurement.

A measure closely related to the quantum discord is the geometric measure of quantum discord and is given by [48]

$$D^g(\rho) = \min_{\chi \in \Omega} \parallel \rho - \chi \parallel^2 \tag{3.14}$$

The minimum in the expression above is taken over Ω which denotes the set of zero discord states and $\parallel \chi \parallel^2 := \mathrm{Tr}(\chi^\dagger \chi)$ is the square of the Hilbert-Schmidth norm of Hermitian operators. The geometric measure of quantum discord for two-qubit states with the X-state structure can be evaluated in an analytical form following the explicit expression given in Ref. [48]. The quantum discord, and its geometric measure, has rich possibilities and potential applications in a non-markovian open quantum system such J-aggregate systems. A detailed examination of the entanglement dynamics of J-aggregate systems via these measures is yet to be performed. Accordingly, results of the quantum discord present in a simple linear chain will be shown in a forthcoming section.

4. The Green's Function Formalism

We begin with the following crystal Hamiltonian [5]

$$\hat{H}_T = \hat{H}_{ex} + \hat{H}_p + \hat{H}_{ep1} + \hat{H}_{ep2} \tag{4.1}$$

$$\hat{H}_p = \sum_q \hbar\omega(q) b_q^\dagger b_q \tag{4.2}$$

$$\hat{H}_{ep1} = N^{-1/2} \sum_{k,q} \left[F(k,q) + \chi(q)\right] B_{k+q}^\dagger B_k (b_{-q}^\dagger + b_q) \tag{4.3}$$

$$\hat{H}_{ep2} = N^{-1/2} \sum_{k,q} \chi(q) B_k^\dagger B_k (b_{-q}^\dagger + b_q) \tag{4.4}$$

where $\hat{H}_{ex}$ is given in Eq. (2.4), $\hat{H}_p$ denotes the phonon energies and $b_q^\dagger(b_q)$ is the creation (annihilation) phonon operator with frequency $\omega(q)$ and wavevector q. Explicit forms for the coupling functions $F(k,q)$ and $\chi(q)$ are given in Ref. [5]. It is important to note the difference in the two coupling functions, while $\chi(q)$ acts to localize an exciton at its original occupation site, $F(k,q)$ operates during dispersion of an exciton to neighboring sites. Thus $F(k,q)$ is influenced by the same factors which determine the exciton bandwidth. In situations where $\chi(q) >> F(k,q)$, self-trapped excitonic effects are expected to dominate while in the case of comparable coupling strengths $\chi(q) \sim F(k,q)$, coexistence of free and trapped excitons and localization effects are expected to prevail. Due to the finite time associated with lattice vibrations, nonlocal effects which govern entanglement properties of the delocalized exciton are likely to experience decoherence effects due to the first coupling term in Eq. (4.3). At very high neighbor transfer energies J, the second dispersive term in Eq. (4.4) acts as a small perturbation on exciton dynamics. We now investigate the effect of these different coupling terms using Green's functions formalism introduced by Davydov [5].

The Green's function for an exciton at time t is given by [5, 49]

$$G(k,t) = -i\langle 0 : n_q | T\{B_k(t) U(t,0) B_k^\dagger(0)\} | 0 : n_q\rangle \tag{4.5}$$

$$B_k(t) = B_k(0)\exp(iE(k)\,t) \tag{4.6}$$

where $|0 : n_q\rangle$ denotes a state with n_q as the population of phonons with wavevector q and zero exciton population. T is the time ordering operator and $U(t,0)$ is the Dyson function

$$U(t,0) = T\exp\left(-i\int_0^t \exp(i\hat{H}_T t')\left[\hat{H}_{ep1} + \hat{H}_{ep2}\right]\exp(-i\hat{H}_T t')dt'\right) \tag{4.7}$$

where $\hat{H}_T$, $\hat{H}_{ep1}$ and $\hat{H}_{ep2}$ are given in Eqs. (4.2), (4.3) and (4.4) respectively. A direct relation between the Green's function $G(k,t)$ in Eq. (4.5) and exciton energy $E(k)$ has been derived by Craig et al. (see Eqs. (4.12) and (4.13) in Ref. [50]) at times larger than the inverse of phonon frequencies,

$$G(E(k),t) = -i\exp\left[-i\left(E(k) + \Delta(k) - N^{-1}\sum_q \frac{|\chi(q)|^2}{\omega(q)^2}\right)t\right] \tag{4.8}$$

$$\times\exp[-\gamma(k)]\exp(-\Lambda) \tag{4.9}$$

where

$$\Delta(k) = N^{-1}\sum_q |F(k,q)|^2\left[\frac{\bar{n}_q}{\Omega^-_{k,q}} - \frac{\bar{n}_q + 1}{\Omega^+_{k,q}}\right] \tag{4.10}$$

$$\gamma(k) = N^{-1}\sum_q |F(k,q)|^2\left[\bar{n}_q\,\delta(\Omega^-_{k,q}) + (\bar{n}_q + 1)\,\delta(\Omega^+_{k,q})\right] \tag{4.11}$$

$$\Omega^\pm_{k,q} = \hbar\omega(q) \pm (E(k+q) - E(k)) \tag{4.12}$$

$$\Lambda = N^{-1}\sum_q \frac{|\chi(q)|^2}{\omega(q)^2}(2\bar{n}_q + 1) \tag{4.13}$$

where $\Delta(k)$ yields the principal value of the term associated with it, and the mean phonon occupation number at temperature T is given by $\bar{n}_q = \{\exp[\hbar\omega(q)/k_B T] - 1\}^{-1}$. We note that Λ is independent of time unlike the other terms, however this only applies in the Markovian approximation.

An expression which is valid for smaller time t and at low temperatures has been obtained as [15, 50]

$$G(E(k), t) = -i \exp\left[-i \left(E(k) + \Delta(k) - \sum_q \frac{|\chi(q)|^2}{N\omega(q)^2}\right) t\right] \tag{4.14}$$

$$\exp[-\gamma(k)t]$$

$$\times \prod_q \sum_{m_1=0}^{\infty} \sum_{m_2=0}^{\infty} \left(\frac{|F(k,q)|^2}{N(\Omega^-_{k,q})^2}\right)^{m_1} \left(\frac{|F(k,q)|^2}{N(\Omega^+_{k,q})^2}\right)^{m_2} \frac{1}{m_1! m_2!}$$

$$\times \exp\left[i(m_1\Omega^-_{k,q} - m_2\Omega^+_{k,q})t\right]$$

where m_1, m_2 are integers. A rigorous treatment in the non-Markovian range using Eq. (4.14) involves intense numerical computations and therefore will not be considered here. Nevertheless the Green's function $G(k,t)$ in Eq. (4.8) will suffice to reveal vital difference in the roles played by the two coupling functions, $\chi(q)$ and $F(k,q)$ in influencing the times of coherence between lattice sites.

The dynamics of a free particle in a one-dimensional system is determined entirely by the Green function $G_{n,m}(t) = \langle n|G(t)|m\rangle$ [51]. which yields the amplitude for an excitation to move from site m at $t = 0$ to site n at time t. In a one-dimensional lattice with nearest-neighbor interaction energy $2V\cos(K)$ and zero lattice vibrations, $G_{n,0}(t)$ is given by [15]

$$G_{n,0}(t) = N^{-1} \sum_k e^{ikn} \exp\left[-i\frac{2Vt}{\hbar}\cos(K)\right] \tag{4.15}$$

Accordingly, the probability $P_{n0}(t)$ that an excitation which originates at site 0 at $t = 0$ and appears at site n at time t can be obtained using $P_{n0}(t) = |G_{n,0}(t)|^2$. To calculate the probability $P_{n0}(t)$ for the exciton in the presence of dispersion and resonance terms $\chi(q)$ and $F(k,q)$ quantifying the exciton-phonon coupling, we use the Green's function $G(E(k),t)$ in Eq. (4.8). For very large $N \to \infty$, we obtain [15]

$$P_{n0}(t) \approx N^{-1} \mathrm{J}_n^2(\frac{2V}{\hbar}t) \exp(-\Lambda) \sum_k e^{-\gamma(k)t} \tag{4.16}$$

where $\mathrm{J}_n(x)$ is Bessel function of the first kind and $\gamma(k)$ is given in Eq. (4.11). In the absence of lattice vibrations, $P_{n0}(t)$ reduces to the well-known form obtained by Merrifield [52] for an excitation propagating on infinite aggregate systems. We have assumed that the transfer interaction term V is independent of the environmental variables associated with lattice vibrations, also known as Condon approximation in Eq. (4.16). Recent experimental results [13] have shown that electron-phonon interactions between off-diagonal elements modulate the couplings between the dipoles, resulting in decoherence of the transfer interaction term V. Here we assume that the transfer interaction term V remains intact, and Eq. (4.16) is a reasonably accurate description of the exciton propagation in the presence of lattice vibrations.

The subtle differences in influence of the coupling functions, $\chi(q)$ and $F(k,q)$ on the probability transfer function $P_{n0}(t)$ is evident from Eq. (4.16). $\chi(q)$ tends to localize the excitation at the original site of excitation rather than introduce decoherence during propagation while $F(k,q)$ retards the tendency for the exciton to move further away from the initial site of propagation during propagation. Eq. (4.16) can be rewritten in terms of dimensionless arguments

$$P_{n0}(t) \approx N^{-1} \mathrm{J}_n^2(ct) \exp(-a^2) \sum_k e^{-b^2 t} \tag{4.17}$$

where $c = \frac{2V}{\hbar\omega(q)}$, $a = \frac{\chi(q)}{\hbar\omega(q)}$ and $b = \frac{F(k,q)}{\hbar\omega(q)}$. Time t is dimensionless as it is scaled by the factor $\omega(q)^{-1}$. It is to be noted that $t > 1$ describes the Markovian limit reasonably well. At the Markovian limit, the flow of information from phonons back to the excitonic system can be neglected as the exciton hopping times are large compared to the phonon reservoir correlation times. This means that two propagating exciton states cannot be distinguished as they remain unaffected by correlations within the phonon reservoir at times larger than phonon correlation times.

In an earlier work [15], explicit results based on Eq. (4.16) were obtained by making a few approximations. Firstly the dispersion in phonon frequencies was neglected so that a single frequency, $\omega(q) \sim 50\text{cm}^{-1}$, representative of the lowest vibrational branch in common molecular crystals [50, 53] is considered. The band averages of the two coupling functions, $\chi(q)$ and $F(k,q)$ are then varied via parameters a and b (given below Eq. (4.17)), together with the electronic transfer energy, quantified by c. The von Neumann entropy measure of the system using $S = \sum_{n=1}^{N} -\mathrm{Tr}_n \rho_n \ln \rho_n$ which enables one to examine the influence of the coupling functions obtained in Eq. (4.17).

Figure 4.1a shows the results in which S versus time t was obtained using Eq. (3.3) for various values of N, and for an initial state, $P_{n0}(0) = \delta_{n,0}$ [15]. At time t =0, average entropy S =0 due to presence of excitation at just one site, however the entropy rises rapidly reflecting the spread of excitation to more sites with constant velocity. Due to the constraints of a finite system, the propagating exciton leaves the system under study after a critical time. Hence the entropy S decreases in all three cases of N, with the decrease most pronounced at small N where there faster exit of the exciton. The parameters a and b are set to zero in Fig. 4.1(a), consistent with the absence phonon-related decoherence processes. This implies that the noticeable drop in S is indeed due to the finite size effects of our chosen model. Results in Fig. 4.1 (b), (c) show the localization tendencies associated with the exciton-phonon interaction, with the fall in S is noticeably steep for high values of b or the coupling function $F(k, q)$. Fig. 4.1(c) [15] shows that high electronic transfer energies or c lead to fast rise in S in the time domain ($0 < t < 6$), which however is not sustained as excitation leaves the system as quickly as it spreads. For lower c, the excitation remains extended for comparatively longer times with noticeable oscillations associated with "death and birth" type excitation and de-excitation process [46]. In Fig. 4.1(d), we note that entropy S of the extended state is higher than all other S values. The results indicate that at large N, the entanglement persist for longer times and S is almost comparable to that of a fully extended system. Figs. 4.1 (a)−(d) based on an earlier work [15] show the interplay of three competing effects: (1) the increase of S due to spreading; (2) subsequent decrease of S due to finite size effects and, (3) decrease of S due to coupling with phonons. The results suggests the large scale energy transport abilities of one-dimensional excitonic systems and highlight the prominence of the J-band in systems of large N. The results also show the usefulness of the entropy measure in capturing the transition from coherent to incoherent energy transfer.

It is expected that similar transition regimes as seen in Figs. 4.1 (a)−(d) can be also observed with other entanglement measures discussed in the earlier sections. Thus far the influence of two exciton phonon coupling terms on the von Neumann entropy measure has been analysed using a simplified model, in which a single excitation is considered to propagate to neighboring sites. In reality, the electronic excitation arises from a single site but does not originate from all N molecular sites of the crystal. Such a model requires an entanglement measure which is able to incorporate the superposition of several lattice sites which are the source of an excitation. A full superposition of amplitudes from all possible sites have to be then taken into account with time as an added dimension to obtain an accurate picture of the excitation dynamics. In order to simplify analysis of the full model where all sites have equal probability of being excited, we consider the Wootters concurrence [39] in the next section.

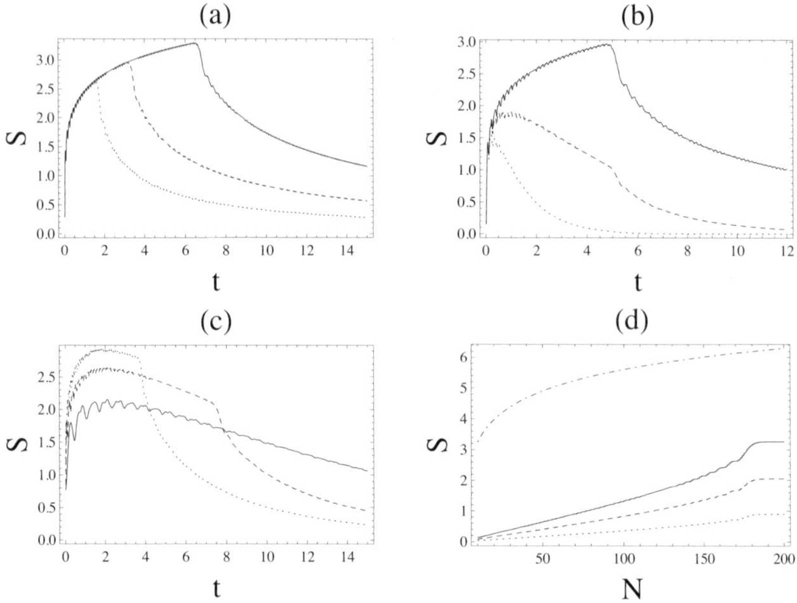

Figure 4.1. (a) The von Neumann entropy S versus time t [15] at $N = 200$ (solid line), $N = 100$ (dashed) and $N = 50$ (dotted) at $a = b = 0$ and $c = 30$. (b) The von Neumann entropy S versus time t at $N = 100$, $a = 0$ and $c = 20$. The three curves correspond to (from top to bottom) $b = 0$, 0.5 and 1. (c) The von Neumann entropy S versus time t at $N = 200$, $a = 0.5$ and $b = 0.3$. The three curves correspond to $c = 40$ (dotted), 20 (dashed) and 5 (solid line). (d) The von Neumann entropy S versus time N at $b = 0$ (solid line), $b = 0.3$ (dashed) and $b = 0.5$ (dotted) at $a = 0$, $c = 30$ and $t = 6$. Dotted-dashed line corresponds to a fully extended state for which $S = \ln N - (N-1)\ln(1 - \frac{1}{N})$ [15] (Copyright, 2011, Institute of Physics).

5. Molecular Cooperativity as a Paradigm of Collective Entanglement

As mentioned earlier, a notable optical response exhibited by one-dimensional excitonic systems is their enhanced rate of spontaneous emission compared to that of the monomer, an effect known as superradiance [54–56]. Mukamel and coworkers [57–60] have shown a connection between cooperative emission mechanisms such as superradiance and "molecular co-operativity", a term used to describe the effective coherent dynamics of a system of delocalized Frenkel excitons. The concept of a group of molecules acting in tandem to produce enhanced features in $\chi^{(3)}$, the third order optical response of one dimensional molecular systems was also shown in an earlier work by Hanamura and coworkers [55]. The oscillator strength of the $K = 0$ exciton state is directly proportional to the number of molecular sites in the absence of decoherence mechanisms. This is the result of the involvement of a large number of molecular dipoles which act collectively in phase to produce giant oscillator strengths. This raises the possibility of harnessing $\chi^{(3)}$ as a quantitative measure of multipartite entanglement. The superradiance of one-dimensional excitonic systems can be interpreted in terms of the entanglement measure, Wootters concurrence [39] (see Sec. 3.1) which provides a measure of entanglement between a pair of qubits.

The average pairwise concurrence $\langle C \rangle$ [61, 62] associated with a system of single-particle states is given by [15]

$$\langle C \rangle = \frac{2}{N(N-1)} [\zeta - 1] \tag{5.1}$$

ζ is a measure of delocalization and is also known as inverse participation ratio (IPR) [63]. Small values of ζ correspond to localized states while larger values are linked with delocalization. At the extreme limit of localization at a single site $\zeta = 1$, $\langle C \rangle = 0$, while for states which are delocalized completely, $\zeta = N$ which yields the maximum entanglement of $\frac{2}{N}$. The scaled concurrence is obtained by dividing Eq. (5.1) with $\frac{2}{N}$ and we obtain

$$\langle C \rangle_s = \frac{1}{(N-1)} [\zeta - 1] \tag{5.2}$$

For extended excitonic system of J-aggregate structures,ζ takes on a similar role as the characteristic coherence size N_c [15]. N_c has been formulated in terms of the density matrix as [60]

$$N_c = \left[N \sum_{mn} |\rho_{mn}|^2 \right]^{-1} \left(\sum_{mn} |\rho_{mn}| \right)^2 \tag{5.3}$$

where N is the total number of molecules or monomers and ρ_{mn} is the reduced density matrix associated with sites m and n. N_c provides an approximate measure by which exciton at site n is entangled with site m, and decays monotonically as a function of $(n - m)$. In the absence of any entanglement with occurence of extreme localization, $\rho_{mn} = N^{-1}\delta_{nm}$, $N_c = 1$ and the scaled concurrence $\langle C \rangle_s = 0$. In the limit of complete delocalization, $\rho_{mn} = N^{-1}$, $N_c = N$ which yields the maximum $\langle C \rangle_s = 1$.

Spano et al. [59] has determined the coherence size N_c using a microscopic definition $G(0, t) \sim \exp(-N\ \gamma_r t)$ where $N\ \gamma_r$ is the radiative rate. In the superradiant limit, excitation associated with individual sites can be taken as acting collectively to yield the unique features of the molecular aggregate such as the giant nonlinear optical properties [56]. In particular, Spano et al. [59] obtained an empirical relation relating N_c and the coupling function $F(k,q)$, temperature T, nearest neighbor transfer energy V and phonon energy, $\hbar\omega(q)$. The empirical relation obtained by Spano et al. [59] can be rewritten in terms of the dimensionless constants b and c (given below Eq. (4.17)) as [15]

$$N_c \approx 2.16 \left(\frac{V^2\, \hbar\omega(q)}{F^2(k,q)\, k_B T} \right)^{\frac{1}{3}} = 2.16 \left(\frac{c^2}{b^2 t} \right)^{\frac{1}{3}} \tag{5.4}$$

where the dimensionless temperature term, $t = \frac{k_B T}{\hbar\omega(q)}$.

Using the defination of concurrence in Sec. 3.1, we can evaluate the concurrence of a linear chain of monomers for a given set of parameters, a, b and c. Fig. 5.1 shows change of the concurrence with time for aggregate system with different speed of propagation c in which number of sites $N = 150$. The figures has features similar to those in Fig. 4.1(c). A sharp rise is noted for the system with the highest speed of energy propagation, which however is not maintained as the excitation leaves the system as quickly as it spreads. It is to be noted that these results are specific to a model of a one dimensional excitonic chain with a finite number of lattice sites. As noted in Fig. 4.1(c), we note that at lower c, the excitation remains extended for comparatively longer times, however the oscillations associated with “death and birth” type excitation and de-excitation process is not prominent in Fig. 5.1.

The quantum discord defined in Sec. 3.2 is evaluated for the density matrixassociated with two selected lattice sites for the linear monomer system, and for a given set of parameters, a, b and c. Fig. 5.2 illustrates the change of the quantum discord with time two monomer sites ($N =$ 0, 8) of the density matrix associated with. The figure highlights the distinct nature of the quantum discord as compared to the von Neumann entropy or concurrence measures. Specifically, the quantum discord stands out as a unique entity as it remains almost almost intact as time increases.There are noticeable oscillations during the initial phase of propagation of the excitation possibly due to excitation/de-excitation processes occurring at a lattice site. Importantly, results the in Fig. 5.2 highlights the potential of J-aggregates to be utilized as channels of non-classical correlations, with the possibility of being harnessed for their superior optical properties.

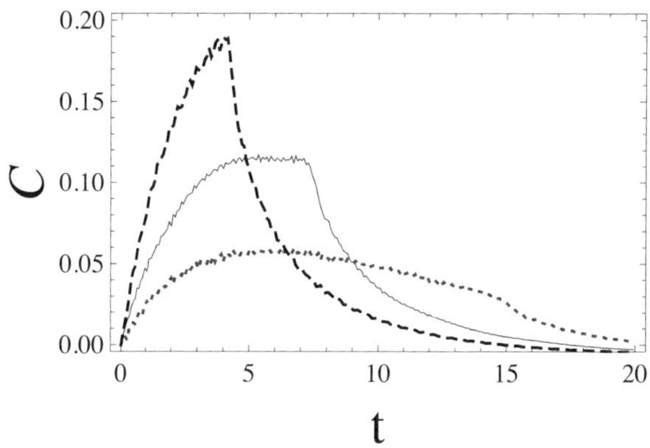

Figure 5.1. Concurrence C versus time for $N = 150$, $a = 0.2$ and $b = 0.4$. The three curves corresponds to $c = 10$ (dotted), 20 (solid line) and 35 (dashed).

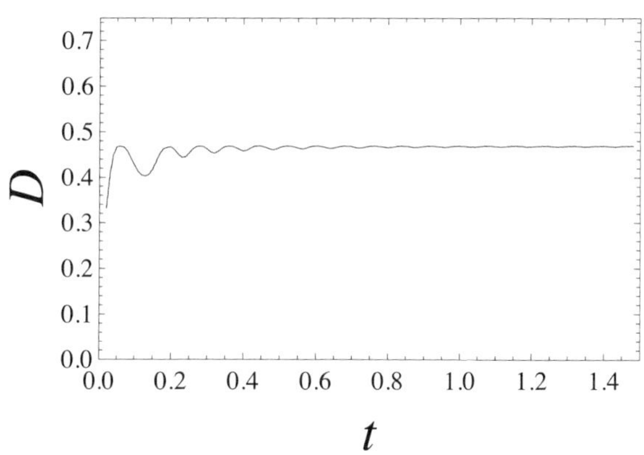

Figure 5.2. Quantum discord D between the excitation at initial site ($N = 0$) and the $N = 8$ site as a function of time. We use $a = 0.04$, $b = 0.02$ and $c = 40$ and the initial coupling parameter is set at $\frac{1}{\sqrt{10}}$.

6. Theory of Two-exciton States

In this section, we examine the optical features of J-aggregates which present experimentally quantifiable measures of multipartite entanglement even if dephasing mechanisms [25] are present. The class of entanglement quantity known as the Wei-Goldbart geometric measure of entanglement [40] may be convenient in analysing multipartite entanglement of J-aggregates based on properties of the third order optical susceptibility, $\chi^{(3)}$. An important feature related to non-linear optical properties of material systems is the existence of one- and two-exciton states. One-exciton states are easily identified as states in which an excitation is present at any one of the N monomers while in a two-exciton state, any two of N monomers are excited while the other monomers remain in the ground state. The two-exciton state here is not the same as the biexciton state which consists of the two coupled exciton pair system that propagates throughout the crystal as a single entity.

Two-exciton states may be identified with entangled states formed due to superposition of two spatially uncoupled excitons, such states having twice the bandwidth of the usually considered one-exciton states. This can be shown by writing the diagonal Hamiltonian of the two-exciton state in K space (without involvement of phonon operators) as [15]

$$\hat{H}_{ex} = \sum_k E_1(k) B^\dagger_{1,k} B_{1,k} + E_2(k) B^\dagger_{2,k} B_{2,k} + T_k(ij) \tag{6.1}$$

$$T_k(ij) = \sum_{m \neq n} \exp\left[ik.(mi - nj)\right] V_{mn}$$

where $E_1(k)$ is energy of the exciton (labelled by subscript 1) with wavevector k and $E_2(k)$ is energy of the second exciton (labelled by subscript 2) with wavevector k. The exciton operators $B^\dagger_{1,k}$, $B_{1,k}$ retain the same meaning as used in earlier sections. V_{mn} denotes the interaction energy coupling of transition moments for molecules located at m, n. Equation (6.1) can be diagonalized further in terms of the exciton operators

$$\begin{aligned} \chi_a(k) &= \cos\beta(k) B^\dagger_{1,k} + \sin\beta(k) B^\dagger_{2,k} \\ \chi_b(k) &= -\sin\beta(k) B_{1,k} + \cos\beta(k) B_{2,k} \end{aligned} \tag{6.2}$$

we obtain

$$\hat{H}_{ex} = \sum_k E^*_a(k) \chi_a(k)^\dagger \chi_a(k) + E^*_b(k) \chi_b(k)^\dagger \chi_b(k) \tag{6.3}$$

where $E^*_a(0)$ and $E^*_b(0)$ are exciton energies associated with Davydov branches at $k = 0$. The asterisk in the superscript denotes energies which are split due to the intermolecular interaction. Equation (6.3) shows that two-exciton states are highly entangled with the degree of entanglement determined by $\cos\beta(k)$ and $\sin\beta(k)$. We next make a connection between two-exciton states and generalized symmetric states via the third order optical susceptibility, $\chi^{(3)}$.

In the presence of lattice vibrations, Eq. (6.3) is modified by the inclusion of the following phonon operators

$$\begin{aligned} \varphi_1(q) &= b_1(-q)^\dagger + b_1(q) \\ \varphi_2(q) &= b_2(-q)^\dagger + b_2(q) \\ \eta(q) &= b_1(q) + b_2(q) \\ \eta(q)^\dagger &= b_1(-q)^\dagger + b_2(-q)^\dagger \end{aligned} \tag{6.4}$$

In order to avoid confusion when dealing with phonon operators associated with two different exciton states, we have adopted a slightly different notation for the phonon operators ($b_1(q)$ and $b_1(q)$) as compared to that used in Eqs.(4.1) and (4.2).

Using the phonon operators in Eq. (6.4), we can extend Eq. (6.3) to a more complicated form as follows

$$\begin{aligned} \hat{H}_{2\text{-}exp} = \sum_{k,q}\sum_{g=a,b}\sum_{r=1,2} & E_r^*(k)\chi_g(k)^\dagger\chi_g(k) + K_g^1(k,q)\chi_g(k+q)^\dagger\chi_g(k)\varphi_r(q) \\ & + K_g^2(k,q)\chi_a(k+q)^\dagger\chi_b(k)\varphi_r(q) + K_g^3(k,q)\chi_b(k+q)^\dagger\chi_a(k)\varphi_r(q) \\ & + K_g^4(q)\chi_g(k)^\dagger\chi_g(k)\varphi_r(q) + K_g^5(q)\chi_a(k)^\dagger\chi_b(k)\varphi_r(q) \\ & + K_g^6(q)\chi_b(k)^\dagger\chi_a(k)\varphi_r(q) + \omega(q)\eta(q)^\dagger\eta(q) \end{aligned} \tag{6.5}$$

where $K_g^i(k,q)$ (i = 1,2,3) are resonance coupling functions associated with exciton-phonon interaction and $K_g^i(q)$ (i = 4,5,6) are dispersion coupling functions. These coupling functions are expected to determine the *critical number of coherently coupled* monomers linked to the superradiance properties of the J-band. At higher temperatures the coupling functions $K_g^i(k,q), K_g^i(q)$ are increased, and the superradiance phenomenon is suppressed. The Hamiltonian in Eq. (6.5) needs to be solved to fully examine the important influence of decoherence mechanisms associated with two-exciton states in determining the unique properties of J-aggregate systems. Currently this task lies beyond the scope of this work due to its complicated form.

6.1 *Third Order Optical Susceptibility, $\chi^{(3)}$ and Entropy Measure*

The third order optical susceptibility, $\chi^{(3)}$ is defined as [56]

$$\chi^{(3)}(-\omega_s;\omega_1,\omega_2,\omega_3) = \sum \left[\chi_c^{(3)}(-\omega_s;\omega_1,\omega_2,\omega_3) + \chi_c^{(3\star)}(-\omega_s;\omega_1,\omega_2,\omega_3)\right] \tag{6.6}$$

where the summation is performed over all possible permutations of frequencies of incident light, ω_1, ω_2, ω_3 and $\omega_s = \sum_i \omega_i$. Expressions for $\chi_c^{(3)}(-\omega_s;\omega_1,\omega_2,\omega_3)$ and its conjugate are lengthy and given in the earlier work by Knoester [56], in terms of $\mu_{0;e1}$ and $\mu_{e2;e2,e3}$. $\mu_{0;e1}$ is the dipole moment associated with transition between the ground state and one-exciton state in which an excitation is present at any one of the N monomers, a generalization of the well known Dicke state $|W\rangle = \frac{1}{\sqrt{N}}\left(|100...0\rangle + |01...0\rangle + ... + |0...01\rangle\right)$. $\mu_{e2;e2,e3}$ is the dipole moment associated with transition between the one-exciton state and the two-exciton state. Using Eq. (6.3) as basis, we consider that the two-exciton state are a manifestation of symmetric states, which are dependent on the number of ground states or 0's [40, 42]

$$|S(N,M)\rangle \equiv \sqrt{\frac{M!(N-M)!}{N!}} \sum_{\text{permutations}} |0\rangle^{\otimes M} \otimes |1\rangle^{\otimes(N-M)} \tag{6.7}$$

The states in Eq. (6.3) are known to have an entanglement entropy measure [40, 42]

$$E_\Lambda(N,M) = -\ln \Lambda(N,M)^2 \tag{6.8}$$

$$\Lambda(N,M) = \sqrt{\frac{N!}{M!(N-M)!}} \left(\frac{M}{N}\right)^{\frac{M}{2}} \left(\frac{N-M}{N}\right)^{\frac{N-M}{2}}. \tag{6.9}$$

For fixed N, the minimum entanglement eigenvalue or geometric measure of entanglement [40] $\Lambda_{\max}$ occurs at $M = N/2$ (even N) and $M = (N\pm1)/2$ (odd N).

We consider that transitions involved in nonlinear optical effects occur via a simplified pathway represented as follows

$$\begin{aligned} |S(N,0)\rangle &\longrightarrow |S(N,1)\rangle \text{ (one-exciton state)} \\ &\longrightarrow |S(N,2)\rangle \text{ (two-exciton state)} \end{aligned} \tag{6.10}$$

The terms of the form given in Eq. (6.7) are assumed to give rise to the observed dependence of $\chi^{(3)}$ on non-local effects associated with energy transfer within various exciton site basis. Accordingly $\chi^{(3)}$ scales with the entanglement entropy E_Λ in Eq. (6.8), however the explicit nature of relation between the two quantities is not immediately clear, this predicament also applies to a possible relation between $\chi^{(3)}$ and the genuine multipartite entanglement involving all N monomers. Hence we keep the analysis tractable by examining the dependence of the intensity of third-harmonic generation (THG) $I_{THG} \propto |\chi^{(3)}(-3\omega;\omega,\omega,\omega)|^2$ on N instead. We simplify the approach via the following proposed relations

$$\begin{aligned} \mu^2_{0;e1} &\approx N\mu^2\zeta_1 \\ \mu^2_{e2;e2,e3} &\approx N\mu^2\zeta_2 \\ \zeta_1 &= \frac{E_\Lambda(N,1)}{E_\Lambda(N,N/2)} \\ \zeta_2 &= \frac{E_\Lambda(N,2)}{E_\Lambda(N,N/2)} \end{aligned} \tag{6.11}$$

where μ is the transition dipole of individual monomers and the entanglement entropy measure $E_\Lambda(n,k)$ is given in Eq. (6.8). ζ_1 and ζ_2 are effective entropy measures associated with the transition efficiencies of one-exciton state and two-exciton state respectively. Figure 6.1(a) shows gradual decrease in these we note that entropy measures with increasing aggregate size N effective. ζ_2 the entropy measure associated with two-exciton band is more resilient than the effective measure associated with the one-exciton band. Due to the wider bandwidth of the two-exciton states, their entanglement persists for longer times. For both types of exciton states, the entropy decrease with N can be interpreted as due to the faster increase of $E_\Lambda(N,N/2)$ with N. While we have considered in some details the specific case of the two-exciton state, the related cases of the K-exciton state ($K > 2$) in which any two of K monomers are excited while the other monomers remain in the ground state, have not been considered here due to the complexities involved in deriving an effective entropy measure. Nevertheless, the K-exciton state is expected to play an important role in the optical properties of J-aggregates.

Using Eq. (6.11) and $\omega = \frac{1}{3}\Delta E$ where the excitation energy ΔE is defined in Eq. (2.3), we write the absolute value of the third order optical susceptibility with damping rate Γ as

$$|\chi^{(3)}(-3\omega;\omega,\omega,\omega)| = N\frac{\mu^2 E_\Lambda(N,1)E_\Lambda(N,2)}{2\Gamma\hbar^3\,(\omega^2-\Delta E^2)} \tag{6.12}$$

Figure 6.1(b) shows gradual increase of $\frac{1}{N}|\chi^{(3)}(-3\omega;\omega,\omega,\omega)|$ (in which a constant term in Eq. (6.12) was factored out) with aggregate size N be-

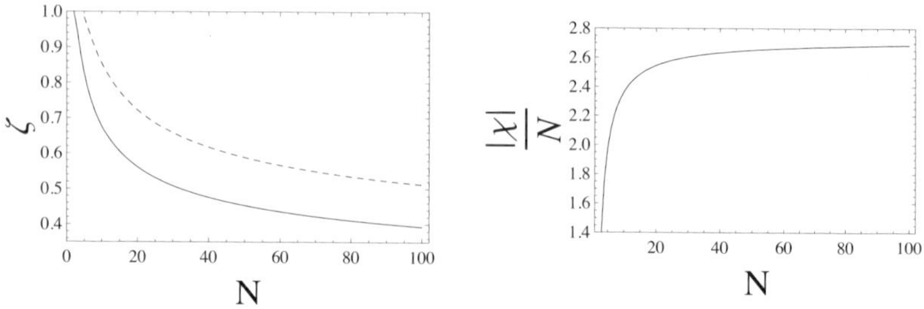

Figure 6.1. (a) Entropy measures ζ_1 (solid line) and ζ_2 (dotted) versus aggregate size N. (b) $\frac{1}{N}|\chi^{(3)}(-3\omega;\omega,\omega,\omega)|$ (see Eq. (6.12)) versus aggregate size N.

fore reaching an almost constant value which is independent of the size of system. The results of Fig. 6.1(b) illustrate the important fact that properties of the J-band is determined by a *critical number of coherently coupled* monomers rather than the total number N of monomers present in the system.

The slope in Fig. 6.1(b) yields the degree of entanglement or multipartite entanglement measure. In this respect, the *onset of zero slope* during change of third order optical susceptibility with the system size N can be viewed as the point of maximum entangled state. The results in Fig. 6.1(b) are consistent with earlier observations [56] which showed the critical role played by the exciton delocalization length. It has been noted that the results in the earlier work [56] was not analysed from a quantum entanglement perspective.

The results obtained in Fig. 6.1(b) show that scaling of the third order optical response $\chi^{(3)}$ with exciton delocalization size may be used as a basis in quantifying multipartite entanglement in J-aggregates or systems composed of excitonic spin chains. The result obtained here is significant considering that a rigorous definition of genuine multipartite entanglement is still lacking in the literature. Thus entanglement measures may be defined based on experimental observations and taking into account decoherence mechanisms [22, 23], exciton-photon interactions and laser detuning processes. In this regard, the results obtained so far has established some basic guidelines for an experimentally demonstrable measure of identifying multipartite entanglement in excitonic systems.

7. Quantum Information Attributes of J-aggregates

It is well known that the quantum correlation quantified by entanglement present opportunities for useful quantum tasks such as quantum teleportation, quantum dense code and and quantum computing [10, 64]. One- and two-qubit logic gate systems are essential in quantum information processing, and it well known that solid-state qubits are based on inter-qubit interactions corresponding to various exchange Hamiltonians such as XY [65], XXZ, or the isotropic Heisenberg exchange model [66]. Qubit system based on J-aggregates appear to have the advantage that when inter-qubit interactions are invoked, there is less chance of creating secondary decoherence channels associated with lattice vibrations. In this regard, quantum logic gates based on dye systems may form reliable quantum logic gates such the CNOT gate which consists of at least *two* iSWAP gates [10]. iSWAP gates are known to be useful in the fabrication of large cluster states and which form the basis of one-way quantum computing, a novel measurement-based approach to quantum computing [67].

7.1 *The Grover-like Search Problem*

In a recent work [16], the physical implementation of a Grover-like search problem by means of Frenkel exciton trapping was proposed. Quantum search via the Grover's algorithm allows combinatorial search to be speeded up considerably over those employing classical rules. The time taken to move from one point of a graph vertex to another is exponentially faster for a walker following quantum rules over one who obeys classical physics. Unlike their classical counterparts, paths in quantum walks can interfere leading to nonexistence of a quantum walker at a vertex.

The exciton transfer mechanisms in J-aggregates can be viewed as similar to that of a quantum walker on a graph in which each vertex point represents a lattice site. The route that an exciton takes as it propagates from one site to another is governed by the edge joining one vertex to another. The number of edges incident on a specific vertex is considerably simplified for one dimensional systems such as J-aggregates. For the quantum search problem, a particular vertex is distinguished from all others and a successful outcome occurs when this vertex is reached by a quantum walker originating from a given initial point. An initially delocalized exciton may be considered to examine all of the sites simultaneously in a superposed form before reaching its target. The exciton lifetime therefore can be seen to yield a measure of the efficiency of search.

In order that the search for the impurity site to be translated into a quantum task, it is imperative that the coherent length associated with the superposed exciton state exceed the number of sites to be searched. This means that the exciton state should retain its coherent form at least for the time needed for the exciton to reach the target impurity site. Moreover it is possible that an exciton is created at an impurity site due to the binding of free charge carriers. This may occur instead of the trapping of an exciton participating in the Grover-like search mechanism. The rate associated with such a process which acts to disrupt the search process is given by [16] $R_f \approx \frac{2n_i m_e^* N^* d_0}{\hbar^2 \xi v} \left[(E_T - E_b)^2 + \Delta p^2\right]$ where N^* and d_0 are the number of molecular sites and inter-molecular distance respectively at an edge of the crystal unit cell. E_T is the transport gap, the minimum energy required for the formation of a free electron and hole pair, E_b is the binding energy, Δp is the trap depth, ξ is mass density of the host molecules, m_e^* is the respective effective mass of the electron, and v denotes the velocity of sound in the crystal. Due to the large binding energies of the Frenkel exciton, participation of phonons of relatively higher energies is needed before excitons can be formed at the impurity site. Thus the process of exciton formation at impurity site is not likely to be significant in J-aggregates.

7.2 *Environment-assisted Quantum Transport*

The results in the earlier Sections have shown the negative influence of the environment on the propagation of excitation in J-aggregates. However recent works [68, 69] have alluded to the possibility of an environment-assisted quantum transport via the quantum Zeno effect. The quantum Zeno effect (QZE) describes the retarded time evolution of a quantum state subjected to frequent measurements [70, 71]. In the limiting case of continuous measurement, the time evolution of the state is expected to come to a standstill. The opposite effect which leads to enhancement in time evolution of the quantum state is known as anti-Zeno effect (AZE). In unstable systems, the occurrence of both QZE and AZE effects depends on critical parameters like measurement frequencies and environmental noise that may be associated with a bath of harmonic oscillators.

In recent years, the concept of Zeno effect has been studied using an adiabatic theorem in which alternative outcomes are eliminated, with the system evolving as a group of exclusive quantum Zeno subspaces within the total Hilbert space. The quantum Zeno effect may be utilized to exclusively direct a quasi-particle such as the exciton to propagate uninterrupted in a pre-selected route, however such directed control via the Zeno effect has yet to be experimentally demonstrated. The high-energy transfer efficiency in biological systems such as chromophoric complexes has been attributed an environment-assisted quantum walk model [69], and in this regard the quantum zero effect has potential applications in bio molecular systems. We next examine a few mechanisms in J-aggregates that may act to disrupt coherent the propagation of exciton.

7.3 *Exciton-exciton Annihilation in J-aggregates*

It is possible that excitons in close proximity can dissociate due to electrostatic interactions, creating free charge carriers. The energy released from recombination of an electron and a hole bounded together as an exciton at one molecular site can lead to dissociation of exciton at another site, with possible involvement of phonons. This process can lead to decoherence of the entangled exciton state in J-aggregates. Exciton-exciton annihilation is a critical process in J-aggregates as it affects the nonlinear properties which influence electronic processes that occur at high excitation intensities. There are various channels through which the annihilation can take place. For instance, two excitons from the same strand of aggregate can be involved in the annihilation process, or two excitons from different strands of the aggregate system can annihilate as well. These different channels of annihilation mechanisms are expected to be characterized by various parameters, one of which is the distance between the two excitons.

We consider the two interacting excitons in the singlet spin state ($\sigma_S = 0$), with the conservation of momentum rule allowing the exciton momentum to be transferred to electron after the annihilation process. The participation of phonons is minimized in this process, therefore the order of magnitude of the annihilation rates are not greatly affected by omission of interaction terms associated with phonons. The eigenvectors of the initial and final states involved in the annihilation process can be written as

$$|i\rangle = B^{\dagger}_{\mathbf{K_1}} B^{\dagger}_{\mathbf{K_2}} |0; n\rangle \tag{7.1}$$

$$|f\rangle = \sum_{\sigma_e, \sigma_h} a^{\dagger}_{\mathbf{k}}(\sigma_e) d^{\dagger}_{\mathbf{k'}}(\sigma_h) |0; n\rangle \tag{7.2}$$

where $a^{\dagger}_{\mathbf{k}}(d^{\dagger}_{\mathbf{k}})$ which denote the electron (hole) creation operators are associated with free charge carriers. The transition from $|i\rangle$ to $|f\rangle$ is enabled by electrostatic interactions V_{mn} between molecules m and n with a transition matrix of the form [16]

$$\langle f|\hat{V}|i\rangle = 2N^{-1}\sum_{\mathbf{l},\mathbf{n}} \exp[\mathbf{k}.(\mathbf{n}-\mathbf{l})] \times \left[\langle \mathbf{0},\mathbf{m};\mathbf{e}+\mathbf{h},\mathbf{l}\,|V_{mn}|\overset{\mathbf{e}}{\underset{\mathbf{h}}{|}},\mathbf{n};\overset{\mathbf{e}}{\underset{\mathbf{h}}{|}},\mathbf{m}\rangle - \frac{1}{2}\langle \mathbf{e}+\mathbf{h},\mathbf{l};\mathbf{0},\mathbf{n}\,|V_{mn}|\overset{\mathbf{e}}{\underset{\mathbf{h}}{|}},\mathbf{n};\overset{\mathbf{e}}{\underset{\mathbf{h}}{|}},\mathbf{m}\rangle\right] \tag{7.3}$$

The first term within the square bracket describes recombination of a free exciton (denoted by $\overset{\mathbf{e}}{\underset{\mathbf{h}}{|}}$) at site $\mathbf{m}$, with the released energy inducing the exciton at $\mathbf{n}$ into a pair of free carriers (denoted by $\mathbf{e}+\mathbf{h}$) at site $\mathbf{l}$. A similar interpretation holds for the second term, with recombination of the free exciton taking place at site $\mathbf{n}$ instead. The initial and final state energies are obtained as

$$E_i = W_0 + 2E_0(K) + \frac{\hbar^2 K^2}{2M^*} \tag{7.4}$$

$$E_f = W_0 + E_e(K) + \frac{\hbar^2 K^2}{2m_e^*} \tag{7.5}$$

W_0 is the total electronic energy of crystal with all molecule in their ground states, $E_0(K)$ is the exciton energy and $E_e(K)$ denotes the free electron energy. Using the Golden Rule formula, Eqs. (7.3) and (7.4), the exciton-exciton annihilation rate R_{an} has been obtained as [16]

$$R_{an} \approx \frac{10\, n_{ex}}{\hbar^4}\left[m_e^{*3/2}\, E_T^{5/2}\, N^{*6}\, d_0{}^6\right] \tag{7.6}$$

where n_{ex} is the concentration of free excitons, and N^* and d_0 are the number of molecular sites and inter-molecular distance respectively at an edge of the crystal unit cell. E_T, the transport gap is used to estimate the square of the transition matrix in Eq. (7.3). Equation (7.6) shows that at high exciton concentration, the annihilation mechanism can dominate the exciton dynamics leading to a breakdown of the coherence of J-aggregate systems.

7.4 *Exciton Fusion and Fission*

There are differences between the exciton-exciton annihilation mechanism described earlier and the exciton fusion process. During exciton fusion, two triplet excitons combine to form a singlet exciton [16]

$$\left(\overset{\mathbf{e}}{\underset{\mathbf{h}}{|}}\right)_T + \left(\overset{\mathbf{e}}{\underset{\mathbf{h}}{|}}\right)_T \longrightarrow \left(\overset{\mathbf{e}}{\underset{\mathbf{h}}{|}}\right)_S + E_v \tag{7.7}$$

The subscripts T and S correspond to the triplet and singlet exciton respectively, with excess energy E_v generally released in the form of phonons. In organic systems, there exists a positive energy difference between two triplet excitons and a singlet exciton, with the triplet exciton energy tending to decrease faster than the singlet exciton. A reverse process in which a singlet exciton undergoes conversion into a pair of triplet excitons is termed exciton fission. Hence in the fusion (fission) process, a singlet (triplet) exciton is formed with all excess energy is released (absorbed) via intra- or inter- molecular vibrations. The calculation of rates of exciton fission or fusion is similar to that for exciton-exciton annihilation for which free charge carriers were created in the final state. Similar to the exciton-exciton annihilation process, the exciton fusion and fission processes are sensitive to n_{ex}, the concentration of free excitons and fusion rates in the picosecond range $\sim 1-10$ ps is not uncommon for organic systems. Thus at low exciton concentration and packing density, exciton fusion and fission processes are less likely to disrupt the entanglement dynamics of J-aggregate systems.

We summarize some quantum information attributes of J-aggregates in the table below.

Table 7.1.

State/Event	Quantum information attributes
Ideal J-aggregate in	
Vacuum state	Null state
Frenkel exciton	Entangled state
Number of monomers over which excitation is localized	coherence length
One-exciton state	Encompasses Dicke States
Two-exciton states	Encompasses Symmetric States
Molecular cooperativity	Paradigm of collective entanglement
Third order optical susceptibility	Effective multipartite entanglement measure
J-aggregate systems	Channels of energy propagation Sources of non-classical correlations
Exciton-exciton annihilation, exciton fusion and fission	Sources of error

In conclusion, we have shown the usefulness of the von Neumann entropy, Wootters concurrence and geometric measure of entanglement in characterizing the quantum information processing attributes in quantum J-aggregate systems. The applicability of another correlation, the quantum discord, has been discussed as well. We identify the one-exciton state as encompassing Dicke states and two-exciton states as incorporating the class of symmetric states. We allude to the possibility that molecular cooperativity phenomenon in J-aggregates is an notable feature of collective entanglement and robust non-classical correlations, which can be exploited for applications involving information processing in material systems.

The entanglement and quantum correlations associated with a delocalized exciton over many monomer sites provides a framework for the understanding of the processes behind the observed characteristics of J-aggregates. However the delocalization of electronic states which give rise to narrowness in the J-aggregate spectrum may be complicated by the fine interplay of excitation transfer mechanism and decoherence associated with a background of phonon bath. With control of the dispersive coupling ($\chi(q)$) and resonance coupling ($F(k,q)$) parameters via temperature, matrix characteristics and the Zeno-effect, J-aggregate systems may be utilized as robust channels of large scale energy propagation.

The scaling of the third order optical response $\chi^{(3)}$ with exciton delocalization size may provide an experimentally demonstrable measure of quantifying multipartite entanglement in Frenkel excitonic systems. The results obtained here suggest that the optical response, $\chi^{(3)}$ provides the possibility of distinguishing between system with low and high fidelities. However the links between the system fidelities and secondary processes such as exciton-exciton annihilation, exciton fusion and fission which may interfere with efficient energy transport in J-aggregates need further investigation. Nevertheless, the results of this work will be useful in understanding the role of many-body correlated dynamics of excitonic systems and in determining the multipartite entanglement in J-aggregate systems.

References

[1] T. Kobayashi, *J Aggregates* (World Scientific, Singapore, 1996).
[2] T. Kobayashi and K. Misawa, *J. Lumin.* **72-74**, 38 (1997).
[3] E.E. Jelley, *Nature* **138**, 1009 (1936).
[4] G. Scheibe, *Angew. Chem.* **50**, 212 (1937).
[5] A.S. Davydov, *Theory of Molecular Excitons* (Plenum, New York, 1971).
[6] E. W. Knapp, *Chem. Phys.* **85**, 73 (1984).
[7] J. S. Briggs and A. Herzenberg, *Mol. Phys.* **21**, 865 (1971).
[8] A. Eisfeld and J. S. Briggs, *Chem. Phys.* **281**, 61 (2002).
[9] J. Roden, A. Eisfeld, W. Wolff, and W. T. Strunz, *Phys. Rev. Lett.* **103**, 058301 (2009).

[10] M. A. Nielsen and I.L. Chuang, *Quantum Computation and Quantum Information* (Cambridge University Press, Cambridge, U.K., 2000).
[11] D. P. Craig and S. H. Walmsley, *Excitons in Molecular Crystals* (Benjamin Inc., New York, 1968).
[12] Y. Toyozawa, *Optical Processes in Solids* (Cambridge, New York, 2003).
[13] F. Milota, J. Sperling, A. Nemeth, and H. F. Kauffmann, *Chem. Phys.* **357**, 45 (2009).
[14] I. Stiopkin, T. Brixner, M. Yang, and G.R. Fleming, *J. Phys. Chem. B* **110**, 20032 (2006).
[15] A. Thilagam, *J. Phys. A: Math. Theor.* **44**, 135306 (2011).
[16] A. Thilagam, *Phys. Rev. A* **81**, 032309 (2010).
[17] P. Schouwink, H. V. Berlepsch, L. Dähne, and R. F. Mahrt, *Chem. Phys. Lett.* **344**, 352 (2001).
[18] A. Pawlik, A. Ouart, S. Kirstein, H.-W. Abraham, and S. Daehne, *European Journal of Organic Chemistry* **2003**, 3065 (2003).
[19] W. T. Simpson and D. L. Peterson, *J. Chem. Phys.* **26**, 588 (1957).
[20] J. Klafter and J. Jortner, *Chem. Phys. Lett.* **50**, 202 (1977).
[21] V. V. Egorov, *Chem. Phys. Lett.* **336**, 284 (2001).
[22] A. Thilagam, *J. Phys.: Condens. Matter* **21**, 045504 (2009).
[23] A. Thilagam and M. A. Lohe, *J. Phys.: Condens. Matter* **20**, 315205 (2008).
[24] A. Thilagam , *J. Phys. A: Math. Theor.* **43**, 155301 (2010).
[25] A. Eisfeld and J. S. Briggs, *Phys. Rev. Lett.* **96**, 113003 (2006).
[26] A. J. Scott, *Phys. Rev. A* **69**, 052330 (2004).
[27] R. Horodecki, P. Horodecki, M. Horodecki, and K. Horodecki, *Rev. Mod. Phys.*, **81**, 865 (2009).
[28] L. Amico, R. Fazio, A. Osterloh, and V. Vedral, *Rev. Mod. Phys.* **80**, 517 (2008).
[29] M. Hein, W. Dür, J. Eisert, R. Raussendorf, M. Van den Nest, and H. J. Briegel. *Entanglement in graph states and its applications*, Proc. of the Int. School of Physics, " Quantum Computers, Algorithms and Chaos", July 2005.
[30] M. B. Plenio and S. Virmani, *Quantum Inf. Comput.* **7**, 1 (2007).
[31] J. Eisert and D. Gross, *Lectures on Quantum Information*, eds. D. Bruis and G. Leuchs (Wiley-VCH, Weinheim, (2007).
[32] D. A. Meyer and N. R. Wallach, *J. Maths. Phys.* **43**, 4273 (2002).
[33] V. Coffman, J. Kundu and W. K. Wootters, *Phys. Rev. A* **61**, 052306 (2000).
[34] J. Niset and N. J. Cerf, *Phys. Rev. A* **74**, 052103 (2006).
[35] H. Ollivier and W. H. Zurek, *Phys. Rev. Lett.* **88**, 017901 (2001).
[36] L. Henderson and V. Vedral, *J. Phys. A* **34**, 6899 (2001).
[37] V. Vedral, *Phys. Rev. Lett.* **90**, 050401 (2003).
[38] A. Kitaev and J. Preskill, *Phys. Rev. Lett.* **96**, 110404 (2006).
[39] W. K. Wootters, *Phys. Rev. Lett.* **80**, 2245 (1998).
[40] T. C. Wei and P. M. Goldbart, *Phys. Rev. A* **68**, 042307(2003).
[41] C. Spitz, J. Knoester, A. Ouart, and S. Daehne, *Chemical Physics* **275**, 271 (2002).
[42] T. C. Wei, K. Nemoto, P. M. Goldbart, P. G. Kwiat, W. J. Munro, and F. Verstraete, *Phys. Rev. A* **67**, 022110 (2003).

[43] T. Werlang, S. Souza, F. F. Fanchini and C. J. Villas-Boas, *Phys. Rev. A* **80**, 024103 (2009).
[44] A. Ferraro, L. Aolita, D. Cavalcanti, F. M. Cucchietti and A. Acin, arXiv:0908.3157 (2009).
[45] S. Luo, *Phys. Rev. A* **77**, 042303 (2008).
[46] A. Thilagam, *J. Phys. A: Math. Theor.* **42**, 335301 (2009).
[47] S. Luo and S. Fu, *Phys. Rev. A* **82**, 034302 (2010).
[48] B. Dakic, V. Vedral and C. Brukner, e-print arXiv:1004.0190 (2010).
[49] A. Suna, *Phys. Rev.* **135**, A111 (1964).
[50] D. P. Craig and L. A. Dissado, *Chem. Phys.* **14**, 89 (1976).
[51] E. N. Economou, *Green's Functions in Quantum Physics* (Springer-Verlag, Berlin, 1979).
[52] R. E. Merrifield, *J. Chem. Phys.* **28**, 647 (1958).
[53] G. S. Pawley and S. J. Cyvin, *J. Chem. Phys.* **52**, 4072 (1970)
[54] T. Takagahara, *Surface Science* , **196**, 590 (1987).
[55] T. Tokihiro, Y. Manabe, and E. Hanamura, *Phys. Rev. B* **47** 2019 (1993).
[56] J. Knoester, *Chem. Phys. Lett.* **203**, 371 (1993).
[57] F. C. Spano, J. R. Kuklinski, and S. Mukamel, *Phys. Rev. Lett.* **65**, 211 (1990).
[58] T. Meier, Y. Zhao, V. Chernyak, and S. Mukamel, *J. Chem. Phys.* **107**, 3876 (1997).
[59] F. C. Spano, J. R. Kuklinski, and S. Mukamel, *J. Chem. Phys.* **94**, 7534 (1991).
[60] O. Dubovsky and S. Mukamel, *J. Chem. Phys.* **95**, 7828 (1991).
[61] A. Lakshminarayan and V. Subrahmanyam, *Phys. Rev. A* **67**, 052304 (2003).
[62] I. Varga and J. A. Mendez-Bermudez, *Physica Status Solidi C.* **5**, 867 (2008).
[63] A. Barelli, J. Bellissard, P. Jacquod, and D. L. Shepelyansky, *Phys. Rev. Lett.* **77** 4752 (1996).
[64] D. Bouwmeester, A. Ekert, A. Zeilinger (Eds.), *The Physics of Quantum Information: Quantum Cryptography, Quantum Teleportation, Quantum Computation*, Springer, Berlin (2000).
[65] J.Q. You and F. Nori, *Phys. Today* **58** (11), 42 (2005).
[66] D. Loss and D.P. DiVincenzo, *Phys. Rev. A* **57**, 120 (1998).
[67] H.J. Briegel and R.Raussendorf, *Phys. Rev. Lett.* **86**, 910, 5188 (2001).
[68] S. A. Gurvitz, *Phys. Rev. Lett.* **85**, 812 (2000).
[69] K. Fujii and K. Yamamoto, *Phys. Rev. A* **82**, 042109 (2010).
[70] B. Misra B and E. C. G. Sudarshan, *J. Math. Phys.* **18** 758, 1977.
[71] P. Facchi and S. Pascazio *J. Phys. A: Math. Theor.* **41** 493001, 2008.

CHAPTER 11

J-AGGREGATES OF PERYLENE DYES

Frank Würthner[a] and Ivan G. Scheblykin[b]

[a)] *Institut für Organische Chemie and*
Röntgen Research Center for Complex Material Systems,
Universität Würzburg, Am Hubland, 97074 Würzburg, Germany
E-mail: wuerthner@chemie.uni-wuerzburg.de

[b)] *Chemical Physics, Lund University, Box 124*
22100 Lund, Sweden
E-mail: ivan.scheblykin@chemphys.lu.se

During the last decade detailed information on the supramolecular organization of perylene bisimide (PBI) dyes in their aggregates in solution has been achieved by spectroscopic methods. Whilst the preferential packing for the parent PBIs, i.e. dyes without core-substituents and specific directing groups, affords sandwich-type aggregates with dominant H-type absorption and strongly bathochromically displaced excimer-type emission bands, construction of dye assemblies with J-type coupling was achieved by supramolecular design. In particular, the combination of contorted PBI cores by bay substituents and directional hydrogen bonding afforded PBI J-aggregates that exhibit significantly bathochromically displaced absorption bands, band narrowing and small Stokes shifts. Owing to the nucleation-induced instantaneous growth of extended and well-defined thermodynamically stable J-aggregates and their high fluorescence quantum yield of almost unity, a variety of spectroscopic investigations could be performed for these J-aggregates. For individual PBI J-aggregates, fluorescence blinking corresponding to collective quenching of up to 100 PBI monomers has been observed that could be related to an exciton diffusion length of up to 70 nm in these assemblies at room temperature

Keywords: exciton migration, J-aggregates, perylene dyes, self-assembly, single molecule spectroscopy.

1. Introduction and General Considerations on the Packing of PBIs

Perylene dyes, i.e. dyes based on the perylene 3,4:9,10 tetracarboxylic acid bisimide (PBI) skeleton (also called perylene tetracarboxylic diimide and abbreviated as PDI or PTCDI) are a most interesting class of colorants with exceptional fluorescence quantum yields up to unity and outstanding photostability.[1–3] Whilst significant efforts during the last two decades aimed at utilization of this class of dyes in organic (opto-)electronic materials[4–7] their major application is still in the field of high performance color pigments with shades ranging from red to violet, maroon and black.[8–10] This variety of shades is attributable to different packing arrangements of the PBI molecules in the crystal and has been rationalized on various levels of theory.[11–16] Figure 1.1 illustrates the results of calculations for the ground state potential energy surface (PES) of PBI dimers with the DFT-D approach which accounts for

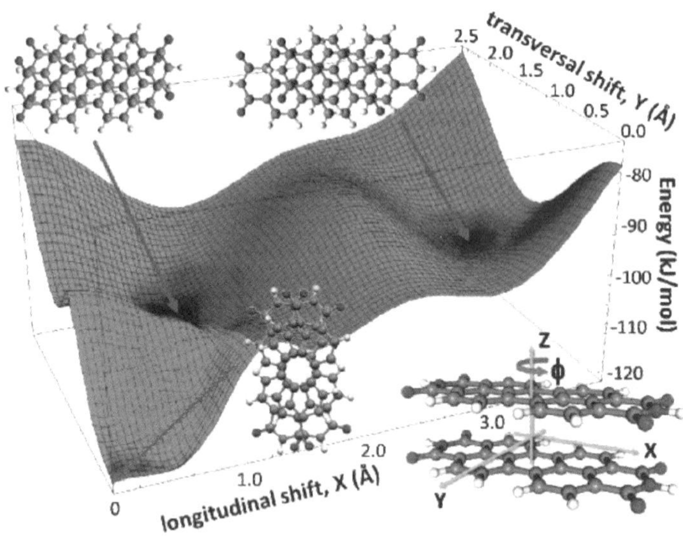

Figure 1.1 Calculated potential energy surface for the ground state of PBI dimers according to reference 13. The major part of the surface with two distinct local minima corresponds to purely translational offsets between the molecules whilst the surface in the front left corner with the deepest energy minimum corresponds to rotational offsets. The definition of the rotational angle ϕ is indicated on the right-hand side. Reprinted with permission from reference 13. Copyright (2008) American Chemical Society.

dispersion effects by adding empirical corrections to the standard density functional energy.[13,17] If these calculations are restricted to translational offsets (PES in Fig. 1.1 with the exception of front left corner), two distinct minima are found which correspond nicely to the majority of PBI crystal structures of red-colored pigments (left case, suggesting the presence of H-type excitonic coupling) and black pigments (right case, suggesting the presence of J-type excitonic coupling). If rotational offsets are allowed, an even deeper energy minimum is found at a rotational angle of ~30° (front left corner of PES).

2. Columnar Stacks of Simple PBIs

2.1 *Flat PBIs without Core Substituents*

Such rotational displacements have only rarely been observed in crystal structures of PBIs.[18] They are, however, the preferred arrangements found for aggregates of core-unsubstituted PBIs in solution and in columnar liquid-crystalline mesophases according to several recent NMR and UV/Vis spectroscopic investigations.[19-23] Concentration- and temperature-dependent NMR and UV/Vis spectroscopic studies on these aggregates further revealed a transition from monomeric into extended π-stacks that can be satisfactorily described by the isodesmic model (equal binding constant for upper and lower π-surface).[24] As shown in Fig. 2, the major absorption band of these aggregates is hypsochromically shifted, revealing predominant H-type excitonic coupling.

As a consequence of the rotational displacement between neighboring dyes, however, the optical transition into the lower energy exciton state becomes partially allowed[25,26] as evidenced by a second weaker absorption band at longer wavelengths, i.e. a J-band. These absorption bands could be described with an excitonic model including a single vibrational coordinate.[27,28] Owing to the fact that similarly good agreement with the experiment was found for dimeric and for trimeric entities, we may conclude that the coherent domain is rather localized despite the extended physical size of these π-stacks. These calculations further revealed that the excitonic coupling element $J = 0.11$ eV is of comparable magnitude to the vibrational spacing of 0.175 eV (intermediate coupling regime).

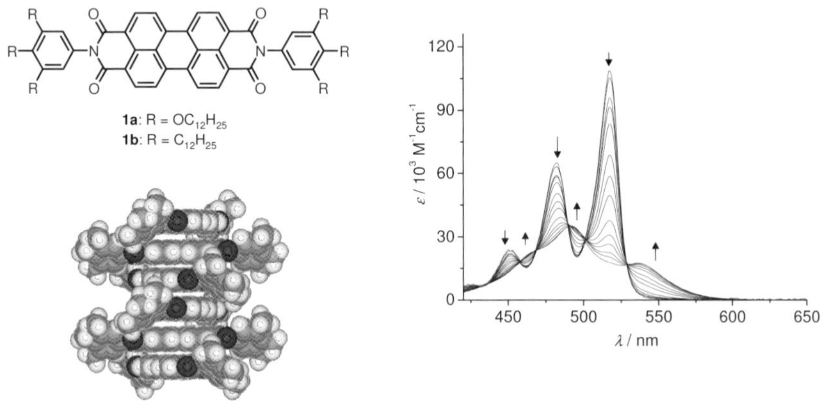

Figure 2.1 Structures of core-unsubstituted PBIs **1a,b** and their columnar aggregates (left) and concentration-dependent (2.1×10^{-7} to 1.1×10^{-3} M) UV/Vis absorption spectra of **1b** in methylcyclohexane (MCH, right). Arrows indicate the spectral changes with increasing concentration. Reproduced with permission from Ref. 21. Copyright (2001) Wiley-VCH Verlag GmbH & Co. KGaA, Weinheim.

The particular arrangement and the concomitant physical properties of columnar PBI stacks appear, however, to be strongly dependent on the respective imide substituents, in particular for the photoexcited states. For PBIs bearing electron-rich substituents like trialkoxyphenyl (**1a**), calixarenes or oligophenylenevinylenes, the emission is totally quenched owing to rapid electron transfer processes from the electron donor substituents to the photoexcited acceptor PBI.[29,30] On the other hand, for electronically rather innocent trialkylphenyl substituents (**1b**), relatively long-lived excited states ($\tau = 30$–40 ns) with appreciably high fluorescence quantum yields have been reported,[21,31] which were attributed to the relaxation of the exciton into an excimer-type state with almost parallel transition dipole moments (i.e. rotational angle close to zero).[13,21]

The preferential packing patterns (Fig.1.1) together with these relaxation pathways may account for the prevalence of non-emissive solid state materials for this class of dyes, despite their outstanding fluorescence properties in the monomeric state. Thus, only few examples for strongly emitting microcrystalline powders of PBI dyes have been reported so far.[32,33] In most cases this fluorescence appears to originate from bulky imide substituents that prohibit the π-systems to approach closely in the ground and photoexcited states. For an exploitation of this highly promising class of chromophores in organic electronics, photovoltaics or in photonic devices, however, closely stacked dye aggregates are required in which excitonic and electronic couplings support the migration of excitons and charge carriers. Because the packing arrangement observed for black PBI crystals (PES in Figure 1.1, right structure, showing the highest energy of the three energy minima) affords bathochromically displaced J-type absorption bands it appeared promising to modify PBI molecules in a way that supports such an arrangement. This can be either achieved by stabilizing this arrangement, e.g. by additional noncovalent forces, giving rise to a more deep energy minimum or by destabilizing the other arrangements, e.g. by introducing bulky bay substituents that interfere with non-slipped sandwich-type packing arrangements.

2.2 *Twisted PBIs Bearing Core Substituents*

In 2001 one of us described the first step in this direction based on the latter approach and reported perylene bisimide dye aggregates that exhibit a predominant bathochromically displaced aggregate band.[19, 34] These J-aggregating PBIs, e.g. **2a**, were equipped with 3,4,5-tridodecyloxyphenyl substituents at the imide N atoms and phenoxy substituents in the four bay positions (Fig. 2.2). Upon self-assembly in low-polarity solvents, a rather broad but clearly red-shifted absorption band was observed. Although this red shift indicates the presence of slipped chromophores, the broadness of the absorption band and the rather large Stokes shift might be taken as an indication for a less ideal system influenced by disorder and phonon coupling[35–37] caused by the flexibility of the bulky phenoxy substituents. In addition, the fluorescence of these compounds was quenched both in the monomeric (ϕ_f = 21%) and the aggregated state owing to a photoinduced electron transfer from the electron-rich trialkoxyphenyl substituents to the electron-poor perylene bisimide.[29] Accordingly, PBI **2b** was synthesized which revealed significantly improved fluorescence properties (ϕ_f = 95 % for monomers in solution) while maintaining the J-type character of the aggregates.[38] Owing to the bulkiness of the core substituents, no extended π-stacks are formed anymore at spectroscopic concentrations. Instead the presence of dimeric aggregate species was confirmed in aliphatic solvents for 0.001 to 0.01 M solutions by UV/Vis spectroscopy and vapor pressure osmometry. Notably, a whole series of trialkyl- and trialkoxyphenyl-substituted PBIs were shown to exhibit columnar liquid crystalline mesophases over a broad temperature range.[19,21,38]

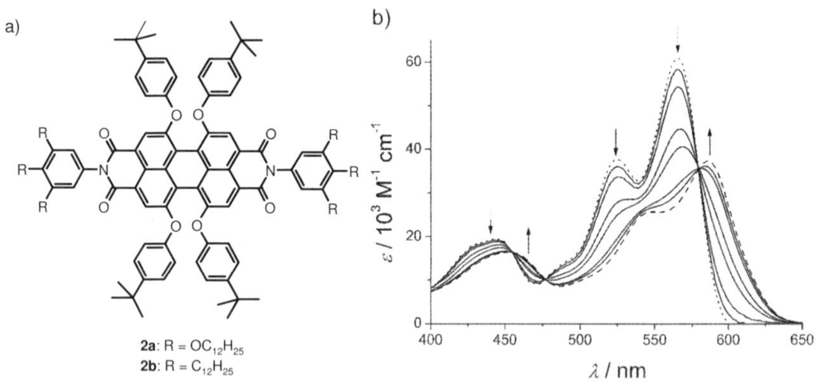

Figure 2.2 a) Structures of PBIs **2a,b** that form aggregates with J-type character. (b) Concentration-dependent UV/Vis absorption spectra of **2a** in methylcyclohexane (MCH). Arrows indicate the spectral changes with increasing concentration. Dotted lines represent spectra for the free and the aggregated chromophore calculated from the respective data set. Reproduced with permission from reference 19. Copyright (2001) Wiley-VCH Verlag GmbH & Co. KGaA, Weinheim.

3. Hydrogen-bond Enforced Columnar Stacks of PBIs

3.1. *Flat PBIs without Core Substituents*

In a serendipitous discovery it was found that also core-unsubstituted PBI derivatives containing trialkoxyphenyl groups attached to the imide positions via amide spacers can form aggregates exhibiting J-type absorption spectra (Fig. 3.1).[39, 40] These amide-functionalized PBIs form long and defined helical nanofibers by hydrogen-bond assisted self-assembly that are able to immobilize a broad variety of solvents giving rise to organogels. Whilst the PBI derivatives **3a,b** with simple n-alkyl chains in the periphery form red-colored H-type aggregates, as expected for core-unsubstituted PBIs, aggregates of dyes **3c,d** bearing branched alkyl chains exhibit a strongly bathochromically displaced broad J-band in solution (Fig. 3.1) and in the organogel state as well. The latter PBI aggregates are almost black and possess even more pronounced gelation ability than the former ones. This feature has been exploited for the preparation of supramolecular bulk heterojunction photovoltaic materials by gelation of triarylamine-based p-type semiconducting polymers.[41] Despite the favorable absorption properties of these black materials and excellent hydrogen bonding-assisted phase separation on the nanoscale only modest photovoltaic performance of these BHJ blends was observed.

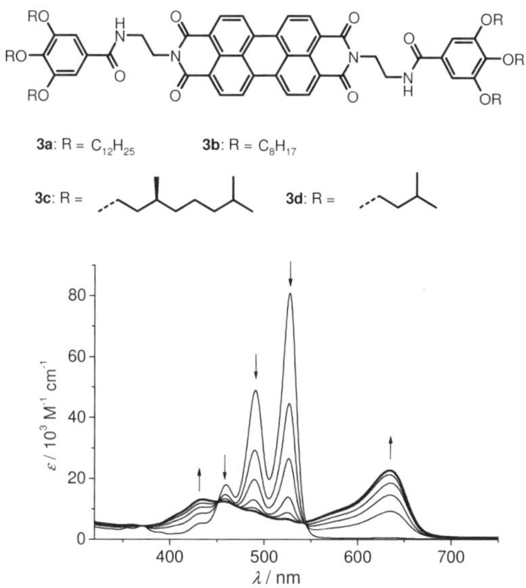

Figure 3.1 Structures of PBIs **3a-d** and solvent-dependent UV/Vis absorption spectra of chiral PBI **3c** in MCH/$CHCl_3$ solvent mixtures from 50:50 to 80:20 at a constant concentration of 10^{-5} M. Arrows indicate spectral changes with increasing amounts of nonpolar MCH. Reproduced with permission from reference 40. Copyright (2008) Wiley-VCH Verlag GmbH & Co. KGaA, Weinheim.

It is indeed very intriguing that the aggregation mode (i.e., H- vs J-type) of PBI chromophores could be altered by a subtle variation of the peripheral substituents. Additional experimental studies supported by molecular modeling revealed that the steric nature of the peripheral side chains dictates the self-assembly code of these core-unsubstituted PBI dyes as derivatives bearing linear alkyl chains (less steric demand) form the usual H-type assemblies, while for dyes containing branched alkyl chains (high steric demand) assemblies with J-coupling prevail.[42] Simplified packing models proposed for the J-type and H-type assemblies of these PBIs are illustrated in Fig. 3.2.

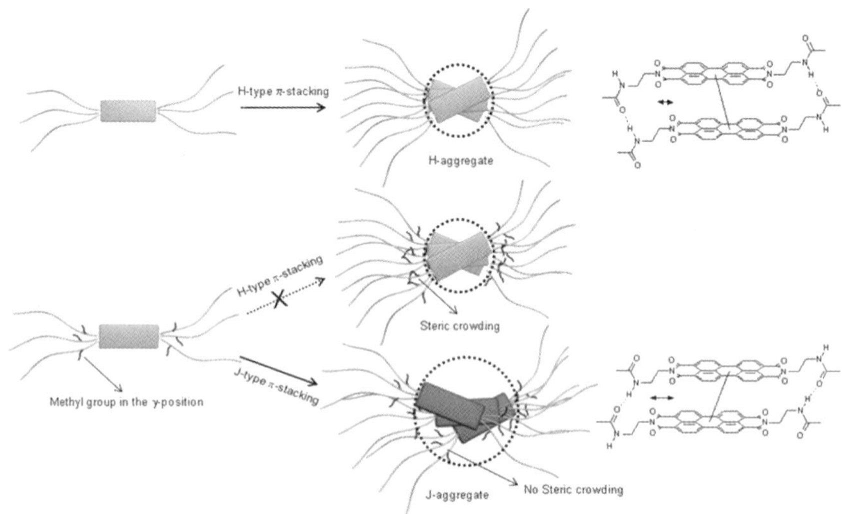

Figure 3.2 (Left) Schematic representation of PBI chromophores (having orange color) with linear (**3a,b**, top) and branched (**3c,d**, bottom) alkyl substituents. (Middle) The transition from red-colored H- (top) to black J-type (bottom) π-stacking with increasing steric demand of the peripheral alkyl side chains. (Right) Proposed packing model for H- (top) and J-type (bottom) π-stacking. In both cases additional rotational offsets are needed to enable both close π-π contact and hydrogen bonding. Reproduced with permission from Ref. 42. Copyright (2008) Wiley-VCH Verlag GmbH & Co. KGaA, Weinheim.

3.2. *Twisted PBIs Bearing Core Substituents*

The same concept as applied before, i.e. alteration of the chromophore **3a** by introducing four phenoxy substituents at all four bay positions, afforded a PBI organogelator **4** with a remarkably sharp J-type absorption band (Fig. 3.3) and high fluorescence quantum yield.[43] Interestingly, whilst this molecule failed to gelate a broad variety of solvents, highly luminescent lyotropic mesophases were observed in aliphatic solvents.

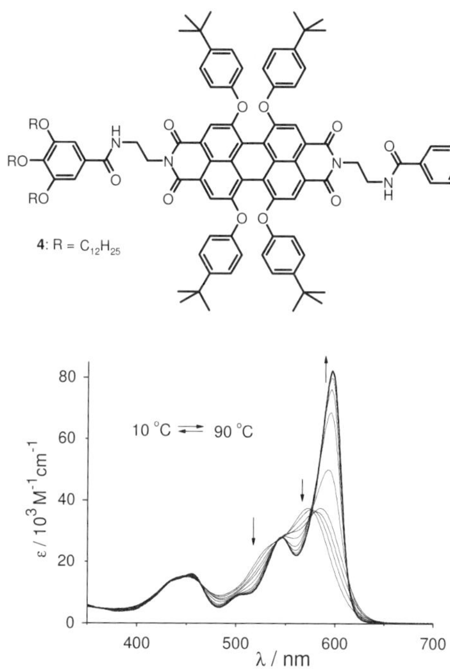

Figure 3.3. Structure of PBI **4** and temperature-dependent transition from monomers (90 °C) to aggregates (10 °C) with J-type character in methylcyclohexane (MCH). Arrows indicate the spectral changes with decreasing temperature at a concentration of 1.1×10^{-4} M. Reproduced with permission from reference 43. Copyright (2008) Wiley-VCH Verlag GmbH & Co. KGaA, Weinheim.

3.3. *Miscellaneous Hydrogen-bonded PBIs*

Few more examples of PBI aggregates with J-type character were reported, which exhibit in general rather broad bathochromically shifted bands comparable to those of PBI **2a** (Fig. 2.2).[44, 45] Not only homoaggregates of PBI dyes, but also heteroaggregates consisting of PBIs and other π-systems, e.g. melamine derivatives, have been reported.[46-48] A triple hydrogen-bonding motif from perylene bisimide to melamine derivatives (ADA-DAD, with A = H-bond acceptor and D = H-bond donor) together with π-π interactions of the adjacent PBIs led to the formation of extended fluorescent PBI–melamine networks.[47] The long alkyl chains of the melamine increase the solubility of the assemblies, and the van-der-Waals interactions between the aliphatic chains further stabilize the network. Absorption spectroscopy revealed a small bathochromic shift of the J-band, assumed to be originated from a weakly slipped displacement of the chromophores in the aggregate.

J-type coupling between PBI dyes was also observed for aggregates composed of hydrogen-bonded and covalent donor–acceptor–donor arrays of bay-substituted PBI acceptors and oligo(p-phenylene vinylene) (OPV) donors bearing chiral side chains by concentration- and temperature-dependent UV/Vis absorption spectroscopy.[49-53] Further investigations by circular dichroism, photoluminescence, and femtosecond pump-probe spectroscopy revealed the formation of chiral assemblies, in which the PBI monomers are arranged in a slipped J-type arrangement (similar to the black J-aggregates shown in Figure 3.1), and ultrafast photoinduced charge transfer from the OPVs to the PBIs.[50, 52] More recently, a related system was reported where the OPV was replaced by azobenzenes.[54] A unique example of a melamine-functionalized, core-unsubstituted PBI was reported by Yagai *et al.* that self-assembles upon addition of 0.5 equivalents of ditopic cyanurate into H-type dimers but, interestingly, addition of 1 equivalent of cyanurate transforms the H-dimers into J-type aggregates with strongly red-shifted absorption band (> 100 nm) compared to that of the corresponding monomeric PBI.[55] Such a concept for the control of H- and J-type aggregation indeed has prospect to accomplish responsive J-assemblies with fluorescence sensor functionality.

4. J-Aggregated PBIs by Supramolecular Design

The most rational concept towards PBI J-aggregates combines core distortion by four bay substituents and hydrogen bond-directed chromophore alignment. Thus, PBI derivatives **5a-c** containing hydrogen atoms at the imide nitrogens formed J-aggregates with very characteristic optical properties that are comparable with those of the well studied classical cyanine J-aggregates, thus representing the first genuine J-aggregating PBIs (Fig. 4.1).[56, 57]

UV/Vis and fluorescence spectroscopy studies of PBI **5a** in nonpolar solvent MCH revealed the reversible formation of these J-aggregates (Fig. 4.1) and the strong narrowing of the red-shifted absorption band from a full-width-at-half-maximum (fwhm) value of 2393 cm^{-1} down to 885 cm^{-1} and that of the fluorescence band from 1660 cm^{-1} to 878 cm^{-1} for the monomer and the aggregates, respectively, concomitant with an increase of the fluorescence quantum yield from 93% to 96%. As expected for J-type coupled chromophores with a significantly enhanced transition dipole for the $S_0 \rightarrow S_1$ transition, the fluorescence lifetime is decreased for the aggregate (2.6 ns) with respect to that of the monomer (6.8 ns).[56] This J-type aggregation of functional perylene bisimide chromophores could be achieved by a design of monomeric building blocks that encode the desired slipped face-to-face arrangement by mutual effects of hydrogen bonding and π–π interaction and, on the other hand, prevent aggregation into columnar stacks owing to their twisted π-conjugated core and sterically demanding substituents in the bay area.

The proposed model for the aggregate formation is illustrated in Fig. 4.2. In depth investigations with the series of PBIs **5a-c** (for structures, see Fig. 4.1) revealed the impeded formation of a dimeric nucleus (Fig. 4.2c) prior to the H-bond supported elongation into double-string cable-like J-aggregates (Fig. 4.2d) with the monomers aligned with translational offset (Fig. 4.2e).[57] This cooperative nucleation-elongation mechanism is in contrast to the isodesmic (or equal K) model which was previously used to describe the aggregation process for common columnar stacks of PBI dyes such as **1a,b** (Fig. 2.1). For PBI **5c** equilibrium constants for dimerization (= nucleation, Fig. 4.2c) of $K_2 = 13\ M^{-1}$ and for elongation of $K = 2.3 \times 10^6\ M^{-1}$ in methylcyclohexane (MCH) were obtained by applying the nucleation-elongation model to concentration-dependent UV/Vis absorption studies. Accordingly, for this system dimeric species do never show up in

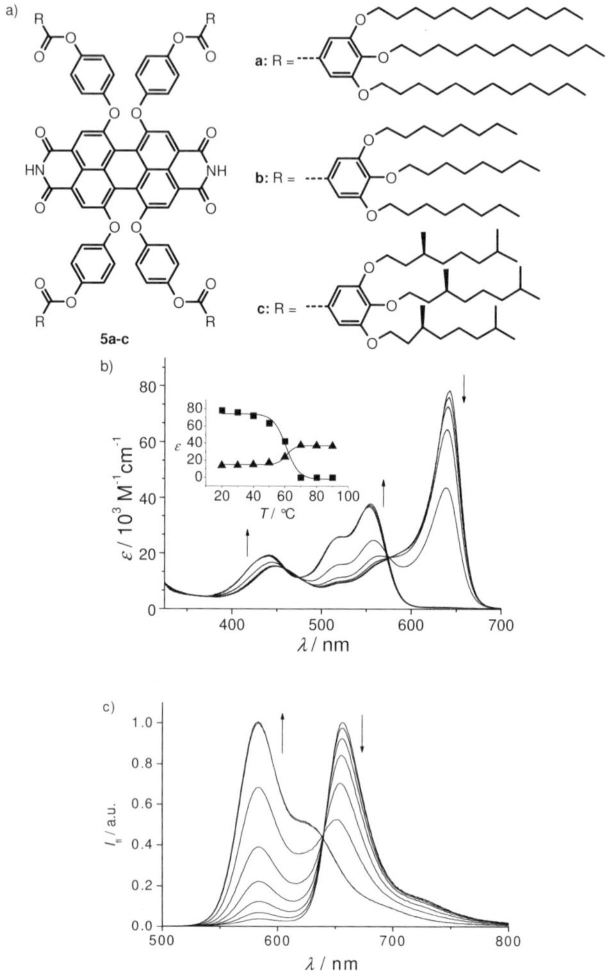

Figure 4.1. (a) Structures of PBI dyes **5a-c**. (b) Temperature-dependent UV/Vis spectra of **5a** in MCH (1.5×10^{-5} M) at 20–90°C. Arrows indicate the spectral changes with increasing temperature. Inset: Changes in absorption at 642 nm (■) and 553 nm (▲). (c) Temperature-dependent fluorescence spectra of **5a** in MCH (6×10^{-7} M, λ_{ex} = 476 nm) at 15–50°C; arrows indicate the spectral changes with increasing temperature. Figures 4.1(b) and (c) are reproduced with permission from Ref. 56. Copyright (2007) Wiley-VCH Verlag GmbH & Co. KGaA, Weinheim.

significant quantity as, e. g., observed for pseudoisocyanine dyes at low concentrations.[58,59] Instead, the instantaneous growth into extended nanofibers takes place at a temperature- and solvent-dependent critical concentration. Furthermore, by CD spectroscopy the nonlinearity of chiral amplification in PBI heteroaggregates composed of **5b** and **5c** directed by sergeants-and-soldiers principle[60] was demonstrated.[57]

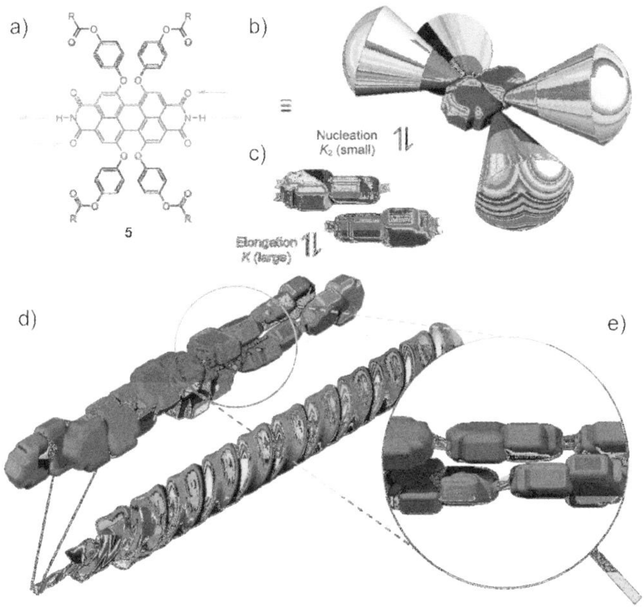

Figure 4.2 Schematic illustration of self-assembly of the perylene bisimide dyes **5** into J-aggregates. (a) Structure of PBI **5** (substituents R are defined in Figs. 4.1(a) and (b) graphical representation of the monomer. (c) Schematic representation of π-stacked dimeric nucleus and (d) that of an extended hydrogen-bonded J-aggregate of **5**. Black (and dark grey) twisted blocks represent the perylene bisimide cores (in the adjacent chains), gray cones represent the bay substituents, and light grey lines represent hydrogen bonds. The dyes **5** self-assemble in a helical fashion as shown in the magnification (substituents are omitted and only the left-handed helical structure is shown for simplicity). (e) The magnifier visualizes the J-type arrangement of the core perylene bisimide units in a double-string cable in the side view. Reprinted with permission from reference 57. Copyright (2009) American Chemical Society.

The same supramolecular strategy for PBI displacement by means of core-twisting and hydrogen-bonding could be applied to diamino-substituted PBIs **6**[61] to afford green-colored "chlorophyll-type" J-aggregates (absorption maximum of the J-band at 822 nm) and to octachloro-substituted PBIs **7**.[62] In the latter case densely packed and poorly soluble crystalline solid state materials are formed that could be characterized by single crystal X-ray diffraction and deposited in vacuum to give organic thin film transistors with exceptional n-type charge transport properties under ambient conditions.[62]

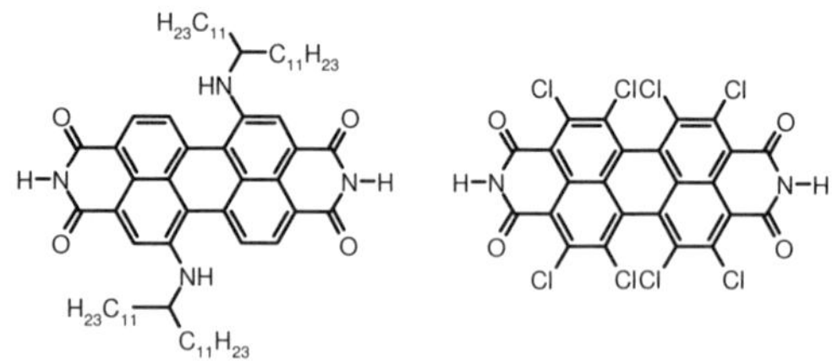

6 7

Figure 4.3. Structures of PBIs **6** and **7**.

The exciton transport along J-aggregates of **5a** has been studied by spectroscopy at low temperatures (from 300 K down to 5 K)[63] and by single-molecule spectroscopy.[64] These investigations provided insight into disorder phenomena and information on the coherence and exciton migration lengths which will be discussed in the following sections.

5. Disorder in PBI J-Aggregates

5.1 *The Coherence Length*

Probably the most important characteristics of any excitonic system and J-aggregates in particular is the so-called exciton coherent or delocalization length which is basically equal to the number of monomers the exciton wavefunction is delocalized over. The larger the delocalization length the narrower spectral bands and shorter radiative lifetimes one would expect. The delocalization length is limited by disorder, which can be static (e.g. differences between excited state energies of monomers as originated by conformational degrees of freedom or differences in interactions between the monomers governed by the supramolecular arrangement) or dynamic (which originates from exciton interaction with different types of phonons). A way to estimate the coherent length N_C is to compare the radiative lifetime of the aggregate τ^0_J and that of the monomer (τ^0_m):

$$\tau_J^0 = \frac{\tau_m^0}{N}(1+Q) \tag{5.1}$$

where the factor of Q takes into account the presence of Davydov splitting of the exciton band.[65] The simple case of $Q = 0$ originates if an aggregate is of a pure J-type, meaning that it has only one exciton band in its absorption spectrum and the fluorescence is associated with the same band.

PBI **5a** aggregates are of such a pure J-type with $Q = 0$ and a fluorescence quantum yield close to 100% meaning that the excited state decays radiatively. With a fluorescence lifetime of 2.6 ns for PBI **5a** J-aggregates at room temperature and a monomer fluorescence decay time of 6.8 ns[57] equation gives a coherent domain of about three monomers. This value is several times smaller than those of most classical cyanine J-aggregates (usually about 10–20 monomers at room temperature).[65]

5.2 *Temperature Dependences of Spectral Properties*

Static and dynamic disorders are also the reasons for spectral broadening. Similar values for the size of the coherent domain (N_C = 3–4) are obtained from the band narrowing for PBI **5a** (Figure 4.1) upon aggregation. The origin of the disorder can be better understood by evaluation of the temperature dependence of spectral lines. For PBI **5a** a bathochromic shift of λ_{max} (shift towards lower energy) for both absorption and fluorescence was observed upon decreasing temperature (Fig. 5.1a). Bathochromic shifts upon cooling are commonly observed for conjugated polymers where, in contrast to J-aggregates, the monomer units are connected by covalent bonds. These spectral shifts are usually explained in terms of increasing conjugation length and increasing refractive index (due to a higher density of the solvent) upon cooling. The same effect is expected for J-aggregates and indeed observed for PBI **5a** aggregates (Fig. 5.1).

Nevertheless, for most of the non-covalently bound J-aggregates (i. e. cyanine dye aggregates) investigated so far a hypsochromic shift of λ_{max} was observed.[66] Because all of these classical J-aggregates are formed in

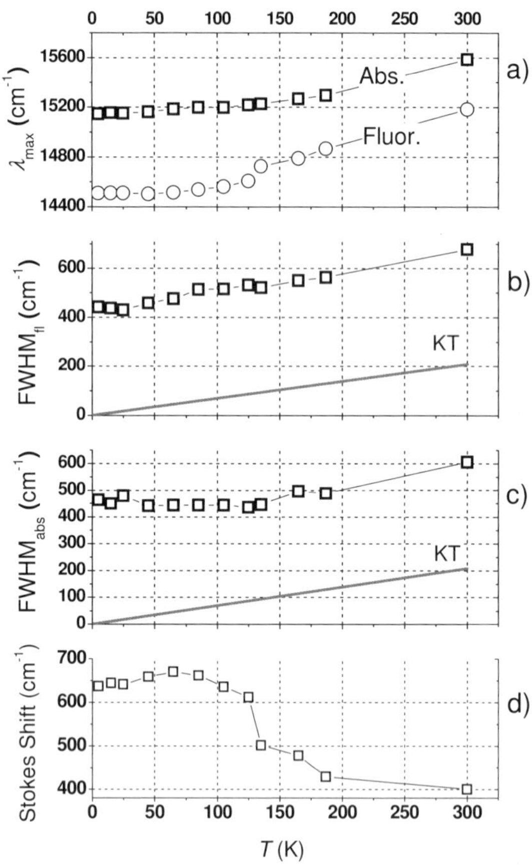

Figure 5.1 Temperature dependence of spectral parameters of PBI **5a** aggregates in MCH ($c = 5.2 \times 10^{-5}$ M). (a) Fluorescence maxima λ_{max} (circles) and absorption maxima λ_{max} (squares), b) fluorescence FWHM, c) absorption FWHM, and d) Stokes shifts. The grey straight lines in (b) and (c) illustrate the thermal energy (kT) changes with temperature. Adapted from Ref. 63. Copyright (2009) American Chemical Society.

Protic solvents like water, where solvophobic effects play a major role, it is likely that specific solvation effects are the reason for their unexpected behavior. In contrast, the present PBI **5a** aggregates are formed in nonpolar organic solvents where specific solvation effects are absent.

Absorption and fluorescence bandwidths of J-aggregates usually decrease upon cooling. This is also the case for PBI **5a** aggregate fluorescence. $FWHM_{fl}$ in MCH decreases monotonically upon cooling, except for the temperature region around 125 K corresponding to the freezing point of the solution. However, cooling from room temperature to 5 K leads to a decrease of the fluorescence bandwidth by a factor of 1.5 only. For J-aggregates formed from cyanine dyes, a much larger effect on the bandwidth (a factor of 3-4) was observed. In reference [67] the reader can find a short review on experimental data on temperature dependences of spectral properties of J-aggregates including references to original publications. Even more striking, the width of the lowest energy absorption band changes very little with temperature in case of PBI **5a**. $FWHM_{abs}$ is very large, even at low temperatures (460 cm^{-1} at 5 K in MCH). This again shows that the present system is quite different from other known J-aggregates, which exhibit up to 15 times smaller $FWHM_{abs}$ values (typically between 160 cm^{-1} and 30 cm^{-1})[67,68] than that of the present assemblies.[63] This clearly indicates that the static disorder (which is temperature independent) is much larger in the case of the PBI **5a** aggregate system compared to most of previously reported J-aggregates. Broad spectra imply small exciton coherent length in agreement with the fluorescence lifetime measurements.

Thus, the main reason for the weak temperature dependence of the bandwidths of PBI **5a** aggregates is the pronounced static disorder as revealed by the large spectral widths of PBI **5a** assemblies at low temperatures (absorption FWHM = 460 cm^{-1} at 5 K). Figure 5.1 illustrates the smaller contribution of temperature-dependent dynamic band broadening including a comparison to the thermal energy kT.

The Stokes shift depends on temperature non-monotonically and reaches its maximum at about 60 K (Fig. 5.1(d)). Similar dependence was reported for thiacarbocyanine dye THIATS[67] where the Stokes shift reaches its maximum at 20 K. It is interesting that THIATS and PBI **5a** aggregates show qualitatively the same dependence of the Stokes shift in spite of being completely different in terms of the disorder level. THIATS aggregates possess five times narrower spectral lines at liquid helium temperatures and probably even larger differences exist with regard to the exciton coherent length. The main effect behind temperature dependence of the Stokes shift at low temperatures appears to be a limited ability of the excitons to sample different states of the exciton manifold. At very low temperatures excitons are not mobile and cannot reach low energy levels situated at the tail of the density of states. When the temperature becomes higher (few tens of Kelvin) excitons are able to reach the lower levels. This leads to an increase of the Stokes shift. Here we refer to the original paper[63] or Chapter 9 of this book where the temperature dependence of the Stokes shift and especially energy migration efficiency will be discussed in detail.

6. Exciton Migration Probed in Individual Aggregates by Single Molecule Spectroscopy

6.1 *Experimental Observation of Fluorescence Blinking of Individual Aggregates on a Glass Surface*

Since about 15 years ago there have been several attempts to study optical properties of individual J-aggregates immobilized on surfaces by scanning near field optical microscopy (SNOM)[69, 70] and fluorescence microscopy at room and low temperatures.[71–73] Although studies of individual aggregates are still quite rare, these recent investigations demonstrate the great potential of this approach.

J-aggregates of PBIs **5a-c** appeared to be particularly promising because their formation via a nucleation-growth mechanism[74] ensures the instantaneous formation of large aggregates (composed of > 100 molecules) at a well-defined critical concentration $c = 1/K$.[57] This feature is illustrated in Fig. 6.1 for the experimentally derived values for PBI **5c** in MCH at room temperature. For PBI **5a** the binding constants were even higher and could not be determined accurately.[57] J-aggregates of PBIs **5a-c** are also sufficiently robust for orientation by shearing forces and transfer to substrate surfaces by spin casting. AFM studies showed J-aggregates of PBIs **5a-c** as separated 1D thread-like structures without tendency to bundle formation.[56,57] As the reader will see below, this unique combination of properties was extremely useful for quantitative characterization of fluorescence intensity of individual aggregates and exciton migration range.

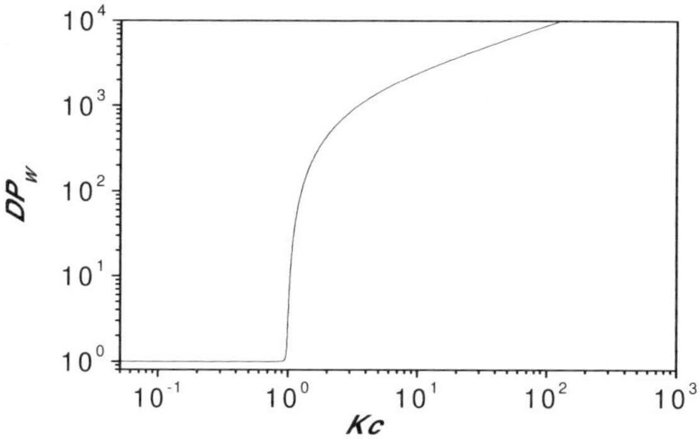

Figure 6.1 Calculated weight-averaged degree of polymerization DP_w for a supramolecular polymer formed by the nucleation-growth mechanism with K_2/K = 5 x 10^{-6} (experimental values for PBI **5c** in MCH at room temperature: nucleation constant $K_2 = 13\ M^{-1}$ and elongation constant $K = 2.3 \times 10^6\ M^{-1}$).[57]

Fluorescence of such individual threads was studied by a wide-field fluorescence microscopy setup based on an inverted microscope and a CCD camera with on-chip multiplication gain. A sample placed in a nitrogen chamber was excited by a circularly polarized CW laser light at 514 nm (for monomers) or 633 nm (for J-aggregates) through an oil-immersion objective lens (N.A. = 1.25).

Samples were prepared by spin casting 10^{-7}–10^{-8} M solutions of PBI **5a** in MCH (freshly diluted from a concentrated stock solution) on microscope cover slips. Such samples showed diffraction-limited fluorescent spots with characteristic fluorescence spectra of J-aggregates. Different individual J-aggregates were found possessing almost identical emission spectra slightly broader than the emission spectrum of the J-aggregates in MCH solution,[56] implying that the aggregates on the surface possess more disordered structure than in solution. Notably, a more disordered structure has also been found for aggregates of THIATS adsorbed on an LB film or polyelectrolyte compared to the same aggregates in solution.[75,76]

Strong fluorescence intensity fluctuations (or fluorescence blinking) were observed in isolated **5a** J-aggregates at excitation power density $I = 50\ W/cm^2$ (Fig. 6.2). This was the first time that the blinking phenomenon was reported for J-aggregates. The reason for blinking could be attributed to photogeneration of long-lived quenchers because it could be suppressed by decreasing the excitation power. The most probable origin of such long-lived quenchers with lifetimes up to several seconds is photoinduced charge-transfer processes involving the dye molecules.[77,78] Below we will show how one can use the blinking effect for obtaining information on excitation migration in J-aggregates.

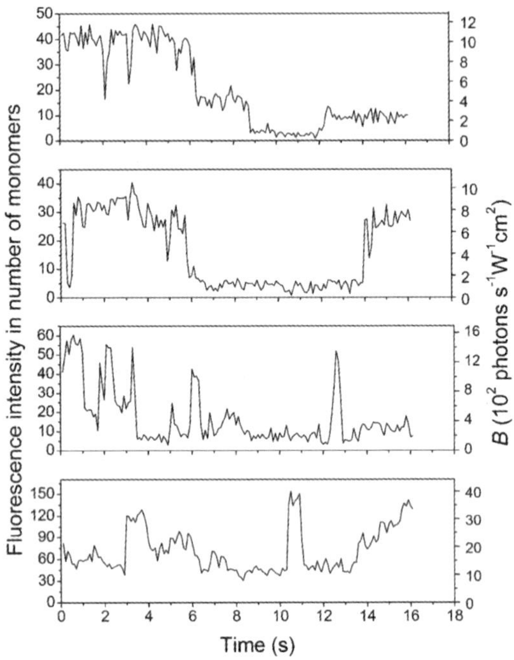

Figure 6.2. Fluorescence intensity trajectories (time resolution 100 ms) for several individual J-aggregates under CW excitation at 633 nm. Fluorescence intensity is shown in numbers of dye monomers contributing to the fluorescence (see the text for details) and in brightness coefficient *B*, i.e., photon counts per second at 1 W/cm^2 of excitation power density. Adapted from reference 64. Copyright (2010) American Chemical Society.

Blinking phenomenon is very general and it is observed not only in single quantum systems like dye molecules and small quantum dots, but also in multichromophoric ensembles such as conjugated polymers,[79] polymer aggregates,[77] and natural light-harvesting complexes.[80] It has been demonstrated that the quencher responsible for blinking at time scale of seconds in conjugated polymers is a hole polaron.[77, 81-83] Fluorescence blinking in multichromophoric systems called also "collective blinking effect" is related to energy exchange between chromophores[84–87] and to exciton "funneling" to a small number of lowest energy emitting sites.

The general idea is that once a photogenerated quencher exists in an ensemble of chromophores, it can quench any singlet exciton created therein. Even if an exciton is not formed in the vicinity of the quencher, it will reach the quencher due to energy migration among the chromophores if the migration is efficient. The size of the ensemble, i.e. the number of chromophores that can be collectively quenched, depends on the excitation migration length and the quenching radius. Because of this dependence, the blinking effect can be used as a tool for monitoring energy transfer in individual macromolecules and self-assembled dye aggregates.

It is important to realize that such collective quenching processes have a direct relation to the operation of light-driven molecular devices like e.g. solar cells.[88] Indeed, in order to use photon energy efficiently all photoexcitations created in a large antenna should be harvested at a reaction center (quencher) to produce electron-hole pairs or to run a desired chemical reaction. Accordingly, collective blinking in single molecule experiments pinpoints the ability of a light-harvesting antenna to funnel energy to a reaction center.

In spite of the fact that collective blinking "off" for up to 5-10 chromophores has been confirmed in small well-defined molecular ensembles like dendrimers and light-harvesting natural antennas,[80, 89] until recently[64,89] there were no direct experiments showing how many chromophores can be actually quenched in a single blinking event in a large system, i.e. individual chains of conjugated polymers or extended dye aggregates.

The challenge for such studies is that just by looking at the relative fluorescence intensities one cannot tell how many chromophores are quenched without knowing how many chromophores are contributing to fluorescence initially, i.e. before the quenching happens.[90] Therefore, measuring the absolute number of collectively quenched chromophores became an important task that was firstly accomplished for J-aggregates of PBI **5a** dye.

6.2 *Fluorescence Brightness of Individual Aggregates and Their Length*

Fluorescence blinking is a result of temporal fluctuation of the fluorescence quantum yield $\Phi = \Phi(t)$. Because the quenchers which cause the blinking are photo-generated, Φ is also parametrically dependent on the excitation power density I [W/cm^2]. In order to see how many monomers were collectively quenched in each blinking event, the fluorescence intensity of a J-aggregate should be "translated" into the number of monomers contributing to the fluorescence $N(t)$, which can be defined as follows:

$$N(t) = N_{total} - N_{quenched} = N_{total}\frac{\Phi(t,I)}{\Phi_a^0} \tag{6.1}$$

where N_{total} is the total number of monomers in the aggregate and Φ_a^0 is the fluorescence quantum yield in the absence of any quenchers, which can be approached by using very low excitation power density ($I \to 0$).

N_{total} in a 1D J-aggregate can be calculated from its length. According to the arrangement of monomers in PBI **5a** aggregates (Figure 6.3a),[56, 57] every monomer adds 0.75 nm to the aggregate length. It was not possible to measure the lengths of those blinking aggregates directly from the fluorescence images because they were obviously shorter than the optical resolution of the microscope. However, using more concentrated solutions for spin casting, it was possible to prepare much longer aggregates with lengths > 1 μm directly measurable from the fluorescence images (Fig. 6.3b).[64] The number of monomers in such long aggregates could then be estimated. Of course, these long aggregates did not show any blinking because they were so large ensembles that independent fluorescence intensity fluctuations of their different parts were averaged out in the overall fluorescence signal. Below it will be shown how fluorescence of such long aggregates can be used as a reference to determine $N(t)$ for blinking aggregates.

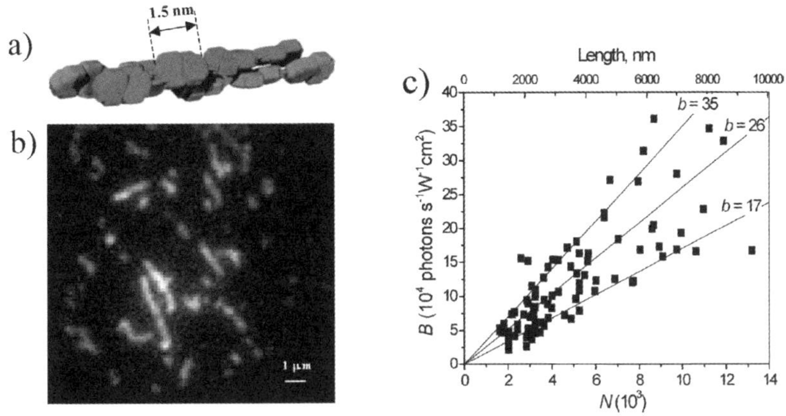

Figure 6.3. (a) The double string structure of the PBI J-aggregates (compare Figure 4.2). The dye molecule has a length of approximately 1.5 nm, however, due to double string geometry every dye molecule adds only 0.75 nm to the J-aggregate length. (b) A typical fluorescence image of a sample containing isolated long J-aggregates with length > 1 μm. (c) Scatter plot of fluorescence brightness coefficient (B_l) as a function of the number of monomers (N_l) for 92 individual long J-aggregates, where N_l was estimated according to the aggregate length measured from the fluorescence images. Fluorescence brightness per monomer $b = B_l/N_l$ was calculated to be 26 ± 9 photons s^{-1} W^{-1} cm^2. Adapted from Ref. 64. Copyright (2010) American Chemical Society.

To characterize the emission ability of a single object it is convenient to use excitation power density independent fluorescence brightness coefficient B, which can be experimentally measured or calculated:

$$B(t) = \frac{F(t)}{I} = \eta \frac{\sigma \Phi(t,I)}{h\nu} \tag{6.2}$$

where $F(t)$ [photons/s] is the detected fluorescence emission rate of the object, η is the efficiency of the fluorescence detection, $\Phi(t, I)$ is the fluorescence quantum yield, σ is the absorption cross-section at the excitation frequency ν, and I is the excitation power density. Note that B is an instrumental-dependent parameter since η varies from instrument to instrument and also depends on the configuration of the sample.

For a J-aggregate the absorption cross-section σ is directly proportional to the length of the aggregate N_{total}. Let us compare the fluctuating brightness $B(t)$ of a small aggregate containing N_{total} monomers with the non-fluctuating (due to ensemble-averaging) brightness B_l of a micrometer-sized long aggregate containing N^l_{total} monomers which can be measured directly from the image:

$$\frac{B(t,I)}{B_l(I_l)} = \frac{\sigma\Phi(t,I)}{\sigma_l\Phi_l(I_l)} = \frac{N_{total}\Phi(t,I)}{N^l_{total}\Phi_l(I_l)} \tag{6.3}$$

where I and I_l are excitation power densities used to excite the small and the long reference aggregate, respectively. Using eq. 6.3 and the definition of $N(t)$ given in eq. 6.1 we get:

$$N(t) = \frac{B(t,I)}{B_l(I_l)/N^l_{total}} \frac{\Phi_l(I_l)}{\Phi^0_a}. \tag{6.4}$$

At low excitation intensity limit:

$$I_l \to 0,\ \Phi_l(I) \to \Phi^0_a \tag{6.5}$$

we obtain:

$$N(t) = \frac{B(t,I)}{B_l/N^l_{total}} = \frac{B(t,I)}{b} \tag{6.6}$$

where $b = B_l/N^l_{total}$ is the fluorescence brightness of J-aggregates per monomer unit measured at low excitation power density conditions. In practice it means that the decrease of the fluorescence yield due to fluorescence blinking should be negligible at this excitation power. This quite obvious equation allowed us to determine the number of monomers in J-aggregates smaller than the diffraction limit.

In order to satisfy the condition (6.5) excitation power density as low as 0.5 W/cm^2 was used (100 times lower that the one used for observation of fluorescence blinking[64]). Contour lengths of 92 long aggregates were measured directly from the corresponding images and their fluorescence brightness coefficients were extracted. There indeed exists a linear relationship (Eq. 6.6) with $b = 26 \pm 9$ photons cm^2 s^{-1}W^{-1} (Fig. 6.3c). The distribution of b is broad due to the inhomogeneous environment around the single J-aggregates and the experimental error in the determination of N^{l}_{total}.

Note that so far we have not considered the possibility that some of the monomers could be permanently quenched (e.g. oxidized) so that not all the monomers are contributing to the fluorescence even at low excitation power density. In order to make sure that the effect of such "inherently dark monomers" is negligible, one can use yet another approach solely based on fluorescence intensity comparison with a reference molecule of known fluorescence yield and absorption cross section.

A good reference molecule should be highly fluorescent and photostable. Single molecules of perylene bisimides have been proven to fit these requirements very well.[91,92] Therefore, the most convenient approach is to use a single molecule of PBI **5a** dye itself as a reference for PBI **5a** aggregates. In order to prepare the reference sample we spin cast ~10^{-10} M PBI **5a** dichloromethane solution (in this solvent hydrogen-bonding and π–π-interactions are pretty weak)[47] on quartz cover slips. If the reference molecule and the molecules which aggregate have the same total transition oscillator strength (which neglects the 20% increase of oscillator strength per molecule upon formation of the J-aggregate observed for PBI **5a**[57]), the fluorescence brightness of a J-aggregate per monomer unit is equal to (see Ref. 64 for details):

$$b = B_{ref} \frac{A_a(\nu_a)\Phi_a^0}{A_{ref}(\nu_{ref})\Phi_{ref}} \frac{\nu_{ref}}{\nu_a} \tag{6.7}$$

where B_{ref} is the brightness of the reference molecule, ν_a and ν_{ref} are excitation frequencies and Φ_a^0 and Φ_{ref} are fluorescence quantum yields for J-aggregate and for the reference molecule. $A_{ref}(\nu_{ref})$ and $A_a(\nu_a)$ are normalized absorption spectra of the reference molecules and J-aggregates ($\int A_a(\nu)\,d\nu = \int A_{ref}(\nu)\,d\nu = 1$). The ratio of the normalized absorption spectra takes into account the difference between the cross sections of the reference dye and the J-aggregate originated from the differences in their absorption spectral shapes.

At first approximation fluorescence quantum yields of the monomeric PBI dye and its J-aggregates in the absence of quenchers were taken as 100 %.[56] Fluorescence intensities for several hundreds of individual PBI **5a** molecules were measured at low excitation power density (to reduce fluorescence blinking and other photoinduced quenching effects) giving $B_{ref} = 11$ photons $s^{-1}\,W^{-1}\,cm^2$ in average. From this value and Eq. (6.7) we get $b = 31$ photons $s^{-1}\,W^{-1}\,cm^2$, which agrees well with the value obtained by the first method (26 ± 9 photons $s^{-1}\,W^{-1}\,cm^2$).

This shows that J-aggregates of PBI **5a** dye immobilized on glass are indeed very good emitters where all monomers contribute to absorption and emission within the limits of experimental errors of the current analysis. Basically it means that fluorescence intensity of J-aggregates scales linearly with its length and matches well the expected values based on simple calculations.

6.3 *Blinking and Energy Migration Distance*

Now Eq. (6.6) with $b = 26 \pm 9$ photons $s^{-1}\,W^{-1}\,cm^2$ can be used to recalculate fluorescence intensity from the experimental photon count rate to the time dependence of the number of monomers $N(t)$ contributing to emission of single J-aggregates.

Examples of abrupt fluorescence intensity changes involving quenching of tens of monomers in single events are presented in Fig. 6.2. By analyzing intensity time trajectories of all studied aggregates a distribution of the maximal values of N(t) (which at first approximation is equal to the total number of monomers) N_{total} was obtained (Fig. 6.4(a)). It allowed us to see the length of aggregates present in the studied sample. The distribution is broad, ranging from 10 to 300 monomers with a peak around 30 monomers. Then the distribution of jump amplitudes was extracted (Fig. 6.4(b)). On average one jump with amplitude of more than ~50 monomers happened every 10 seconds (integral of the distribution from 50 to 120 is equal to 0.8 [jumps/10 s]) of CW excitation. Every 100 seconds a jump of more than 90 monomers could be observed. In some rare cases jumps of more than 100 monomers were detected.

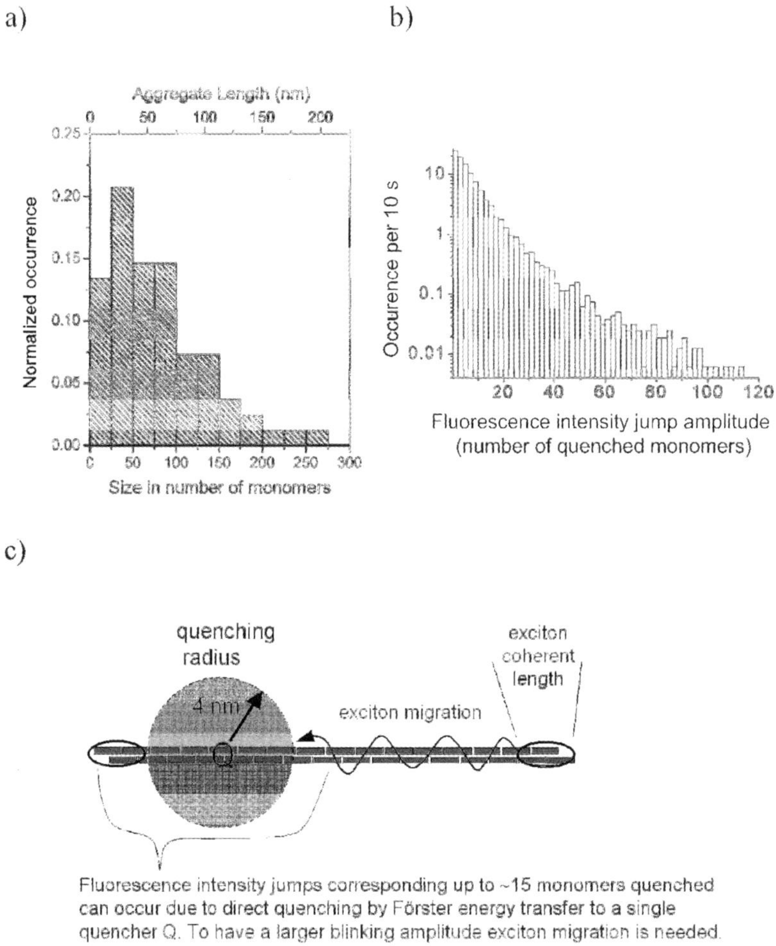

Figure 6.4 (a) The aggregate length distribution for the investigated sample. (b) Distribution of fluorescence intensity jump amplitudes in single J-aggregates of PBI **5a**, represented by the average occurrence per 10 s under excitation. (c) Schematic illustration of the effect of quenching radius, exciton coherent length, and migration distance on the jump amplitude (number of collectively quenched monomers) in PBI J-aggregates. Adapted from reference 64. Copyright (2010) American Chemical Society.

Figure 6.4(c) schematically shows how the jump amplitude (number of collectively quenched monomers) is dependent on the quenching radius, the exciton coherence length and the exciton migration length.

The exciton coherence length is about three monomers according to the emission bandwidth of the J-aggregates (see discussion in 5.1. of this chapter).[57] For the quenching radius one can assume the typical Förster distance of about 4 nm for an energy transfer process between a donor–acceptor pair with reasonably good spectral overlap and oscillator strength (as given for PBI dyes). Accordingly, direct quenching by a single quencher[93] can account only for jump amplitudes involving up to 15 monomers. To explain the much larger blinking amplitudes observed experimentally, one has to take into account an exciton migration along the 1D aggregate chain. For the experimentally observed collective quenching of 100 monomers the migration distance should be as large as 70 nm. It is noteworthy that a similar value for the exciton diffusion length (~ 100 nm) has recently been obtained for PBI **4** (Fig. 3.3) J-aggregates from exciton-exciton annihilation dynamics observed in femtosecond spectroscopy.[94] Note that both of these J-aggregates originate from the same tetraphenoxy-PBI building block but that their self-assembled architectures are dictated by different supramolecular codes imprinted in the imide substituents.

One of the possible reasons for the quite low probability of large blinking amplitudes is that among the probed aggregates those containing more than 100 monomers were substantially fewer than those containing e.g. 20 monomers (Fig. 6.4(a)). Moreover, the J-aggregate structure is expected to possess large heterogeneity on the scale of tens of nanometers because of the fragility of non-covalent bonds (Fig. 4.2) under the shearing forces applied during the sample preparation. This heterogeneity is also reflected by the distribution of the data points on the B vs. N graph (Fig. 6.3(c)). Heterogeneity may result in very different efficiency of energy transport in different aggregates or even in different parts of the same aggregate. It is the single molecule spectroscopy approach that allowed us to find the most intact self-assembled J-aggregates and to reveal the limit of ~ 70 nm for one-dimensional exciton diffusion in PBI **5a** aggregates which is of similar magnitude to that recently reported for crystallites of diindenoperylene molecules.[95] This shows that well-organized self-assembled aggregates can compete in energy transfer efficiency with the best crystalline organic materials developed so far for solar cell applications. According to our judgement, for highly ordered J-aggregate systems with larger coherent domains exciton migration lengths significantly beyond 100 nm should be feasible.

7. Conclusions and Outlook

In this article we have discussed the different aggregate types found for perylene dyes upon self-assembly in aliphatic solvents. Whilst the most common core-unsubstituted PBI derivatives prefer to form self-assembled columnar π-stacks with predominant H-type character, our recent synthetic efforts resulted in a variety of extended dye assemblies that exhibit J-coupling. Owing to the utilization of directional intermolecular hydrogen-bonding interactions unprecedented control of the dye packing could be achieved for these aggregates. Accordingly, we could establish proper relationships between supramolecular ordering and spectroscopic features for this class of dyes. The combination of contorted PBI cores (achieved by bay substituents) and directional hydrogen bonding afforded the most valuable J-aggregates that exhibit significantly red-shifted absorption bands, band narrowing, small Stokes shifts and fluorescence quantum yields close to unity. Owing to a nucleation-induced instantaneous growth of extended and well-defined thermodynamically stable J-aggregates a variety of spectroscopic investigations could be performed for these J-aggregates. Most importantly, fluorescence blinking corresponding to collective quenching of up to 100 PBI monomers has been observed for individual J-aggregates of these dyes that could be related to exciton diffusion lengths of up to 70 nm in these assemblies at room temperature. Due to the fact that rather large static disorder is present in these aggregates based on temperature-dependent spectroscopic data we envision that exciton migration lengths significantly beyond 100 nm for highly ordered J-aggregate systems with larger coherent domains should be achievable. Accordingly, it will be of interest to construct J-aggregates composed of more rigid PBI derivatives, to elucidate their photophysical properties and to apply them in devices where efficient exciton transport is required.

Acknowledgments

We are deeply indebted to our co-workers and collaboration partners who have diligently contributed to this research, and whose names are mentioned in the respective literature cited. Generous financial support by the Deutsche Forschungsgemeinschaft (DFG), Volkswagen-Stiftung, Alexander von Humboldt-Stiftung, Swedish Research Council and Knut & Alice Wallenberg foundation are gratefully appreciated. Special thanks are given to Profs. Mark Van der Auweraer, Fred Brouwer and Boris Rybtchinski for their careful reading of this chapter and helpful suggestions.

References

1. H. Langhals, *Heterocycles* **46**, 477 (1995).
2. F. Würthner, *Chem. Commun.,* 1564 (2004).
3. A. Y Avlasevich, C. Li and K. Müllen, *J. Mater. Chem.* **20**, 3814 (2010).
4. K.-Y. Law, *Chem. Rev.* **93**, 449 (1993).
5. P. Peumans, S. Uchida and S. R. Forrest, *Nature* **425**, 158 (2003).
6. M. R. Wasielewski, *J. Org. Chem.* **71**, 5051 (2006).
7. J. E. Anthony, A. Facchetti, M. Heeney, S. R. Marder and X. Zhan, *Adv. Mater.* **22**, 3876, (2010).
8. W. Herbst and K. Hunger, *Industrial Organic Pigments*, 2nd edition (VCH, Weinheim, 1997).
9. E. B. Faulkner and R. J. Schwartz, Eds., *High Performance Pigments*, 2nd Edition (Wiley-VCH, Weinheim, 2009).
10. P. Erk and H. Hengelsberg, in *Handbook of Porphyrins and Phthalocyanines*, Vol. 19, Eds.: K. Kadish, R. Guillard and K. M. Smith (Elsevier, Amsterdam, 2003).
11. F. Graser and E. Hädicke, *Liebigs Ann. Chem.* 483 (1984).
12. P. M. Kazmaier and R. J. Hoffmann, *J. Am. Chem. Soc.* **116**, 9684 (1994).
13. R. Fink, J. Seibt, V. Engel, M. Renz, M. Kaupp, S. Lochbrunner, H.-M. Zhao, J. Pfister, F. Würthner and B. Engels, *J. Am. Chem. Soc.* **130**, 12858 (2008).
14. H. M. Zhao, J. Pfister, V. Settels, M. Renz, M. Kaupp, V. C. Dehm, F. Würthner, R. F. Fink and B. Engels, *J. Am. Chem. Soc.* **131**, 15660 (2009).
15. J. Vura-Weis, M. A. Ratner and M. R. Wasielewski, *J. Am. Chem. Soc.* **132**, 1738 (2010).
16. M. C. Ruiz Delgado, E.-G. Kim, D. A. da Silva Filho and J.-L. Brédas, *J. Am. Chem. Soc.* **132**, 3375 (2010).

17. S. Grimme, J. Antony, T. Schwabe and C. Mück-Lichtenfeld, *Org. Biomol. Chem.* **5**, 741 (2007).
18. P. Zugemeier, J. Duff and T. L. Bluhm, *Cryst. Res. Technol.* **35**, 1085 (2000).
19. F. Würthner, C. Thalacker, S. Diele and C. Tschierske, *Chem. Eur. J.* **7**, 2245 (2001).
20. F. Würthner, Z. Chen, V. Dehm and V. Stepanenko, *Chem. Commun.*, 1188 (2006).
21. Z. Chen, V. Stepanenko, V. Dehm, P. Prins, L. D. A. Siebbeles, J. Seibt, P. Marquetand, V. Engel and F. Würthner, *Chem. Eur. J.* **13**, 436 (2007).
22. M. R. Hansen, T. Schnitzler, W. Pisula, R. Graf, K. Müllen and H. W. Spiess, *Angew. Chem. Int. Ed.* **48**, 4621 (2009).
23. A. Wicklein, A. Lang, M. Muth and M. Thelakkat, *J. Am. Chem. Soc.* **13**, 14442 (2009).
24. Z. Chen, A. Lohr, C. R. Saha-Möller and F. Würthner *Chem. Soc. Rev.* **38**, 564 (2009).
25. E. G. McRae and M. Kasha, *J. Chem. Phys.* **28**, 721 (1958).
26. M. Kasha, R. Rawls and M. A. El-Bayoumi, *Pure Appl. Chem.* **11**, 371 (1965).
27. J. Seibt, P. Marquetand, V. Engel, Z. Chen, V. Dehm and F. Würthner, *Chem. Phys.* **328**, 354 (2006).
28. J. Seibt, V. Dehm, F. Würthner and V. Engel, *J. Chem. Phys.* **126**, 164308-1 (2007).
29. E. H. A. Beckers, S. C. J. Meskers, A. P. H. J. Schenning, Z. Chen, F. Würthner and R. A. J. Janssen, *J. Phys. Chem. A* **108**, 6933 (2004).
30. C. Hippius, I. H. M. van Stokkum, E. Zangrando, R. M. Williams and F. Würthner *J. Phys. Chem. C* **111**, 13988 (2007).
31. A.Tolkki, E. Vuorimaa, V. Chukharev, H. Lemmetyinen, P. Ihalainen, J. Peltonen, V. Dehm and F. Würthner, *Langmuir* **26**, 6630 (2010).
32. F. Würthner, T. E. Kaiser and C. R. Saha-Möller, *Angew. Chem.* **123**, 3436; *Angew. Chem. Int. Ed.* **50**, 3376 (2011).
33. S. Nakazono, Y. Imazaki, H. Yoo, J. Yang, T. Sasamori, N. Tokitoh, T. Cédric, H. Kageyama, D. Kim, H. Shinokubo and A. Osuka, *Chem. Eur. J.* **15**, 7530 (2009).
34. First evidence for J-aggregating PBIs in a lyotropic liquid crystalline phase was reported by L. B.-Å. Johansson and H. Langhals, *Spectrochim. Acta* **47A**, 857 (1991).
35. A. Eisfeld and J. S. Briggs, *Chem. Phys. Lett.* **446**, 354 (2007).
36. A. Eisfeld and J. S. Briggs, *Phys. Rev. Lett.* **96**, 113003-1 (2006).
37. A. Eisfeld and J. S. Briggs, *Chem. Phys.* **281**, 61 (2002).
38. Z. Chen, U. Baumeister, C. Tschierske and F. Würthner, *Chem. Eur. J.* **13**, 450 (2007).
39. X.-Q. Li, V. Stepanenko, Z. Chen, P. Prins, L. D. A. Siebbeles and F. Würthner, *Chem. Commun.* 3871 (2006).
40. F. Würthner, C. Bauer, V. Stepanenko and S. Yagai, *Adv. Mater.* **20**, 1695 (2008).

41. A. Wicklein, S. Ghosh, M. Sommer, F. Würthner and M. Thelakkat, *ACS Nano* **3**, 1107 (2009).
42. S. Ghosh, X.-Q. Li, V. Stepanenko and F. Würthner, *Chem. Eur. J.* **14**, 11343 (2008).
43. X.-Q. Li, X. Zhang, S. Ghosh and F. Würthner, *Chem. Eur. J.* **14**, 8074 (2008).
44. X. Yang, X. Xu and H.-F. Ji, *J. Phys. Chem. B* **112**, 7196 (2008).
45. B. Jancy and S. K. Asha, *Chem. Mater.* **20**, 169 (2008).
46. F. Würthner, C. Thalacker and A. Sautter, *Adv. Mater.* **11**, 754 (1999).
47. F. Würthner, C. Thalacker, A. Sautter, W. Schärtl, W. Ibach and O. Hollricher, *Chem. Eur. J.* **6**, 3871 (2000).
48. C. Thalacker and F. Würthner, *Adv. Funct. Mater.* **12**, 209 (2002).
49. A. P. H. J. Schenning, J. v. Herrikhuyzen, P. Jonkheijm, Z. Chen, F. Würthner and E. W. Meijer, *J. Am. Chem. Soc.* **124**, 10252 (2002).
50. F. Würthner, Z. Chen, F. J. M. Hoeben, P. Osswald, C.-C. You, P. Jonkheijm, J. von Herrihuyzen, A. P. H. J. Schenning, P. P. A. M. van der Schoot, E. W. Meijer, E. H. A. Beckers, S. C. J. Meskers and R. A. J. Janssen, *J. Am. Chem. Soc.* **126**, 10611 (2004).
51. E. H. A. Beckers, S. C. J. Meskers, A. P. H. J. Schenning, Z. Chen, F. Würthner, P. Marsal, D. Beljonne, J. Cornil and R. A. J. Janssen, *J. Am. Chem. Soc.* **128**, 649 (2006).
52. E. H. A. Beckers, Z. Chen, S. C. J. Meskers, P. Jonkheijm, A. P. H. J. Schenning, X.-Q. Li, P. Osswald and F. Würthner, *J. Phys. Chem. B* **110**, 16967 (2006).
53. F. J. M. Hoeben, J. Zhang, C. C. Lee, M. J. Pouderoijen, M. Wolffs, F. Würthner, A. P. H. J. Schenning, E. W. Meijer and S. De Feyter, *Chem. Eur. J.* **14**, 8579 (2008).
54. S. Yagai, S. Hamamura, H. Wang, V. Stepanenko, T. Seki, K. Unoike, Y. Kikkawa, T. Karatsu, A. Kitamura and F. Würthner, *Org. Biomol. Chem.* **7**, 3926 (2009).
55. S. Yagai, T. Seki, T. Karatsu, A. Kitamura and F. Würthner, *Angew. Chem.* **120**, 3415 (2008); *Angew. Chem, Int. Ed.* **47**, 3367 (2008).
56. T. E. Kaiser, H. Wang, V. Stepanenko and F. Würthner, *Angew. Chem.* **119**, 5637 (2007); *Angew. Chem. Int. Ed.* **46**, 5541 (2007).
57. T. E. Kaiser, V. Stepanenko and F. Würthner, *J. Am. Chem. Soc.* **131**, 6719 (2009).
58. G. Scheibe, *Kolloid-Z.* **82**, 1 (1938).
59. B. Kopainsky, J. K. Hallermeier and W. Kaiser, *Chem. Phys. Lett.* **83**, 498 (1981).
60. A. R. A. Palmans and E. W. Meijer, *Angew. Chem.* **119**, 9106 (2007); *Angew. Chem. Int. Ed.* **46** 8948 (2007).
61. H. Wang, T. E. Kaiser, S. Uemura and F. Würthner, *Chem. Commun.* 1181 (2008).
62. M. Gsänger, J. H. Oh, M. Könemann, H. W. Höffken, A.-M. Krause, Z. Bao and F. Würthner, *Angew. Chem.* **122**, 752 (2010); *Angew. Chem. Int. Ed.* **49**, 740 (2010).
63. T. E. Kaiser, I. G. Scheblykin, D. Thomsson and F. Würthner, *J. Phys. Chem. B* **113**, 15836 (2009).

64. H. Lin, R. Camacho, Y. Tian, T. E. Kaiser, F. Würthner and I. G. Scheblykin, *Nano Lett.* **10**, 620 (2010).
65. I. G. Scheblykin, M. M. Bataiev, M. Van der Auweraer and A. G. Vitukhnovsky, *Chem. Phys. Lett.* **316**, 37 (2000).
66. I. Renge and U. P. Wild, *J. Phys. Chem. A* **101**, 7977 (1997).
67. I. G. Scheblykin, O. Y. Sliusarenko, L. S. Lepnev, A. G. Vitukhnovsky and M. Van der Auweraer, *J. Phys. Chem. B* **105**, 4636 (2001).
68. M. A. Drobizhev, M. N. Sapozhnikov, I. G. Scheblykin, O. P. Varnavsky, M. Van der Auweraer and A. G. Vitukhnovsky, *Chem. Phys.* **211**, 455 (1996).
69. P. J. Reid, D. A. Higgins and P. F. Barbara, *J. Phys. Chem.* **100**, 3892 (1996).
70. D. M. Eisele, J. Knoester, S. Kirstein, J. P. Rabe and D. A. Vanden Bout, *Nat. Nanotechnol.* **4**, 658 (2009).
71. M. Vacha, K. Hashizume and T. Tani, *J. Luminescence* **87-9**, 730 (2000)
72. M. Vacha, M. Saeki, O. Isobe, K. Hashizume and T. Tani, *J. Chem. Phys.* **115**, 4973 (2001).
73. E. Lang, A. Sorokin, M. Drechsler, Y. V. Malyukin and J. Köhler, *Nano Lett.* **5**, 2635 (2005).
74. T. F. A. De Greef, M. M. J. Smulders, M. Wolffs, A. P. H. J. Schenning, R. P. Sijbesma and E. W. Meijer, *Chem. Rev.* **109**, 5687 (2009).
75. N. Vranken, P. Foubert, F. Köhn, R. Gronheid, I. Scheblykin, M. Van der Auweraer and F. C. De Schryver, *Langmuir* **18**, 8407 (2002).
76. E. Rousseau, M. M. Koetse, M. Van der Auweraer and F. C. De Schryver. *Photochem. Photobiol. Sci.* **1**, 395 (2002).
77. I. Scheblykin, G. Zoriniants, J. Hofkens, S. De Feyter, M. Van der Auweraer and F. C. De Schryver, *ChemPhysChem* **4**, 260 (2003).
78. R. Zondervan, F. Kulzer, M. A. Kol'chenko and M. Orrit, *J. Phys. Chem. A* **108**, 1657 (2004).
79. J. Yu, D. H. Hu and P. F. Barbara, *Science* **289**, 1327 (2000).
80. M. A. Bopp, Y. W. Jia, L. Q. Li, R. J. Cogdell and R. M. Hochstrasser, *Proc. Natl. Acad. Sci. USA* **94**, 10630 (1997).
81. P. Dyreklev, O. Inganas, J. Paloheimo and H. Stubb, *J. Appl. Phys.* **71**, 2816 (1992).
82. P. R. Hania and I. G. Scheblykin, *Chem. Phys. Lett.* **414**, 127 (2005).
83. J. Yu, N. W. Song, J. D. McNeill and P. F. Barbara, *Isr. J. Chem.* **44**, 127 (2004).
84. F. C. De Schryver, T. Vosch, M. Cotlet, M. Van der Auweraer, K. Müllen and J. Hofkens, *Acc. Chem. Res.* **38**, 514 (2005).
85. J. Hofkens, M. Maus, T. Gensch, T. Vosch, M. Cotlet, F. Kohn, A. Herrmann, K. Müllen and F. De Schryver, *J. Am. Chem. Soc.* **122**, 9278 (2000).
86. D. A. Vanden Bout, W. T. Yip, D. H. Hu, D. K. Fu, T. M. Swager and P. F. Barbara, *Science* **277**, 1074 (1997).
87. A. J. Gesquiere, Y. J. Lee, J. Yu and P. F. Barbara, *J. Phys.Chem. B* **109**, 12366 (2005).

88. S. E. Gledhill, B. Scott and B. A. Gregg, *J. Mater. Res.* **20**, 3167 (2005).
89. M. Cotlet, R. Gronheid, S. Habuchi, A. Stefan, A. Barbafina, K. Müllen, J. Hofkens and F. C. De Schryver, *J. Am. Chem. Soc.* **125**, 13609 (2003).
90. H. Z. Lin, Y. X. Tian, K. Zapadka, G. Persson, D. Thomsson, O. Mirzov, P. O. Larsson, J. Widengren and I. G. Scheblykin, *Nano Lett.* **9**, 4456 (2009).
91. H. Uji-i, S. M. Melnikov, A. Deres, G. Bergamini, F. De Schryver, A. Herrmann, K. Müllen, J. Enderlein and J. Hofkens, *Polymer* **47**, 2511 (2006).
92. E. Lang, R. Hildner, H. Engelke, P. Osswald, F. Würthner and J. Köhler, *ChemPhysChem* **8**, 1487 (2007).
93. O. Mirzov, F. Cichos, C. von Borczyskowski and I. G. Scheblykin, *Chem. Phys. Lett.* **386**, 286 (2004).
94. H. Marciniak, X.-Q. Li, F. Würthner and S. Lochbrunner, *J. Phys. Chem A*, **115**, 648 (2011).
95. D. Kurrle and J. Pflaum, *Appl. Phys. Lett.* **92**, 133306 (2008).

CHAPTER 12

NON-COVALENT APPROACHES TO FACILE SYNTHESIS OF DIMENSION-SPECIFIC H AND J-AGGREGATES

Archita Patnaik

Department of Chemistry
Indian Institute of Technology Madras
Chennai- 600 036, India
E-mail: archita59@yahoo.com

Facile and co-operative 2D syntheses of new functionals (E)-1-(3-chloro-4-(alkoxy)-phenyl)-2-phenyl-diazene and their stoichiometrically ordered assembly of parallel-stacked unit mixed-H type aggregates as building blocks are described upon neat addition of charged fluorescent units of Rh B to the interfacial photochromes. For these integrated assemblies, molecular order was encoded in the shape and chemical functionality and was governed by the strength and directionality of secondary interactions. The solution phase H- type aggregates from the amphiphiles required long range orientational order; transformation into more stable tubular architectures with controlled molecular alignment at 2D was energetically favorable solely on account of surface-induced phenomena. Rh B influenced π–π interaction and anisotropic polarizability in addition to electronic interactions, while the diazene derivatives supported coulombic binding.

For the same amphiphiles to form J aggregates, a two component interfacial matrix stabilization technique was necessitated and devised. Langmuir isotherms of (E)-1-(3-chloro-4-(alkyloxy)phenyl)-2-phenyl-diazene (C_nCD, n = 8, 10, 12) with stearic acid (STA) as the two dimensional matrix revealed a phase separation of the C_nCD from the matrix at a compression pressure of 10 mN/m. While a one component C_nCD failed to form a 2D monolayer, the STA supported C_nCD binary system crossed a compression induced mixed monolayer phase with formation of highly ordered matrix stabilized and phase separated C_nCD J-aggregate upon proper tuning of STA- CnCD ratio, suggesting a '*know-how*' for the preparation of J-aggregates from hydrophobic functionals.

The dimeric association of a Fullerene (C_{60})-N,N-dimethylaminoazobenzene dyad (DPNME), a donor-bridge-acceptor hybrid has been explored. The study started by presenting a comprehensive DFT-study on the dyad, continued by investigation of its tunable self assembly as a function of solvent polarity in a more exploratory manner and finished with an extensive discussion on its dimerization, where the electronic coupling of the azo chromophore with the substituted 3D Fullerene C_{60} had a predominant impact on the structure organization. For a binary solvent dielectric constant $7.56 \leq \varepsilon \leq 66$, a solvent polarity dependent dynamic equilibrium between the monomer $\leftrightarrow$ J-aggregate $\leftrightarrow$ H-aggregate was established. A synergetic effect of the intermolecular donor-acceptor interaction between the electron deficient fullerene ring and the electron rich N,N-dimethylamino substituted aromatic ring governed the process of aggregation with predominant van der Waals and π-π interaction between a pair of dimers. The linear geometry of the donor and the conformational rigidity of the Fullerene C_{60} acceptor played crucial roles in envisaging the resulting π-extended network. An important part of the work comprised of our emphasis on the observed finite oscillator strength of the hitherto forbidden anti-parallel H-aggregates. The results established a novel chromophoric system in fullerene (C_{60})-azobenzene hybrid for the construction of supramolecular assemblies with opto-electronically active constituents. Importantly, substitution of fullerene C_{60} with the J-aggregate forming AZNME chromophore yielded the anti-parallel H-dimer. The overall process of H and J aggregate formation represented significant advance towards controlled molecular self assembly with predictable shapes by controlling the surface molecular density and bulk phase chemical structure and solvent polarity. The results exposed interesting new model systems for dye aggregates from which laws of cluster formation could be deduced and the ability to organize the building blocks in a specified direction could provide insight for nano-electronics design.

Keywords: Functional amphiphiles, H and J aggregates, Electronic transitions, Donor-bridge-acceptor dyads, Molecular mechanics, Surface pressure, Solvent polarity

1. Introduction

Molecular H and J aggregates are of particular interest in view of their unique electronic and spectroscopic properties. The J-aggregate is a one-dimensional molecular arrangement in which the transition moments of individual monomers are aligned parallel to the line joining their centers[1a] (end-to-end arrangement) and the H-aggregate is a one-dimensional array of molecules in which the transition moments of individual monomers are aligned parallel to each other, but perpendicular to the line joining their centers (face-to-face arrangement)[1b]. Depending on whether the unit dimers are from the same or different molecular units, they are termed as mixed aggregates. The characteristic J-aggregate exhibits a narrow J-band, red-shifted in the absorption spectrum with respect to the monomer absorption, while the absorption spectrum of the H-aggregate consists of a blue-shifted band, not as narrow as the J-band with respect to the monomer absorption, as depicted in Scheme 1.1. The energy shift of the absorption bands of the aggregates has been explained by exciton theory.[1c]

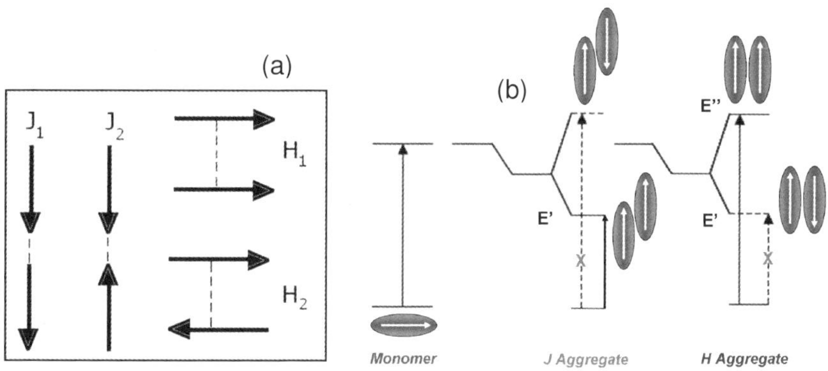

Scheme 1.1 (a) Schematic for J and H dimers; (b) Electronic energy level representation showing the splitting of energy levels of J and H dimers.

The molecular orbital overlap between two chromophores that brings interaction between the excited states leads to their splitting described by a detailed quantum mechanical treatment along with approximation of electrostatic interaction of the transition dipoles in the excited state. For aggregates composed of N monomers, the spectral shift in wave number from the monomer peak has been given as,

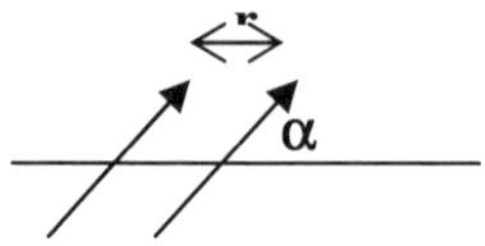

$$\Delta v = \left(\frac{2}{hc}\right)\left(\frac{N-1}{N}\right)\frac{\mu^2}{r^3}(1-3\cos^2\alpha) \qquad (1.1)$$

Where, h is the Planck's constant, *c* is the velocity of light, μ is the dipole moment, *r* is the distance between the dipoles and α, the angle between the chromophore axes and the chromophore center-to-center line. Arrangement for a typical H dimmer defines the angle α. For $\alpha = 0^{o}$, the chromophores are arranged in a head-to-tail fashion, yielding the J-aggregates and with $\alpha = 90^{o}$, the chromophores are stacked in a parallel fashion giving rise to the H-aggregates. The shift from J to H occurs at 54.7^{o}.

1.1 *H and J Aggregation in Azobenzene-based Photochromic Systems*

Azobenzene containing materials have been of immense interest for a variety of linear and nonlinear optical applications.[2–4] Their fascinating properties are largely based on the reversible photoisomerisation of the azobenzene molecule. The optical and photophysical properties of azobenzene molecules depend strongly on the substituents.[5,6] In optical applications, the pseudo-stilbene-type molecules, where the azobenzene is substituted with electron-donor and electron-acceptor groups have been often used. Pseudo-stilbenes are characterised by short lifetime of the cis isomer, a property that is useful in optical information storage.[7] Moreover, the asymmetric electron distribution gives rise to large ground-state dipole moment and hyperpolarizabilities, making them suitable for nonlinear optical applications.[8] In order to obtain high response to the optical fields, a high chromophore concentration was required where the intermolecular interactions between the dyes warranted attention. For high-dipole-moment chromophores, such interactions changed the spectroscopic properties of the material and the overall response of the system. Azobenzenes are known to form J- and H-aggregates. Characteristic molecular organization, aggregation and *cis* $\leftrightarrows$ *trans* isomerization of azobenzene derivatives in solution as well as in their confined monolayers have revealed changes in the molecular packing and surface potential, implying presence of both monomers and H-aggregates.[9] Polar and aromatic groups have promoted the formation of H aggregates. In contrast, J aggregates have formed upon carbonyl and aliphatic substitution.

Azo polymers have been considered as highly versatile materials due to their photo induced activity when irradiated with polarized laser light.[10] Various azo-polymers bearing amino-nitro substituted azobenzene units have been synthesized whose absorption maxima have been reported close to those for similar push-pull azo-compounds.[11–13] In these materials, both J and H-type aggregation have been observed in cast films. Synthesis, characterization and optical properties of a series of novel amphiphilic amino-nitro substituted azobenzene bearing end capped oligo (ethyleneglycol) side chains (RED-PEGM-n series) have been reported. Morales-Saavedra[14] showed the formation of H and J aggregates from pnPEGMAN to be a strong function of the number of PEG units.

1.2 *Azobenzene based Dendritic Chromophores and Dye Aggregates*

Organic materials have emerged as viable alternatives to conventional inorganic crystalline materials in active photonic components such as electro-optic modulators in view of their lower dielectric constant and higher electro-optic (EO) coefficient. Incorporating azo-benzene based push pull chromophores in dendritic architectures as active materials for electro-optic applications was introduced by Padma Gopalan *et. al.*[15] Compared with polyene or stilbene-based systems, the advantages of using azo-benzene based dendritic chromophores were simple, high-yielding synthetic steps and increased photo stability in air.[16,17] Azo-benzene containing dendritic benzyl aryl ethers have been extensively studied. Use of three dimensional molecules or dendrimers for control of functionality on nanometer length scales has been one of the fastest growing areas of research; attributes such as spherical shape, increased solubility, low viscosity, and monodisperse nature of dendritic molecules have been exploited in nano-electronics, drug delivery systems, and non-linear optics.[18]

Well-ordered, two-dimensional, amalgamation-type J-aggregates consisting of two different kinds of structurally analogous anionic cyanine dyes were self-assembled[19] on an amino-functionalized Au(111) electrode surface. Similar binary J-aggregates, though considerably less ordered in either one- or three-dimensional structure, could also be grown easily in a concentrated (1.5M) aqueous solution of NaCl and were characterized by a singly peaked J-band associated with a unique molecular exciton shared by the two different dye components coherently. This cooperative binary excitonic state could live with substantial energy gaps between the HOMO and LUMO levels of the two mixed dyes, where the energy gaps allowed the binary exciton to relax rapidly to an intra-aggregate charge-separated state, in good competition with the metal-induced quenching of the excited- state energy. The enhanced charge separation thus led to a high-efficiency photocurrent generation with a simple binary J-aggregate-coated Au(111) electrode in a reversible redox solution. Recent interest on supramolecular aggregates of organic compounds have focused onto ordered J and H aggregate structures, ever since their discovery by Jelley[20] for their potential in opto-electronics, light harvesting systems, and non-linear properties.[21,22]

1.3 *Langmuir and Langmuir-Blodgett Strategies in Controlled Fabrication of H and J aggregates*

The formation of stable monolayers of several cyanine dyes at the air-water interface was first described by Kaufer and Scheibe.[23] 5,6-5'6-Dibenzo-pseudoisocyanine, only slightly soluble in water, has been studied in details from its absorption spectrum by means of multiple passes of the light beam through the floating monolayers. Upon compression, a reversible formation of the Scheibe aggregate was observed at surface pressures above 12 mN/m. The monolayers of azobenzene-containing urea amphiphiles have shown carbon parity determined from the even-odd effect of alkylene spacer length to have affected the packing state of the azobenzene-urea monolayers; the urea moiety associated via intermolecular bifurcated NH---O=C hydrogen bonds afforded a robust but a highly directional intermolecular joint where the carbon number altered the aggregate state attributed to the long axis of the chromophore and a change in the molecular orientation was predicted from the intensity ratios of the π-π^* band.[24,25] Composite LB films of a cyanine dye with amphiphilic azobenzene formed irreversible J aggregates with circular domains changing to fractal-ones. Amphiphilic cyanine and merocyanine dyes that have long hydrocarbon substituents instead of the usual methyl or ethyl groups on the nitrogen have formed stable monolayers at the air-water interface and have been widely used for investigation of energy[26] and photo-induced electron transfer processes in the monolayer assemblies.[27]

Mixed Scheibe aggregates have been prepared in monolayers using amphiphilic cyanine dyes that showed a continuous shift of the position of the aggregate band with composition. As in mixed aggregates in solution, the dye couple needed to meet geometrical requirements. Compared to solution, controlling the molar ratio in monolayers for preparation of mixed aggregates was easier.

Optical anisotropic domains have been observed in monolayers of J-aggregates from transmission between crossed polarizers, fluorescence microscopy and Brewster angle microscopy. Depending on the dye and preparation conditions, the geometry of the single crystalline domains was governed. Nonlinear optical techniques have provided new insights into the exciton dynamics in aggregates.[28,29] The spatial correlation of molecular disorder could be determined by measuring the dependence of the pump-probe spectrum on pump frequency. Model experiments in monolayer assemblies using amphiphilic pseudoisocyanine as donor, 1, I1-dioctadecyl-4,4-bipyridiiiium perchlorate as acceptor and various amphiphilic molecules as super sensitizers have clearly shown charge transfer /delocalization in the aggregates of the donor.[30] A new azo containing rare earth amphiphilic coordination compound, (E)-l-methyl-4-(2-(4-(dihexadeeylamino) phenyl)azo) pyridinium tetrakis (l-phenyl-3-methyl-4- benzoyl-5-pyrazolonato) lanthanum(III) in its stable Langmuir Blodgett monolayer on hydrophilic quartz and calcium fluoride has exhibited very large second order molecular hyperpolarizability ~ 1.4×10^{-47} C m s V^{-2}, one of the largest values for the azo-based non-linear optical materials reported.[31] The molecular orientation and H and J dimer formation in Crystal Violet dye monomolecular layers were investigated through polarized reflection spectroscopy with horizontal transition dipole moments.[32] Compared to a broad band for an H dimer, the H aggregate tetramer showed a blue shifted sharp band from p-polarized spectral measurement exclusively, without any feature for the s-polarized one. The condensed phase ($40mN.m^{-1}$) 2D dye monolayer H aggregates showed peak shifting in reference to the LE phase ($10mN.m^{-1}$) aggregates owing to an increase in the slip angle, explained in terms of the extended dipole model.

2. The Experimental Technique: Langmuir Monolayers and the Langmuir–Blodgett Technique

Langmuir–Blodgett (LB) films invite attraction owing to intrinsic control of the internal layer structure down to a molecular level and the precise control of the resulting film thickness and orientation. With the advent of sophisticated LB troughs and characterization techniques, it has become possible to process vivid materials of different functionalities and offer possibilities to tune the layer architecture according to the demands of the desired molecularly engineered thin-film devices. Langmuir monolayers at the air–water interface represent quasi-two dimensional model systems possessing rich phases and structures, and have been extensively studied. Sophisticated surface analytical tools such as X-ray reflection and scattering techniques, together with Fluorescence and Brewster Angle Microscopy (BAM) have provided a detailed picture of the general phase diagrams, structure and morphology of such Langmuir monolayers formed at the air-water interface.[33,34] Structured and oriented LB films have been constructed by transferring floating monolayers from the air-water interface to a solid substrate[35–39] for potential fabrication of nano films with controlled structure.

Amphiphilic molecules when sprayed on the water surface orient with their hydrophilic heads towards water and hydrophobic tails in the air region. A flat sheet of molecules of single molecule thickness thus formed is the Langmuir monolayer, formed with different degrees of hydrophilicity. For any given substance, it depends on the ratio of total area of the water surface available to the number of molecules spread. It is indeed this property that is exploited in the Wilhelmy method where the change in surface pressure with respect to the change in the area is followed. The method uses a rectangular platinum plate or a filter paper of defined thickness and area in contact with the water surface (Fig. 2.1) where the forces acting on the plate consist of the gravity and surface tension downward, and buoyancy due to displaced water upward.

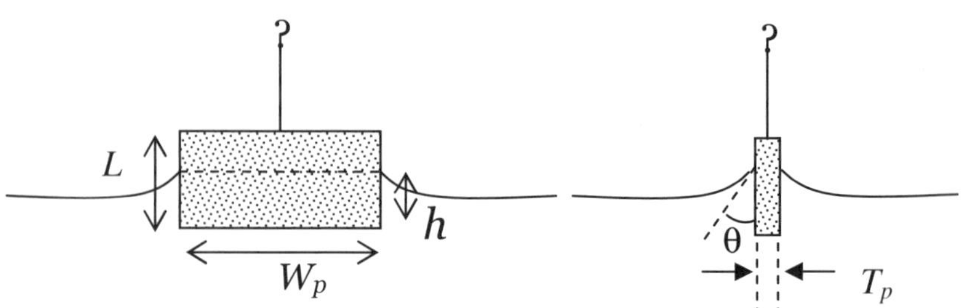

Figure 2.1. Geometries under consideration for a Wilhelmy plate immersed in water.

For a rectangular plate of dimensions L_p, W_p and T_p, of material density ρ_p immersed into a depth h_l in a liquid of density ρ_l , the net downward force is given as,

$$F = \rho_p g\, L_p\, W_p\, T_p + 2\gamma(T_p W_p)(\cos\theta) - \rho_l\, g\, T_l\, W_l\, h_l \qquad (2.1)$$

where 'γ' is the liquid surface tension, 'θ ' is the contact angle of the liquid on the solid plate and 'g' is the gravitational constant. The surface pressure is then determined by measuring the change in F for a stationary plate between a clean surface and the same surface with a monolayer present. For complete wetting of the plate, (i.e. $\cos\theta = 1$) the surface pressure is obtained from the following equation:

$$\Pi = -\Delta\gamma = -\left[\frac{\Delta F}{2(T_P + W_P)}\right] = -\left[\frac{\Delta F}{2(W_P)}\right] \text{ (if } W_p >> T_p) \qquad (2.2)$$

The force in this way is determined by measuring the changes in the mass of the plate, which is directly coupled to a sensitive balance. A plot of surface pressure vs. area, the Pressure-Area isotherm, is the common output of any Langmuir monolayer technique, measured at a constant temperature. Figure 2.2 represents the (π-A) isotherm of stearic acid at 25°C.

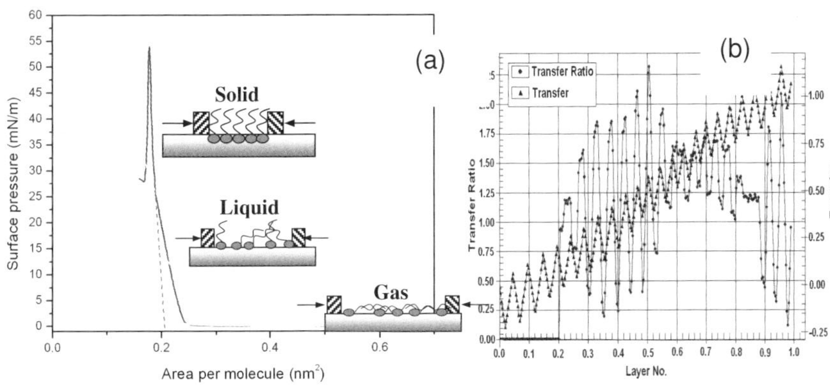

Figure 2.2 (a). (π-A) isotherm of Stearic acid at 25°C on water subphase. The inset schematics show the orientation of molecules at different compression intervals corresponding to Gas, Liquid and Solid phases. (b) A typical transfer ratio plot for fabrication of a Langmuir-Blodgett film on a designed substrate.

The surface tension of pure water is 72 mN/m. Hence the surface tension when an amphiphile is spread at the water surface is calculated as,

$$\pi = (\gamma_0 - \gamma) \tag{2.3}$$

where γ_0 is the surface tension of pure water. When the surfactant forms a thin sheet at the interface, the Gibbs equation is given as,

$$d\gamma = -\Gamma_s \, d\mu \tag{2.4}$$

where Γ_s is the surface excess defined as the excess of the amphiphile accumulated at the interface and μ is the chemical potential of the system. In terms of molar concentration, Eq. (2.4) becomes,

$$d\gamma = -RT \, \Gamma_s \, d \ln C \tag{2.5}$$

or

$$\left(\frac{\partial \gamma}{\partial C}\right) = -\frac{RT \, \Gamma_s}{C} \tag{2.6}$$

Amphiphilic molecules at the ultrapure water (Resistivity 18 MΩ cm, pH ≈ 5.9) surface upon compression undergo gas→liquid→solid phase transitions akin to gaseous systems, effected by increase in the surface molecular density to form a quasi-two dimensional solid film. The thermodynamic equation of state holds good here, but is restricted to two dimensions. Thus, the Langmuir monolayers can be represented by the 2D gas equation $\pi A = kT$. The film deposition is characterized by a deposition/transfer ratio, T_R given by Langmuir *et al.*[40] as, $T_R = A_L / A_S$, with A_L as the decrease in the area occupied by the monolayer on the water surface held at constant pressure and A_S, the coated area of the solid substrate. A typical transfer ratio plot is shown in Fig. 2.2(b) where the transfer ratio is plotted against the layer number. The films are transferred onto substrates by vertically immersing the latter at a fixed surface pressure and at a constant speed (Fig. 2.3) where, a transfer ratio of 0.9–1.0 represents an efficient transfer.

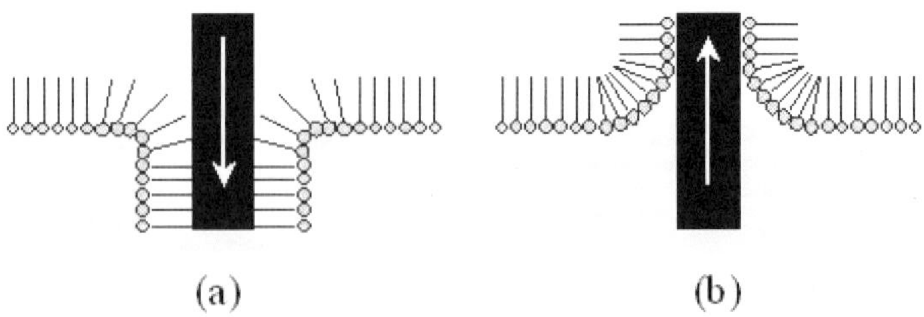

Figure 2.3. Schematic representation of transfer of Langmuir monolayers from the air-water interface to fabricate a Langmuir-Blodgett film. (a) hydrophobic substrate (immersion) to yield a Z-type film, (b) hydrophilic substrate (emersion) to result in a Y-type film.

3. Dimension-specific Self-assembly of Azobenzene-based Photochromic Amphiphiles

3.1. *Synthesis of 2D Dye-capped Azobenzene Mixed H-Aggregates*

Engineering organic crystals at a molecular level in the nano / meso scale utilizes diverse intermolecular interactions comprising of electrostatics, van der Waals forces, hydrogen bonding etc., and the secondary interactions of all categories. [41]Azobenzene derivatives form enthalpically driven pre-micellar ordered J- and H-aggregates. Characteristic molecular organization, aggregation and cis ⇆ trans isomerization of these derivatives in solution as well as in their hydrophobically confined monolayers have revealed changes in the Brewster angle reflectivity, molecular packing and surface potential, implying presence of both monomers and H-aggregates. In realizing functionality-encoded molecular architectures, fabrication of dimensionally dictated molecular J and H-aggregates from photochromic amphiphiles have been attempted,[42, 43] as shown in Scheme 3.1.

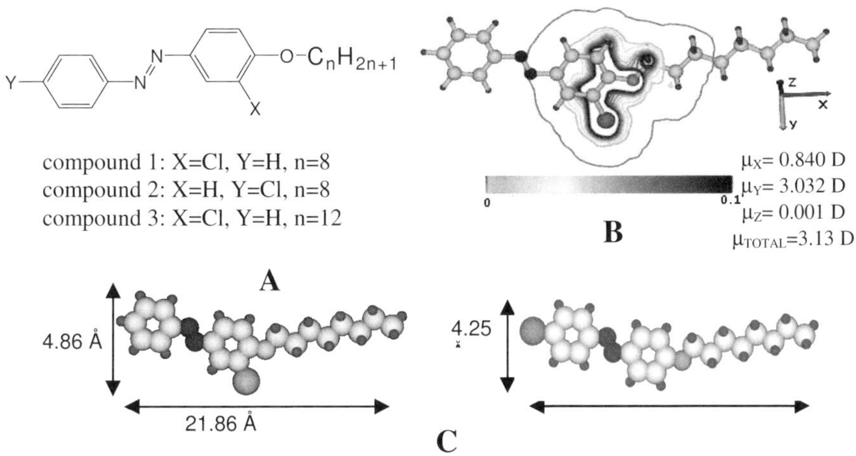

Scheme 3.1 (A) Azobenzene molecules under study. (B) The electron density contour cross section for the probable orientation of compound 1 at the air-water interface show localized electron density over chlorine and oxygen atoms. (C) Compound 1 and compound 2, geometry optimized with *ab-initio* HF calculations with 6-31+G* level of theory.[42]

A "one-shot approach" of facile and co-operative 2D synthesis (vide Fig. 3.1(I)) of a highly ordered hybrid assembly was achieved at room temperature when charged fluorescent units of Rhodamine B (RhB) were added to (3-chloro-4-octyloxy-phenyl)-phenyl-diazene (referred as ortho-C_8CD). Continuous tuning of the surface molecular density of ortho-C_8CD in association with the fluorescent dye formed highly oriented cylindrical micro-tubules with hollow centers running the entire distance of the assembly (Fig. 3.1(II)). In contrast, a planar crystalline morphology with irregular topology was observed for the (4-chlorophenyl)-4-octyloxy-phenyl)-diazene (referred as para-C_8CD). The surface tension driven self-organized structures were evidenced from electronic absorption and steady state fluorescence spectroscopy in conjunction with optical, polarizing and epi-fluorescence microscopy and micro-spectroscopy. The degree of birefringence of the tubes Δn, given by the formula $\Delta n = n\ L - nS$ (where nL and nS represent the refractive index along the long axis and short axis of the tube respectively) was calculated. In the absence of estimation of exact magnitude, the sign of Δn was arrived at using the Michel-Levy interference color chart.[44] The interference colors of the tubular structures depended upon the azimuthal angle from the extinction direction; the tubes inclined right appeared yellow while those inclined left appeared blue, vide Fig. 3.2. Δn throughout the tubular structure was found to be negative with $nS > nL$ and was attributed to large electronic polarization via strong electronic interaction between the π systems of the mixed H-type aggregates. Here the long molecular axis of the azobenzene derivative was oriented perpendicular to the long axis of the tube where refractive index was positively correlated to the electronic polarizability through the Lorenz-Lorenz equation.[45] Yao *et.al.* reported[46] mesoscopic string-like J aggregates from a solution phase thiocyanine dye that spontaneously transformed into an energetically more favorable rod-like morphology. The polarization properties of the later showed negative birefringence. In view of the dipole maximum perpendicular to the long molecular axis for compound 1

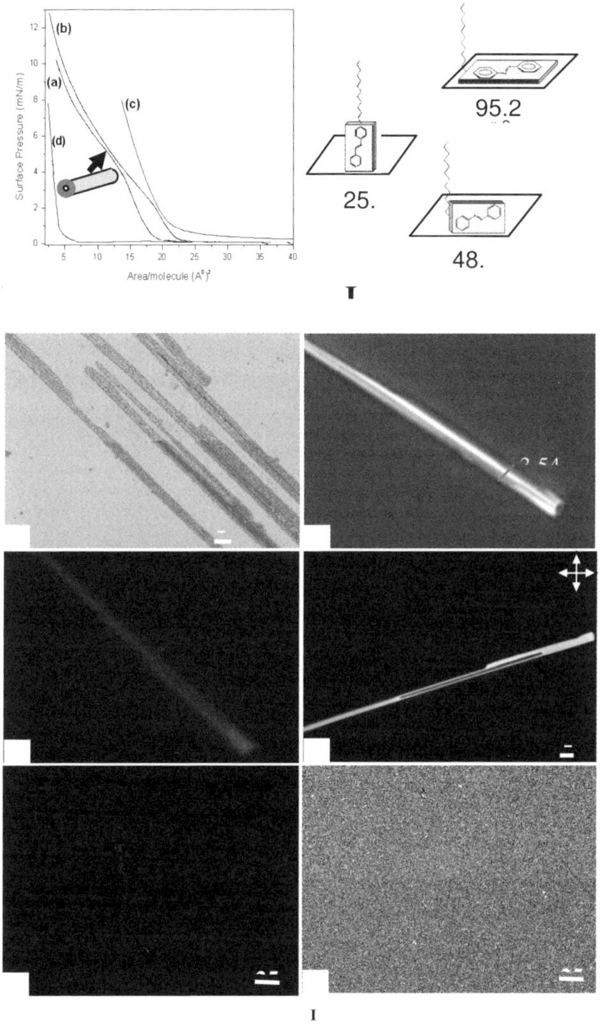

Figure 3.1 I. Pressure-Area isotherms of (a) compound 1, (b) compound 1 with Rh B in 1:0.01 molar ratio at 25°C exhibiting a distinct gas, liquid expanded, liquid condensed and a condensed phase, without collapse, (c) compound 2, (d) compound 3. The pointer indicates the pressure at which the film was transferred: Figures (A), (B), (C) represent the possible orientations of compound 1 at the air-water interface with three basic cross-sectional areas, 25.2 Å^2 along the short axis, 49.0 Å^2 along the long axis, and 95.2 Å^2 in the plane of the ring system. **II.** Optical and epifluorescence micrographs of compound 1 and compound 2 showing (a) tubular / cylindrical bundles, (b) an expanded single tube showing opening at the end, (c) a fluorescent tube of compound 1, (d) tubes under crossed polarizers, (e) Non-fluorescent crystallites of compound 2, (f) Crystallites of compound 2. (vide Ref. 42).

vide (Scheme 3.1), the dipoles in the tubes were oriented parallel to the axis of the tube. Thus in a cylindrical structure, a staircase arrangement of the dipoles was visualized. The apparent rod-like morphology was thus identified to be a tubular architecture with a monomolecular wall and open ends from optical and polarized light microscopies and fluorescence microscopic studies. The tubes from ortho-C_8CD: Rh B were smooth and unflocculated having a uniform width of 2–3 μm and were epifluorescent at λ_{ex} = 510 nm. In contrast, the planar crystallite assemblies formed from the para-C_8CD were non-fluorescent.

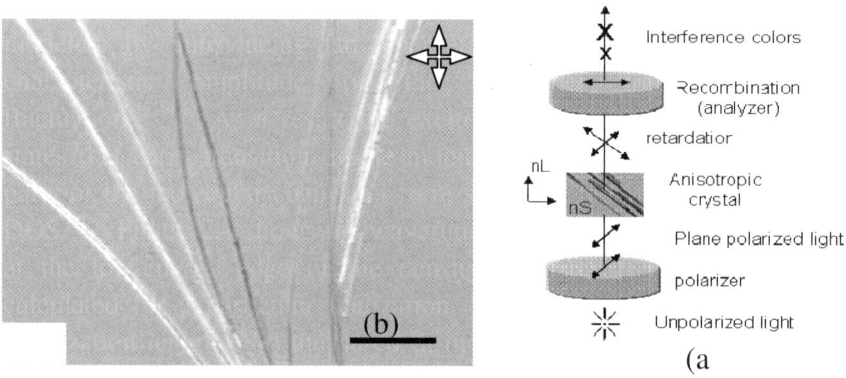

Figure 3.2 (A) Polarizing optical micrograph of tubular mesostructures with 530 nm retardation plate. The cross represents the polarizers in crossed geometry. (B) Sketch showing the phenomena of interference through retardation.

3.1.1 *Structure Elucidation for Mixed-H Aggregates: UV-Visible Absorption Spectral Characteristics in Bulk Solution and Characteristics of 2D Mesostructures*

The pristine compound 1 showed absorption maxima at 350 and 432 nm for $\pi \rightarrow \pi^*$ solvent sensitive transitions and $n \rightarrow \pi^*$ transitions respectively, while Rh B showed a characteristic prominent absorption at 550 nm. Upon addition of Rh B solution to compound 1 in increasing molar ratio, growth of the new peak centered at 319 nm (Fig. 3.3 (a)) with concomitant decrease in the monomer absorption band implied formation of an ion-association complex between compound 1 and Rh B, upon ion-dipole interaction. The trans isomer of azobenzene is planar in solution where the transition is regarded localized in the phenyl rings as a result of interaction between the π orbitals of the chromophore and the phenyl rings. Upon addition of the cationic dye to the azobenzene surfactant, appearance of the new feature, blue shifted with respect to the $\pi \rightarrow \pi^*$ band evidenced formation of mixed H-type aggregates. These unit mixed aggregates effectively gave rise to the H-type aggregate from a parallel orientation and favorable van der Waals interaction, as evidenced from the observed blue shift, falling in the ambit of the exciton theory.

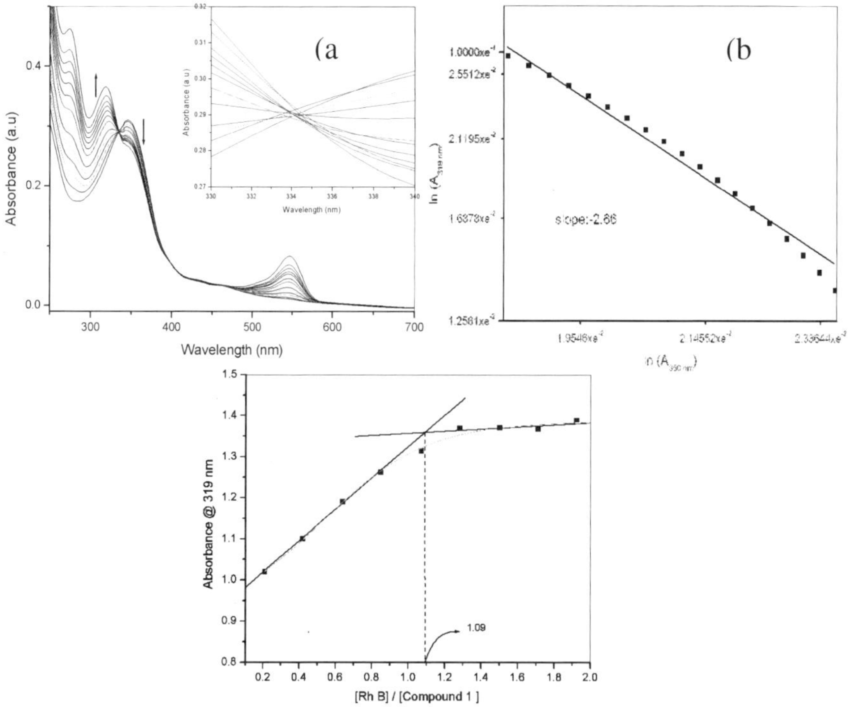

Figure 3.3 (a) UV-Vis absorption spectra of compound 1 in chloroform on titrating with Rh B over a molar ratio range of 0.0025 – 0.03, showing decrease in intensity at 346 nm, characteristic of $\pi - \pi^*$ feature of compound 1. Inset shows the cross over point instead of an isosbestic point near 334 nm, suggesting the formation of mixed aggregates; (b) Variation of monomeric absorbance of compound 1 vs. aggregate absorbance at 319 nm; (c) Plot of aggregate absorbance at 319 nm against mole ratio showing ~1:1 stoichiometry between Rh B and compound 1.[42]

The formation of a homo H-dimer/aggregate was ruled out from the nonlinearity observed in the Benisi-Hildebrand kinetic plot in Fig. 3.3(b) for the monomer $\leftrightarrow$ aggregate equilibrium:

$$\text{Aggregate} \Leftrightarrow n(\text{Monomer})$$

Thus,

$$K = [\text{Monomer}]^n \big/ [\text{Aggregate}] \quad (3.1)$$

$$\ln K = n\, ln[\text{Monomer}] - \ln[\text{Aggregate}] \quad (3.2)$$

Where, K is the equilibrium constant of the above process. The equation demonstrated the relationship between the monomer concentration to that of the aggregate in solution using 'n', the aggregation number. Applying Benisi-Hildebrand kinetics in terms of A_{agg} and A_{mono} as the absorbances of the aggregates and the monomer respectively at equilibrium we have,

$$\ln\left(A_{agg}\right) = n \ln\left(A_{mono}\right) + B \quad (3.3)$$

Where, B is a constant. For an equilibrium of the type shown above, a plot of $\ln(A_{agg})$ vs. $\ln(A_{mono})$ in Fig. 3.3(b) should have yielded a straight line with a slope = n . Deviation from linearity in the Figure discarded the monomer-aggregate equilibrium and hence formation of any homo H-aggregate of compound 1 induced by Rh B. With a slope $n > 2$ (2.66), any induced homo H-aggregate was ruled out; further from the appearance of a cross-over point instead of a sharp isosbestic point, the mixed H-aggregate formation was confirmed. Koti *et. al.*[47] reported formation of mixed aggregates from a 1:4 stiochiometric meso-tetrakis (4-sulfonatophenyl) phorphine dianion and cationic 3,3′-dihexyloxacarbocyanine iodide upon electrostatic binding with a binding constant 7.7 x10^{22} M^{-1}. On the otherhand, 3, 3'-diethyloxadicarbocynine iodide was found to induce J-aggregation in tetrakis (4-sulfonatophenyl) phorphine dianion. In Fig. 3.3 (c), the aggregate absorbance vs. the mole ratio plot of Rh B: compound 1 indicated the saturation threshold at 1:1.09, thus confirming a 1:1 stoichiometrically mixed aggregate in solution with the following equilibrium:

$$\text{compound 1} + \text{Rh B} \underset{}{\overset{K_b}{\rightleftarrows}} (\text{compound 1 - Rh B})$$

$$K_b = [\text{compound 1 - Rh B}] \big/ [\text{compound 1}][\text{Rh B}] \quad (3.4)$$

From titration experiments, K_b, the binding constant of the mixed aggregate was estimated to be 7.56 x 10^7 M^{-1} with [compound1-RhB], [compound 1] and [Rh B] as the equilibrium concentrations of the corresponding species in solution.

Figure 3.4 shows the UV-Visible absorption profiles of the mesostructures on quartz substrates with the characteristic absorption at 319 nm similar to the one in bulk solution, implying the existence of 2D mixed H-type aggregates. Linear dichroic measurements produce polarizability tensors related to tilt angles, defining molecular orientation. The tilt angle direction of the azobenzene chromophores in the multilayer mixed aggregates was visualized assuming π–π* transition directed along the molecular axis of the compound 1. When the dipping direction was rotated by 90° for parallel and perpendicular configurations, the films exhibited linear dichroism implying a regular and periodic arrangement of the constituent molecules. An enhanced intensity in the figure signified a more tilted configuration towards the surface normal when $\vec{E}$, the electric vector component of the radiation interacted with the film in parallel dipping direction.

The absorption profiles of the spin coated films on pretreated hydrophilic quartz plates at 2500 rpm showed only the presence of pristine components at 355 nm. The observation uniquely established 2D surface pressure to be the driving force for the self assembled mixed H-aggregate structures in Langmuir films. A report by Acharya *et.al.*[48] substantiated the prevalent two dimensional surface pressure to govern the coalescence of nanorods into nanowires near room temperature. The micro fluorescence spectrum of the tubes excited by 520 nm argon ion laser with emissions collected in the reflection mode showed the tubes yellowish red when excited. The spectrum showed a distinct Rh B emission band at 582 nm, close to its isotropic phase intense fluorescent band at 550 nm, when excited at 510 nm, while the pristine compound 1 did not show any fluorescence at that excitation wavelength. The 582 nm emission feature could aid in inferring that the tubes were bilayered with an outer Rh B and an inner layer of compound 1.

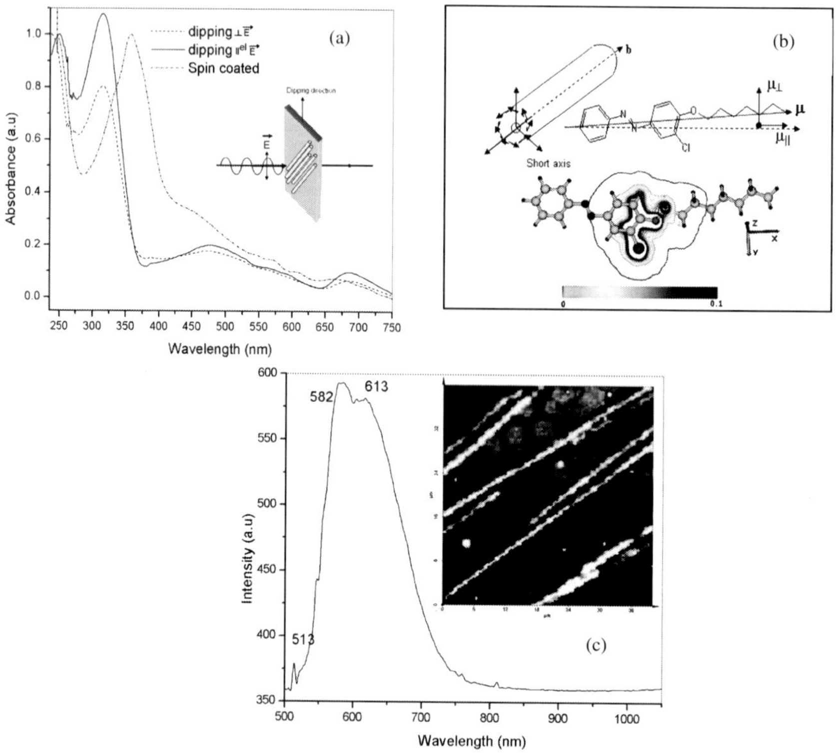

Figure 3.4. (a). UV-Visible absorption characteristics of the formed mesostructures on quartz substrates. Inset explains the enhanced absorbance with respect to perpendicular dipping direction where, $\vec{E}$ the electric vector component of the radiation interacts with the more tilted alignment in the direction to the surface normal; (b) Arrangement of dipoles in the tubules with the cross sectional electron density contour map of ortho-C_8CD. (c) Micro epifluorescence spectrum of 2D tubular mesostructure; inset shows the micro- epifluorescene image, mapping 40μm X 40μm of the tubules with 520 nm laser excitation.[42]

3.1.2 *Dependence of Aggregate Geometry on Alkyl Chain Length, Temperature and Substituent Effect*

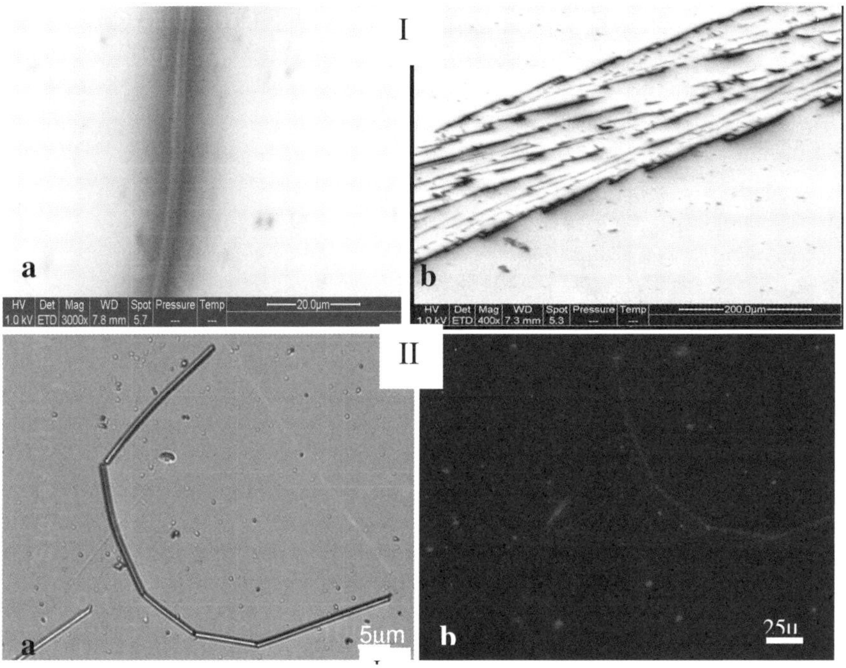

A proper balance between the hydrophobic and electrostatic interactions was accrued upon increasing the chain length to C_{12} (compound 3), that directed the final geometry of the aggregate in anti-parallel serrate assembly as shown in Fig. 3.5(Ib). Temperature controlled supramolecular synthesis of the hollow tubules was demonstrated when the interfacial tubules formed at 40°C assembled in an end-to-end fashion against parallel stacking at room temperature, as a result of increase in the line tension between the condensed phase and the fluidic phase.[49]

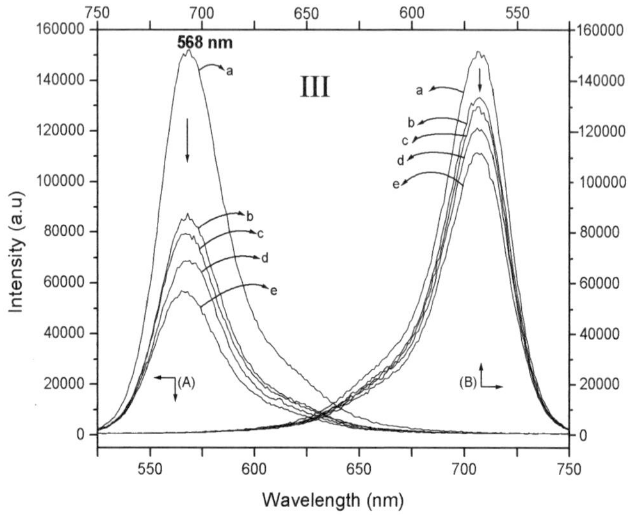

Figure 3.5 I. Scanning electron micrograph of (a) compound 1 with Rh B (b) compound 3 showing anti-parallel serrate arrangement; II. Optical micrographs of compound 1 at 40°C showing end to end 2D interracial self-assembly. (b) Epifluorescence micrograph of the end -to-end assembled tubes; III. A. Fluorescence quenching of Rh B by compound 1, B. compound 2: with varying concentration of Rh B. Curve (a) Rh B 100% ; (b–e) 2–8% v/v of azobenzene derivative (Excitation wavelength = 510 nm).[42]

Shape transitions as a result of enhanced in-plane dipole moment at low temperature have been revealed by quantum mechanical calculations. Self-assembly depends critically on thermal energy where the resultant motion allows structures to attain the optimum positions for highly ordered structures with thermally induced self-assembly, typically requiring repulsive or only very weakly attractive interactions.[50] A compact shape would have been favored with decreasing temperature, whereas elongated shapes would have arisen with increasing temperature. Steady state fluorescence measurements of solutions of Rh B mixed with the compound 1 in Figure 3.5 III showed quenching of the 550 nm emission band of Rh B while the compound 2 did not, indicating absence of any interaction of the latter with Rh B. The formation of an electrostatically associated complex between Rh B and compound 1 was attributed to the fact that the lone pair of electrons on the alkoxy oxygen facilitated the coulombic binding of ion-dipole type with the net positive charge existing on the quaternary nitrogen of Rh B, as depicted in Scheme 3.2(a).

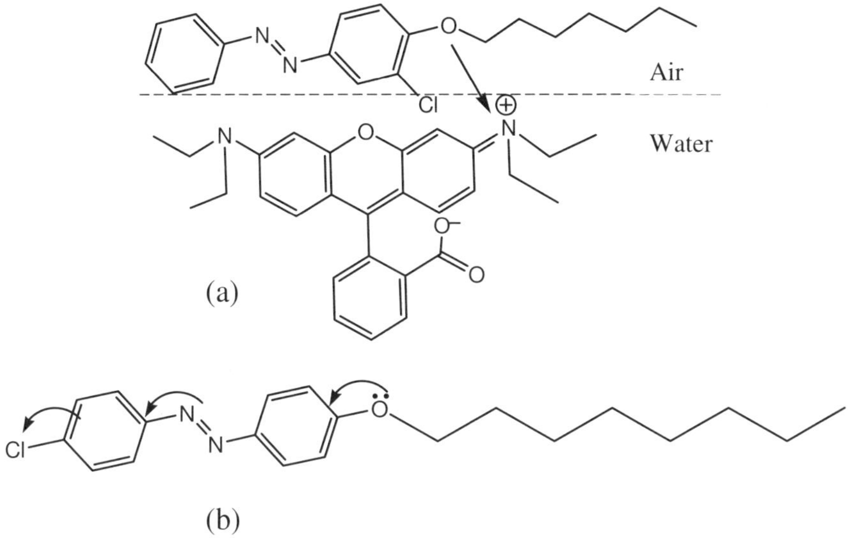

Scheme 3.2 (a) Ion-dipole interaction between ortho-C_8CD and Rh B at the air-water interface. (b) The electron withdrawing effect in para-C_8CD responsible for non-fluorescent crystallites.

For compound 2, the electro negativity of the chloro group situated para to the –N=N– decreased the coordinating ability of the alkoxy oxygen. The specificity of chlorine moiety on the molecular skeleton of compound 1 played a vital role in the structure formation; absence of chlorine in compound 1 as well as in compound 2 did not form such ordered structures. A probable explanation was attributed to the electron density localization by the electronegative chlorine substituent, facilitating the ion-dipole interaction between compound 1 and Rh B. The above mechanism of self assembly suggested that the linear configuration of the azobenzene moiety, the *trans* isomer, alone favored such a self-assembly, further validated by the fact that photoisomerization of compound1 did not serve as a trigger to induce self-organization when the formation of the meso H-aggregate structures was specific to 2D. A stair-case arrangement of dipoles was conceived from the negative birefringence of the tubules.

The model proposed below for the tubular assembly of the dimension specific H- aggregates in the amphiphilic system based on the above investigations thus revealed the aggregates to exist in different morphologies depending on the substitution pattern of the azobenzene chromophores in presence of the Rh B dye. The sensitivity of the aggregate morphology to slightest chemical modifications in the molecular framework has thus been implied. Tubule formation has been modeled,[51,52] where the anisotropy of films due to symmetry breaking in the molecular packing has led to formation of tubular structures. Added to this, the energy contribution from stacking of the rigid π-system of Rh B and its electronic interaction accounted for the observed shape. The hydrophobic tails attached to the chromophores favored an aggregation of the building blocks, the associated complex between compound 1 and Rh B, into bilayer sheets which could roll into a tubular assembly with variation of surface tension as the driving force. With the direction of the molecular alignment in the rod aggregate evaluated from the polarization excitation under the 530 nm retardation plate, the formation mechanism for the tubular assembly is shown in Scheme 3.3.

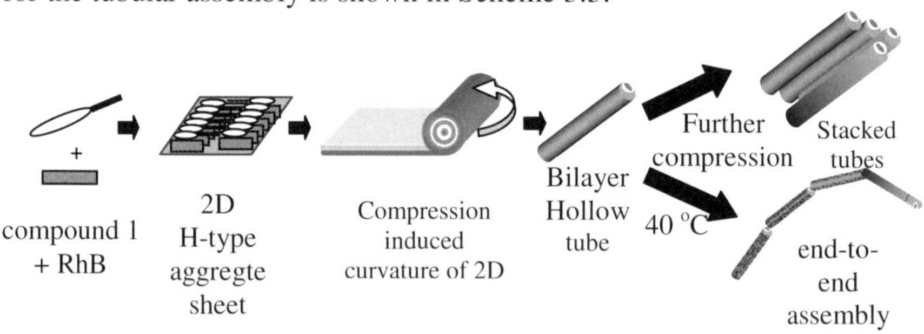

Scheme 3.3 Mechanism of surface driven self assembly leading to tubular aggregates of compound 1 and Rh B in two dimensions.[42]

3.2 Matrix stabilized J-aggregates of (E)-1-(3-chloro-4-(alkoxy)-phenyl)-2-phenyldiazenes (C_nCDs) at the Air-Water Interface

The underlying assumption in the LB method states the Langmuir monolayers to serve as simple building blocks resembling the features of pre-formed monolayers at the air–water interface. Oriented molecules in such films exhibit properties different from that of the same when in bulk or in three dimensions. This exciting behavior in 2D makes LB films to be extended to ultra-thin film devices of the current generation. Electronically excited molecules in LB films have properties quite different from those of the solution or the bulk. Highly ordered thin films of amphiphilic dye molecules with specific orientation and restricted motion of the chromophores have shown large optical non-linearity.[53]

Single component films formed at the air-water interface have required scaffolds to stabilize the monolayer. Recent reports on such mixed monolayers as multi-component systems with vivid properties have established them in par with the pure ones.[54–56] Mixing a non-amphiphilic system with a long chain acid or an inert polymer matrix has resulted in the formation of high quality LB films of such materials.[57–59] Martin *et al.*[60] studied the two dimensional ordering of non-amphiphilic cationic/anionic porphyrins by electrostatic interaction with amphiphilic anchors. A more recent work[61] emphasized on H-bonded monolayers initiating interdigitated multilayers at the gas-liquid interface. Host-guest monolayers of cobalt complexes of octaalkoxyphthalocyanines with nematic liquid crystals have been reported by Seki *et al.*[62] A phospholipid matrix has been used to form water soluble porphyrin aggregates at the air-water interface.[63] Ocko[64] reported the structure and phase behavior of mixed monolayers of saturated and unsaturated fatty acids.

The behavior of mixed monolayers of cholesterol and poly(methylmethacrylate) (PMMA) with molecular weights of Mw 120000g/mol and Mw 15000g/mol was investigated at the air/water interface.[65] With existence of two collapses in the mixed monolayers, the molecular area-mole fraction plots demonstrated the immiscibility of the film components at the interface. The application of Crisp phase rule to the phase diagram of PMMA-Cholesterol mixed monolayers helped to explain the existence of a biphasic system, regardless of their composition and surface pressure. The high reflectivity heterogeneous cholesterol domain when immersed in a homogeneous polymer separate phase with low reflectivity, extended a candid application to contact lense use at PMMA–Cholesterol tear interface. Behavior of sulfatide/ cholesterol mixed monolayers at the air/water interface was investigated.[66] Thermodynamic analysis indicated negative deviation of the excess molecular area and surface excess Gibbs energy from the ideal behavior at various molar ratios, suggesting an attractive interaction between sulfatide and cholesterol in the monolayers as compared with the pure component monolayers. The behavior of sulfatide was thought to be involved in lateral domain formation in biological membranes, mimicking "lipid rafts" domains. Structural characterization of a mixed LB film of a merocyanine dye derivative MS 18 and a deuterated arachidic acid C20-d binary system was achieved.[67] Influence of successive hydrothermal treatment in the liquid phase (HTTL) on the film was investigated by polarized UV-Visible and IR absorption spectroscopy. The variations in the conformation and orientation of the MS 18 hydrocarbon chain and the orientation of the C20-d hydrocarbon chain changed from ordered and disordered structures to more disordered and ordered ones respectively, where the former was mainly caused by the priority action of thermal energy and the latter by hydrophobic effect due to the presence of warm water. Consequently, a correlation between the degree of structural order for both hydrocarbon chains during the HTTL treatment was established.

Monolayers of stearic acid (STA) and octadecylamine proved highly rigid with low compressibility and with strip-like collapse morphology, observed from Brewster angle microscopy.[68] Mixed monolayers at the air–water interface of the cationic surfactant, hexadecyltrimethylammonium bromide (CTAB), with alkanes showed a first-order freezing transition as a function of temperature. Surface freezing was more favorable in mixed monolayers at the air–water interface than at the bulk alkane–water interface for the same surfactant concentration.[69] Studies on mixed Langmuir monolayers are relevant to a variety of scientific and industrial areas such as membrane biophysics, lungs function, pharmaceutical science, food technology, and detergent technology. The interaction between deuterated dipalmitoylphosphatidylcholine (DPPC-d_{62}) and palmitic acid (PA) in mixed Langmuir monolayers was studied using vibrational sum frequency generation (VSFG) spectroscopy, which revealed palmitic acid to have increased the chain ordering of DPPC and had promoted the liquid expanded (LE) phase with phase separation. The presence of palmitic acid decreased the chain tilt angle of DPPC and increased the orientational ordering of DPPC chains.[70]

Apart from the fundamental studies, mixed-monolayer technique could be conveniently used in controlled-aggregation phenomena. The method became vital when functional J aggregates were formed in ultra thin films with potential applications in light harvesting systems as well as in transport of energy and/or charge. Two dimensional mixed J-aggregates of structurally and spectrally analogous anionic cyanine dyes, co- adsorbed on amino alkanethiolate self-assembled monolayers on gold electrode showed unique light-absorbing behavior, yielding large cathodic photo current.[71] Comparison between the organization of dye molecules into J-aggregates from a thiacarbocyanine dye in the presence of an oppositely charged polyelectrolyte and the polymer bound J-aggregates in solution have been studied with respect to the influence of the polyelectrolyte and the structure of the J-aggregate. Bathochromic shifts in the absorption and emission spectra of mixed benzfluoranthene and STA LB films revealed the formation of ordered aggregates in the former.[72] In the mixed monolayers of 4-n-octyl-4′-cyanobiphenyl (8CB) and STA,[73] STA stabilized 8CB without any steric hindrance or condensation; the alignment of 8CB depended on the surface pressure, but not on its concentration. Interaction between the excitonic molecular aggregate monolayer[74] and surface plasmons[75] of the metallic nanoparticles drew current interest in view of their application in biological and chemical sensors[76,77] as well as in surface enhanced Raman scattering.

3.2.1 *Matrix-assisted Interfacial Monolayers, Phase Behavior, Effect of Chain Length on Ordering from C_nCDs*

For the mixed surfactant systems, relation between the surface pressure and molecular orientation of the amphiphile in the monolayer and the existing interaction between the constituent molecules play an important role, thereby demanding a better understanding. The mixed Langmuir films consisted of two molecular species with different chemical properties, as shown in Scheme 3.4.[43]

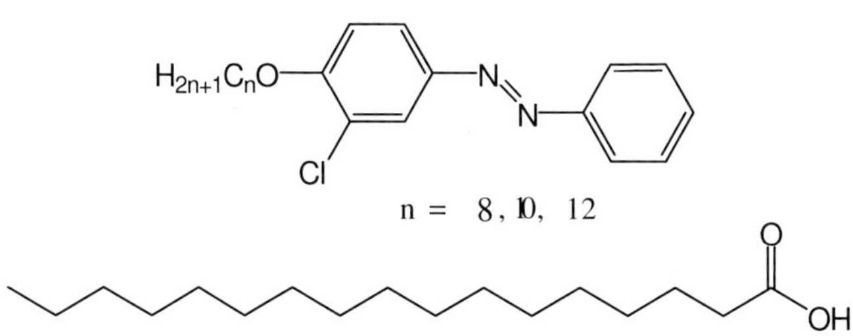

Scheme 3.4 Chemical structures of amphiphiles for the preparation of matrix stabilized molecular aggregates.[43]

The molecules were geometry optimized with density functional calculations at the B3LYP/6-31G level of theory. For the pure C_nCD and the C_nUD (the underivatized amphiphile with $n = 12$), when spread at the air–water interface, resulted in aggregates. An undefined collapse pressure and an area/molecule < 22 Å^2 (ideal STA monolayer) discarded the formation of ideal monolayers of sprayed C_nCDs.

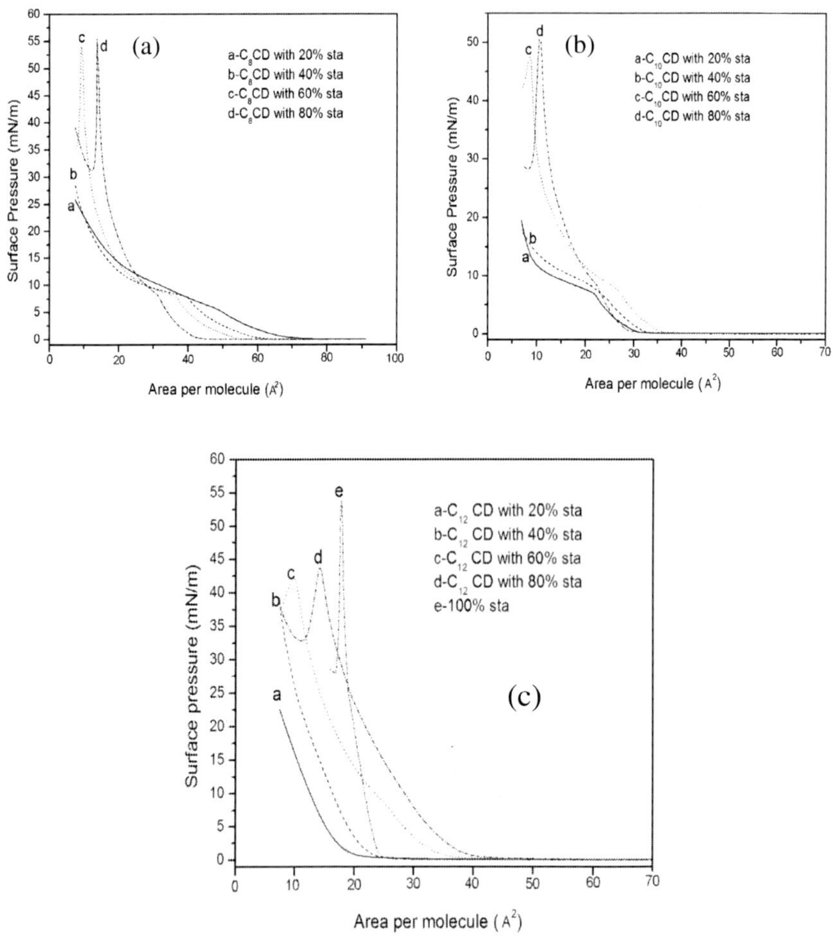

Figure 3.6. Surface pressure–area isotherms of (a) C_8CD, (b) $C_{10}CD$ and (c) $C_{12}CD$, mixed with stearic acid at different ratios at 25°C.[43]

With STA as the molecular matrix at the air–water interface, C_nCDs formed stable compressible monolayers, withstanding high surface pressures, ~ 55 mN/m, as shown in Figures 3.6(a)–(c). With increase in the mole fraction of STA, a decrease in the area per molecule suggested accommodation of the compound within the two dimensional STA matrix. The general shape of the isotherm was thus a region where the surface pressure was almost zero with a slowly increasing surface pressure region upon compression, and a region with a plateau in the isotherm. From the molecular models of the *trans*-azobenzene,[78] the probable monolayer orientation could be discussed. The molecular dimensions of azobenzene moiety with length, width and thickness as 13.6 Å, 7.0 Å and 3.6 Å respectively,[79] three basic cross-sectional areas with corresponding orientations: 25.2 $Å^2$ along the short axis parallel to water surface, 48.9 $Å^2$ along the short axis perpendicular to the water surface, and 95.2 $Å^2$ in the plane of the ring could be visualized, and are shown in Fig. 3.7. The area per molecule from $A_{\pi\rightarrow 0}$ ~50 $Å^2$ for 40% STA in Fig. 3.6 (a) suggests C_8CD moiety to be oriented in a sideways fashion according to Fig. 3.7c, with the short axis perpendicular to the water surface. Upon increasing the STA content as well as increasing compression, the decreased average area implied close packing of 2D STA monolayer with the formation of aggregated C_nCD domains. In the two-component mixed monolayer systems, three types of interactions could prevail: (i) STA-C_nCD (ii) STA-STA and (iii) C_nCD-C_nCD. Among such molecular interactions, a predominant C_nCD-C_nCD interaction could form the 2D aggregated domains. From the isotherms of the mixed systems, the average area per molecule at 10 mN/m when plotted for various molefractions of C_nCDs (Fig. 3.8), a positive deviation from ideality was observed for $C_{10}CD$ and $C_{12}CD$. The results illustrated in Fig. 3.8 show C_8CD with a negative deviation. The positive deviation indicated a repulsive interaction between the components, while a negative deviation for C_8CD on the other hand implied the interaction to be attractive. The additivity rule was tested for the probable existence of ideal mixing[80] and for immiscible monolayers.

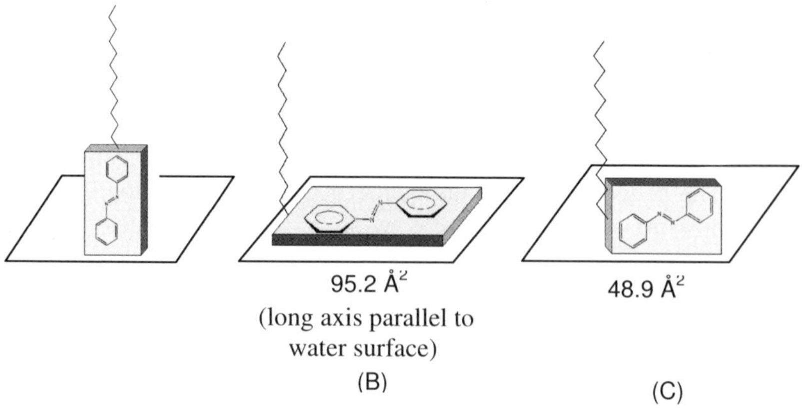

Figure 3.7 Possible orientations of azobenzene moiety of C_nCDs at the air–water interface.[43]

The rule assumes each molecular species to occupy an area at a certain pressure (π), independent of the other molecules in the film as,

$$A(\pi) = A_{CnCD}(\pi).X_{CnCD} + A_{sta}(\pi).X_{sta} \quad (3.5)$$

Where, X is the mole fraction of the species 'i' in the mixed film, A (π) and A_i (π) are the area per molecule in the mixed film and the single components respectively, as a function of the surface pressure. Since the pristine C_nCDs did not form a well defined monolayer, the possibility of miscibility was ruled out. Thus, at higher compressions the whole space at the air–water interface was occupied only by STA molecules, implying the azobenzene molecules to be squeezed out of the water surface. The phenomenon was supported by the steep increase in the surface pressure at higher concentrations of STA or at increased compression with a decrease in the area/molecule, tending towards 20Å^2. The isotherm implied a phase separation where the C_nCDs were completely removed from the mixture owing to their insolubility in water and consequently formed aggregated domains on top of the STA monolayer.

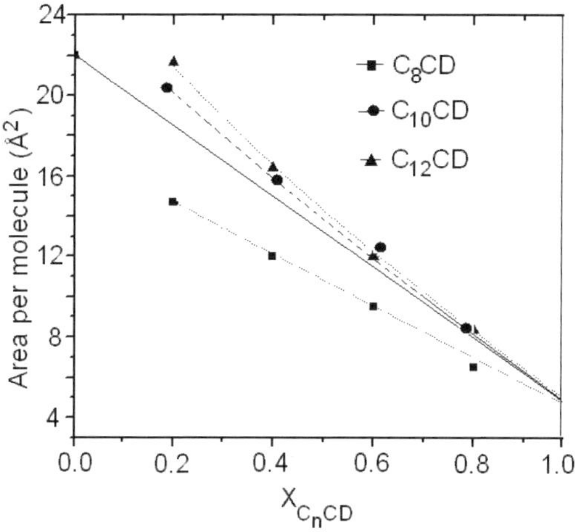

Figure 3.8 Area/composition plots (at π =10mN/m) of $C_{10}CD$ and $C_{12}CD$ showing positive deviation and C_8CD representing negative deviation from ideality, shown as the solid line.[43]

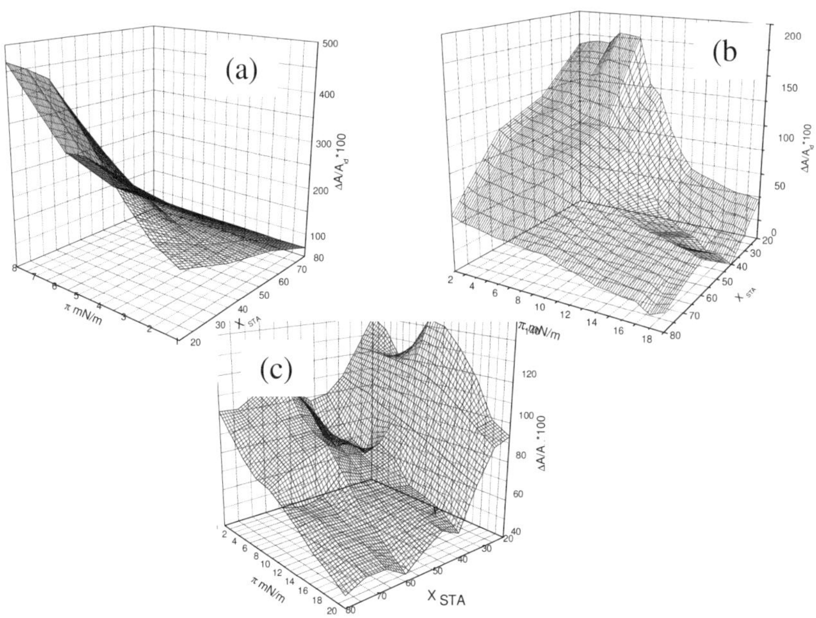

Figure 3.9 [$\Delta A/A_{id}$] relative excess areas in (%) of (a) C_8CD, (b) $C_{10}CD$ and (c) $C_{12}CD$ in the STA matrix with various compositions.[43]

An analysis for the excess area of mixing (ΔA) = $A_{(CnCD,sta)}$ – [$A_{CnCD}(\pi).X_{CnCD} + A_{sta}(\pi).X_{sta}$] also evidenced (Fig. 3.9) phase separation in the mixed monolayers, where $A_{(CnCD,sta)}$ is the average surface area per molecule in the mixed monolayer and A_{CnCD} and A_{sta} are the molecular areas of single component monolayers at the same surface pressure and X_{CnCD} and X_{sta}, the mole fractions of the components in the mixture. With increasing X_{sta}, ΔA decreased with increase in the surface pressure, finally leading to a complete phase separation. Thus, mixing was less favored at higher surface pressures, where the film was in a compressed state.

A proper balance of hydrophobic and hydrophillic interactions among the monolayer components is critical in obtaining a stable film at the air-water interface. Increasing the chain length from C_8 to C_{12} resulted in a drastic change in the phase behavior of monolayers. The plateau representing the two phase coexistence of liquid condensed and liquid expanded phases in the isotherms in Figures 3.6(a)–(c) decreased from C_8 to C_{10} as a result of enhanced hydrophobic interaction in the molecular skeleton, finally disappearing in $C_{12}CD$. In the most hydrophobic system of $C_{12}CD$, aggregated domains were formed after compression beyond the gas phase region in the STA matrix. Figures 3.9(a)–(c) represent the changes in the relative areas [$\Delta A/A_{id}$] with A_{id} as the ideal area, as the specific area changes along the isotherm for different mole fractions of STA. C_8CD showed a better mixing with STA while $C_{10}CD$ and $C_{12}CD$ showed greater repulsion in two dimensions because of an increased hydrophobic nature. However, for a proper combination of surface pressure and surface molecular density of the monolayer components, a minimum could be observed for $C_{12}CD$ case, implying formation of a stable film at the air-water interface. Thus, aggregation at the interface could be controllable by adopting suitable experimental conditions in conjunction with appropriate chemical structures.

3.2.2 *Matrix Stabilized J-Aggregates from C_nCDs*

The Langmuir film transfer onto hydrophilic quartz substrates (vertical dipping, compression speed 5 mm/min at 10 mN/m) followed a Y-type deposition with transfer ratios between 1.1 and 1.2, suggesting an efficient transfer of the condensed phase of the C_nCDs onto the substrate. UV-visible absorption profiles of the transferred LB films in Fig. 3.10 showed a red shift of the $\pi \rightarrow \pi^*$ transition from 347 nm to 393 nm and of the $n \rightarrow \pi^*$ transition from 430 to 459 nm, indicating the formation of J-aggregates of C_nCDs. J-aggregates are characterized by their absorption bands shifted to longer wavelengths with respect to the monomer absorption and a sharp emission band with very little Stokes shift.

When the alignment of the dipoles in the aggregates was such that the molecular tilt angle followed $0 < \theta < 54.7^{\circ}$, the exciton band was located energetically below the monomeric state, resulting in a red shifted absorption spectrum. The topography and nanoscale domains of J-aggregates have been well probed using SPM techniques.[81,82] The AFM images of 1:1 C_8CD and $C_{12}CD$ with STA shown in Figure 3.10 represent the respective LB films with densely packed, superimposed globular islands of ~50 nm.

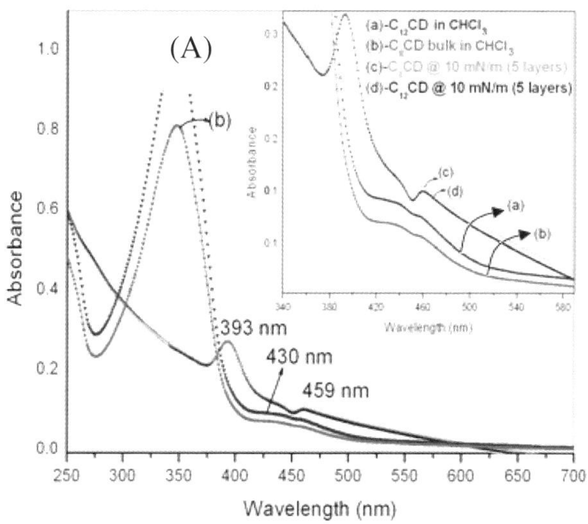

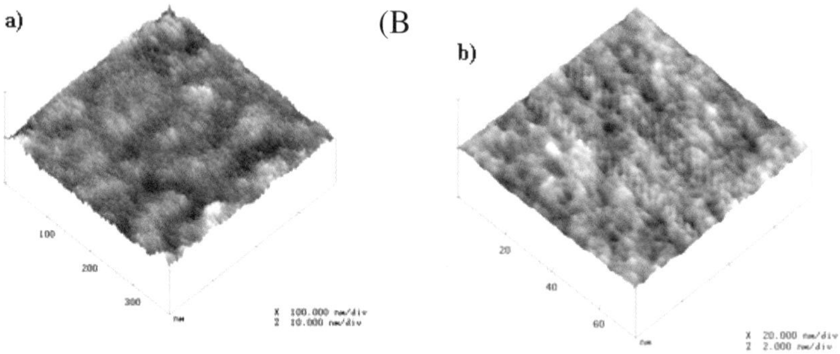

Figure 3.10 A. UV-visible absorption profiles of (a) and (b) C_8CD and $C_{12}CD$ in bulk. (c) and (d) the LB films transferred at10 mN/m and in bulk solution. Inset shows the expanded regions of the absorption profiles. B. Atomic force microscopic images of (a) 1:1 C_8CD: STA and (b) 1:1 $C_{12}CD$:STA monolayers on Si(100) transferred at 10 mN/m.[43]

Raman spectroscopy as a versatile tool for analyzing molecular aggregates was employed. The surface Raman spectroscopy of azobenzene containing fatty acid films[83] has shown strong Raman bands in the regions 1135–1150 cm^{-1}, 1400–1450 cm^{-1} and 1580–1600 cm^{-1} from azobenzenes and their derivatives. The micro-Raman spectra of the LB films acquired from two different spatially resolved domains in Figure 3.11 depict the –N=N– Raman stretching frequency at 1443 cm^{-1}, characteristic of the surface adsorbed azobenzenes and the Raman stretching frequencies of –N=N– and phenyl –N=N– groups shift to 1443 cm^{-1} and 1127 cm^{-1} from 1448 cm^{-1} and 1153 cm^{-1} respectively. Raman spectral analysis of J-aggregates of Congo-red azo dye incorporated in a polyion matrix has been investigated by He *et al.*,[84] where the surface adsorbed aggregates showed a similar shift towards the lower wave numbers. Raman images generated by mapping with 1443 cm^{-1} showed the presence of aggregated domains of C_nCDs (Figs. 3.11 (d) and (e) and the image mapped with 1603 cm^{-1} showed an image complementary to that of Fig. 3.11 (e).

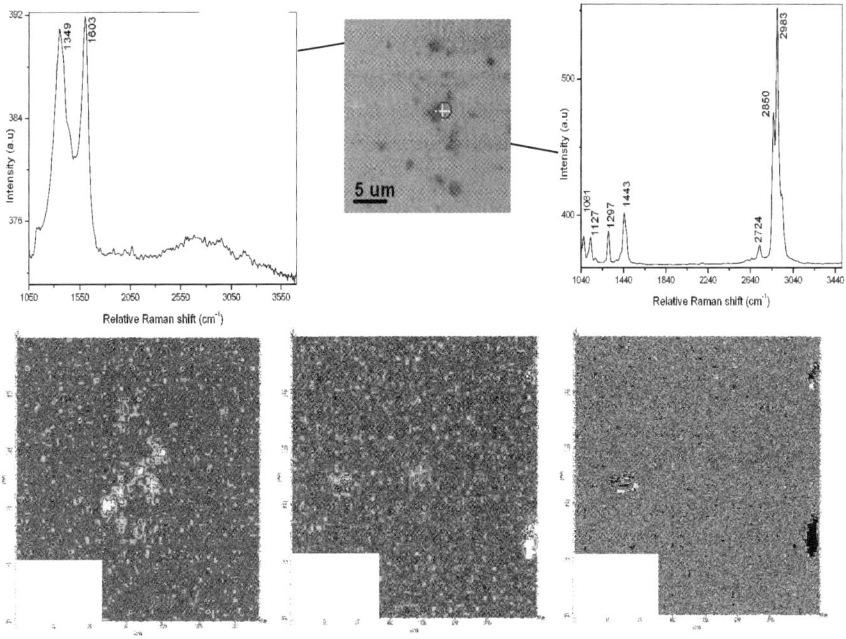

Figure 3.11 (a) and (c) Micro-Raman spectra of the 1:1 C_8CD:STA LB film acquired from two spatially different domains of (b). Raman image of (d) 1:1 C_8CD:STA (20 μm x 20 μm) and (e) 1:1 C_{12} CD:STA (25 μm X 25 μm) transferred at 10 mN/m. (f) Image mapped with 1603 cm^{-1} showing the complementary image of (e).[43]

The aggregated areas appeared dark in the Fig. 3.11(e), suggesting the presence of phase separated C_nCD domains. The background showed the carbonyl stretching frequency at 1603 cm^{-1} (Fig. 3.11(a)) with the appearance of a broad feature around 2750 cm^{-1}, corresponding to the alkyl chains of STA matrix. However, the sharp 2983 cm^{-1} Raman band in Fig. 3.11(c), representative of the asymmetric alkyl stretching frequency, implied an ordered environment in terms of molecular orientation and packing in the J-aggregated domains.

4. Donor-Bridge-Acceptor Dyads: Modulated 3D and 2D J and H Aggregation

Azobenzenes of donor–acceptor and electronically symmetric type[85,86] have been found to form characteristic aggregates depending on the solvent composition, polarity and the solute concentration. With the aim of developing photo-modulated phthalocyanine J aggregates, α-aryl/alkoxy-substituted zinc phthalocyanine dyads (3-azo-ZnPc and 4-azo-ZnPc) have been synthesized by Li Hong Niu *et.al.*[87] It was found that 3-azo-ZnPc in chloroform could be effectively photo-controlled in a wide range owing to changes in the geometry and dipole moment of azobenzene during the photo-isomerization process, whereas, 4-azo-ZnPc proved not to have this ability at all, with or without UV light illumination. The relative positions of the oxygen atoms to which the aryl/alkoxy substitution was attached essentially controlled the extent of aggregation. Supramolecular aggregates of azobenzene phospholipids and related compounds in bilayer assemblies have been studied for structure, properties, and photoreactivity.[88] Strong evidence of H aggregate formation in the pure and mixed dispersions has been indicated. On the basis of simulations and studies with similar stilbene phospholipids and the induced circular dichroism signals from the aggregate, a chiral "pinwheel" unit aggregate structure, similar to that for several aromatics was proposed.

Solvent characteristics have played important roles in controlling the formation of J and H aggregates;[89] zeaxanthin aggregates in hydrated ethanol showed a characteristic J band at 530nm, and that of an H band at 400nm. Three parameters of importance, namely, 1. pH, referring to ability to form a hydrogen bond, 2. the initial concentration of zeaxanthin, that essentially controlled the distance between zeaxanthin molecules and 3. solvent polarity, the ratio of ethanol/water. To create H-aggregates, the ability to form hydrogen bonds was crucial, while J-aggregates were preferentially formed when hydrogen-bond formation was prevented. Further, the formation of J-aggregates required a high initial zeaxanthin concentration and a high ethanol/water ratio, while H-aggregates were formed under the opposite conditions. Time-resolved experiments revealed excitation of the 530-nm band of J-zeaxanthin to produce a different relaxation pattern than excitation at 485 and 400 nm, showing that the 530-nm band was not a vibrational band of the S2 state but a separate excited state formed by J-type aggregation. H-type aggregation promoted a new relaxation channel that formed the zeaxanthin triplet state.

4.1 *A Novel Fullerene (C_{60})-N, N-dimethylaminoazobenzene Dyad: Solvent Polarity Regulated Aggregation:*

Novel molecular hybrids could be constructed utilizing the fascinating 3D Fullerene C_{60}'s optical and electronic properties[90–93] when coupled with photoactive azobenzene chromophore with novel photo, electrochemical and electrical properties. The covalent linkage of azobenzenes with a strong absorption coefficient in the visible region could aid to understand 2D and 3D molecular and supramolecular assembly formation. The only report on non-covalent interaction-induced self-assembly of fullerene C_{60} and azobenzenes by Min Li *et al.*[94] showed the site specific accommodation of fullerene C_{60} guest in the tetra-acidic azobenzene molecule's 'Kagome' open networks of specific size and symmetry. In spite of several recent reports on the covalently linked fullerene – azobenzene hybrids[95–100], tunable and discreet molecular assembly of specific types upon controlled experimental conditions seemed important when viewed at par with a range of aggregate forming systems[101–110].

Characteristic aggregation of Fullerene C_{60}-N, N-dimethylamino azo benzene dyad (DPNME) in a binary THF/ water solvent medium as well as at the air–water interface was investigated by Kumar and Patnaik.[111] In doing so, self-assembly of its donor and acceptor constituents AZNME and NMFP respectively (vide Scheme 4.1) were investigated from the resultant spectroscopic signatures, single crystal packing pattern, semi-empirical computations for insight into the electronic structure associated with specific aggregate-type, electronic absorption spectral co-relationships and the differences invoked on the monomer's electronic structure upon molecular association. Scheme 4.1 depicts the chemical structures of the donor, the acceptor and the dyad molecules.

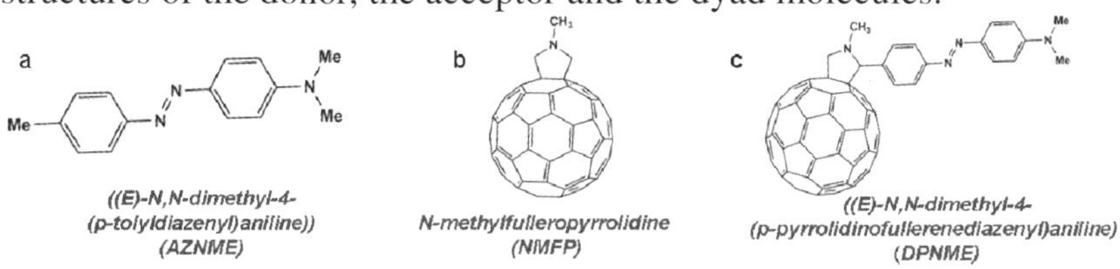

Scheme 4.1 Chemical structures of (a) AZNME, the donor (b) NMFP, the acceptor and (c) DPNME, the dyad

4.1.1 *Molecular Self-assembly of the Donor (AZNME) in THF/ Water Binary Solvent of Varying Polarity*

Molecular association in AZNME was studied in a THF/ Water binary solvent with varying polarity according to Eq. (4.1).

$$\varepsilon_{\text{mix}} = f_{\text{THF}}\varepsilon_{\text{THF}} + f_{\text{H}_2\text{O}}\varepsilon_{\text{H}_2\text{O}} \quad (4.1)$$

'ε' as the dielectric constant 7.52 and 79.52 respectively was used for pure THF and water solvents with '*f*' as the volume fraction. Gradual increase in the volume fraction of water in which AZNME was insoluble, increased the dielectric constant (ε_{mix}) of the binary solvent and facilitated intermolecular association between AZNME moieties. Figure 15 depicts the 407 nm absorption maximum for $^1(\pi\rightarrow\pi^*)$ electronic transition with the transition moment parallel to the long axis of the azo chromophore, whereas, the transition moment parallel to the short axis of the azo chromophore was observed at a lower $\lambda = 252$ nm. The 440 nm feature signified the monomeric $^1(n\rightarrow\pi^*)$ absorption. With gradual increase in solvent polarity, the absorption maximum was bathochromically shifted without any distinct peaking. At $\varepsilon_{mix} = 72.32$, the Gaussian deconvoluted absorption spectrum showed splitting of the absorption band accompanied by growth of a new feature at 449 nm, attributed to head-to-tail associated AZNME J-dimer, in accordance with the exciton coupling theory. The percentage weights in the figure established a solvent polarity controlled dynamic equilibrium between Monomer ↔ J aggregate in the system. Steady state fluorescence spectra ($\lambda_{excitation}$= 407 nm) implied AZNME in pure THF to be non-emissive, whereas, in more polar THF/ water binary solvents, orders of magnitude enhanced emission was observed. In spite of a negligible quantum yield $\sim 10^{-7}$–10^{-5} from azobenzene solutions,[112] a few exceptional bilayer aggregates of azobenzene-containing amphiphiles[113–114] and azobenzene-functionalized dendrimers[115] in aqueous solution have exhibited fluorescence emission at ~ 600 nm.

Other exceptions to non-emissive character of H-aggregates have been reported under special conditions,[115a] at low temperature in frozen solution[115b] or for dye aggregates embedded in Langmuir–Blodgett layers.[115c] Tethered mero- and hemicyanine chromophores could fold into fluorescent H-type aggregates with two face-to-face-stacked strongly dipolar dyes, where dimerization was driven by electrostatic interactions,[115d] related to the magnitude of their dipole moments with Gibbs binding energy being proportional to the square of the ground-state dipole moment $\Delta G^0 \propto \mu_g^{\ 2}$.[115e] This unique example of a merocyanine H-aggregate exhibiting a well-resolved fluorescence spectrum and a significantly longer fluorescence lifetime relative to its monomer was rationalized within the concept of exciton theory by taking into account the small transition probability caused by a slight rotation of the two coupled dyes in the excited state.

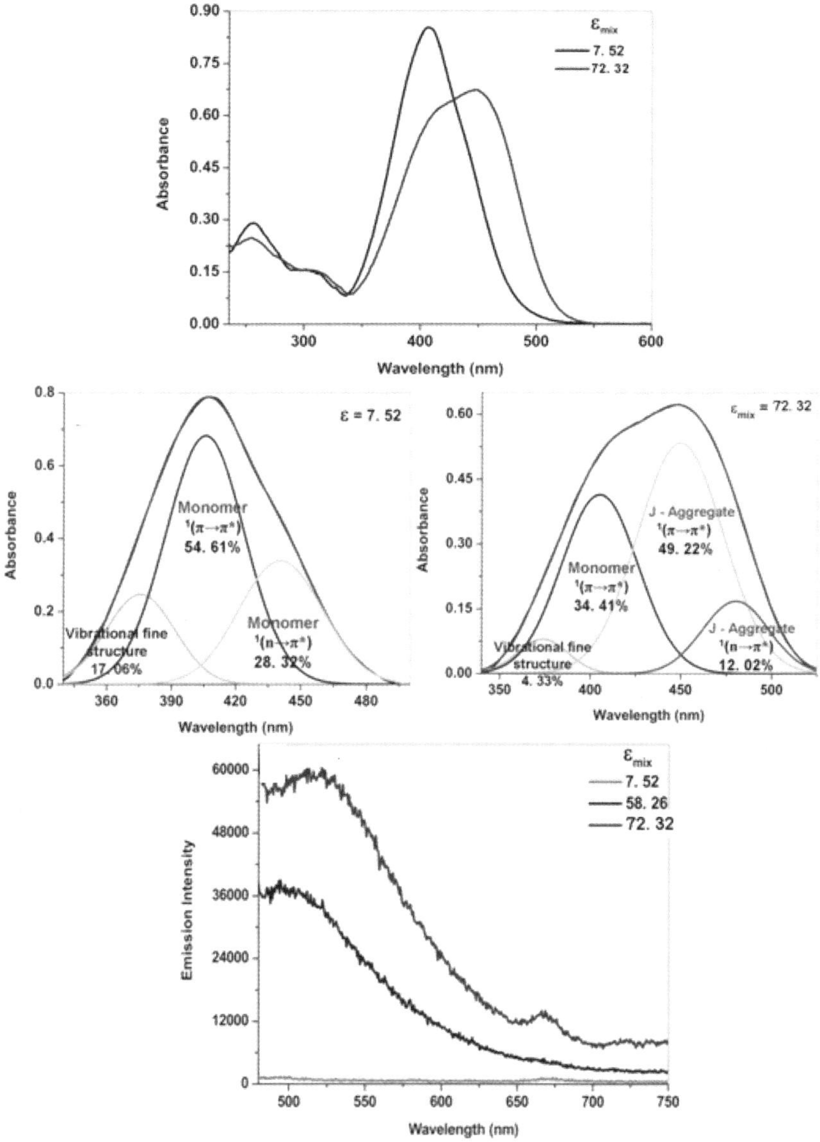

Figure 4.1 Solvent polarity dependent, baseline corrected and Gaussian deconvoluted UV-Vis absorption spectra of AZNME with the individual weights of the characteristic monomeric and J-aggregate $^1(\pi\rightarrow\pi^*)$ and $^1(n\rightarrow\pi^*)$ transitions. The steady state fluorescence emission spectra of 1×10^{-5} M AZNME for the corresponding excitation wavelengths.[111]

4.1.2 *Structure Elucidation of AZNME J Aggregates: Unit Cell Packing and Electronic Structure*

The single crystal X-ray diffraction packing diagram in Fig. 4.2 bears a direct relevance to the intermolecular interactions existing along head–to–tail arrangement of the molecules along with partial inter-digitation along the b axis. Across the c axis, molecules are stacked with an inter-molecular distance of 6. 153 Å. The packing diagram depicts a layered pattern, stacked along the crystallographic a axis with each layer comprising of a directionally distinct yet similar head-to-tail molecular orientation in a parallel fashion. In each layer, intermolecular interactions were facilitated through two CH–π type short contacts with a distance of 2.883 Å. The solid state molecular orientation of

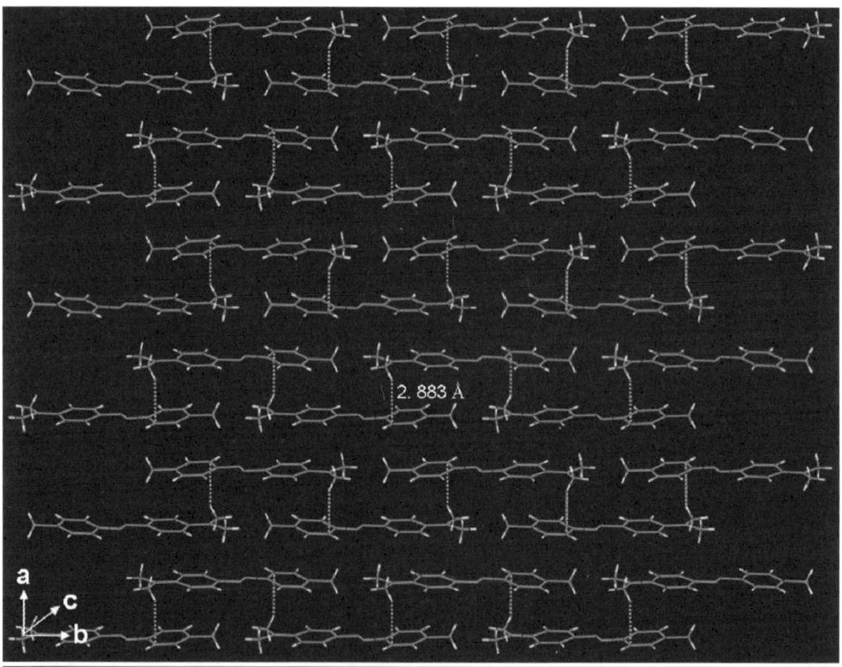

Figure 4.2 Crystal packing of AZNME viewed through c axis. Orthorhombic, space group= $P_{na}2_1$, a = 15.3708 (4) Å, b = 14.0139(04) Å, c = 6.1563(02) Å, α = 90°, β = 90°, γ = 90°, V = 1326.1(7) $(Å)^3$, cell formula units = 4, T = 273(2) K (standard deviation in parentheses), Mo Kα radiation (λ = 0.71073 nm).[111]

AZNME corroborates well the exciton coupling theory predicted red shifted peaking of the new absorption band in Fig. 4.1 upon increased head to tail inter-molecular interactions.

The electronic structure of the AZNME J dimer providing best agreement with the solution phase UV-Vis absorption spectrum was unraveled with molecular mechanics and semi-empirical ZINDO calculations, as illustrated in Fig. 4.3. Here, the electronic absorption spectrum of the monomeric AZNME was generated using the semi-empirical method, as implemented in the Gaussian 03 suite of programmes[116], where the crystal coordinates obtained from the single crystal X-ray diffraction data were used. The dimer conformation for calculation was chosen with strong facial interactions, facilitated through dispersive forces as observed for linear aromatics[117-119] in view of the absence of a large dipole moment and a weak donor–acceptor character of AZNME. The minimum energy J-dimer depicted in the figure showed pronounced intermolecular facial interaction between the electron rich N, N-dimethylaniline ring and the relatively electron deficient methyl substituted aromatic ring. For this dimeric arrangement, ZINDO predicted a dominant absorption feature at ~ 428 nm with an oscillator strength of ~ 2.32, double that of the predominant monomeric transition calculated at ~ 405 nm with an oscillator strength of 1.07. Several aspects of azobenzene systems have been studied by theoretical and computational modeling;[120,121] the semi-empirical methods have reached the desired quality, but with the cost of eventually compromising the generality.

The $^1(\pi \rightarrow \pi^*)$ transition of the J–dimer involved promotion of an electron from the HOMO to LUMO + 1 orbitals that have large amplitudes of N=N group, and were delocalized throughout the molecule and have finite contribution from both the AZNME moieties, rather than the predominantly contributing one AZNME. A reasonable extent of exciton coupling therefore explained the bathochromic shifting of the absorption maximum as expected for a head to tail dimeric arrangement. The calculation rightly predicted the energy lowering for the $^1(\pi \rightarrow \pi^*)$ transition of the J-dimer with a comparable magnitude as observed experimentally and its close proximity with the $^1(\pi \rightarrow \pi^*)$ transition of the monomeric AZNME. Further proof for reliability of the above calculations was unraveled by performing ZINDO calculations on the H-dimer of AZNME,[111] constructed by a similar methodology as that for the J dimer. The hypsochromic shifting of the H dimer's absorption maximum vs. the 405 nm monomeric band revealed the predictive utility of the computational methodology adopted in this study.

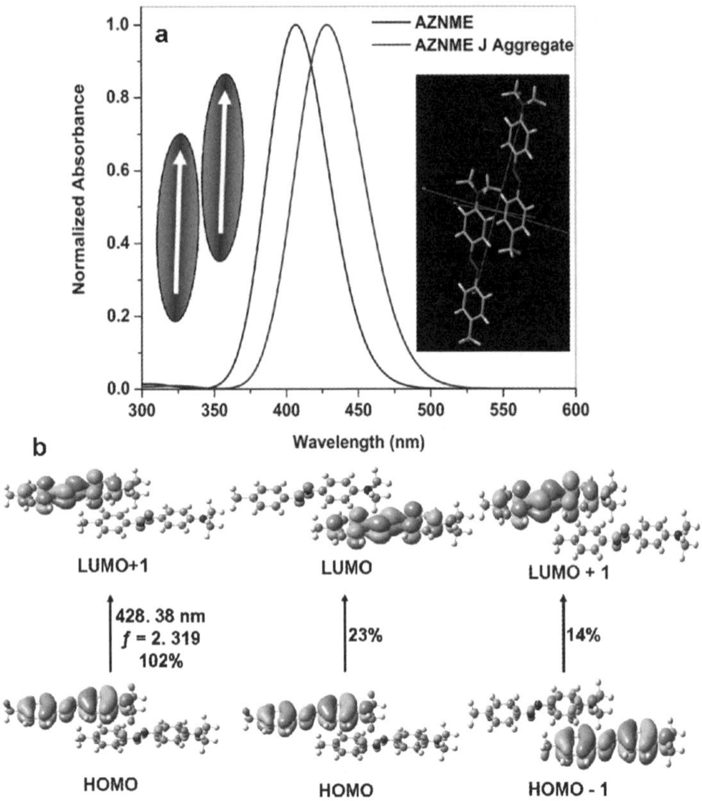

Figure 4.3 (a) ZINDO calculated UV-Vis absorption spectra of monomeric AZNME and J aggregate of AZNME. Inset shows the molecular mechanics (MM) optimized J-aggregate structure. (b) Molecular orbitals involved in the ZINDO calculated electronic transitions of AZNME J-dimer showing characteristic transitions, vide Ref. 111.

4.1.3 *Molecular Self-assembly of the Dyad (DPNME) in THF/Water Binary Solvent of Varying Polarity*

Functionalized Fullerene C_{60} derivatives with intrinsic hydrophobicity and amphiphilicity have shown geometrical possibilities for self-organization and have been studied intensely in view of their unique electronic and optical properties.[122–124] Asanuma *et.al.*[125] coined 'Supramolecular polymorphism' from an atypical C_{60} based surfactant exhibiting exotic molecular assemblies upon changing the chemical structure design and conditions of assembly. C_{60} functionalized with hydrophobic groups have been seen to self-organize into structures, soluble in polar solvents.[126-128] Nakanishi[129] devised new and novel architectures of Fullerene C_{60} in non-polar media by eliminating hydrophilic units, but maintaining sufficient amphiphilicity for improved solubility. A Fulleropyrrolidine with 3,4,5-(hexadecyloxy) phenyl assembled into various super structures[130] depending on the nature of the solvent; vesicles originated in 2-propanol/toluene mixture, while 1D fibrous structures of few μM resulted in 1-propanol. In an equimolar THF/ H_2O, cone shaped structures with a hole diameter ~ 60 nm were predominant.

Aggregation of the Fullerene–azobenzene hybrid, the DPNME dyad, signified the effect of substitution of a 3D fullerene moiety on a 2D azobenzene chromophore and yielded fascinating results for structure governing of dimer formation.[111] Exciton theory proposed for the molecular dimer formation was invoked and the experimentally observed results were interpreted with the help of molecular mechanics and semi-empirical ZINDO calculations, as explained in the following sections. Figure 4.4 shows absorption spectra of DPNME in pure THF with the $^1(\pi\rightarrow\pi^*)$ electronic transition of the azo moiety at 417 nm along with the minor, but characteristic 433 nm N-methylfulleropyrrolidine (NMFP) based peak. Increase in the dielectric constant of the medium led to gradual red shifting of DPNME $^1(\pi\rightarrow\pi^*)$ transition upto an ε_{mix} of 43.79. Further increase in solvent polarity led to appearance of a new 469 nm feature at the expense of the pristine one. More importantly, the disappearance of the characteristic NMFP absorption maximum at 433 nm implied aggregation of DPNME. Experiments performed with NMFP acceptor constituent of DPNME (figure not shown) indicated monomer to aggregate conversion with structureless, enhanced absorption[111] in the visible region at $\varepsilon_{mix} \geq 43.79$. Here, mixing of the $^1(n\rightarrow\pi^*)$ feature of AZNME with the dyad J–dimer centered electronic transition was observed with loss of true $^1(n\rightarrow\pi^*)$ character of the monomer at the expense of the system tending to aggregate. However, at this polarity, a finite DPNME monomeric weight retained with the characteristic 433 nm band (cf. Fig. 4.4(b)). At a solvent dielectric constant of 51.01, a skewed absorption profile in Fig. 4.4(c) implied loss of 433 nm NMFP feature indicating the onset of J aggregation of the dyad; the larger intensity 471 nm feature was attributed to the J type DPNME dimer. At increased ε_{mix} of ~ 58 and ~ 65, an interconversion between the J to H

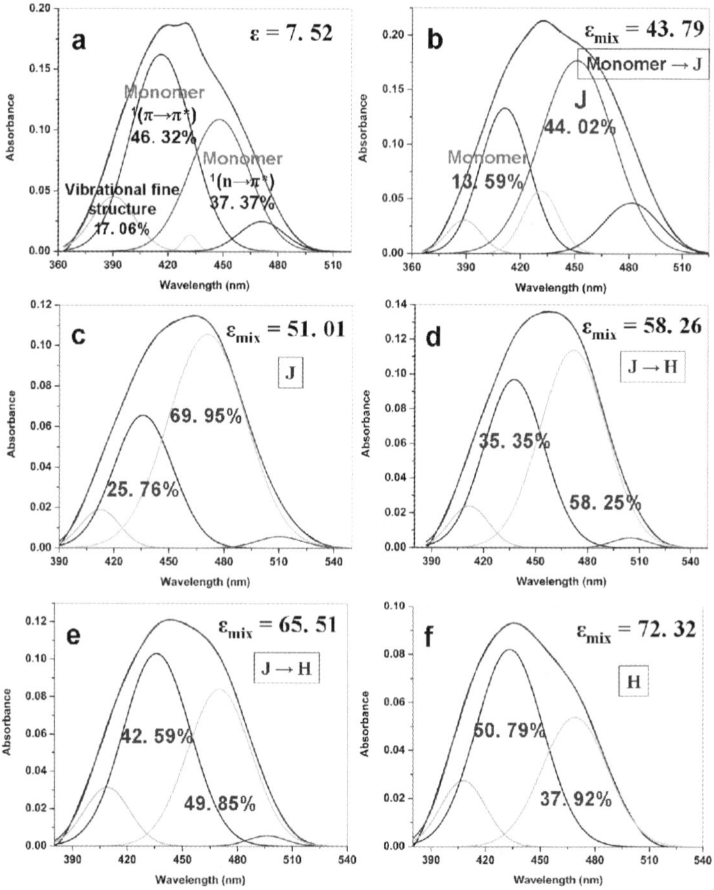

Figure 4.4 Baseline corrected and Gaussian deconvoluted electronic absorption spectra (a)-(f) of DPNME dyad as a function of solvent polarity, depicting the sequential monomer-to-J-to-H-aggregate conversion.[111]

dimer was imminent with the equilibrium shifting more towards the H dimer, as evident from the relative weights of the blue shifted 434 nm and red shifted 471 nm features, vide Figs. 4.4(d) and (e). At ε_{mix} = 79.5, the peak maximum at 435 nm was 18 nm bathochromically shifted in reference to that of the DPNME monomer in pure THF at ε = 7. 52, implying the predominant existence of H-dimers with its respective $^1(\pi\rightarrow\pi^*)$ and $^1(n\rightarrow\pi^*)$ transitions coupled with fullerene centered transitions.

The overall structural changes involved in DPNME aggregation are depicted in the absorption versus ε_{mix} plot in Fig. 4.5 illustrating dimerization behavior as a function of solvent polarity. With a critical dielectric constant 41.9 for monomer to aggregate conversion, co-existence of J and H-type dyad dimers was established between 58.26 ≤ ε_{mix}≤ 65.95. An increased probability of H-dimer characteristics was observed beyond this polarity. Thus, solvent polarity dictated a dynamic equilibrium between the aggregates due to different intermolecular structural ordering in solutions of varying dielectric constant.

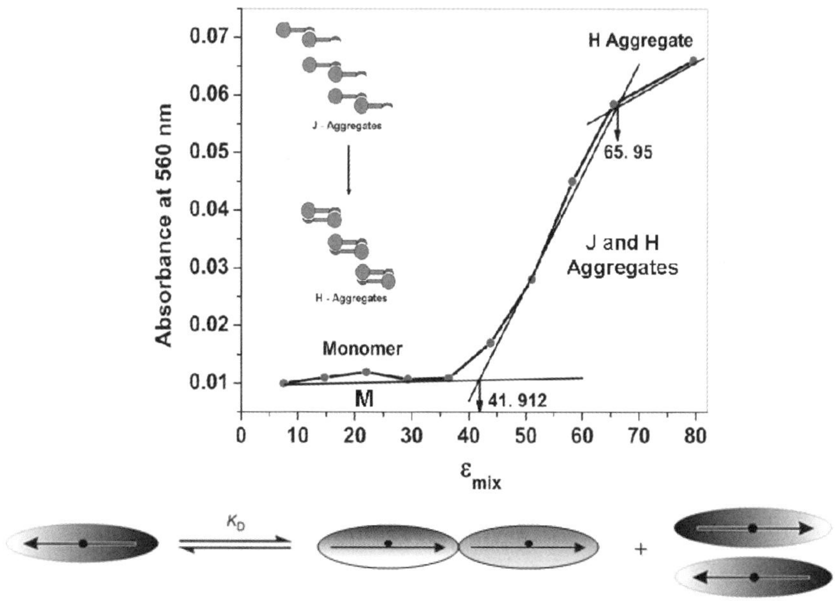

Figure 4.5 Variation of absorbance as a function of ε_{mix} of the THF-Water binary solvent establishing a dynamic equilibrium between the Monomer→ J aggregate + H aggregates till an ε_{mix}= 65.9, beyond which only H- dimers were formed.

In compliance with the exciton theory and the results obtained for AZNME, the 471 nm peak could be attributed to the DPNME J dimer. However, attribution of 435 nm feature in Figure 18(c) (red shifted in reference to the monomer $^1(\pi \rightarrow \pi^*)$ band at 417 nm) to the H dimer of DPNME seemed unacceptable under the ambit of exciton theory,[1c] since only electronic transitions for the face to face, parallel arranged molecular dipoles constituting an H dimer are allowed with a hypsochromically shifted absorption maximum vs. the monomer band. The observed results in retrospect could be due to electronic transitions of the H dimer for its anti-parallel orientation of the dipoles (vide Fig. 4.5), though cited forbidden. The validation for this attribution was followed through the single crystal structure and semi-empirical calculations in the two subsequent sections.

4.1.4 *Structure Validation of Dyad H Aggregates: Unit Cell Packing and Electronic Structure*

Single crystals of dyad grown from a saturated solution of chloroform with slow diffusion of cyclohexane into the solution were analysed; the crystal structure in Fig. 4.6 represents the first ever crystal structure of a fullerene–azobenzene family.[111] Single crystal analysis showed DPNME to belong to the monoclinic crystal system with the space group C2/c. The unit cell comprised of four pairs of dimers with each pair arranged in an inter-digitated manner with figures (b) and (c) depicting part of the unit cell viewed through b and c axes. A view through the b axis of the unit cell yielded one-dimensional arrays of stacked molecules, extended along the crystallographic c axis through (i) two similar CH–π interactions between the aromatic hydrogen atoms of the AZNME part of DPNME and the electron rich 6,6-ring carbons of fullerene with a distance of 2.696 Å and, (ii) the inter-pair interactions between two fullerene moieties through the 6,5-ring carbon junctions with a distance of 3.227 Å, vide Figure 4.6 (c). The most interesting feature of the structure was the intra-dimer packing between fullerene and the AZNME

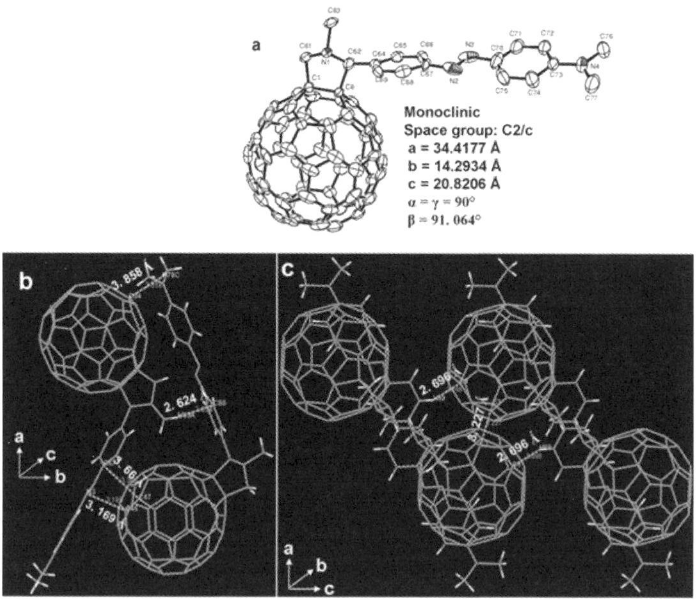

Figure 4.6 a. ORTEP diagram of the dyad crystallized from chloroform – cyclohexane solvent with 50% thermal ellipsoids along with important crystal parameters; solvent molecules chloroform and cyclohexane were omitted for clarity and only a single layer of molecules is shown. (b) Crystal packing in the monoclinic unit cell with a view from b axis, and (c) two pairs of dimers visualized through b axis. [111]

moieties. As depicted in Figure 4.6 (b), the 6,6 ring carbon atoms of the Fullerenes made two intermolecular contacts, one with the azo nitrogen and the other with the aromatic carbon of the closely interdigitated AZNME part of DPNME. The intra-dimer interactions were further strengthened by the two CH–π interactions.

A rare packing pattern of DPNME, unlike the porphyrin[130–133] or pthalocyanin[134] bowl-shaped electron donors, was thus visualized; the linear geometry of the donor AZNME along with the dominant Fullerene–Fullerene intermolecular π–π interactions have brought out the structural transformations. The geometrical constraints imposed by rigid C_{60} on the dyad framework could contribute substantially towards the resultant dyad arrays. Importantly, substitution of fullerene C_{60} with the J-aggregate forming AZNME chromophore yielded the anti-parallel H-dimer, evidenced from X-ray diffraction pattern in Fig. 4.7(A).

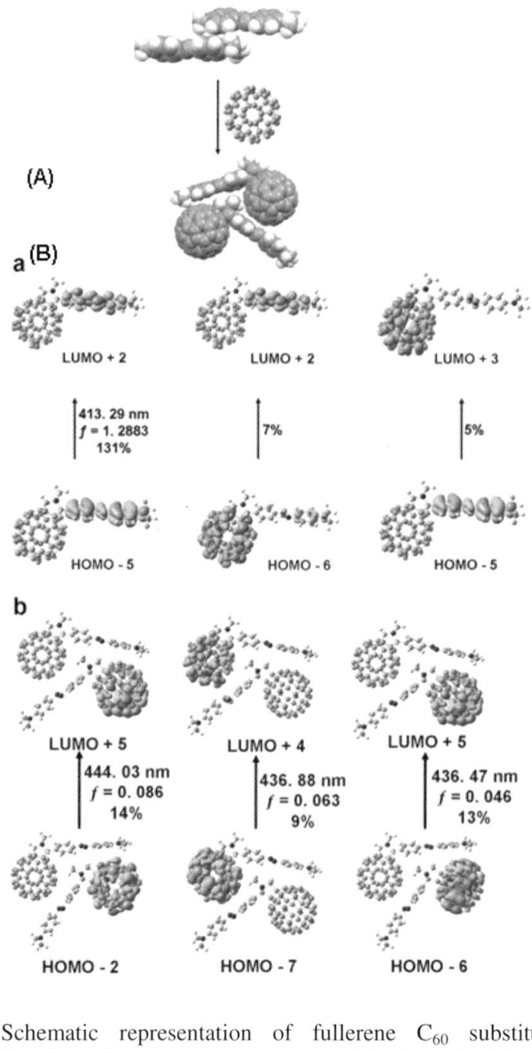

Figure 4.7 (A) Schematic representation of fullerene C_{60} substitution-induced H-aggregate formation from the J-aggregate forming AZNME. (B) Molecular orbitals involved in the ZINDO calculated electronic transitions of (a) DPNME monomer and (b) its H-dimer (structure as in single crystal XRD unit cell) depicting transition types and oscillator strengths.[111]

The calculated monomer absorption maximum at 413.29 nm tallied with the experimentally observed absorption maximum of 417 nm. The predicted $^1(\pi\rightarrow\pi^*)$ transition essentially involved electronic transition from HOMO-5 to LUMO+2 orbital, as shown in Fig. 4.7(B)a. For the dimer, the XRD single crystal unit cell structure when used as the input for ZINDO calculation, the few observable transitions in Fig.4.7(B). b showed very low oscillator strengths involving transitions with molecular orbital coefficients concentrated only on fullerenes, contradictory to the experimental spectra obtained in Fig. 4.4 with predominant AZNME-based absorptions. In achieving the electronic absorption characteristics of the experimental J to H equilibrium weights and the ultimate H dimer, MM and ZINDO calculations were carried out for a varied range of DPNME dimer conformations, vide Fig. 4.8(A) and (B). The head to tail or J dimer's energy optimized conformation in Fig. 4.8(A).a showed fullerene-fullerene and fullerene–arene distances of 3.14 Å and 3.29 Å respectively, in very good agreement with the literature reported values and distances obtained from the X-ray crystal structure analysis.

The molecular orbitals in Fig. 4.8(A).b revealed a predominantly azo centered nature of the electronic transition at 450 nm with a reasonable electronic coupling in the electronic excited state. However, the higher intensity transition at 424 nm involved contributions from both azo and fullerene moieties with a remarkable coupling in the excited state. These transition characteristics mapped well with the experimental absorption profiles in the ε_{mix} range between 43 to 51 and could be attributed to DPNME J dimer's characteristic absorption. The discrepancies between the predicted oscillator strengths and the experimentally observed intensities could be attributed to coupling of J dimer's absorption characteristics with the monomeric $^1(n \rightarrow \pi^*)$ transition of the azo part of the dyad, the increasing fullerene-fullerene interactions, and the usage of a semi-empirical ZINDO method with its in-built approximations. For the dimeric arrangement with hypsochromic shifting of the absorption maximum upon increasing binary solvent polarity, DPNME moieties were arranged in a dimeric conformation with the phenyl ring carrying the electron donating N, N – dimethylamino groups in π-stacking interaction with the fullerene rings of the other DPNME moiety (Fig. 4.8(B).a) in an inter-digitated manner. In view of the 3D spatial orientation of Fullerene C_{60}, an anti-parallel arrangement of transition dipoles was thus observed, unlike that in classical H aggregate forming materials. The 3D spatial orientation along with the cream of electronic cloud laced above the alternative double and single bonds, and an electron deficient core have been the main reason for fullerene aggregation in mono and binary solvents.[135,136] This characteristic

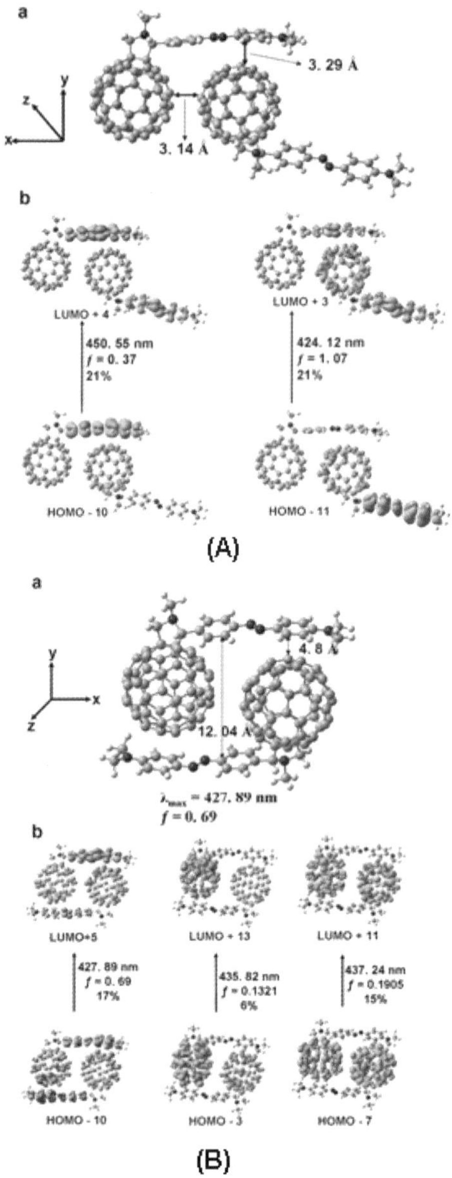

Figure 4.8 (A). (a). Molecular structure of the J dimer of DPNME. (b) Molecular orbitals involved in the ZINDO calculated electronic transitions of J - dimer of DPNME depicting transition types and oscillator strengths. (B). (a) Molecular structure of the anti- parallel H dimer of DPNME. (b) Molecular orbitals involved in the ZINDO calculated electronic transitions of the anti-parallel H-dimer of DPNME depicting transition types and oscillator strengths.[111]

electron density distribution has prevailed even for functionalized fullerene derivatives towards aggregate formation and structures of varying size and dimensions, emphasizing on the electronic and van der Walls interactions between the neighboring molecules.

ZINDO calculations in Fig. 4.8(B).b predicted an electronic transition of maximum oscillator strength at 427 nm, predominantly involving contributions from the azo as well as fullerene parts with excellent electronic coupling and in very good agreement with the experimentally observed 435 nm band at an ε_{mix} = 72.32. The two new transitions predicted at ~ 436 nm further showcased the intermolecular interactions between the fullerene moieties constituting the DPNME dimer. For the DPNME molecule with a donor–acceptor structure and with an aggregation-philic fullerene moiety, the predominant intermolecular interactions could be, (i) the electronic interaction between the electron rich part of the molecular skeleton of one moiety with the electron deficient part of the other, as observed in the case of fullerene–porphyrin co-crystals[137–139] and (ii) the strong intermolecular interaction between the fullerene moieties in close proximity to each other. The molecular orientation in solution or in solid state therefore could have been governed by such interactions with each type trying to maximize its influence in the assembly process. Apart from the strong π–π and van der Walls interactions between the fullerene moieties, the donor–acceptor type interactions between the fullerene moiety and N, N-dimethylaniline substituted aromatic ring of the other DPNME moiety gains importance; for the J-aggregate with a linear head to tail arrangement (cf. Fig. 4.8(A)(b)), only one such interaction was possible showing a much lesser degree of stabilization in comparison to the H-dimer in Fig. 4.8(B)(b)). When the two DPNME moieties were arranged in an interdigitated manner with their dipole moments pointing diametrically opposite to each other, the donor and acceptor fragments of each DPNME could interact with its opposite number without compromising the fullerene–fullerene interaction, as shown in Fig. 4.9 (A).b. This enhanced possibility of intermolecular donor–acceptor interaction was the sole origin of formation of the interdigitated H-dimer with increasing solvent polarity, which enhanced the hydrophobic/ van der Waals fullerene–fullerene interaction. The semi-empirical ZINDO calculations on the monomeric and the H-dimer of DPNME had excellent agreement with the experimental observations and the results are shown in Fig. 4.9(B). More importantly, the ZINDO calculated absorption maximum of the DPNME H-dimer at 427 nm was in good agreement with the experimental absorption maximum at 436 nm, but appeared with a much reduced oscillator strength (f = 0.69) in comparison with the monomeric DPNME (f = 1.29) against an ideal strength of double that of the monomer. For H aggregates, exciton theory predicted parallel oriented dipoles with the higher energy state, where transition from the singlet ground state was allowed; the forbidden transition was the lowest energy split state with anti-parallel alignment of dipoles. In view of the meager oscillator strength for a dimer, the electronic transition attributed to the H-dimer of DPNME though could be forbidden, was experimentally observable. The H-dimer of DPNME thus could bring in a conceptually important molecular system. The close agreement between the experimental and ZINDO calculated UV-Vis absorption spectral features could prove well the assignment of the calculated dimeric structure to the dimeric H-aggregate existing in the

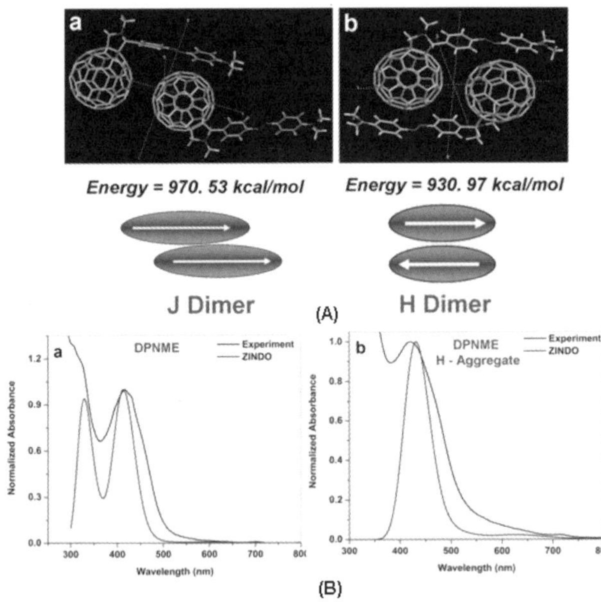

Figure 4.9 (A) Molecular Mechanics geometry optimized structures of (a) J- dimer and (b) Anti -parallel H- dimer of DPNME showing relative stabilities. (B) The semi-empirically calculated absorption spectra of (a) DPNME and (b) H dimer of DPNME, showing a comparison with the experiment.[111]

THF/water binary solvent medium, where strong intermolecular van der Walls and π–π electronic interaction between the two hydrophobic fullerene moieties, augmented by the presence of water molecules of large dielectric constant could have been the driving force for such molecular association

4.1.5 *Self-Assembly of DPNME Dyad at the Air–Water Interface: Molecular Orientation and Supported H and J Aggregates*

The Langmuir monolayer profiles from DPNME dyad concentrations ranging from 0.1 to 1 mM on sub-phase pH 5.9 in Fig. 4.10 (a) implied stable molecular layers with appreciable surface pressure. At 1mM concentration, compression induced bilayers of DPNME with high surface pressure, similar to that reported for pristine fullerene C_{60}[140-143] was noted indicating structural reorganization of the floating film at the interface. From the molecular model depicted in Fig. 4.10 (b), a flat-lie orientation of the dyad in the monolayer with ~ 102 $Å^2$ was inferred at 0.1 mM concentration. The characteristic interfacial molecular organization of the dyad was further studied by transferring Langmuir films onto pre-treated and hydrophilized Si(100) substrates by a single upstroke as Langmuir-Blodgett (LB) films. AFM images of the LB films transferred from a pH 5.9 subphase at 10 mN/m surface pressure revealed layered 2D vesicular structures with vertically stacked layers, as depicted in Fig. 4.10(c) and (d). At higher resolutions, the cohesive nature of the constituent smaller architectures led to the formation of extended assemblies, dictated by strong intermolecular van der Waals and hydrophobic interactions. The UV-Vis absorption characteristics of the transferred film on a quartz substrate in Fig. 4.11 (A) upon Gaussian deconvolution revealed 428 and 468 nm features, mimicking those of the THF–water binary solvent media and implied presence of both the J and H-type dimer-based aggregates, but with predominance of H aggregates. The obtained 2D architecture could be visualized from the molecular modeling of hydrophilic (AZNME)–hydrophobic (C_{60})/head–tail–type bilayer arrangement of the DPNME H-dimers, as shown in

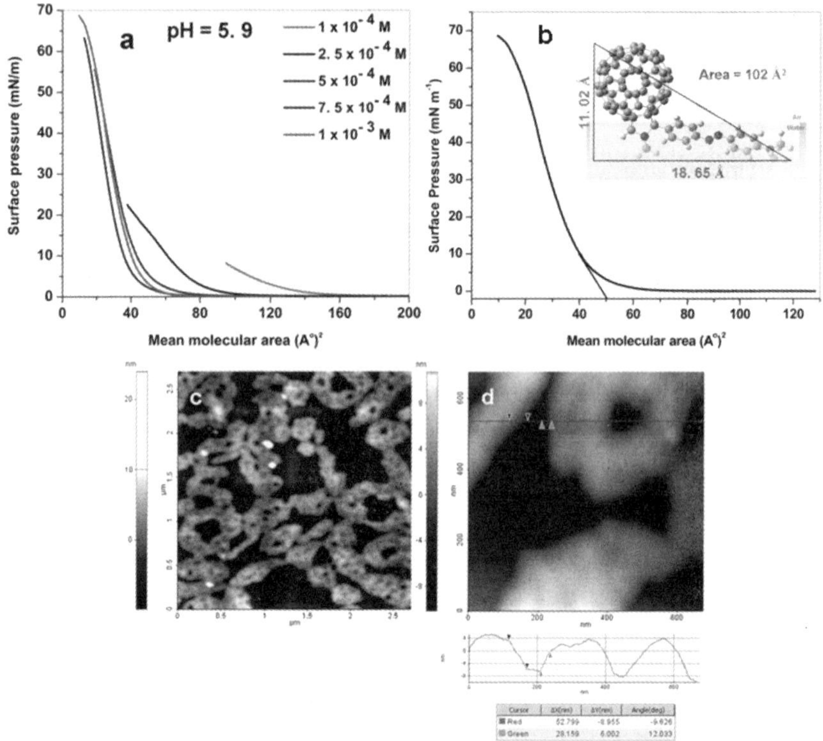

Figure 4.10 (a) Langmuir isotherms of DPNME in Chloroform spread over pH 5.9 sub-phase at varying concentrations. (b) Probable flat-lie orientation of a single DPNME molecule at the air – water interface for a 0.1 mM concentration. The molecular geometry was full optimized with DFT–B3LYP/ 6-31g (d, p) level of theory. The total area occupied by the DPNME molecule was deduced by considering it as a right angle triangle as, $\left(\frac{1}{2} \text{x } 18.65\right) \text{x } 11.02 = 102 \text{ A}^{o2}$. (c) and (d) AFM images of DPNME LB films transferred onto pretreated Si(100) substrates at a surface pressure of 10 mN/m from a pH 5.9 sub-phase.

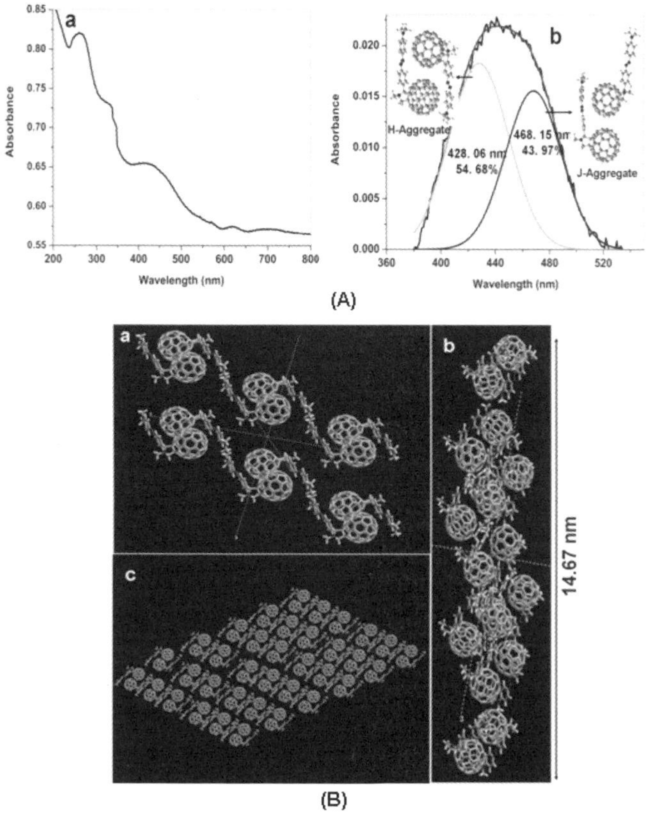

Figure 4.11 (A) UV–Vis absorption spectrum of 1 x 10^{-3} M DPNME transferred onto a quartz substrate at a surface pressure of 10 mN/m. (b). Baseline corrected and Gaussian deconvoluted 360 to 550 nm absorption region of the spectrum showing the presence of 2D H and J aggregates. (B) Schematic representation of 2D DPNME H-dimer arrangement. (a) Two rows of H dimers arranged parallel, resulting from AZNME interdigitation and hydrophobic C_{60} – C_{60} interactions, (b) Molecular mechanics full geometry optimized 10 pairs of DPNME H dimers stacked vertically. The 14.67 nm height of this assembly matched well with the AFM height analysis depicted in Figure 24. (c) A model molecular assembly built from 36 pairs of DPNME H dimers from full geometry optimization with molecular mechanics computational method employing AMBER force field. [111]

Fig. 4.11 (B). Upon full structure optimization with molecular mechanics (MM), the schemes represented the 2D molecular architectures depending upon a number of internal factors, such as, the chemical structure of the molecule, the competing intra-molecular forces, the donor–acceptor and the intermolecular interactions. The relative magnitudes of the attractive and repulsive forces associated with molecular organization have resulted into an ultimate equilibrium structure.

Acknowledgments

This work was supported by Council of Scientific and Industrial Research, New Delhi, India and Department of Science and Technology, New Delhi, India. The author would like to thank Dr. B. Vijai Shankar and Dr. K. Senthil Kumar, whose doctoral work has been cited in this chapter. United States- India Educational Foundation, New Delhi is gratefully acknowledged for having granted the fellowship, during which the chapter was written.

References

1. a. E. E. Jelley, *Nature* **138**, 1009 (1936); b. G. Schiebe, *Angew. Chem,* **49**, 567 (1936); c. M. Kasha, H. R. Rawls, M. A. El-Bayoumi, *Pure Appl. Chem.* **11**, 371 (1965).
2. J. A. Delaire, K. Nakatani, *Chem Rev.* **100**, 1817 (2000).
3. A. S. Matharu, S. Jeeva, P. S. Ramanujam, *Chem Soc Rev.* **36**, 1868 (2007).
4. D. R. Kanis, M. A. Ratner, T. J. Marks, *Chem Rev.* **94**, 195 (1994).
5. H. Rau, Photoreactive Organic Thin Films. In: Sekkat Z, Knoll W, Eds. San Diego, USA: Academic Press 2002.
6. C. Bosshard, K. Sutter, P. Prêtre et al. Organic nonlinear optical materials. Basel: Gordon and Breach Science Publishers SA 1995.
7. A. Natansohn, P. Rochon, *Chem. Rev.* **102**, 4139 (2002).
8. S. K. Yesodha, C. K. S. Pillai, N. Tsutsumi, *Prog Polym Sci.* **29**: 45 (2004).
9. J. Maack, R. C. Ahuja, and H. Tachibana, *J. Phys. Chem.*, **99**(22), 9210 (1995).
10. A. Natansohn and P. Rochon, *Chem. Rev.* **102**, 4139 (2002).
11. S. Freiberg, F. Lagugn´e-Labarthet, P. Rochon, and A. Natansohn, *Macromolecules* **36**, 2680 (2003).
12. H. Rau, In J. K. Rabek Editor, Photochemistry and Photophysics,Vol II. (Boca Raton, Florida: CRC Press; 1990).
13. D. M. Shin, K. S. Schanze, and D. G. Whitten, *J. Am. Chem. Soc.***111**, 8494 (1989).
14. O. G. Morales-Saavedra, *REVISTA MEXICANA DEF´ ISICA* **56** (6), 449 (2010).
15. Padma Gopalan, Howard E. Katz, David J. McGee, Chris Erben, Thomas Zielinski, Danielle Bousquet, David Muller, John Grazul, Ylva Olsson, *J. Am.Chem.Soc.***126**,1741 (2004).
16. A. Galvan-Gonzalez, K. D. Belfield, G. I. Stegeman, *Appl. Phys. Lett.* **14**, 2083(2000).
17. A. Galvan-Gonzalez, M. Canva, G. I. Stegeman, *Appl.Phys. Lett.* **21**, 3306(1999).

18. H. Ma, B. Chen, T. Sassa, L. R. Dalton, A. K.-Y Jen, *J.Am.Chem.Soc.***123**, 986((2001).
19. Mitsuo Kawasaki, Satoshi Aoyama, and Eishi Kozawa, *J. Phys. Chem. B* **110**, 24480((2006).
20. E. E. Jelley, *Nature,* **138**, 1009 (1936) .
21. P. J. Meadows, E. Dujardin, S. R. Hall, and S. Mann, *Chem. Commun.*, **29**, 3688 (2005).
22. T. Hasobe, S. Fukuzumi, and P. V. Kamat, *J. Am. Chem. Soc.*, **127**, 11884 (2005).
23. H. Kaufer, G. Scheibe, *J. Elektrochem.* **59**, 584 (1955).
24. T. Ko bayashi, T. Seki, and K. Ichimura, K. *Chem. Commun.*, **13**, 1193(2000).
25. T. Kobayashi (Ed.) *J-aggregates*, World Scientific, Singapore, 1996.
26. H. Biicher, K. H. Drexhage, M. Fleck, H. Kuhn, D. Mobius, F. P. Schafer, J. Sondermann, W. Sperling, J. Wiegand, *Mol. Cryst.* **2**, 199 (1967).
27. H. Kuhn, D. Mobius, In Investigations of Surfaces and Interfaces (Eds: B. W. Rossiter. R. C. Baerzold), Wiley, New York 1993, pp.375-542.
28. F. C. Spano, S. Mukamel, *Phys. Rev. A* **40**, 5783 (1989).
29. H. Fidder. J. Knoester. D. A. Wiersma, *J. Chem. Phys.* **98**, 6564 (1993).
30. T. L. Penner, D. Mobius, *J. Am. Chem. Soc.* **104**, 7407 (1982).
31. L. H. Gao, K. Z. Wang, C. H. Huang', Y. F. Zhou, T. K. Li, J. M. Xu, X. S. Zhao and X. H. Xia, *Thin Solid Films* **286**, 237(1996)
32. Akio Miyata, Yukuta Unuma and Yoshiyuki Higashigaki, *Bulletin Chem. Soc. Japan* 64, 2786(1991).
33. A. Ulman, An introduction to ultrathin organic films: From Langmuir-Blodgett to self-assembly, Academic Press 1991
34. H. Motschmann, H. Möhwald, Langmuir Blodgett Films: Handbook of Applied Surface and Colloid Chemistry, edited by Krister Holmberg, Wiley, New York, 629-648 (2001)
35. Andreas Ziegler, Joachim Stumpe, Ali Toutianoush, Bernd Tieke, Photoorientation of azobenzene moieties in self-assembled polyelectrolyte multilayers, *Colloids Surf. A* **198-200**, 777 (2002).
36. J. A. Zasadzinski, R. Viswanathan, L. Madsen, J. Garnaes, and D.K. Schwartz, *Science*, **263**, 1726 (1994)
37. R. Viswanathan, L. L. Madsen, J.A.N. Zasadzinski, and D. K. Schwartz, *Science* , **269**, 51 (1995)
38. D. K. Schwartz, *Surf. Sci. Reports*, **27**, 245 (1997).
39. D. K. Schwartz, in Encyclopedia of Materials: Science and Technology, K.H.J. Buschow et al. eds. Elsevier, Oxford (2001) pp. 4392-4399.
40. I. Langmuir, V. K. Schaefer, and H. Sobotka, *J. Am. Chem. Soc.,* 59, 1751 (1937).
41. C. J. Faul, and M. Antonetti, *Chem. Eur. J.* **(8)**12**,** 2764 ((2002).
42. B .Vijai Shankar and Archita Patnaik, *Langmuir* **22**, 4758 (2006).
43. B. Vijai Shankar, Archita Patnaik, *Journal of Colloid and Interface Science* **302**, 259 ((2006).
44. T. J. Kato, *Geol. Soc .Jpn.* **107**, 64 (2001).
45. J. N. Israelachvili, *Intermolecular and Surface Force,* Academic Press: London. 1991.
46. H. Yao, C. A. Michaels, S. J. Stranick, T. Isohashi, and K. Kimura, *Letters in Organic Chemistry*, **1**(3), 280(2004).
47. A. S. R. Koti, and N. Periasamy, *J. Mater. Chem.***,** **12**(8*),* 2312 (2002).

48. S. Acharya, S. J. Efrima, *J. Am. Chem. Soc.*, *127*, 3486 (2005).
49. H. M. McConnell, *Annu. Rev .Phys. Chem.*, ***42***, 171 (1991).
50. L. Ramos, T. C. Lubensky, N. Dan, P. Nelson, D. A. Weitz, *Science* **286**, 2325 (1999).
51. J. M. Schnur, *Science **262***, 1669 (1993).
52. G. W. Orr, L. J. Barbour, J. L. Atwood, *Science* **285**, 1049(1999).
53. E. O. Potma, D.A. Wiersma, *J. Chem. Phys.* **108**, 4894(1998).
54. J. M. Chovelon, K. Wan, and N. J. Renault, *Langmuir,* **16**(15), 6223 (2000).
55. T. H. Chou, and C. H. Chang, *Colloid Surf. B.,* **17**(2), 71(2000).
56. I. H. Ku, Y. L. Lee, C. H. Chang, Y. M.Yang, J. R. Maa, *Colloids Surf. A* **191**(3), 223 (2001).
57. F. E. Antunes, E. F. Marques, R. Gomes, K. Thuresson, B. Lindman, and M. G. Miguel, *Langmuir* **20**(11), 4647(2004) .
58. P. A. Antunes, C. J. L. Constantino, R. F. Aroca, and J. Dufft , *Langmuir* **17**(10), 2958((2001).
59. S. Y. Arzhantsev, S. Takeuchi, and T. Tahara, *Chem. Phys. Lett.* **330**(1-2), 83 ((2000).
60. M. T. Martin, I. Prieto, L. Camacho, and D. Mobius, *Langmuir,* **12**(26)**,** 6554 (1996).
61. S. M. Martin, K. Kjaer, M. J. Weygand, I. Weissbuch, and M. D. Ward, *J. Phys. Chem. B*, **110**(29), 14292(2006).
62. T. Seki, and K. Ichimura, *Langmuir,* **13** (6), 1361(1997).
63. J. M. Pedrosa, M. Perez, I. Prieto, M. T. Martin-Romero, D. Mobius, and L. Camacho, *Phys. Chem. Chem. Phys.*, **4(11)**, 2329 (2002).
64. B. M. Ocko, and M. S. Kelley, *Langmuir* **18**(25), 9810 (2002).
65. M. Minones Conde, O. Conde, J. M.Trillo, and J.Minones, Jr., *J. Phys. Chem. B*, **114**(33), 10774 (2010).
66. C. Hao, R. Sun, J. Zhang, Y. Chang, C. Niu., *Colloids Surf. B Biointerfaces.* **69**(2), 201 (2009).
67. Hirano Yoshiaki, Yamazaki Asuka, Maio Ari, Kitahama Yasutaka, and Ozaki Yukihiro, *J. Phys. Chem. B* **114**(33), 10782 (2010).
68. Y. L. Lee, Y. C. Yang, and Y. J. Shen, *J. Phys. Chem.B.,* **109**(10), 4662((2005).
69. K. M. Wilkinson, L. Qunfang, and C. D. Bain, *Soft Matter*, **2**(1), 66 (2006).
70. Gang Ma and H. C. Allen, *Langmuir* **23**(2), 589(2007).
71. M. Kawasaki, and S. Aoyama, *Chem. Commun.,* **8**, 988 (2004).
72. S. Acharya, D. Bhattacharjee, and G. B. Talapatra, *J. Phys. Chem. Sol.* **64**(4)**,** 651(2003).
73. Th-Enderle, A. J. Meixner, and, I. Z. Granacher, *J. Chem. Phys.* **101**(5) 4365 (1994).
74. G. P. Wiederrecht, G. A. Wurtz, and J. Hranisavlijevic, *Nano Lett.* **4**(11), 2121 (2004).
75. T. Uwada, R. Toyota, H. Masuhara, and T. Asahi, *J. Phys. Chem. C.,* **111**(4), 1549 (2007).
76. C. Sönnichsen, B. M. Reinhard, J. Liphardt, and A. P. Alivisatos, *Nat. Biotechnol.*, **23(6)**, 741((2005).
77. E. Dulkeith, M. Ringler, T. A. Klar, J. Feldmann, A. M. Javier and W. J. Parak, *Nano Lett.* **5,** 585 (2005).
78. H. Nakahara and K. Fukuda, *J. Coll. Interface Sci.,* **93(2),** 530(1983).

79. C. J. Brown, *Acta Cryst.* **21**, 146(1966).
80. M. Aston, *Chem. Soc. Rev.* **22**, 67(1993).
81. N. Vranken, P. Foubert, F. Kohn, R. Gronheid, I. Scheblykin, M.V. der Auweraer, F.C. de Schryver, *Langmuir* **18,** 8407(2002).
82. S. Ozcelik, M. M. Demir, B. Birkan, *J. Phys. Chem.B.* **108,** 4679(2004).
83. Y. M. Jung, H. Sato, T. Ikeda, H. Tashiro and Y. Ozaki, *Spectrochim. Acta A.,* **60**(8-9), 1941(2004).
84. J. A. He, S. Bian, L. Li, J. Kumar, S. K. Tripathy, and L. A. Samuelson, *J. Phys.Chem. B.,* **104**(45), 10513 (2000).
85. C. Hohle, U. Hofmann, S. Schloter, M. Thelakkat, P. Strohriegl, D. Haarerb, S. J. Zilkerb, *J. Mater. Chem.* **9,** 2205 (1999).
86. E. Rivera, M. P. Carreo – Castro, I. Buendia, G. Cedillo, *Dyes Pigm.* **68**, 217 (2006)
87. Li Hong Niu, Cheng Zhong, ZiHui Chen, Zhi Zhang and ZhongYu Li, *Chinese Science Bulletin* **54 (7)**, 1169 (2009)
88. X. Song, J. Perlstein, D. G. Whitten, J. Am. Chem. Soc. **119**, 9144 (1997).
89. Helena Horvin Billsten, Villy Sundstrom, and Tomas Polıvka, *J. Phys. Chem.A* **109**, 1521 (2005).
90. H. W. Kroto, J. R. Heath, S. C. O'Brien, R. F. Curl, and R. E. Smalley, *Nature.* **318**, 162 (1985).
91. D. M. Guldi, *Chem. Commun.*, 321 (2000).
92. L. Echegoyen, L. Echegoyen, *Acc. Chem. Res.*, **31**(9), 593 (1998).
93. J. P. Hare, H. W. Kroto, R. Taylor, *Chem. Phys. Letts.*, **177**(4-5), 394 (1991).
94. M. Li, K. Deng, S-B Lei, Y-L Yang, T-S Wang, Y-T Shen, C-R Wang, Q-D Zeng, C. Wang, *Angew. Chem.Int. Ed.* **47**(35), 6717 ((2008).
95. Oh-ishi. K., Okamura. J., Ishi-i. T., Sano. M., Shinkai. S. (1999) Large Monolayer Domain Formed by C_{60}-Azobenzene Derivative, Langmuir, 15, 2224 – 2226.
96. T. Sasaki, J. M. Tour. *Org. Lett.* **10** (5), 897 (2008)
97. D. I. Schuster., K. Li., D. M. Guldi, A. M. Palkar, L. Echegoyen., C. Stanisky, R. J. Cross, M. Niemi, N. V. Tkachenko, H. Lemmetyinen, *J. Am. Chem. Soc.*, **129**, 15973 ((2007).
98. Y. Shirai, T. Sasaki, J. M. Guerrero, B. C. Yu, P. Hodge, J. M. Tour , *ACS Nano.*, **2** (1), 97 (2008).
99. K. S. Kumar, A. Patnaik, *J.Comput. Chem.* **31**(6), 1182(2010).
100. K. S. Kumar, A. Patnaik, *Chem. Phys.Chem.***11**(17), 3645 (2010).
101. Z-Z Gu, S. Hayami, Q. B. Meng, T. Iyoda, A. Fujishima, O. Sato, *J. Am. Chem. Soc.* **122**, 10730 (2000).
102. Z. Tian, Y. Chen, W. Yang, Y. Yao, L. Zhu, Z. Shuai, *Angew. Chem. Int. Ed.*, **43**, 4060(2004).
103. C. J. Faul, M. Antonietti, *Advanced Materials*, **15**(9), 673(2003)
104. D. G. Whitten, L. Chen, H. C. Geiger, J. Perlstein, X. Song, *J. Phys. Chem. B.*, **102** (50), 10098(1998).
105. S. Yagai, T. Kinoshita, M. Higashi, K. Kishikawa, T. Nakanishi, T. Karatsu, A. Kitamura, *J. Am. Chem. Soc.*, **129** (43), 13277 (2007).
106. E. Arunkumar, C. C., Forbes, B. D. Smith, *Eur. J. Org. Chem.***19**, 4051(2005).
107. J. A. A.W. Elemans, A. E. Rowan, R. J. M. Nolte, *J. Mater. Chem.***13**, 2661(2003).
108. H-A Klok, K. A. Jolliffe, C. L. Schauer, L. J. Prins, J. P. Spatz, M. Möller, P. Timmerman, D. N. Reinhoudt, *J. Am. Chem. Soc.***121** (30), 7154(1999).

109. S. S. Gayathri, A. Patnaik, *Langmuir* **23**(9), 4800(2007).
110. I. Ramakanth, B. Vijai Shankar and A. Patnaik, *Chem Commun.*, 2008, 4081
111. K. S. Kumar and A. Patnaik, Chem. European J. 2011(In Press)
112. C. G. Morgantea, W. S. Struve, *Chem. Phys. Letts.* **68**(2-3), 267(1979).
113. M. Han, M. Hara, *J. Am. Chem. Soc.*, **127** (31), 10951(2005).
114. M. Shimomura, T. Kunitake, *J. Am. Chem. Soc.***109** (17), 5175(1987).
115. J. Nithyanandhan, N. Jayaraman, R. Davis, S. Das, *Chemistry – A European Journal*,**10**(3), 689(2004).
115a. U. Roesch, S. Yao, R. Wortmann, and F. Wurthner, *Angewandte Chemie Int. Ed.* **45**,7026(2006).
115b. M. vander Auweraer, G. Biesmans, F. C. DeSchryver, *Chem.Phys.* **119**, 355(1988).
115c. M. vander Auweraer, B. Verschuere, F. C. DeSchryver, *Langmuir* **4**, 583(1988).
115d. L. Lu, R. J. Lachicotte, T. L. Penner, J. Perlstein, D. G. Whitten, *J.Am.Chem.Soc.* **121**, 8146 (1999).
115e. F. W. Mrthner, S. Yao, T. Debaerdemaeker, R. Wortmann, *J.Am.Chem.Soc.* **124**, 9431(2002).
116. M. J. Frisch et al. Gaussian 03, Revision C.02, Gaussian, Inc., Wallingford CT, 2004.
117. M. L. Waters, *Current Opinion in Chemical Biology* **6**(6), 736(2002).
118. C. A. Hunter, *Chem. Soc. Rev.* **23**, 101(1994).
119. C. A. Hunter, K. R. Lawson, J. Perkins, C. J. Urch, *J. Chem. Soc., Perkin Trans.* **2**, 651(2001).
120. R. Zales'ny, K. Matczyszyn, A. Kaczmarek, W. Bartkowiak, P. Cysewski, *J. Mol. Model.* **13**, 785(2007).
121. L. Briquet, D. P. Vercauteren, J. M. André, E. A. Perpète, D. Jacquemin, *Chem. Phys. Letts.* **435**, 257((2007).
122. S. S. Gayathri and A. Patnaik, *Chem. Commun.*, 1977 (2006).
123. S. Campidelli, J. Lenoble, J. Barberá, F. Paolucci, M. Marcaccio, D. Paolucci, R. Deschenaux, *Macromolecules*, **38** (19), 7915(2005).
124. F. Wudl, *J. Mater. Chem.*, **12**, 1959(2002).
125. H. Asanuma, H. Li, T. Nakanishi, H. Moehwald, *Chem. Eur. J.*, **16**(31), 9330 (2010).
126. T. Nakanishi, K. Ariga, T. Michinobu, K. Yoshida, H. Takahashi, T. Teranishi, H. Moehwald, D. G. Kurth, *Small* **3**(12), 2019 (2007).
127. T. Nakanishi, W. Schmitt, T. Michinobu, D. G. Kurth, K. Ariga, *Chem. Commun.*, 5982 (2005).
128. S. S. Gayathri and A. Patnaik ., *J. Chem. Phys.* **124**(1), 131104 (2006).
129. T. Nakanishi, *Chem. Commun.*, **46** (20), 3425 (2010).
130. H. Nobukuni, Y. Shimazaki, H. Uno, Y. Naruta, K. Ohkubo, T. Kojima, S. Fukuzumi, S. Seki, H. Sakai, T. Hasobe, F. Tani, *Chemistry – A European Journal*, **16**(38), 11611(2010).
131. H. Yamada, K. Ohkubo, D. Kuzuhara, T. Takahashi, Atula., S. D. Sandanayaka, T. Okujima, K. Ohara, K., O. Ito, H. Uno, N. Ono, S. Fukuzumi, *J. Phys. Chem. B.* **114** (45), 14717 (2010).
132. Y. Sun, T. Drovetskaya, R. D. Bolskar, R. Bau, P. W. D. Boyd, C. A. Reed, *J. Org. Chem.*, **62** (11), 3642 (1997).
133. H. Nobukuni, F. Tani, Y. Shimazaki, Y. Naruta, K. Ohkubo, T. Nakanishi, T. Kojima, S. Fukuzumi, S. Seki, *J. Phys. Chem. C*, **113**(45), 19694 (2009).

134. M. S. Rodrguez-Morgade, M. E. Plonska-Brzezinska, A. J. Athans, E. Carbonell, G. D. Miguel, D. M. Guldi, L. Echegoyen, T. Torres, *J. Am. Chem. Soc.*, **131** (30),10484 (2009).
135. A. D. Bokare and A. Patnaik, *J. Phys. Chem. B* **109** (1)**,** 87 (2005).
136. A. D. Bokare and A. Patnaik , *J. Phys. Chem. B* **107** (25)**,** 6079 (2003).
137. P. D. W. Boyd, C. A. Reed, *Acc. Chem. Res.*, **38** (4), 235 (2005).
138. P. Bhyrappa, K. Karunanithi, J. *Inorg. Chem.*, **49**(18), 8389 (2010).
139. G. Gil-Ramrez, S. D. Karlen, A. Shundo, K. Porfyrakis, Y. G. Ito, A. D. Briggs, J. J. L. Morton, H. L. Anderson, *Org. Lett.*, **12** (15), 3544 (2010).
140. M. Matsumoto, H. Tachibana, R. Azumi, M. Tanaka, T. Nakamura, G. Yunome, M. Abe, S. Yamago, E. Nakamura, *Langmuir* **11**, 660(1995).
141. D. Felder, J-L. Gallani, D. Guillon, B. Heinrich, J-F. Nicoud, J. F. Nierengarten, *Angew. Chem. Int. Ed.* **39**, 201 (2000).
142. S. Burghardt, A. Hirsch, N. Medard, R. Abou, K. D. Aussere, M. P. Valignat, J. L. Gallani, *Langmuir* **21**, 7540(2005).
143. A. Honciuc, A. Jaiswal, A. Gong, K. Ashworth, C. W. Spangler, L. R. Peterson, L. R. Dalton, R. M. Metzger, *J. Phys.Chem. B*, **109**, 857(2005).

CHAPTER 13

MESOSCOPIC MORPHOLOGY, OPTICAL ANISOTROPY AND SPECTROSCOPIC PROPERTIES OF CYANINE J-AGGREGATES

Hiroshi Yao

Graduate School of Material Science, University of Hyogo
3-2-1, Koto, Kamigori-cho, Ako-gun, Hyogo 678-1297, Japan
E-mail: yao@sci.u-hyogo.ac.jp

Fabrication of highly organized systems such as J-aggregates from photofunctional molecular building blocks opens up new vistas for the exploration of photonic nanodevice concepts. This review delineates formation and direct observation of J-aggregates with various mesoscopic structures in "solution phases" (in bulk solution and at solution/solid interfaces). Fluorescence microscopy, polarized-light microscopy (PLM) and atomic force microscopy (AFM) are used to characterize their morphologies which include quasi-one-dimensional string, tubular-rod and two-dimensional sheet structures as well as three-dimensional multilayers. The optically anisotropic nature of such J-aggregates is quantitatively evaluated. The J-aggregates carrying the outstanding spectroscopic properties with efficient exciton coupling represent a versatile rationale to realize structurally complex nanosystems with specific shape and functions, so the study on the nanoscopic/mesoscopic structures of such aggregates will be significant in relation to future supramolecular engineering.

Keywords: Mica/water interface; Three-dimensional structure; Morphology transformation; Atomic force microscopy (AFM); Polarized-light microscopy (PLM); Optical anisotropy; Tubular structure; Monomolecular sheet

1. Introduction

One of the well-known chromophore self-assemblies caused by the enthalpically-driven attractive molecular interaction is the J-aggregates. J-aggregates were discovered by Jelley and Scheibe in the late 1930s,[1,2] and characterized by a narrow and intense electronic absorption band (J-band) that shows a bathochromic shift with respect to the relevant monomer band.[3] Because of the striking optical characteristics, J-aggregates are often used as spectral sensitizers in photography.[4] Recent interest has focused on the ability of the aggregates to exhibit coherent excitation phenomena, such as superradiance, which generates large optical nonlinearities.[5,6] It should be noted that many concepts of coherent energy transfer developed for J-aggregates can be applied to light-harvesting complexes.[7]

J-Aggregates have been found in bulk liquids,[8–11] in Langmuir–Blodgett films,[12,13] or on solids such as silver halides.[14,15] Hence it is considered that a J-aggregate has a one or two-dimensional side-by-side arrangement of molecules in which the individual transition moments are parallel to each other and parallel to the line joining them. Various theoretical models have been proposed to describe the relationship between the physical/optical properties and the molecular arrangement within the J-aggregates.[16–19] In a simple exciton model,[19] a bathochromic spectral shift of the J-aggregate comes from the small slip angle α (the slip angle defines the angle between the line-of-centers of a column of dye molecules and the long axes of any one of the parallel molecules). It is generally accepted that the bathochromic shift is the result of $\alpha <$ ~$32°$.[3] In any model, they show that the aggregate structure strongly impacts its spectroscopic properties such as the spectral line shape and peak energy.[19] Despite the multitude of spectroscopic and theoretical studies of J-aggregates, the knowledge of the detailed supramolecular structure of the aggregates is still fragmentary and controversial. Therefore, detailed investigations of the structure and morphology of the J-aggregates are of primary importance, and consequently, direct observation of the aggregates has gained interest.

So far, the structure of J-aggregates has been extensively studied for "dry" films by using STM (scanning tunneling microscopy),[20] AFM (atomic force microscopy),[21–23] and NSOM (near-field scanning optical microscopy).[24] In contrast, the structure of J-aggregates assembled in "solution phases" (which includes bulk solutions and solution/solid interfaces) has been rarely explored, probably due to the difficulty in imaging their fragile structures in "solutions". In recent years, however, studies of the solution phase J-aggregates characterized by cryogenic transmission electron microscopy (cryo-TEM) have proved various nanoscopic/mesoscopic morphologies such as twisted ropes or fibers.[25–28] Although this technique provides highly resolved images of the aggregates in their native environment, it requires very specialized apparatus and can neither distinguish distributions nor dynamic processes (morphological transformation) of the aggregates *in situ* due to the vitrifaction processes.

In this chapter, I would review direct observations of mesoscopic J-aggregates of some cyanine dyes produced at solution/solid interfaces or in bulk solution by various microscopic techniques such as fluorescence microscopy, polarized-light microscopy (PLM) and atomic force microscopy (AFM) as well as by useful spectroscopic measurements. The observed J-aggregates are not vitrified upon drying. We successfully find various mesoscopic morphologies of the J-aggregates having quasi-one-dimensional string, tubule, or two-dimensional sheet-like structures in bulk solution, and those of the interfacial J-aggregates possessing three-dimensional multilayer structures. Optically anisotropic or birefringent properties of the J-aggregates are also described.

2. Direct Observations of J-Aggregates

2.1 Fluorescence Microscopy

I here describe briefly some important microscopic techniques for investigating mesoscopic J-aggregates. The first one is fluorescence microscopy.[29] Fluorescence microscopy requires that the objects should fluoresce. Therefore, if the molecular assembly of interest fluoresces strongly (such as J-aggregates), this technique can be very suitable. In fluorescence microscopy, the limits of detection are generally governed by the darkness of the background with little or no emission; in other words, only the emission should reach the detector so the resulting fluorescent structures are superimposed with high contrast against a very dark background in a properly configured microscope. Note that the fluorescence microscope cannot provide spatial resolution below the diffraction limit of specimen features, but the detection of fluorescing molecules below such limits can be successfully achieved.

To view a fluorescent sample in the microscope, (i) a bright light source (typically, mercury arc lamp) with proper (monochromatic) wavelengths for excitation, and (ii) specific band-pass filters to isolate the excitation and emission lights from the specimen, are required. In particular, a dichroic beam splitter, which reflects shorter wavelengths of light (that is, excitation) and allows longer wavelengths (that is, emission) to pass, plays a significant role in isolating the emission from the excitation light. In "epi-illumination" mode, the most frequently used set-up, the objective acts as a focusing lens for the excitation beam as well as a gathering lens for emission. It is worth noting that fluorescence is the only mode in optical microscopy where the specimen produces its own light.

2.2 Polarized-Light Microscopy (PLM)

Fluorescence microscopy can be applied only for the fluorescent specimens. A regular molecular alignment in the J-aggregate is capable of bringing about their optical anisotropy or birefringence, so the polarized-light microscopy (PLM) should be a useful technique that gives contrast-enhancing imaging.[30] PLM is designed to observe specimens that are visible primarily due to their optically anisotropic character. The microscope should be equipped with both a "polarizer", positioned in the light path before the specimen, and an "analyzer" (a second polarizer), placed in the optical pathway between the objective and the detector in the cross-polarized position. Image contrast arises from the interaction of plane polarized-light in a birefringent specimen to produce two individual wave components that are each polarized in mutually perpendicular planes. The velocities of these components, which are referred to as "ordinary" and "extraordinary" rays, are different and vary with the propagation direction through the specimen.

Hence the light components become out of phase, but are recombined with constructive and destructive "interference" in the analyzer, yielding an image contrast or interference color under white-light illumination in the microscope. Note that, in the case where incident rays impact the crystal in a direction that is parallel to the optical axis, they behave as ordinary light rays and are not separated into individual components by an anisotropic birefringent material.

The interference can be utilized to quantitatively draw conclusions about birefringence or retardation. The retardation (Γ), which is associated with the difference in speed between the ordinary and extraordinary rays, is expressed by "$\Gamma = \Delta n \times d$, where Δn and d are the birefringence and the thickness of a specimen, respectively. In an anisotropic crystal, the two independent refractive indices for the ordinary and extraordinary rays are quantified in terms of birefringence, a measure of the difference in refractive indices (Δn), defined as

$$\Delta n = |n_e - n_o|$$

where n_e or n_o is the refractive index experienced by the extraordinary or ordinary rays, respectively. Note that the degree of birefringence of a specimen will vary with the orientation of the crystal relative to the incident angle of illumination, but we often focus on the maximum Δn value. In addition, the interference colors caused by the retardation are commonly summarized as the Michel-Lévy chart.[31]

When the sample is placed between crossed polarizers in the microscope and rotated to a position of maximum brightness, the observed interference color can be traced down on the retardation axis to find the retardation between the ordinary and extraordinary ray in the sample. The maximum degree of brightness will be achieved when the specimen optical axis (that is, direction in which a ray of transmitted light suffers no birefringence) is oriented at an angle of 45° to the polarizer axis, and the greatest degree of extinction will be observed when the two axes (= optical and polarizer axes) coincide, indicating that, during rotation of the specimen over a range of 360°, its visibility will oscillate between bright and dark four times in 90° increments. In any case, PLM exploits the optical properties specific to anisotropy or birefringence and reveals detailed information concerning the structure, molecular arrangement, or composition of materials that are invaluable for identification and diagnostic purposes.

2.3 Atomic Force Microscopy (AFM)

An atomic force microscope (AFM) has the advantage of imaging almost any type of surface at the nanoscopic scale, including polymers, inorganics, organics, and biological samples.[32] The AFM tip fabricated from Si or Si_3N_4 contacts the surface directly with the interatomic van der Waals forces providing the interaction mechanism. Importantly, the AFM has a variety of operation modes. In "contact mode," the tip is in constant contact with the surface. The *z*-piezo moves the cantilever up and down to maintain constant contact force. Dragging the tip across the surface can be sometimes damaging to soft samples. On the other hand, in "tapping-mode", which is also referred to as "intermittent-contact" mode, the cantilever is driven at near its resonant frequency so that the cantilever vibrates the tip with a certain amplitude. As the tip approaches the surface, damping occurs and the amplitude of the vibration decreases. When the tip scans the surface, the *z*-piezo adjusts the height of the cantilever to maintain a constant amplitude of vibration. The advantage of tapping the surface is improved lateral resolution on "soft" samples and preferable to imaging molecular J-aggregates. Lateral forces such as drag, common in contact mode, are virtually eliminated. In AFM observations, a vertical resolution of 0.01 nm (~1/20 height of atom) can be achieved with the laser sensor system. The horizontal resolution of AFM is limited by the radius of curvature of the tip since commercial tips have a radius of about 10 nm.

3. Pseudoisocyanine (PIC) J-Aggregates at A Mica/Water Interface

3.1 Three-Dimensional Multilayer Structures of PIC J-Aggregates at a Mica/Water interface

It is well known that cationic dye 1,1'-diethyl-2,2'-cyanine (pseudoisocyanine: PIC, chemical structure is shown in Fig. 3.1 produces J-aggregates at a high dye concentration in aqueous solution, and the aggregate shows an intense absorption band (J-band) at about 572 nm at room temperature.[33] It has been proposed that the J-aggregate with a peak of 572 nm (termed as J_S-type) would be a thread-like or ribbon-like structure.[17,34] A recent study showed that the aggregate had a nanoscopic cylinder morphology consisting of six stacks of two adjacent PIC molecules.[26] Meanwhile, the PIC dye also produces J-aggregates at solution/solid interfaces. In the early work by Scheibe, the J-aggregates were observed at a mica/solution interface as revealed by absorption spectroscopy.[35] The peak of the J-band was further red-shifted (termed as J_L-type) compared to that of the J_S-band, and a monolayer arrangement of the dye molecules in the aggregates was proposed. Later, on the basis of quantum mechanical calculations and epitaxial match between the dye and the lattice of a mica substrate, Kuhn and co-workers have proposed a two-dimensional (2D) brick stone work arrangement of the dye molecules at the mica/solution interface.[36] The present author has also found J_L-type aggregates of PIC at a soda lime glass/water interface at room temperature.[33] It is important to note that J_L-aggregate formation can be induced by the presence of anionic sites on a solid surface.[33,37]

Since spectroscopic properties of J-aggregates are influenced by the structures represented by the aggregation number, packing dimension and dye molecular orientations,[19] the more red-shift absorption of J_L-aggregates observed at solution/solid interfaces, in comparison to the position of the J_S-band, is ascribed to a difference in their internal structures or morphologies. Hence detailed investigations of the J_L-type aggregates are of primary importance to elucidate characteristic features of the J-aggregates produced at solution/solid interfaces. In recent years, AFM has been used to study the J-aggregate structures under dry conditions;[21–23] however, a complementary study on the J-aggregates *at solution/solid interfaces* using AFM has never been explored. In this section, I describe characteristic morphologies of the PIC J-aggregates produced at a mica/water interface, which are investigated by *in situ* tapping-mode AFM and fluorescence microscopy as well as by spectroscopic studies such as absorption, fluorescence, and fluorescence lifetime measurements.[38]

3.1.1 *Spectroscopy*

Figure 3.1a shows an optical path length dependence of the absorption spectrum of an aqueous PIC solution (2.0 mM) in a custom-made cell. The sample in the cell was prepared by placing an aliquot of the aqueous PIC solution between mica and hydrophobic glass plates.[38] The spectra showed a sharp and intense J-band (or J_L-band; 580 nm) in addition to the monomer (525 nm) and dimer (480 nm) bands. The peak wavelength of the J-band was red-shifted compared to that of the well-known J_S-aggregate (~572 nm) in a solution phase. Note that no J-band could be detected when the absorption measurement was conducted using a cell composed of only hydrophobic glasses at the concentration. Thus, J-aggregate formation is originated from the interactions between PIC molecules and mica surface.[35,38] Clearly, the absorbance of the monomer or dimer band increases with an increase in the optical path length, whereas that of the J_L-band is independent of the path length, suggesting that the J-aggregates are confined at the mica/water interface.

A concentration dependence of the absorption spectrum of PIC was studied to examine the growth of the J_L-aggregate. Figure 3.1(b) shows

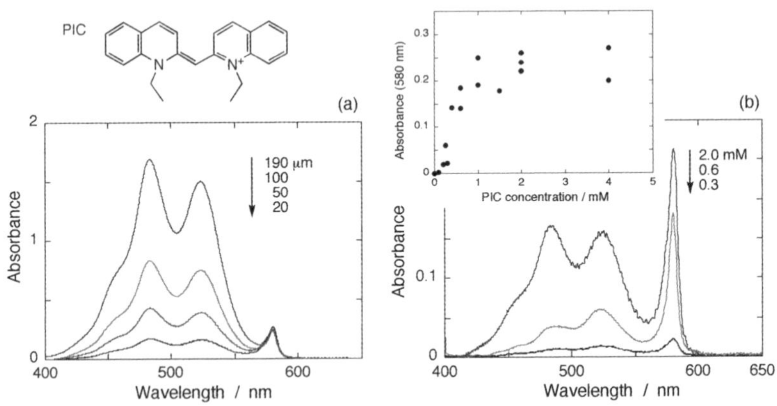

Figure 3.1 (a) Optical path length dependence of the absorption spectrum for aqueous PIC solution (2.0 mM) between mica and hydrophobic glass plate. (b) Absorption spectra of the PIC J-aggregate produced at a mica/solution interface at various PIC concentrations. The inset shows the concentration dependence of the J_L-band (580 nm). Chemical structure of pseudoisocyanine (PIC) dye is also shown.

the absorption spectrum of an aqueous PIC solution at a different concentration (2.0, 0.6, or 0.3 mM). The absorption peak wavelength and line width (HWHM; ~120 cm^{-1}) of the J-band were constant at any concentration (0.1–4.0 mM), indicating the coherence size of the produced J-aggregates which determines such optical properties is unchanged.[16,17] The concentration dependence of the absorbance at 580 nm showed a quasi-adsorption behavior (inset in Fig. 3.1(b)); the J-band appeared at above 0.1 mM, and the absorbance increased with an increase in the PIC concentration and saturated at > ~1.0 mM. The results also suggest that J-aggregate formation took place at the mica/water interface.

The J_L-aggregates showed a sharp fluorescence spectrum (HWHM; ~120 cm^{-1}) as shown in Fig. 3.2(a). The fluorescence spectrum did not show a Stokes shift as a characteristic of that of the J-aggregate. To examine the effect of the growth of the J-aggregates on the excitonic state, fluorescence lifetime measurements were conducted at various PIC concentrations. Fig. 3.2(b) shows a typical fluorescence decay profile (2.0 mM) together with an instrumental response function (IRF). Although a very fast component was dominant in the decay, the profile was best fitted by a double exponential function probably due to inhomogeneity in interactions between the J-aggregates and the mica surface. Thus, we estimated an average fluorescence lifetime ($\langle \tau \rangle$) as is often used for fluorescence decay analysis of a molecule in an inhomogeneous adsorbed state;

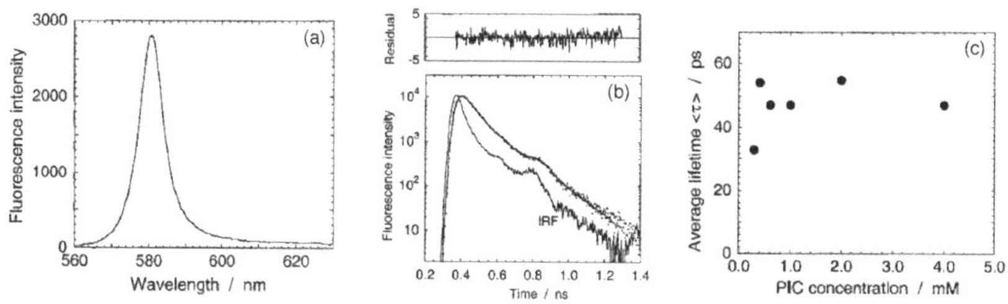

Figure 3.2. (a) Fluorescence spectrum of the J-aggregate at a mica/ solution interface ([PIC] = 2.0 mM). (b) Fluorescence decay profile of the J-aggregate excited at 570 nm. The double exponential fit is also shown. (c) Concentration dependence of the average lifetime $\langle \tau \rangle$.

$$\langle\tau\rangle = \sum_i a_i \tau_i^2 / \sum_i a_i \tau_i$$

where a_i and τ_i are the pre-exponential factor and the fluorescence lifetime of the i-th component, respectively. This approach is advantageous in determining $\langle\tau\rangle$ with a very high reproducibility even when a_i or τ_i is somewhat scattered from one measurement to another.[39] The solid curve in Fig. 3.2(b) is the best fit of the decay profile with the parameters of $a_1 = 30.5$, $\tau_1 = 44$ ps, $a_2 = 4.0$, and $\tau_2 = 96$ ps. The average fluorescence lifetime $\langle\tau\rangle$ was then calculated to be 55 ps. A short fluorescence lifetime compared to that in solution has been generally observed for the J-aggregate adsorbed on surfaces.[40] Figure 3.2(c) shows the PIC concentration dependence of $\langle\tau\rangle$, indicating almost no dependence on [PIC] and $\langle\tau\rangle$ was almost constant at 45–50 ps at any concentrations. The results mean that homogeneity in the exciton state of the J-aggregate does not change even upon the growth of the aggregates.

3.1.2 *Microscopy*

Since the J-aggregate possesses a highly emissive property but monomeric PIC exhibits no emission,[37] fluorescence microscopy is a very useful technique to determine the mesoscopic structures of the J-aggregates produced at the mica/water interface. Figures 3.3(a) or 3.3(b) shows a typical fluorescence micrograph of the interfacial J-aggregates at 0.4 or 0.8 mM, respectively. At [PIC] $\geq$ ~0.8 mM, all of the interface regions were brightly emissive, indicating that the mica/solution interface was completely covered with the PIC J-aggregate islands. In addition, we found that the interface was composed of a complicated mesoscopic domain that is reminiscent of a piece of a jigsaw puzzle. When the [PIC] decreased, however, the coverage by the emissive J-aggregates also decreased, resulting in the discrete distribution of the aggregate islands at the interface.

In an attempt to further explore the interfacial J-aggregate structures, AFM was carried out. Figures 3.4(a) – 3.4(c) show typical AFM images in the concentrations region where the absorbance of the J_L-band increases drastically. The surface of mica remained unchanged and was atomically flat until the J-band appeared (< ~0.1 mM). At [PIC] =

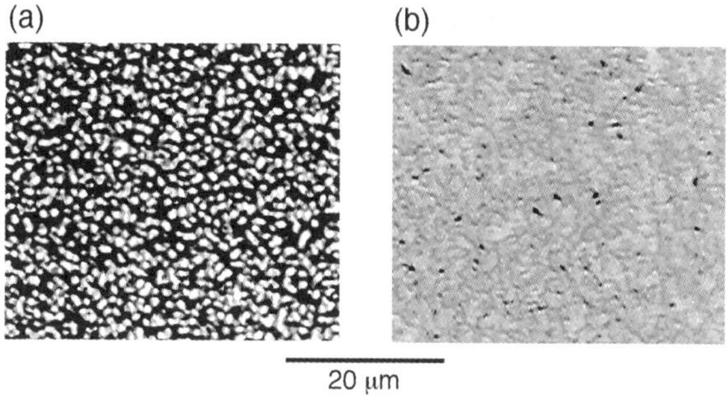

Figure 3.3 Fluorescence micrographs of the PIC J-aggregates at the mica/water interface. The PIC concentration is (a) 0.4 mM or (b) 0.8 mM.

0.1mM where J-aggregate formation took place, leaf-like islands appeared abruptly (Fig. 3.4(a)). Thus, the islands were considered to be the J-aggregates. The size of these islands ranged between ~400–600 nm in length, ~80–100 nm width, and ~4 nm in height (see the cross-sectional profile in Fig. 3.4(a). Importantly, our AFM topographs revealed that the J-aggregate formed at a mica/water interface has a three-dimensional (3D) disk-like structure, but not a 2D monolayer structure. On the assumption of the edge-on arrangement of PIC molecules on mica, the height of the islands corresponds to several molecular layers (multilayers). Such a structure formed by the interactions between the substrate and molecules should bring about different optical properties compared to those of the J_S-aggregates produced in bulk solution, where the latter has been reported to posses a thread-like cylinder structure.[26,34] With an increase in the dye concentration, the number density of the leaf-like islands also increased, and then, they coalesced into complex larger domains with the apparent surface area of ~2–4 μm^2 (0.3 mM; Fig. 3.4(b)). In contrast, the height of the islands was independent of the solution concentration (~3–6 nm). At 3.0 mM, the surface of mica was almost completely covered with islands (Fig. 3.4(c)), suggesting a quasi-adsorption behavior as already shown in Fig. 3.1(b). These morphological changes clarified by *in situ* AFM suggest that the J-aggregates are grown at the mica/solution interface by a Volmer-Weber type growth process; spatially discrete nucleation on mica and successive growth to the islands.[41] The growth process

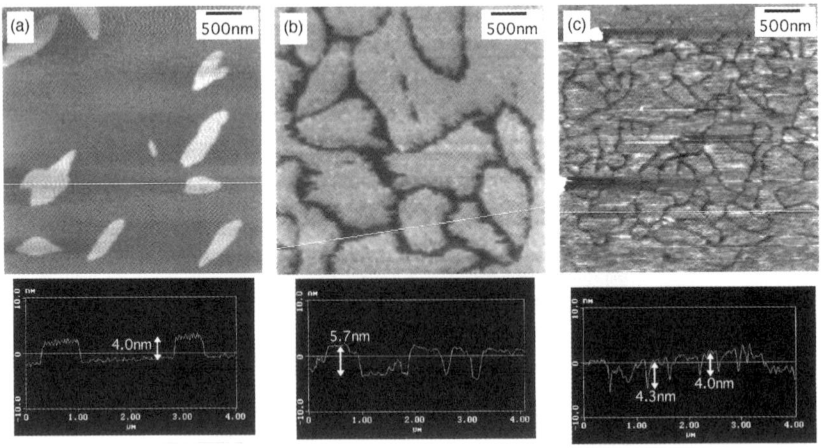

Figure 3.4 AFM images of the PIC J-aggregates at the mica/water interface and cross-sectional profiles along the lines indicated in each figure. The PIC concentrations for the images (a), (b) and (c) are 0.1, 0.3 and 3.0 mM, respectively.

indicates that flocculation of molecules at the anion sites on mica could rise the adsorption energy, leading to formation of discrete nuclei, and successive growth to larger J-aggregate islands. The constant thickness of the J-aggregates (3–6 nm) would be determined by the balance between the adsorption/aggregation and dissolution energies. This issue will be discussed later again.

On a closer inspection of the AFM images, the anisotropic growth of the J-aggregate islands is obvious.[42] We then determined the orientation of the long axis of the islands to model the J_L-aggregate structure. The orientation of the islands with respect to the surface structure of mica was analyzed as follows: First, we imaged the lattice structures of the mica surface in air and confirmed a three-fold symmetric arrangement of negatively charged holes (Fig. 3.5(a)).[43,44] When a mica is cleaved, interlayer K^+ ions are divided onto two mica surfaces, resulting in the formation of negative holes left by K^+ ions dissociating. As the ordering of the K^+ ions is hexagonal between silicate layers in mica, the ordering of the induced negative holes is also hexagonal.[44] Black spots in Fig. 3.5(a) would then correspond to the positions of K^+ ions. The orientation of the hole alignments, namely, direction neighboring two holes, was labeled as A, B, or C as shown in Fig. 3.5(d). Next, we imaged the same surface of mica in the solution at a PIC concentration region where the J-aggregates islands form (Fig. 3.5(b)), and measured angles (θ) between the long axis of the islands and the line in direction "A". Finally, the obtained angles were plotted in a histogram as shown in Fig. 3.5(c). Although the histogram of θ between 0 and 180° showed some scatter, Fig. 3.5(c) clearly indicates that the long axes of the islands are anisotropically oriented relative to the alignment of the holes formed by dissociating K^+ ions.[43,44] A schematic model is also presented in Fig. 3.5(d).

This anisotropic growth suggests the existence of the epitaxial interaction between the dye molecules and the lattice of a mica substrate during aggregation. The highly probable epitaxial interaction is the one proposed by Scheibe, namely, that the most positively-charged N atoms in the PIC molecules were placed at the negative holes left by the dissociating K^+ ions.[35] According to this epitaxial interaction, there are two possible alignments of the dye molecules in the island, that is, the long axis of the dye molecules are parallel or 60°-inclined relative to the long axis of the islands (Fig. 3.6). Barbara and co-workers found that fibrous J-aggregates have been produced by combining PIC molecules with polymer chains,[24] and explained that the dye molecules are in a brick stone work and parallel to the thread. This means that this kind of alignment is the preferable for producing the J-aggregates. Therefore, the dye molecules would grow so that the long axis of dye molecules is parallel to the long axis of the islands (Fig. 3.6). In this model, the slip angle between PIC molecules shows 30°, which is smaller than ~32° obtained on the basis of theoretical calculations for the J-aggregates.[3] This structure is different from that of bulk crystal,[17,26,45,46] and inherent to a mica/solution interface.

The absorbance of the J-band increased in accordance with the island growth of the aggregates. However, the spectroscopic properties of the J-aggregate such as the absorption and fluorescence peaks or lifetime were almost unchanged upon the growth of the aggregates. Considering that the spectroscopic properties would be determined by the coherence size of the aggregates, the coherence size is much smaller than the physical size of the aggregates detected by AFM.

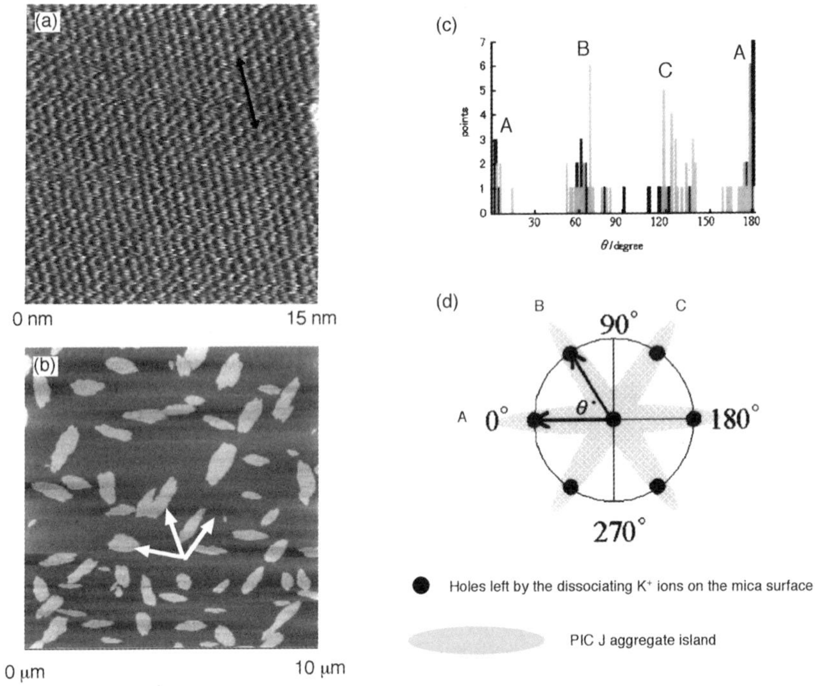

Figure 3.5 (a) Atomic scale AFM image of a mica surface in air. Dark spots correspond to the alignment of holes left by the dissociating K^+ ions on the surface of mica. (b) AFM image of leaf-like islands of J-aggregates at the mica/water interface. Arrows show the periodic orientation (three-fold symmetry) of negative holes as shown in (a). (c) The histogram of the angles θ between 0 and 180°. (d) Schematic representation of the relation between a lattice structure of mica surface (negative holes) and PIC J-aggregate islands.

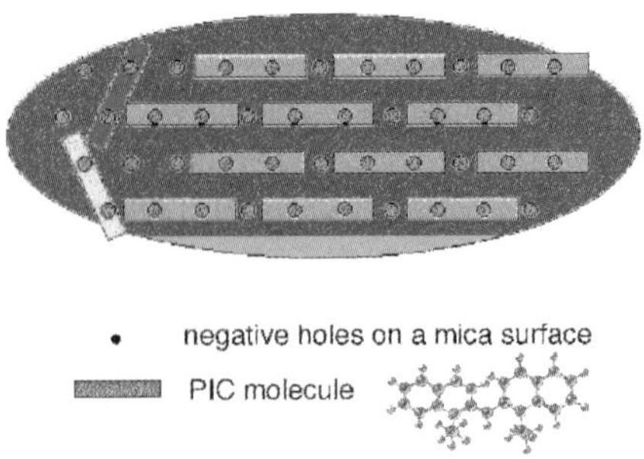

Figure 3.6. Model of the alignment of PIC molecules in the J-aggregate islands formed at the mica/water interface.

3.2 Effect of organic solvents on PIC J-aggregation at a mica/water interface: 3D-to-2D Morphological transition

In the assembly of surfactant systems, cylindrical morphology of CTAB (cetyltrimethylammonium bromide) on a mica substrate has been observed in pure aqueous solution. It is interesting to note that addition of organic solvents can change the adsorption behaviors of the surfactant, yielding different morphology on mica: For example, addition of alcohol decreases the adsorption of CTAB and structural transition from cylinders to spheres has been found.[47] The observed transition can be best explained by hydrophobic interactions between CTAB and alcohol chains. In aqueous phase J-aggregates, on the other hand, spectral and size modifications have been observed for a carbocyanine dye bearing 1,1′-dialkyl substituents upon addition of organic solvents such as alcohols;[48] however, to the best of our knowledge, no data are available for the effect of organic solvents on the interfacial dye assemblies. Contrary to the hitherto accepted model that the J-aggregates at a solution/solid interface have a 2D monolayer structure,[35,36] we have clarified a 3D multilayer structure of the J-aggregates.[37,42] In the case of the PIC J-aggregate at the mica/water interface, if pure water is replaced by aqueous mixed solvents (water/organic mixed solvents), their structures, distributions, or dimensions would be dramatically modified by the organic components. This will provide a new insight on the origin of the 3D structure of interfacial dye assemblies.

In this section, I describe the effect of addition of organic solvents on the 3D morphology of PIC J-aggregates at the mica/solution interface. Co-solvent-*dependent* morphological transitions as well as co-solvent-*independent* spectroscopic properties of the J-aggregates are found, proving that optical properties of the J-aggregates are determined by the 2D coherent interactions within the aggregate.[49]

3.2.1 *Solvent Effect on the Spectroscopic and Morphological Properties*

The organic solvent used was 1-propanol (PrOH) or 1,4-dioxane (Dox), which is completely water-miscible. The boiling point of PrOH or Dox is 97.2 or 101.3 °C, respectively, at normal atmospheric pressure, so the noticeable evaporation of the organic solvents is avoided. Binary mixed solvents were made by adding the organic solvent (PrOH or Dox) of 5 or 10 vol% into water. For concise, the aqueous solution containing *k* vol% PrOH or Dox is refereed to as PrOH(*k*)/water or Dox(*k*)/water, respectively. The PIC dye concentration was fixed at 0.8 mM. For spectroscopic measurements, samples were prepared by placing an aliquot of the PIC solution of mixed solvents between a freshly cleaved mica and a hydrophobic glass plate (optical path length of about 30 μm). Panel A in Fig. 3.7 shows the absorption spectra of PIC (0.8 mM) in various mixed solvents in the presence of mica surface. As noted before, at the mica/pure water interface, the spectrum showed a sharp and intense J_L-band at 580 nm (see Fig. 3.1).

Upon addition of the organic solvent of PrOH or Dox into water, peak positions of the J-band were unchanged (580 nm), whereas its intensity decreased with increasing the amount of organic co-solvent. In addition, the spectral line widths of the J-band were also almost identical with one another (110–120 cm^{-1}; HWHM). In a simple exciton model, the delocalization range of the electronic states determines the sharpness of the J-band that comes from the cooperative effects in the optical response,[5,6,16,17] so that these results indicate that the molecular alignment in the aggregate and the delocalization or coherent size of the aggregate were not influenced by these organic co-solvents. Note again that the broad absorption peak observed at 525 or 480 nm is mainly ascribed to the PIC monomer or dimer, respectively. To better understand the relationship between the spectral and structural properties of the J-aggregates, we next examined their characteristic morphology using fluorescence microscopy and AFM.

Panel B in Fig. 3.7 shows fluorescence micrographs of the PIC J-aggregates in the presence of organic co-solvents. In comparison to the micrograph in Fig. 3.3(b) (pure water; [PIC] = 0.8 mM), the morphology of the aggregate domains altered with increasing the amount of the co-solvent. In PrOH(5)/water, the mica/solution interface was still largely covered by the J-aggregates. With increasing the amount of PrOH to 10 vol% (Panel B; a-2), however, the coverage of the J-aggregate decreased. In the case of Dox addition, the coverage of the J-aggregates was slightly lower than that in the case of PrOH addition at the same volume fraction

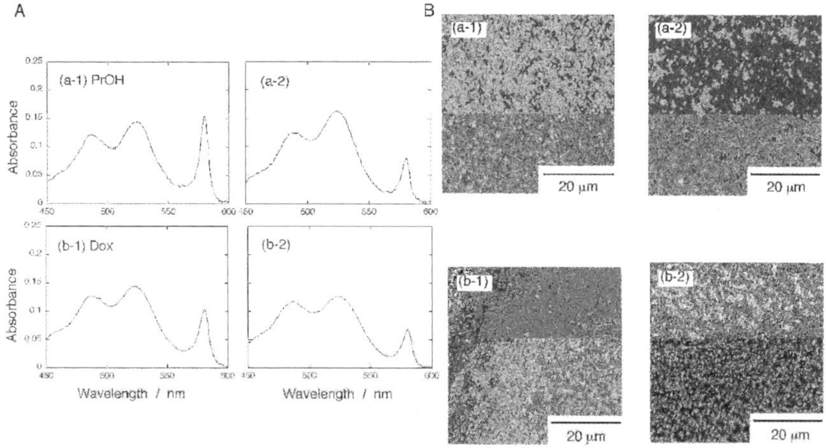

Figure 3.7 Panel A: Absorption spectra of PIC J-aggregates (0.8 mM) at the interfaces of mica/solution of mixed solvents. The mix solvents are (a-1) PrOH(5)/water, (a-2) PrOH(10)/water, (b-1) Dox(5)/water, and (b-2) Dox(10)/water. PrOH or Dox denotes 1-propanol or 1,4-dioxane, respectively. Panel B: Fluorescence micrographs of the PIC J-aggregates at the mica/solution of mixed solvents interfaces. The mixed solvents are (a-1) PrOH(5)/water, (a-2) PrOH(10)/water, (b-1) Dox(5)/water, and (b-2) Dox(10)/water.

(Panel B; b-1 and b-2). Although the absorption intensity of the J-band somewhat scattered from measurement to measurement, this trend approximately agreed with the spectroscopic behavior for the J-band intensity. Hence it is expected that binary phase of mixed solvents might change the structure of the interfacial J-aggregates, so we conducted tapping-mode AFM to visualize the aggregates in higher resolution.

Figure 3.8 demonstrates the dependence of the adsorbed layer morphology (AFM image) of the J-aggregates on the organic co-solvents (PrOH and Dox) and their composition. In both cases for PrOH/water and Dox/water, similar lateral morphologies could be found, and the size and coverage of the J-aggregate islands decreased with an increase in the amount of the co-solvent. The result can be rationalized by the lower binding efficiency of cationic PIC molecules onto mica with more organic solvents. In PrOH(5)/water or PrOH(10)/water, the cross-sectional profile reveals that multilayer islands of the J-aggregates are readily apparent (the thickness of the aggregate domains is ~1.8–2.1 nm).

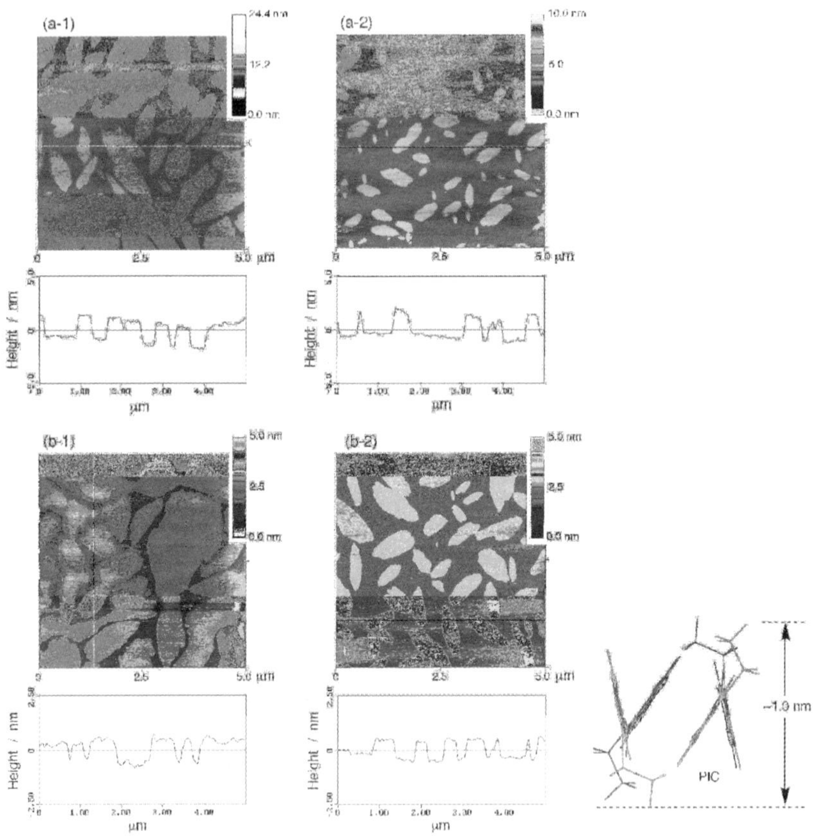

Figure 3.8 AMF images of the PIC J-aggregates at the mica/solution of mixed solvents interfaces. The mixed solvents are (a-1) PrOH(5)/water, (a-2) PrOH(10)/water, (b-1) Dox(5)/water, and (b-2) Dox(10)/water.

On the other hand, when the solution of Dox/water was used, the measured height of the aggregate domains diminished to be ~0.8–1.1 nm irrespective of the Dox content within 5–10 vol%. The thickness data are summarized in Table 3.1. The greatest emphasis is placed on the fact that the thickness of the aggregate islands is determined by the type of organic solvent coexisted, not by the coverage (nor the size of the islands). This thickness or height obtained in the Dox/water systems was very similar to that of the PIC molecular breadth.

Table 3.1 Thickness of the PIC J-aggregate islands at the mica/solution interface.

Solution	Thickness (nm)
Pure water	3.0~6.0
PrOH(5)/water	2.01 ± 0.08
PrOH(10)/water	2.05 ± 0.07
Dox(5)/water	1.05 ± 0.07
Dox(10)/water	0.91 ± 0.09

It is quite reasonable that a combination of π–π stacking and electrostatic interactions between the anionic mica surface and cationic PIC molecules is primarily responsible for the characteristic dye organization at the interfaces.[38] The interactions would be further stabilized via an epitaxial match if the molecules are adsorbed on their molecular plane perpendicular to the surface. Hence PIC lying flat on the surface is unlikely,[50] and thus provides side-on type molecular adsorption. Using this adsorption scheme (that is, the molecular plane perpendicular to the surface), adsorbed PIC molecules form a monomolecular layer of J-aggregates in the mixed solvent of Dox/water, whereas those probably form bilayer J-aggregates in the PrOH/water mixture. Note that the thickness of the monolayer of the interfacial J-aggregate appears slightly larger than that of the breadth of PIC molecule (0.7–0.8 nm). We suppose that, on the basis of the molecular packing in crystallographic data,[26,51] dye molecules can be arranged within the layers with their *N*-ethyl groups distributed above and below the layer plane, providing the effective layer thickness of about ~0.9–1.0 nm (see also the bottom-right in Fig. 8). Organic solvents such as PrOH or Dox would more solubilize the dye molecule and therefore reduce the driving force for adsorption, decreasing the amount of the interfacial dye molecules in both lateral and vertical dimensions.

3.2.2 *Interlayer Interactions in the Multilayer J-Aggregates of PIC at the Mica/Solution Interface*

Addition of organic solvents did not influence the optical properties of the interfacial J-aggregates of PIC (precisely, the J-band width and its peak position), whereas their morphology in the thickness-direction was changeable from multilayer to monolayer. The fact strongly suggests that the spectroscopic nature of the PIC J-aggregates produced at the mica/solution interface is determined by the first monolayer of the dye molecules, and the higher layers of the aggregates do not contribute to their optical responses. It is more difficult to know the influence of the higher layers on the intensity of the J-band because the absorbance somewhat fluctuated from measurement to measurement; however, on the basis of the data on pure water and Dox(5)/water, for example, absorbance of the J-band in Dox(5)/water was about half of that in pure water, despite the fact that the layer thickness was decreased about one-third to one-fifth under a less coverage condition, implying that the higher layers make a very small contribution to the intensity of the J-band. It is not possible to develop an unequivocal model for the packing structure of the second or higher layer of the aggregates, but they would have a large disorder in the molecular alignments that will lead to a strong broadening of the exciton spectral width (note that the thickness of the domain islands was always very uniform, so the amorphous-like arrangement of dye molecules in higher layers seems unlikely.). Interestingly, a bilayer structure of PIC J-aggregates on silver surface in the presence of halide ions has been reported;[52] in the study, the second layer showed a different molecular packing structure of PIC from the first adsorption layer, but absorption spectrum exhibited a single peak of the J-band. Here the layer images of the J-aggregates on silver have been obtained in ambient (that is, dry film prepared by washing away the surface with water, followed by blowing the substrate by nitrogen) using STM.[52] This similar result to our observations supports the possibility of the conceivable adsorption behavior.

We believe weak attractive forces such as CH/π interactions may be the possible candidate for the origin of interlayer interactions in the multilayer J-aggregate at the interfaces.[53] The CH/π interaction is a weak attractive force working between CH groups and π-systems.[53] This includes crystal packing[54,55] and/or self-assembly of organic compounds.[56] To participate in a CH/π interaction, the hydrogen atom should be placed above a π-plane, but the plane-to-plane stacking of aromatic rings is unnecessary. In the crystal structure of PIC-Br, each quinoline ring of the dye cation is stacked in a slipped deck-of-cards manner, and, an H atom (tip of CH_3 in the *N*-substituted ethyl group of PIC) and a quinoline ring can be proximate.[51] Therefore, if some of the *N*-ethyl groups in PIC are distributed above the first monolayer adsorption layer plane on mica, which has been rationalized by the multilayer observation using AFM, CH/π interaction would be plausible. The dimensional transition (from 3D to 2D) of the interfacial aggregates would be due to the balance of solubility of the interfacial aggregates, because the PIC dye has higher solubility in some organic solvents such as alcohols than that in water.[49] We then expect that the controllable dimension of the aggregates of π-systems will provide a useful method for predicting and designing specific mesoscopic architectures at solution/solid interfaces.

4. Thiacyanine (TC) J-Aggregates in Aqueous Solution

4.1 *Quasi-One-Dimensional Morphologies of TC J-Aggregates*

In this section, I describe a direct observation of *single* thiacyanine J-aggregates in bulk aqueous solutions. Although structural and/or morphological differences are expected between the J-aggregates at a solution/solid interface and those produced in bulk solution,[3] details have been poorly understood because it is very hard to *in situ* observe a real structure of the aggregates in solution. Direct observations of single thiacyanine J-aggregates in bulk solution are successful by fluorescence microscopy, PLM and AFM. Spontaneous morphology transformation from strings to tubules and its possible mechanism are presented.[57]

4.1.1 *String and Tubular Rod Morphologies*

An anionic thiacyanine dye, 5,5'-dichloro-3,3'-disulfopropyl thiacyanine (abbreviated as TC, chemical structure is shown in Fig. 4.1(b), was used for the construction of mesoscopic J-aggregates in aqueous solution. A typical dye concentration described here was 0.05 mM, and the electrolyte NaCl was added (11 mM) to promote J-aggregation (This procedure facilitates to obtain clear fluorescence images of the individual J-aggregate under the microscope.).

Figures 4.1(a) and 4.1(b) show the absorption and fluorescence spectra, respectively, for TC in aqueous NaCl solution. The narrow and intense J-band at around 464 nm in the absorption is the evidence that J-aggregates are produced. The absorption peaks at 430 and 408 nm are ascribed to the TC monomer and the dimer bands, respectively.[57] The absence of a Stokes shift in the fluorescence spectrum (Fig. 4.1(b)) is also characteristic of the J-aggregate. The appearance of the J-band was accompanied by a distinctive opalescence of the solution. Hence, fluorescence microscopy was conducted to characterize the mesoscopic structures of the TC J-aggregates *in situ*. Interestingly, we could observe two different morphologies of the J-aggregates; Fig. 4.1(c) shows typical fluorescence micrographs of the solution sample, clearly presenting mesoscopic string or rod structures. The length of the string or rod was several tens of micrometers. Note that the string structures were obtained for as-prepared solution samples, whereas the rod morphology was created upon aging the sample solution at room temperature for at least several weeks, inducing dynamic morphology transformation. There was a period of time in which both of the string and rod were coexisted. During this morphology transformation, however, absorption and fluorescence spectra (peak wavelength and line width) did not change, so that a coherent domain in the aggregates is small and not influenced by the nanoscopic rearrangements in the aggregate. The observed quasi-one-dimensional (1D) morphologies are due to the anisotropic interactions between TC molecules in the aggregate. The single string is likely to form an arc, suggesting that the string J-aggregate is polycrystalline-like. On the other hand, the rod is straight, indicating a different orientation of TC molecules in the rod from that in the string aggregates.

The direction of the molecular alignment in the string or rod aggregate can be evaluated by examining the fluorescence image of each aggregate taken under linearly polarized light excitation, because excitation efficiency depends on the direction of transition moments of the J-aggregate and this direction relates to that of the molecular

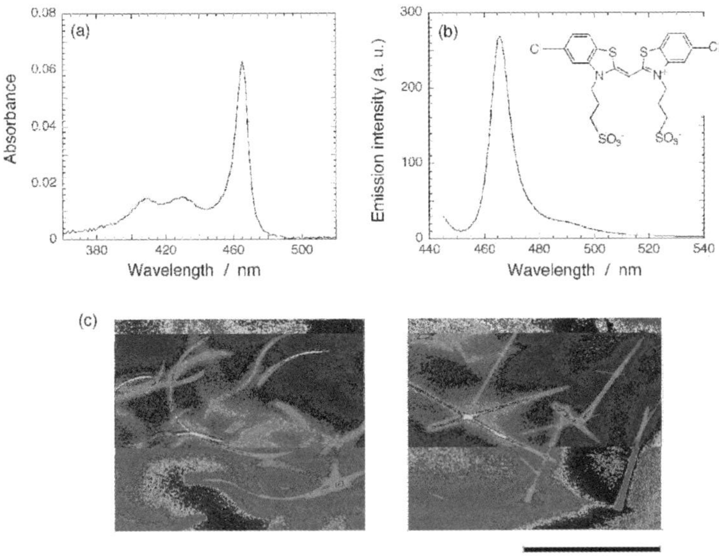

Figure 4.1(a) Absorption and (b) fluorescence spectra of thiacyanine (TC) dye in aqueous NaCl solution ([TC] = 0.05 mM, [NaCl] = 11 mM). The optical path length was about 70 μm. (c) Fluorescence micrograph of mesoscopic TC J-aggregates in solution. We have detected two different morphologies; left and right images show string and rod-like structures, respectively. The scale bar in (c) corresponds to 50 μm.

alignment.[45] Figure 4.2(a) and 4.2(b) represents typical fluorescence micrographs of the aggregates (Figs. 4.2(a) and 4.2(b) for string aggregates, and Figs. 4.2(c) and 4.2(d) for rod aggregates) obtained by using an excitation with linearly polarized light. The orientation of the excitation polarization is indicated by the black arrow at the top right in each figure. A bright fluorescence image of the string (indicated by white arrows) was obtained when the direction of excitation polarization was parallel to that of the long axis of the string, whereas a dark image was obtained when the long axis of the string was perpendicular (compare Figs. 4.2(a) with 4.2(b). These results indicate that, in the molecular alignment, the long axis of each of the TC molecules was approximately parallel to that of the string. In contrast to the string J-aggregates, fluorescence images of the rods had a weak dependence on the excitation polarization (compare Figs. 4.2(a) with 4.2(b)). (Strictly, a slightly brighter image could be taken when the direction of excitation polarization was perpendicular to the long axis of the rod.) Incoherent energy transfer, namely, a site-to-site energy hopping between the domains can also contribute to the isotropic fluorescence images of the rod. However, this effect should be very small because the string-like aggregates with similar size and similar spectroscopic properties exhibited distinct anisotropic nature.

Due to the same spectroscopic properties between the strings and rods, herringbone molecular arrangements in the rod-like J-aggregate can be ruled out.[8] To further elucidate the internal structure of the rod-like J-aggregates, polarized light microscopy (PLM) was carried out since PLM is a useful technique to sense regular molecular alignments in the aggregate exhibiting strong birefringence. Figure 4.2€ shows a typical PLM image with crossed polarizers in the absence (upper image) or presence (lower image) of a 530-nm retardation plate, respectively. The PLM images clarified the presence of strong birefringence throughout. Note that the retardation plate has a fixed optical path difference (530 nm) and insertion of the retardation plate produces a highly accurate analytical instrument that can be employed to determine the retardation (Γ) between the orthogonal wavefronts (ordinary and extraordinary), which are introduced by the sample birefringence. In Fig. 4.2(e), we can see an isotropic region (background in the image) as magenta with the 530-nm retardation plate. Despite an ill-contrast image due to small retardation of the rod, rods inclined right-handed showed deep red as an interference color whereas those inclined left-handed were faint blue. In addition, we could know that the rod-like J-aggregate showed a straight extinction; namely, a rod-like image disappeared when the long axis of the rod was parallel to the extinction direction (i.e., the image *y*-axis in Fig. 4.2(e). These characteristic features indicate that the rod aggregate possesses a uniaxial crystalline nature.

We can also evaluate the "sign" of birefringence (Δn) of the rod-like J-aggregates by the PLM image. Here, $\Delta n = n_L - n_S$, where n_L or n_S represents the refractive index along the long or short axis of the rod, respectively. The sign of Δn is estimated by examining the interference

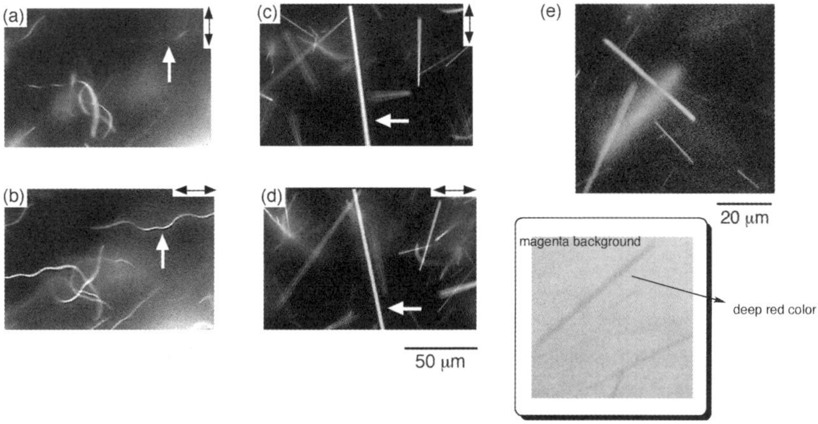

Figure 4.2 Fluorescence microscopy images of string-like TC J-aggregates ((a) and (b)) and rod-like J-aggregates ((c) and (d)) obtained by using linearly polarized light excitation. The direction of excitation polarization is indicated by the black arrow shown at the top right in each figure. (e) Representative polarized-light micrographs of the rod-like J-aggregates with crossed polarizers. The lower PLM image was taken in the presence of a 530-nm retardation plate. A regular molecular alignment in the aggregate is obvious.

color of the sample that depends on the azimuthal angle from the extinction direction.[57] With a help of the interference color chart (Michel-Lévy chart),[31] negative birefringence of the rod-like aggregate is proved on the basis of the PLM image in the presence of the retardation plate (Fig. 4.2(e)); namely, $n_L < n_S$. Because the refractive index is positively correlated with electronic polarizability via the Lorenz-Lorentz relation,[58] a larger refractive index is caused by larger electronic polarization through strong electronic interaction between π-systems of the TC molecules. The negative birefringence in the rod structure suggests that TC molecules are likely to orient or incline along the short axis direction of the rod.

4.1.2 *Structural Model of the Tubular Rod Aggregate*

AFM measurements were conducted at ambient conditions to clarify the string and rod structures of the TC J-aggregate. Samples for the measurements were prepared by placing an aliquot of TC solution on freshly cleaved mica. Before the measurements, we confirmed that the observed fluorescence images were similar to those obtained in solution. Figure 4.3 shows typical AFM images and its cross-sectional profiles for the string-like and rod-like J-aggregates. Figure 4.3(a) indicates that a thick string J-aggregate (~100-300 nm thick) was a bundle of thin strings of ~10-30 nm. This can be confirmed by the fluorescence micrographs (Fig. 9a) because some of the tips of the strings were branched. For the rod-like J-aggregates, we have observed the "edge" image of a rod (Fig.

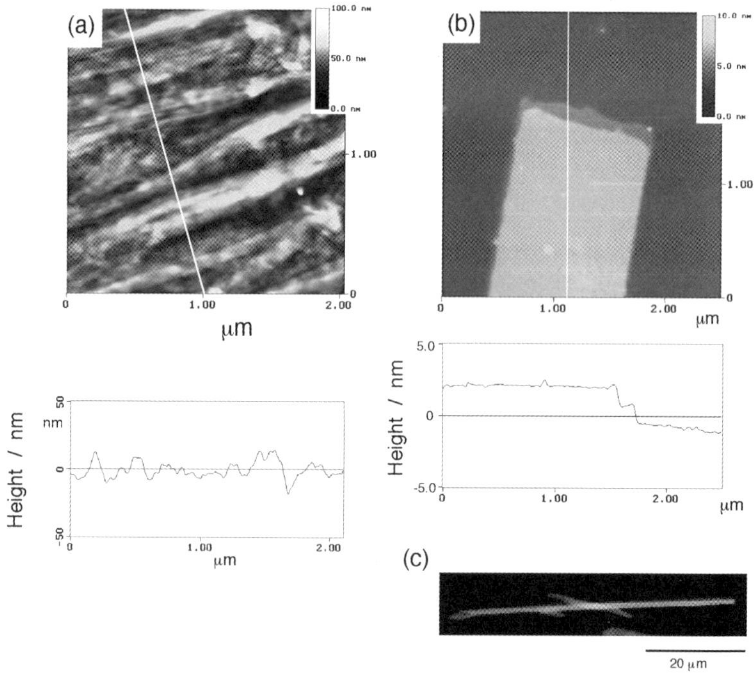

Figure 4.3 (a) AFM topographic image of the J-aggregate strings on mica. Cross-sectional profile along the line is also shown. (b) AFM image and a cross-sectional profile of a rod-like TC J-aggregate on mica. (c) Fluorescence micrograph of a rod aggregate having a deformed open end incidentally obtained in solution.

4.3(b)). The rods were ~1–2 μm wide and ~3 nm high. An important fact is that the cross-sectional profile clearly shows two steps of a TC aggregate monolayer. (Note that a TC molecule possesses a π-electron chromophore of about 0.5 nm and sulfopropyl groups of about 0.6 nm. For symmetry reasons, the dye molecules are presumably arranged within the columns with their sulfopropyl groups distributed above and below the column plane, providing the effective layer thickness of about 1.7 nm).[26] As a result, the rod morphology should be a *bilayer* structure of TC molecules with open ends on mica. Considering that the rods do not have a flat sheet-like shape in solution according to the fluorescence or PLM images, the observed rod morphology must be a *tubular* architecture consisting of a TC monomolecular wall with open ends. On this basis, with very careful observations of the rod-like J-aggregates in solution by fluorescence microscopy, we have incidentally succeeded in catching a rod whose end is deformed and thus open (Fig. 4.3(c)). This fluorescence micrograph clearly proves that an apparent rod structure is a tubule. If a tubular rod having a monomolecular wall is pressed onto a flat surface such as mica, it can be considered as bilayer composed of upper and lower surfaces (halves) of a TC monolayer.

By combining the results of fluorescence microscopy, PLM and AFM, we propose a detailed structure of the tubular rod J-aggregate. The sketch of a proposed model is shown in Fig. 4.4. In the model, a domain with well-ordered molecular alignments is large enough to form a micrometer-sized single sheet phase. When the planar sheet rolls up to produce a tubule, the top view shows a rigid rod morphology of a quasi-bilayer (upper and lower surfaces) structure including water inside the tubule. If molecular orientations are inclined with respect to the short axis (see Fig. 4.4; the molecular orientation in the J-aggregate is assumed to be a staircase pattern),[59] the transition dipole moments of the J-aggregate would be crossed between the upper and the lower surfaces of the tube. Inclination of the molecular orientations against the short axis of the tubule might depend on an each tubule on the basis of the fluorescence micrographs obtained under linearly polarized light excitation. This model can explain little or weak dependence of the fluorescence images on the excitation polarization direction. Therefore, we conclude that a real structure of the apparently rod-like aggregate

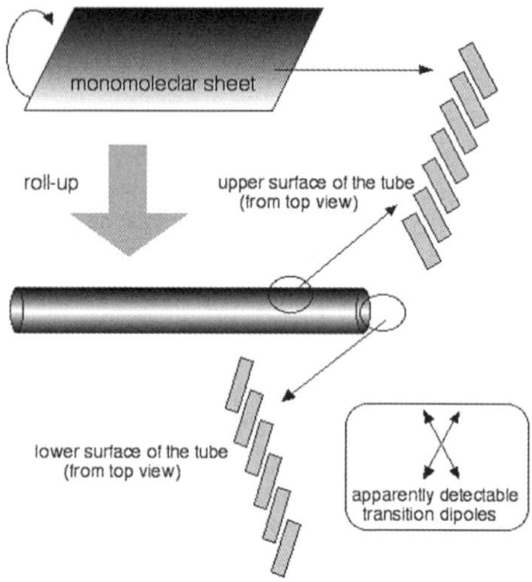

Figure 4.4 Schematic representation of the proposed structure of monomolecular-walled tubular rod J-aggregates of TC dye. The dye molecule is thought of as a rectangle. When the molecular alignment in a monomolecular wall is inclined with respect to the short axis of the tubular rod or sheet, observation of different transition dipole directions can be expected from the top view between the upper half and the lower half of the tubule.

involves a tubule, that is, a tubular rod. If the length of the circumference of the tubule cross section is unchanged during fixation onto a mica substrate, the diameter of the tube is calculated to be ~600 nm based on the AFM image showing the removal of water inside the tube.

4.1.3 *Possible Mechanism for the Morphology Transformation*

We have detected two different, string and tubular rod, morphologies of the TC J-aggregates in solution at a relatively low dye concentration, and found their shape changes as a function of elapsed time. We can understand the observed string-to-tubule morphology transformation in terms of the similarity in polymorphism of crystals.[60,61] When a metastable crystalline solid is in contact with its saturated solution, it is known that a phase transition can occur by one of two routes.[62] In the first, molecules in the metastable structure are able to rearrange themselves into the new stable structure while remaining in the solid state. In the second, the solvent helps this process by allowing dissolution of the metastable phase and re-crystallization of new crystals in the stable phase (solvent-mediated process).[63] In the TC J-aggregate system examined, the second route would be plausible because it is found that fragmentation of elongated strings facilitates the string-to-tubule morphology transformation,[57c] implying that the transformation involves dissolution of the strings and growth of the tubular rods. We here assume a common phase diagram showing that the solubility of phase-S (x_S) is larger than that of phase-R (x_R) at temperature T_x.[61] It is clear that, for a solution having composition x (> x_S and x_R) and temperature T_x, precipitation can yield both the "metastable" phase-S and the "stable" phase-R. In such a situation, however, the empirical observations of Ostwald[64] suggest that the combined nucleation and growth kinetics favor the initial appearance of the metastable phase (Ostwald's Law of Stages).[61] Because the as-prepared mesoscopic structure of the J-aggregates was a metastable string, the phase-S and phase-R would correspond to the string and the tubular rod structures, respectively, in the system. Thus, the starting point for the transformation should be the phase-S (string) in contact with a saturated solution containing nuclei of phase-R (tubular rod). Under the condition, as these nuclei grow, the solution composition falls below the solubility of a string and the solution becomes undersaturated with respect this phase. The strings thus dissolve producing supersaturation for the continued growth of tubular rod, resulting in the morphology transformation. This dissolution-growth process continues until all strings have disappeared and the solution composition has fallen into the solubility of the tubular rods. These considerations strongly support the fact that fragmentation of a metastable string-like J-aggregate accelerated the growth into the stable tubular rod architecture probably due to a faster dissolution of the string fragments.[57c]

4.2 *Two-Dimensional Nanosheet Morphology of TC J-Aggregates*

The TC dye formed mesoscopic string and/or tubular rod J-aggregates in aqueous solution when the dye concentration was relatively low. In this section, I describe 2D sheet-like morphology of the TC J-aggregates produced under high dye concentrations. Investigations by fluorescence microscopy, PLM and AFM reveal the monomolecular nature of the single sheets of the J-aggregate.[65] Quantitative birefringent or optically anisotropic properties of the J-aggregate is discussed in detail.

4.2.1 *Dye Concentration-Dependent String-to-Sheet Morphology Transformation*

Figures 4.5(a) – 4.5(c) show typical fluorescence micrographs of the TC J-aggregate at different TC concentrations ([TC] = 1.0 mM for Fig. 4.5(a), and 7.0 mM for Figs. 4.5(b) and 4.5(c); no electrolyte is added in the solutions), clearly exhibiting the appearance of sheet morphologies. Note that measurements were carried out after ~2 days of storage. The peak position and the line width of the J-band were unchanged in the dye concentration range of 0.5–7.0 mM, indicating that the internal molecular arrangement is also essentially unchanged. The sheet phase was present just under the high dye concentrations. The sheet had an elongated rectangular shape, and its short-axis length was about 4–9 μm. At [TC] < 1.0 mM (a relatively low concentration), only string morphologies were observed as shown in Fig. 4.1(c) (left image). At [TC] = 1.0 mM, a sheet-like J-aggregate appeared in the solution along with a large number of the strings. With an increase in [TC], the number of the sheet increased accompanying by a decrease in that of the string. Then, when the concentration of TC was higher than ~5.0 mM, the sheet morphology could be solely detected (see Figs. 4.5(b) and 4.5(c)). The results indicate that mesoscopic string-to-sheet morphology transformation occurred as a function of the dye concentration. Note that similar morphological changes could be observed as a function of the dye concentration whether an added electrolyte was present or not.[65]

The orientation of molecular alignments of TC in the sheet-like aggregates was evaluated. Figures 4.5(b) and 4.5(c) are the fluorescence micrographs of the sheet-like aggregates obtained under excitation with linearly polarized light in the same view area. The orientation of the excitation polarization is indicated by the arrow in each figure (at top right). Interestingly, a bright fluorescence image was obtained when the direction of excitation polarization was approximately perpendicular to that of the long axis of the sheet, whereas a dark image was seen when it was parallel to the long axis of the sheet (see the region indicated by a white circle in each figure). With respect to the molecular arrangement in the sheet, these results indicate that the long axis (molecular axis) of each of the TC molecules was approximately perpendicular to that of the sheet. As previously mentioned, the string J-aggregate had the molecular alignment in which the long axis of the constituent TC molecules was approximately parallel to that of the string, so that the sheet is not the agglomerate of the strings. It should be noted that a small amount of the sheets showed a reverse polarization behavior in the fluorescence images under linearly polarized light excitation.

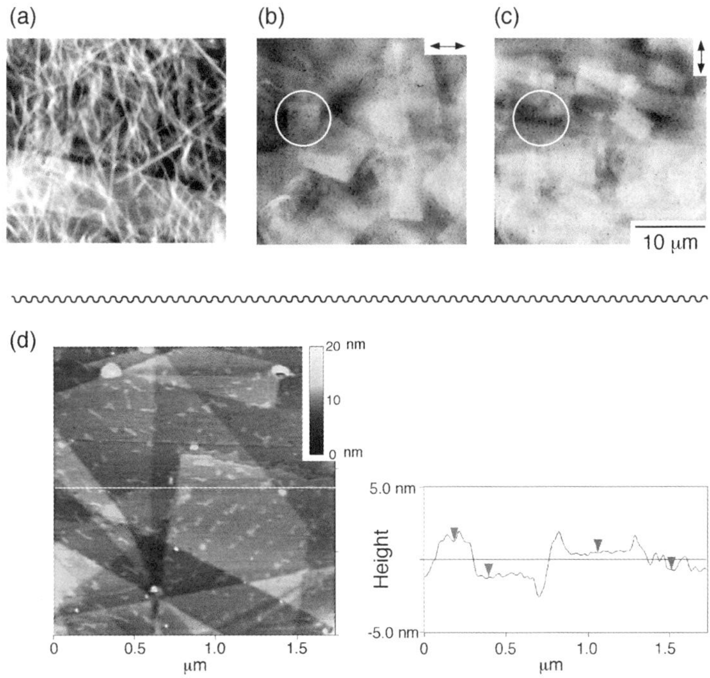

Figure 4.5 Fluorescence micrographs of TC J-aggregates in aqueous solutions. (a) [TC] = 1.0 mM, (b) and (c) [TC] = 7.0 mM. Images (b) and (c) show the same area obtained with linearly polarized light excitation. The direction of excitation polarization is indicated by the arrow at the top right in each figure. (d) AFM topographic images of the sheet-like J-aggregates. Cross-sectional profile along the white line is also shown.

AFM measurements were carried out at ambient conditions to clarify the detailed sheet structure of the J-aggregate. Figure 4.5(d) shows a typical AFM topograph and its cross-sectional profile of the sheet-like J-aggregates. The cross-sectional profile clarifies that the sheets are 1–2 nm high; a monomolecular layer structure. The thicker region shows a turn up of a single sheet. Hence we can conclude that the sheet-like J-aggregates consist of a 2D monomolecular layer of TC molecules, whose long axes lie approximately parallel to the short axis of the sheet structure.

The observed concentration-dependent string-to-sheet morphology transformation of the TC J-aggregates is a similar phenomenon to that observed for surfactant or lipid dispersion.[66] This type of phase transition arises from repulsive interactions between aggregates, where the aggregates are trying to stay as far apart as possible within a confined volume of solution.[58] The repulsive electrostatic interactions between string-like aggregates might be justified by negatively-charged sulfopropyl groups of TC molecules within the aggregate. Under the repulsive interactions between the strings, it is energetically unfavorable to make the aggregates come closer together in solution at high concentrations. To keep their surfaces away from each other and to reduce the internal energy, the string-like aggregates will convert to form a 2D layer or nanosheet as shown in Figs. 4.5(b) and 4.5(c).

4.2.2 *Determination of the quantitative birefringence of a single J-aggregate nanosheet in solution*

When the J-aggregate is composed of regularly aligned dye molecules, its optical anisotropy (or birefringence) is strongly expected. Figure 4.6(a) shows a typical polarized-light micrograph of the TC J-aggregate nanosheets with crossed polarizers. The nanosheet of interest, shown by the white arrow, is oriented at northwest-southeast orientation (maximum level of birefringence). The bright image (caused by interference color) clearly showed an elongated rectangular sheet-like morphology of the J-aggregate with a large degree of birefringence. The size of the sheet was 30–40 μm in length and 4–15 μm in width, indicating that the morphology observed by polarized-light microscopy corresponds to that by fluorescence microscopy.

The nanosheet showed a straight extinction; a dark image was obtained when the long axis of the sheet was parallel to the direction of polarization of the analyzer (extinction direction, the image *y*-axis in Fig. 4.6). This character indicates that the sheet aggregates possess a uniaxial crystalline nature. When observed in the presence of the 530-nm retardation plate (Fig. 4.6(b)), the color of the sheet transformed into faint azure surrounded by magenta of an isotropic area. From the image colors in Fig. 4.6(b), *negative* birefringence ($n_L < n_S$) of the sheet-like aggregates is proved.[31] Here, n_L or n_S again represents the refractive index along the long or short axis of the rectangular sheets, respectively.[31,57] The negative birefringence of the sheet structure

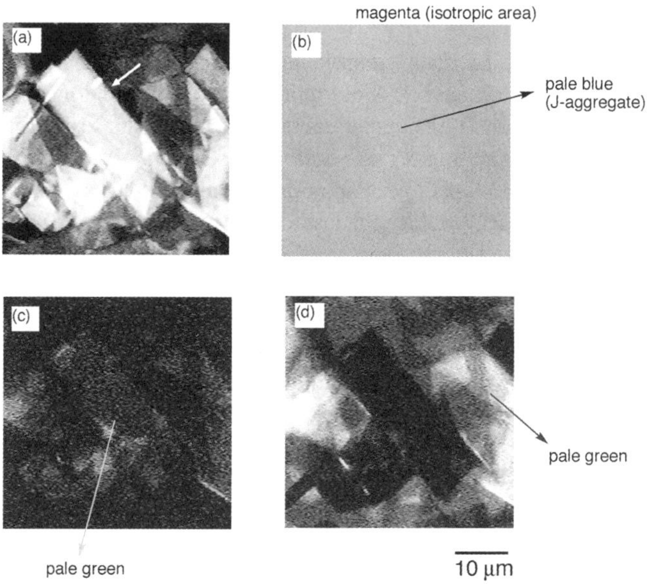

Figure 4.6 Representative polarized-light micrographs of the sheet-like J-aggregates with crossed polarizers. The rectangular J-aggregate nanosheet of interest is oriented at northwest-southeast orientation (maximum level of birefringence). (a) Image observed in the absence of any retardation plate. Bright species correspond to the TC J-aggregates. (b) Image observed in the presence of a 530-nm retardation plate. (c) Image observed in the presence of de Sénarmont compensator at the analyzer angle of 0°. (d) Image observed in the presence of the de Sénarmont compensator at the analyzer angle of 0.7°.

suggests that TC molecules are likely to orient along the *short* axis direction of the sheet.

It is worth noting that the alignment of TC molecules with respect to the structural principal axis switched from parallel to perpendicular during the string-to-sheet transformation. Moreover, one should recall that the molecular alignment also changed during the string-to-tubule transition. In a tubular rod structure, the molecules tend to align toward the slanting direction with respect to the short axis of the tubule. By taking these observations into account, "spatially anisotropic" interaction between the constituent molecules would play a significant role in the characteristic morphology formation of the solution-phase J-aggregates.

The well-defined sheet-like morphology allows us to determine the degree of birefringence of the aggregate quantitatively. De Sénarmont compensator, for introduction of bias retardation, inserted into the polarized-light microscope can be used for quantitative retardation measurements over the range of approximately 550 nm when observed with a monochromatic 550-nm line. By using this compensator, we can estimate the relative retardation of a single J-aggregate nanosheet. Note that the sign of birefringence and orientation of the specimen slow axis must be known before the measurements. For the J-aggregate nanosheet, the slow axis corresponds to the short axis of the rectangular sheet. At the northwest-southeast orientation of the rectangular J-aggregate with negative birefringence, we can see a pale green image of the sheet on the dark background (Fig. 4.6(c)). Compensation was determined by rotating the analyzer from its crossed position through a measured angle (Θ, in degree) until the intensity of the sheet was minimized (extinction; shown in Fig. 4.6(d). The surrounding is brightened from dark to pale green as the analyzer is rotated to obtain extinction. In the case of this J-aggregate nanosheet, $\Theta = 0.7 \pm 0.1°$ was determined. Upon a monochromatic light illumination of 550 nm, the relative retardation (Γ) can be calculated by the equation: $\Gamma = (550 \times \Theta) / 180$. Therefore, $\Gamma = 2.1 \pm 0.3$ nm was obtained.

The degree of (negative) birefringence (Δn) determined as $-\Gamma/d$, where d is the thickness of the J-aggregate nanosheet, depends on the accuracy of d. Although some errors might be included in the obtained thickness of nanosheets, we can reasonably determine d because the nanosheet must be the monomolecular layer of TC molecules. Then the thickness of the monomolecular nanosheet is ~1.7 nm, and up to ~2.0 nm at most. Therefore, $-\Delta n = -1.2 \pm 0.2$ was obtained. Despite the fact that the probe wavelength (550 nm) is off-resonant of 3370 cm^{-1} from the peak energy of the TC J-band, the aggregate nanosheet still exhibited large optical anisotropy. Such a very large birefringence is caused by anisotropic well-ordered molecular alignments within the aggregates, similar to that for heptamethine dye organic crystals with a strong J-band.[67] We believe that this approach using PLM will provide significant information on the optical anisotropy of single, supramolecular mesoscopic aggregates.

5. Conclusions and Outlook

Today's materials science deals increasingly with nanostructures ranged from 1 nm to several tens nanometers, so one of the major challenges is to find a way to organize molecules at a nano/mesoscopic level and to directly analyze/view them with precise structural considerations. This article has then focused on formation of mesoscopic J-aggregates that can be regarded as supramolecular self-assembled structures of excitonically coupled organic dye molecules and on their direct observations by using fluorescence microscopy, polarized-light microscopy (PLM), and atomic force microscopy (AFM) along with spectroscopic techniques. Various mesoscopic morphologies of J-aggregates of pseudoisocyanine (PIC) dye at solution/solid interfaces and those of thiacyanine (TC) dye in bulk solution have been examined.

In case of the PIC J-aggregates at the mica/water interface, they possessed three-dimensional (3D) island (multilayer) structures. The height of the J-aggregate islands was almost invariant irrespective of the dye concentration. The growth of J-aggregates occurred via a Volmer-Weber type process. In our experimental conditions, the absorption peak wavelength of the J-band, line width, and fluorescence lifetime were independent of the growth processes of the aggregates, so the physical aggregation size was considered to be much larger than the coherence size. The 3D multilayer structure of the J-aggregates was influenced by addition of water-miscible organic solvents such as 1,4-dioxane to convert a monolayer structure, whereas spectroscopic properties of the J-band were still unchanged upon addition of the organic solvents. This means that molecular packing in the J-aggregates was unchanged, and the interlayer interaction within the 3D J-aggregate was small. Importantly, characteristic feature of the J-band was then determined by the first monomolecular layer of the interfacial J-aggregates.

In case of the TC dye in bulk aqueous solution, when the dye concentration was relatively low, the aggregates formed quasi-one-dimensional string and tubular rod morphologies that were several tens of micrometers in length and sub-micrometers in width. The aggregates showed spontaneous morphology transformation from string to tubular rod when they were aged in solution. The string J-aggregate was a bundle of thinner strings, whereas the tubular rod had an architecture with a monomolecular wall (single-wall) having open ends. With an increase in the TC dye concentration, string-to-sheet morphological transition of the J-aggregates was found. The rectangular sheet-like J-aggregate consisted of a two-dimensional (2D) monomolecular layer of molecules, whose long axes lie approximately parallel to the short axis of the sheet morphology. The well-defined nanosheet morphology allows us to evaluate its optical anisotropy or birefringence with PLM. The rectangular nanosheet of J-aggregates showed very large negative birefringence of about –1.2 (at 550 nm).

Due to the ability of efficient energy migration and/or photo-induced electron transfer from the aggregate to other species in the J-aggregates, J-aggregates are very promising candidates for designing model systems of artificial light harvesting complexes. The gain in free energy by electron transfer reactions and the self-assembled architecture can be widely varied by their molecular structure of the constituent dyes, which will make the high reaction yields possible. Besides the supramolecular chemistry to assemble artificial light harvesting complexes, extensive static and/or time resolved spectroscopy as well as detailed clarification of internal structures of the aggregates will be further necessary in order to gain deeper understanding of the photophysics in the J-aggregates in future.

Acknowledgments

This work was supported by the Japan Society for the Promotion of Science (JSPS), the Ministry of Education, Culture, Sports, Science & Technology (MEXT), Hokkaido University, and University of Hyogo.

References

1. (a) E. E. Jelley, *Nature* **138**, 1009 (1936). (b) E. E. Jelley, *Nature* **139**, 631 (1937).
2. G. Scheibe, *Angew. Chem.* **49**, 563 (1936). (b) G. Scheibe, *Angew. Chem.* **50**, 212 (1937).
3. K. Norland, A. Ames and T. Taylor, *Photogr. Sci. Eng.* **14**, 296 (1970).
4. A. H. Herz, in *The Theory of the Photographic Process*; 4th Ed., T. H. James, Ed. (Macmillan, New York, 1977).
5. E. Hanamura, *Phys. Rev.* **B37**, 1273 (1988).
6. F. C. Spano, J. R. Kuklinski and S. Mukamel, *J. Chem. Phys.* **94**, 7534 (1991).
7. R. Monshouwer and R. Van Grondelle, *Biochim. Biophys. Acta.* **1275**, 70 (1996).
8. I. G. Scheblykin, M. A. Drobizhev, O. P. Varnavsky, M. Van der Auweraer and A. G. Vitukhnovsky, *Chem. Phys. Lett.* **261**, 181 (1996).
9. M. A. Drobizhev, M. N. Saposhnikov, I. G. Scheblykin, O. P. Varnavsky, M. Van der Auweraer and A. G. Vitukhnovsky, *Chem.* Phys. **211**, 455 (1996).
10. F. Fink, E. Klose, K. Teuchner and S. Dähne, *Chem. Phys. Lett.* **45**, 548 (1977).
11. H. Fidder, J. Terpstra and D. A. Wiersma, *J. Chem. Phys.* **94**, 6895 (1991).
12. N. Vranken, M. Van der Auweraer, F. C. De Schryver, H. Lavoie, P. Bélanger and C. Salesse , *Langmuir* **16**, 9518 (2000).
13. H. Kuhn, D. Möbius and H. Bücher, in *Physical Methods of Chemistry*, A. Weissberger and B. Rossiter, Eds. (Wiley, New York, 1972, p. 577).
14. L. Dähne, J. Tao and G. Mao, *Langmuir* **14**, 565 (1998).
15. J. E. Maskasky, *Langmuir* **7**, 407 (1991).
16. E. W. Knapp, *Chem. Phys.* **85**, 73 (1984).
17. P. O. J. Scherer and S. F. Fisher, *Chem. Phys.* **86**, 269 (1984).
18. E. O. Potma and D. A. Wiersma, *J. Chem. Phys.* **108**, 4897 (1998).
19. T. Kobayashi, Ed., *J-Aggregates* (World Scientific, Singapore, 1996).
20. M. Kawasaki and H. Ishii, *Chem. Lett.* , 1079 (1994).
21. L. Wolthaus, A. Schaper and D. Möius, *Chem. Phys. Lett.* **225**, 322 (1994).
22. H. Saijo and M. Shiojiri, *J. Imaging Sci. Technol.* **40**, 111 (1996).
23. L. Dähne, J. Tao and G. Mao, *Langmuir* **14**, 565 (1998).
24. D. A. Higgins, J. Kemiro, D. A. Vanden Bout and P. F. Barbara, *J. Am. Chem. Soc.* **118**, 4049 (1996).
25. H. von Berlepsch, C. Bötcher, A. Ouart, C. Burger, S. Dähne and S. Kristein, *J. Phys. Chem.* **B104**, 5255 (2000).

26. H. von Berlepsch, C. Bötcher and L. Dähne, *J. Phys. Chem.* **B104**, 8792 (2000).
27. H. von Berlepsch and C. Bötcher, *J. Phys. Chem.* **B106**, 3146 (2002).
28. H. von Berlepsch, S. Kirstein, R. Hania, A. Pugžlys and C. Böttcher, *J. Phys. Chem.* **B111**, 1701 (2007).
29. Becker, E. *Fluorescence Microscopy; Principles, Instruments, Applications* (Ernst Leitz Wetzlar Gmbh, Wetzler Germany, 1985).
30. W. C. McCrone, L. B. McCrone and J. G.Delly, *Polarized Light Microscopy* (Ann Arbor Science, Ann Arbor, MI, 1978).
31. T. Kato, *J. Geol. Soc. Jpn.* **107**, 64 (2001).
32. E. Meyer, *Progress Surf. Sci.* **41**, 3 (1992).
33. H. Yao, H. Ikeda and N. Kitamura, *J. Phys. Chem.* **B102**, 7691 (1998).
34. E. Daltrozzo, G. Scheibe, K. Gschwind and F. Haimerl, *Photogr. Sci. Eng.* **18**, 441 (1974).
35. G. Scheibe, *Angew. Chem.* **52**, 631 (1939).
36. V. Czikkely, H. D. Försterling and H. Kuhn, *Chem. Phys. Lett.* **6**, 11 (1970).
37. H. Yao, R. Kawabata, H. Ikeda and N. Kitamura, *Phys. Chem. Chem. Phys.* **1**, 4629 (1999).
38. H. Yao, S. Sugiyama, R. Kawabata, H. Ikeda, O. Matsuoka, S. Yamamoto and N. Kitamura, *J. Phys. Chem.* **B103**, 4452 (1999).
39. R. K. Bauer, P. de Mayo, E. R. Ware and K. C. Wu, *J. Phys. Chem.* **86**, 3781 (1982).
40. A. A. Muenter, D. V. Brumbaugh, J. Apolito, L. A. Horn, F. C. Spano and S. Mukamel, *J. Phys. Chem.* **96**, 2783 (1992).
41. (a) M. Volmer and A. Weber, *Z. Phys. Chem.* **119**, 277 (1925). (b) J. P. Hirth and G. M. Pound, *Condensation and Evaporation* (Macmillan: New York, 1963).
42. S. S. Ono, H. Yao, O. Matsuoka, R. Kawabata, N. Kitamura and S. Yamamoto, *J. Phys. Chem.* **B103**, 6909 (1999).
43. H. Schmidt, *J. Vac. Sci. Technol.* **A8**, 388 (1990).
44. B. Drake, C. B. Prater, A. L. Weisenhorn, S. A. C. Gould, T. R. Albrecht, C. F. Quate, D. S. Cannel, H. G. Hansma, and P. K. Hansma, *Science* **243**, 1586 (1989).
45. A. P. Marchetti, C. D. Salzberg, and E. I. P. Walker, *J. Chem. Phys.* **64**, 4693 (1976).
46. Von B. Dammeier, and W. Hoppe, *Acta Cryst.* **B27**, 2365 (1971).
47. J. F. Wall and C. F. Zukoski, *Langmuir* **15**, 7432 (1999).
48. H. von Berlepsch, S. Kirstein and C. Böttcher, *Langmuir* **18**, 7699 (2002).
49. H. Yao, Y. Morita and K. Kimura, *J. Colloid Interface Sci.* **318**, 116 (2008).
50. S. Özcelik, M. M. Demir and B. Birkan, *J. Phys. Chem.* **B108**, 4679 (2004).
51. H. Yoshioka and K. Nakatsu, *Chem. Phys. Lett.* **11**, 255 (1971).
52. R. W. Owens and D. A. Smith, *Langmuir* **16**, 562 (2000).
53. (a) M. Nishio and M. Hirota, *Tetrahedron* **15**, 7201 (1989). (b) M. Nishio, M. Hirota, Y. Umezawa and Y. Takeuchi, *Tetrahedron* **51**, 8665 (1995).

54. N. N. L. Madhavi, A. K. Katz, H. L. Carrell, A. Nangia and G. R. Desiraju, *Chem. Commun.* , 1953 (1997).
55. Y. Umezawa, S. Tsuboyama, K. Honda, J. Uzawa and M. Nishio, *Bull. Chem. Soc. Jpn.* **71**, 1207 (1998).
56. S. J. Cantrill, J. A. Preece, J. F. Stoddart, Z.-H. Wang, A. J. P. White and D. J. Williams, *Tetrahedron* **56**, 6675 (2000).
57. (a) H. Yao, M. Omizo, N. Kitamura, *Chem. Commun.*, 739 (2000). (b) H. Yao, S. Kitamura and K. Kimura, *Phys. Chem. Chem. Phys.* **3**, 4560 (2001). (c) H. Yao, C. A. Michaels, S. J. Stranick, T. Isohashi and K. Kimura, *Lett. Org. Chem.* **1**, 280 (2004). (d) H. Yao, T. Isohashi and K. Kimura, *Chem. Phys. Lett.*, **419**, 21 (2006).
58. J. N. Israelachvili, *Intermolecular and Surface Forces* (Academic Press: London, 1985).
59. H. Yao, S. Kitamura and K. Kimura, *Chem. Lett.* , 884 (2001).
60. M. Kobayashi, *Crystallization and Polymorphism of Fats and Fatty Acids*, N. Gartri and K. Sato, Eds., (Mercel Dekker, New York and Basel, 1988, p139).
61. R. J. Davey, P. T. Cardew, D. McEwan, and D. E. Sadler, *J. Cryst. Growth* **79**, 648 (1986).
62. K. Sato, J. Jpn. Assoc. Cryst. Growth **11**, 157 (1984).
63. P. T. Cardew and R. J. Davey, *Proc. Roy. Soc. (London)* **A398**, 415 (1985).
64. M. Kahlweit, *Adv. Colloid Interface Sci.* **5**, 1 (1975).
65. (a) H. Yao, Y. Kagoshima, S. Kitamura, T. Isohashi, Y. Ozawa and K. Kimura, *Langmuir* **19**, 8882 (2003). (b) H. Yao, T. Isohashi and K. Kimura, *Chem. Phys. Lett.* **396**, 316 (2004) .
66. K. Shinoda, T. Nakagawa, B. Tamamushi and T. Isemura, *Colloidal Surfactants* (Academic Press: London, 1963).
67. L. Dähne, K. Kamiya, J. Tanaka, *Bull. Chem. Soc. Jpn.* **65**, 2328 (1992).

CHAPTER 14

J-AGGREGATES IN THE LANGMUIR AND LANGMUIR-BLODGETT FILMS OF MEROCYANINE DYES

Yasuhiro F. Miura and Keiichi Ikegami[†]

Graduate School of Engineering, Toin University of Yokohama,
1614 Kurogane-cho, Aoba, Yokohama, 225-8502, Japan
E-mail: yfmiura@cc.toin.ac.jp

[†]Nanosystem Research Institute, AIST
Tsukuba Central-2, Umezono, Tsukuba 305-8568, Japan

Studies on structure and optical properties of J-aggregates of amphiphilic merocyanine dyes formed in thin film systems fabricated by Langmuir-Blodgett (LB) technique have been reviewed. After giving a theoretical framework for treating structure and optical properties of the dye thin-film systems, nucleation and aggregation number characterized by optical spectroscopy are presented. Studies on reorganization of the J-band in a merocyanine (DS)-arachidic acid (C_{20}) binary LB system using hydrothermal treatment, which is an annealing under relative humidity of 100%, have been also introduced. Furthermore, as a method to induce the H-aggregate formation in the binary systems of the DS-C_{20} LB system, addition of *n*-alkane as a third component has been touched upon. Finally a study on structure of J-aggregate in a merocyanine dye LB film determined by grazing incidence X-ray diffraction by Kato *et al.* has been introduced; it has been found that the interaction between static dipole moments upon excitation contributes significantly to the band shift as well as the transition-dipole interaction.

Keywords: Merocyanine Dye, Polymethine Dye, J- and H-aggregate, Langmuir-Blodgett (LB) technique

1. Introduction

1.1. *Merocyanine Dyes as Exotic J-Aggregate Forming Material*

Merocyanine dyes, which were developed by Kendaal in Ilford and Brooker in Eastman Kodak in 1930's, are classified into the polymethine dye families. As well as other polymethine dyes, such as cyanine and oxonol dyes, merocyanine dyes have been widely used as photosensitizers in the conventional silver halide photographic engineering, which is based on photochemical reduction $Ag^+ \rightarrow Ag$ in AgBr and other silver halide crystallites[1]. Furthermore, it has been also well known in the photographic engineering that the spectral sensitivity is closely correlated with the aggregation state of dye molecules adsorbed at the surface of AgBr.[2] In this respect, a vast literature has been accumulated on the studies of J-aggregates.

The chromophore of cyanine (oxonol) dyes is a polymethine chain consisting of odd number of carbon atoms terminated by two nitrogen (oxygen) atoms [Figs. 1.1(a) and (b)]. Because of its symmetry, the two canonical structures of $N–C=C\cdots C–C=N^+$ and $N^+=C–C\cdots C=C–N$ equally contribute to the resonant structure, resulting in delocalization of the highest occupied molecular orbital (HOMO). This gives the system a large transition dipole moment and optical functions as a dye, but at the same time take the static dipole moment away.

The chromophore of the merocyanine dyes has a form in which a half of cyanine chromophore and that of oxonol are interconnected [Fig. 1.1(c)]. As in the cyanine case, with the help of auxochromes connected to the nitrogen and oxygen atoms, in the merocyanine case the two canonical structures of $N–C=C\cdots C–C=O$ and $N^+=C–C\cdots C=C–O^-$ contribute to the resonant structure with nearly equal weights and the system gains optical functions as a dye. In contrast to the cyanine case, as a result of the resonance, the merocyanine chromophore inevitably bear a large static dipole moment along its long axis.[3] In other words, the merocyanine chromophore is a conjugated system sandwiched by electron withdrawing and repelling groups and inevitably has a large static dipole moment along the long axis.

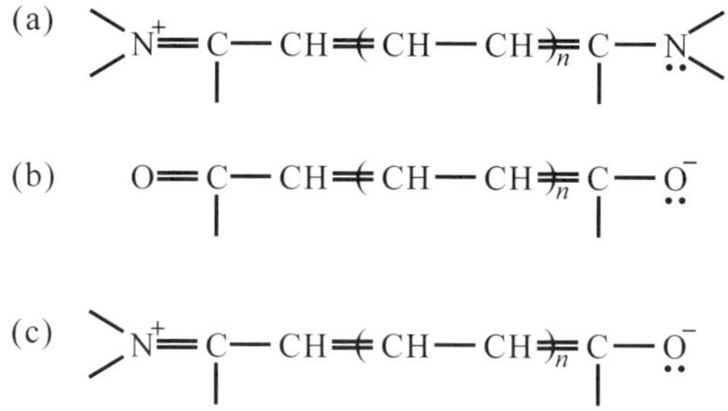

Fig. 1.1. General formulae of (a) cyanine, (b) oxonol, and (c) merocyanine chromophores. Pair of dots stand for lone pairs.

Since J-aggregate typically consists of dye molecules based on symmetrical chromophores, such as cyanine dyes, merocyanine dye is an exceptional and "exotic" constituent for forming J-aggregates. It should be also noted that most merocyanine J-aggregates are formed in a special environment, *i.e.*, under presence of water or water vapor, as in the case of those found at the air/water interface.[4-11]

Therefore, investigation on the formation mechanism of J-aggregate consisting of the "exotic" dyes will widen and deepen our knowledge about the J-aggregates and other self-organized molecular nanostructures. This is the main "raison d'être" of this chapter in this book.

In this chapter, we focus on Langmuir (L) films and Langmuir-Blodgett (LB) films (transferred by LB and LS techniques)[12-15] based on the long-chain derivatives of merocyanine dyes and discuss the structure and properties of J-aggregates formed therein. Figure 1.2 shows the molecular structures of the merocyanine dyes treated in this chapter. A long alkyl chain and carboxylic group are introduced in order to bind the molecules at the air/water interface with restricted molecular orientation.[4,5]

DO:	X=O, R=H
DS:	X=S, R=H
DSe:	X=Se, R=H
6MeDS:	X=S, R=Me

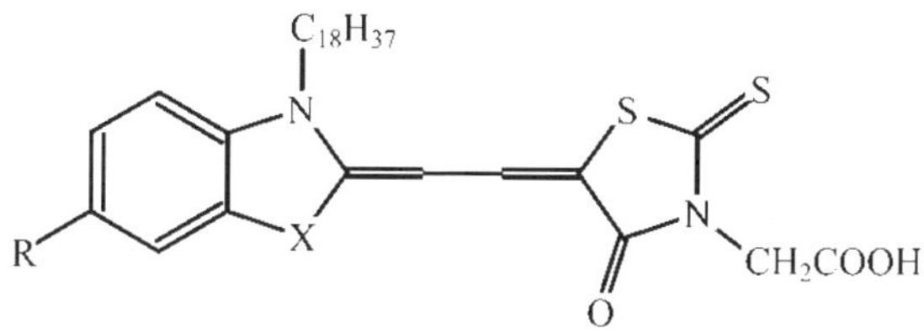

Fig. 1.2. Molecular structures of the merocyanine dyes treated in this chapter.

1.2. *Langmuir-Blodgett (LB) and Langmuir-Shaefer (LS) technique*

The Langmuir-Blodgett technique is a powerful technique that allows us to assemble organic molecules, amphiphilic ones typically, into tailored two dimensional (2D) molecular sheets. As shown in Fig. 1.3(a), amphiphilic molecules, which have both water-loving portion (*e.g.*, hydrophilic head group) and water-hating portion (alkyl chains), are spread on the water surface using a volatile solvent such as chloroform, benzene or toluene. After the volatile solvent is evaporated, the molecules on the water surface are compressed using a barrier, as shown in Fig. 1.3(b). The monomolecular film at the air/water interface is often referred to as Langmuir film or simply L film.

Two methods have been well known for transferring the Langmuir film onto solid substrates. One is by the vertical dipping of the substrate to the film, which is abbreviated as Langmuir-Blodgett (LB) technique, as shown in Fig. 1.4(a). The other is by horizontal lifting technique, which is abbreviated as Langmuir-Schaefer (LS) technique, as shown in Fig. 1.4(b). Sometimes both of the vertical dipping and the horizontal lifting methods are categorized into the common term of Langmuir-Blodgett (LB) method. There have been tremendous amount of modifications and revisions as for both of the LB and LS technique, but for further details, *e.g.*, please refer to Refs. 12-15.

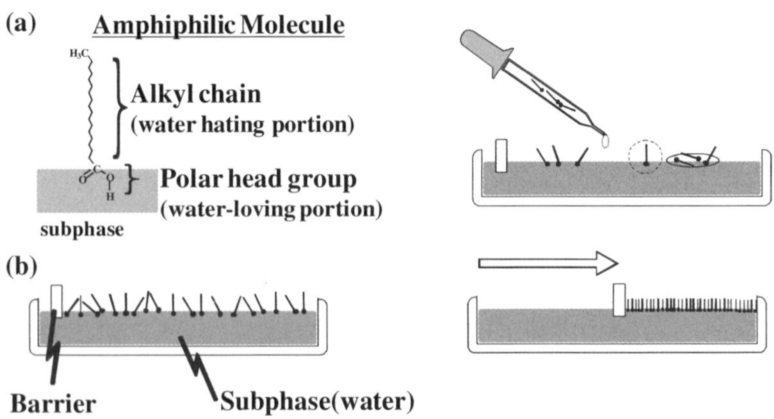

Fig. 1.3. (a) The way to spread amphiphilic molecules at the air/water interface; (b) compression of the interface by a barrier.

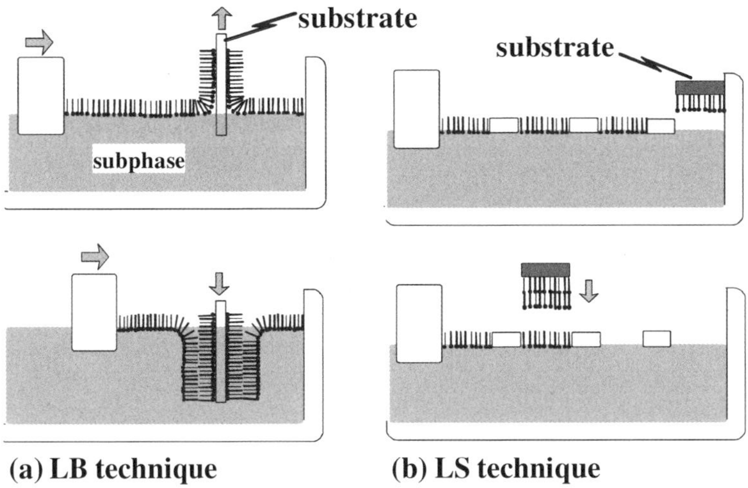

Fig. 1.4. (a) Langmuir-Blodgett (LB) and (b) Langmuir-Schafer (LS) techniques.

During the vertical-dipping process using a conventional LB trough, the longer axes of molecular aggregates, elongated in shape, at the air-water interface tend to be aligned along the dipping and raising direction of the substrate. This is due to the appreciable velocity gradient in the floating films which is inevitably induced by the vertical dipping process. The orientation of the molecular aggregates is then governed by the velocity gradient and the rotatory Brownian motion at the air-water interface during the deposition process. The relative contribution of the rotatory Brownian motion of the molecular aggregate decreases with increasing the length of the longer axis of the one-dimensionally-developed aggregates, since the flow orientation effect strongly depends on the rotatory friction coefficient of the aggregate, which is roughly proportional to the square of the aggregate length. Consequently, the tendency to be aligned parallel along dipping direction of the substrate is more prominent for the longer aggregates.

The orientation effect during the vertical dipping process has been quantitatively characterized by the flow orientation model proposed by Minari and coworkers.[16-18] The model formulates the transfer process of the dye aggregate formed at the air-water interface by introducing a complex velocity potential as representing the flow of the monolayer and the dicrhioic behaviors of the red-shifted and the blue-shifted bands of the merocyanine-dye containing LB films have been well-explained by the model. For a comprehensive review for the flow orientation model, see Ref. 19.

In Sec. 2, a theoretical framework for treating the optical properties of J-aggregates will be given. In Sec. 3, optical properties of the J-aggregate based on 6MeDS dyes confined in the metal-free pure Langmuir film will be given. In Sec. 4, we will introduce a reorganization process of J-aggregate called hydrothermal treatment (HTT), in which the LB samples are annealed at relative humidity of 100 %. In Sec. 5, studies on control of J- and H-aggregate formation in LB films based on DS dyes will be touched upon. In Sec. 6, studies on determination of the structure of J-aggregate in merocyanine dye LB films by grazing incidence X-ray diffraction (GIXD), which have been performed by Kato and coworkers, have been introduced.

2. Theoretical Framework for Simulating J-shifts

In this section, a brief theoretical review of the optical properties of J-aggregates is given. Here, we intend to provide a basis for numerical simulation that will be discussed later. More rigorous and general discussion can be found in the literature.[20,21] In the following, the theory for dimers will be derived at first and then extended to aggregates of larger size.

2. 1. Excitation of Dimer

Let us consider a dimer consisting of equivalent two dye molecules, labeled *A* and *B*. For simplicity, we assume that the molecule has only two state, *i.e.*, the ground and excited states, which of the molecule A (B) are denoted by $|g_A>$ ($|g_B>$) and $e_A>$ ($|e_B>$), respectively, in the followings. Adopting simple products of the molecular wavefunctions as the basis, the Hamiltonian with zero intermolecular interaction $\boldsymbol{H}_0$ is:

$$H_0 = \begin{bmatrix} 0 & 0 & 0 & 0 \\ 0 & \Delta E_0 & 0 & 0 \\ 0 & 0 & \Delta E_0 & 0 \\ 0 & 0 & 0 & 2\Delta E_0 \end{bmatrix}, \tag{2.1}$$

where ΔE_0 = excitation energy of the molecule.

Considering that $|e_A g_B\rangle$ and $|g_A e_B\rangle$ are degenerated, the lowest-order perturbation Hamiltonian accounting for the intermolecular interaction $\boldsymbol{H}_{agg}$ can be written as

$$H_{agg} = \begin{bmatrix} K_{gg} & 0 & 0 & 0 \\ 0 & K_{eg} & J_{AB} & 0 \\ 0 & J_{BA} & K_{ge} & 0 \\ 0 & 0 & 0 & K_{ee} \end{bmatrix}, \tag{2.2}$$

Here,

$$J_{AB} = J_{BA} = \langle e_A g_B | V | g_A e_B \rangle \ , \tag{2.3}$$

$$K_{eg} = \langle e_A g_B | V | e_A g_B \rangle, \quad K_{ge} = \langle g_A e_B | V | g_A e_B \rangle, \tag{2.4}$$

and

$$K_{gg} = \langle g_A g_B | V | g_A g_B \rangle, \quad K_{ee} = \langle e_A e_B | V | e_A e_B \rangle. \tag{2.5}$$

The singly excited states of the dimer can be elucidated by diagnalizing the central 2 × 2 square matrix of $H_0 + H_{agg}$. If the dimer is symmetric and□therefore $K_{eg} = K_{ge}$ (or if $|K_{eg}|, |K_{ge}| \ll |J_{AB}|$), the two eigenvalues (*i.e.*, the excitation energies of the dimer) and the corresponding eigenstates are as follows:

$$E_+ = \Delta E_0 + K_{eg} + J_{AB}, \quad |+\rangle = 2^{-1/2}\left(|e_A g_B\rangle + |g_A e_B\rangle\right). \tag{2.6}$$

$$E_- = \Delta E_0 + K_{eg} - J_{AB}, \quad |-\rangle = 2^{-1/2}\left(|e_A g_B\rangle - |g_A e_B\rangle\right). \tag{2.7}$$

When a dimer at the ground state, $|G\rangle = |g_A g_B\rangle$, is brought to the singly excited states, $|\pm\rangle$, under irradiation of light expressed as $\exp(i\omega t)\boldsymbol{E}$, the transition probability of each excitation is proportional to the square of its absolute transition dipole moment as follows:

$$\left| e\sum_{m,j}\langle G|\exp(itK_{gg}/\hbar)\exp(i\omega t)\sum_{m}\sum_{j}\mathbf{E}\cdot\mathbf{r}_{m,j}(-itE_{\pm}/\hbar)|\pm\rangle\right|^2$$

$$=2^{-1}\delta(\hbar\omega-E_{\pm}+K_{gg})e^2\left|\langle G|\sum_{j}\mathbf{E}\cdot\mathbf{r}_{A,j}|e_A g_B\rangle\pm\langle G|\sum_{j}\mathbf{E}\cdot\mathbf{r}_{B,j}|g_A e_B\rangle\right|^2, \quad (2.8)$$

$$=2^{-1}\delta(\hbar\omega-E_{\pm}+K_{gg})\left|\mathbf{E}\cdot(\mu_{\mathrm{A}}^{\mathrm{tr}}\pm\mu_{\mathrm{B}}^{\mathrm{tr}})\right|^2$$

where e = electron charge, $\hbar=h/2\pi$ (h=Planck constant) and $\boldsymbol{\mu}^{\mathrm{tr}}{}_{\mathrm{A}}$ and $\boldsymbol{\mu}^{\mathrm{tr}}{}_{\mathrm{B}}$ are transition dipole moments of molecules A and B, respectively.

The above calculation is summarized schematically in Fig. 2.1, where angles between $\boldsymbol{\mu}^{\mathrm{tr}}{}_{\mathrm{A}}$ and $\boldsymbol{\mu}^{\mathrm{tr}}{}_{\mathrm{B}}$ is defined as 2θ $(0\leq\theta\leq\pi/4)$, $\Delta E_1=\Delta E_0+K_{eg}-K_{gg}$, and $\boldsymbol{x}$ and $\boldsymbol{y}$ axes are chosen as the directions of $\boldsymbol{\mu}^{\mathrm{tr}}{}_{\mathrm{A}}+\boldsymbol{\mu}^{\mathrm{tr}}{}_{\mathrm{B}}$ and $\boldsymbol{\mu}^{\mathrm{tr}}{}_{\mathrm{A}}-\boldsymbol{\mu}^{\mathrm{tr}}{}_{\mathrm{B}}$, respectively.

(i) When the electric vector of the light $\boldsymbol{E}$ is parallel to $\boldsymbol{x}$-axis, the light $\hbar\omega=\Delta E_1+J_{AB}$ is absorbed and the absorption intensity is proportional to $2(\mu^{\mathrm{tr}})^2\cos^2\theta$ $(|\boldsymbol{\mu}^{\mathrm{tr}}{}_{\mathrm{A}}|=|\boldsymbol{\mu}^{\mathrm{tr}}{}_{\mathrm{B}}|=\mu^{\mathrm{tr}})$.

(ii) When the electric vector of the light $\boldsymbol{E}$ is parallel to $\boldsymbol{y}$-axis, the light $\hbar\omega=\Delta E_1-J_{AB}$ is absorbed and the absorption intensity is proportional to $2(\mu^{\mathrm{tr}})^2\sin^2\theta$.

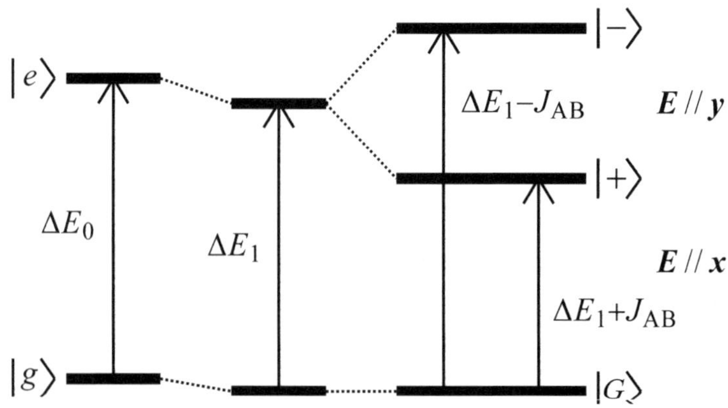

Fig. 2.1. Change of excitation energy when equivalent two dye molecules are dimerized.

Therefore, the absorption band of the dye molecule splits into two bands separated by 2 J_{AB} and polarized to axes orthogonal to each other. The phenomenon is known as Davydov splitting[22] and $K_{eg} - K_{gg}$ is called the "monomer shift".

2.2. *Point Dipole Model and Extended Dipole Model*

Since magnetic interaction is small and negligible, the molecular interaction, J_{AB}, can be expressed solely by electron-electron Coulomb repulsion as follows:

$$J_{AB} = \frac{e^2}{\varepsilon'} \left\langle e_A g_B \left| \sum_j \sum_{j'} \frac{1}{\left| \mathbf{r}_{A,j} - \mathbf{r}_{B,j'} \right|} \right| g_A e_B \right\rangle \tag{2.9}$$

(in CGS-esu unit), where ε' = relative dielectric constant.

If the intermolecular distance is large enough compared to the chromophore size, J_{AB} is approximated by the dipole-dipole interaction as follows:

$$J_{\mathrm{AB}} = \frac{1}{\varepsilon'} \left[\frac{\boldsymbol{\mu}_A^{tr} \cdot \boldsymbol{\mu}_B^{tr}}{\left| \mathbf{r}_{\mathrm{AB}} \right|^3} - \frac{3\left(\boldsymbol{\mu}_A^{tr} \cdot \mathbf{r}_{\mathrm{AB}} \right)\left(\boldsymbol{\mu}_B^{tr} \cdot \mathbf{r}_{\mathrm{AB}} \right)}{\left| \mathbf{r}_{\mathrm{AB}} \right|^5} \right] \tag{2.10}$$

where the inter-chromophore distance is defined as $\boldsymbol{r}_{\mathrm{AB}}$ [Fig. 2.2(a)].

The transition dipole moment, μ^{tr}, can be obtained experimentally from the integral of the solution spectrum, *i.e.*,

$$\left| \mu^{tr} \right|^2 = \frac{3 \cdot 10^3 \cdot \ln 10 \cdot ch}{8\pi^3 N_A} \int \frac{\varepsilon(\nu)}{\nu} d\nu \,, \tag{2.11}$$

where c = velocity of light, N_A = Avogadro's number and $\varepsilon(\nu)$ = molar extinction coefficient. From Eqs. (2.10) and (2.11), J_{AB} can be estimated for a given geometrical arrangement of molecules. The method above is called the "point dipole model" approximation.

Considering that the dye molecules interact with each other via π-π overlaps, it is often not realistic to assume that the inter-chromophore distance is large enough compare to the chromophore size. Then, Kuhn and coworkers proposed the "extended dipole model",[20, 23] in which the point dipole is replaced by a classical dipole consisting of a pair of

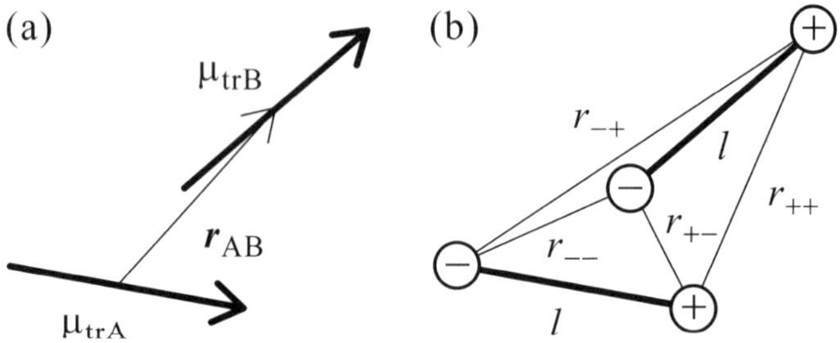

Fig. 2.2(a) Point dipole model and (b) extended dipole model

charges $\pm q$ separated by the length l, as shown in Fig. 2.2(b). Here, the values of q and l are connected to the magnitude of the transition dipole moment as

$$ql = \left|\mu^{tr}\right|. \tag{2.12}$$

Then, J_{AB} can be expressed as follows:

$$J_{\mathrm{AB}} = \frac{q^2}{\varepsilon'}\left(\frac{1}{r_{++}} + \frac{1}{r_{--}} - \frac{1}{r_{+-}} - \frac{1}{r_{-+}}\right). \tag{2.13}$$

Although the transition dipole moment is a virtual quantity unlike the permanent dipole moment that corresponds to the actual charge distribution of a molecule, J_{AB} can be estimated more accurately using this model compared to the point dipole model. In addition, Eq. (2.12) is much easier to handle than Eq. (2.10) on estimating J_{AB} by a computer simulation for dye aggregates with given molecular arrangements, at least when an appropriate value of l is known, because it contains no vectorial calculations.

2.3. *Excitation of Aggregate of Infinite Size*

Let us consider a one-dimensional aggregate of infinite size. As shown in Fig. 2.3, we assume a structure where one single dye molecule is in a unit cell. Adopting wavefunctions in which only the m-th dye molecule is excited $|e_m\rangle = |\cdots g_{m-2} g_{m-1} e_m g_{m+1} g_{m+2}\cdots\rangle$, the Hamiltonian of the singly excited state is given as

$$\mathbf{H}_0 + \mathbf{H}_{\text{agg}} = \begin{bmatrix} \ddots & \vdots & \vdots & \vdots & \vdots & \vdots & \iddots \\ \cdots & J_{-1} & \Delta E_1 & J_1 & J_2 & J_3 & \cdots \\ \cdots & J_{-2} & J_{-1} & \Delta E_1 & J_1 & J_2 & \cdots \\ \cdots & J_{-3} & J_{-2} & J_{-1} & \Delta E_1 & J_1 & \cdots \\ \iddots & \vdots & \vdots & \vdots & \vdots & \vdots & \ddots \end{bmatrix}. \tag{2.14}$$

Here, J_n is defined as the off diagonal matrix element between $|e_m>$ and $|e_{m+n}>$ and the energetic origin is shifted so that $<G|H_0 + H_{\text{agg}}|G> = 0$. By operating this matrix to a column vector whose m-th element is $\exp(ikm)$ and using $J_{-n} = J_n$, we obtain the original vector again with a scalar coefficient as follows:

$$\begin{aligned} &\Delta E_1 e^{ikm} + J_1\left[e^{ik(m+1)} + e^{ik(m-1)}\right] + J_2\left[e^{ik(m+2)} + e^{ik(m-2)}\right] + \cdots \\ &= \Delta E_1 e^{ikm} + 2J_1 e^{ikm}\cos k + 2J_2 e^{ikm}\cos 2k + \cdots \\ &= \left(\Delta E_1 + 2\sum_{n>0} J_n \cos nk\right)\exp(ikm) \end{aligned} \tag{2.15}$$

Thus, using the number of molecules N ($\rightarrow\infty$), the eigen function, $|k>$, and the corresponding eigenvalue, $E(k)$, are obtained as:

$$|k\rangle = N^{-1/2}\sum_m \exp(ikm)|e_m\rangle, \tag{2.16}$$

$$E(k) = \Delta E_1 + 2\sum_{n>0} J_n \cos nk. \tag{2.17}$$

To calculate $E(k)$, sometimes using negative n may be more convenient:

$$E(k) = \Delta E_1 + \sum_{n\neq 0} J_n \cos nk. \tag{2.17'}$$

The transition dipole moment concerning the photoexcitation from the ground state to the state $|k>$ is then obtained as

$$\begin{aligned}\boldsymbol{\mu}_{\rm tr}(k) &= \langle G|e\sum_m\sum_j \mathbf{r}_{m,j}|k\rangle = N^{-\frac{1}{2}}\left(\sum_m \exp(ikm)\langle g_m|e\sum_j \mathbf{r}_{m,j}|e_m\rangle\right)\\ &= N^{-\frac{1}{2}}\boldsymbol{\mu}_{\rm tr}\sum_m \exp(ikm)\\ &= N^{\frac{1}{2}}\boldsymbol{\mu}_{\rm tr}\quad (\text{for } k=0)\quad \text{or}\quad 0\quad (\text{for } k\neq 0)\end{aligned} \quad . \tag{2.18}$$

Therefore, there is only one absorption band, $\hbar\omega = \Delta E_1 + 2\sum_{n>0} J_n$, and the polarization is parallel to the transition dipole of the dye molecule. Here, it should be noted that the sign of the term $2\sum_{n>0} J_n$ depends on the geometry of the aggregate. When the molecules align in the "head-to-tail" manner like displayed in Fig. 2.3(a), it is negative and a red shift of the absorption band will be observed. In this case, the aggregate is called J-aggregate. When the molecules align in the "side-by-side" manner [Fig. 2.3(b)], on the contrary, a blue shift will be observed and the aggregate is called H-aggregate.

The discussion above is of one dimensional (1D) case, but we can extend it easily to two dimensional (2D) and three dimensional (3D) cases because all the molecules are equivalent in infinite aggregates that have no edges.

Indexing molecules in an 2D aggregate by two integers of m_x and m_y, the Hamiltonian is expressed using Kronecker's δ as

$$\left\langle e_{m_x,m_y}\middle|\mathbf{H}_0 + \mathbf{H}_{agg}\middle|e_{m'_x,m'_y}\right\rangle = \Delta E_1 \delta_{m_x,m'_x}\delta_{m_y,m'_y} + J_{m_x-m'_x,m_y-m'_y} \tag{2.19}$$

Then, the eigenfunction and the eigenvalue are,

$$\begin{aligned}|k_x,k_y\rangle &= (N_x N_y)^{-1/2}\sum_{m_x}\sum_{m_y}\exp(ik_x m_x)\exp(ik_y m_y)\left|e_{m_x,m_y}\right\rangle,\\ E(k_x,k_y) &= \Delta E_1 + \sum_{(n_x\ \ n_y)\neq(0\ \ 0)} J_{n_x,n_y}\cos k_x n_x \cos k_y n_y,\end{aligned} \tag{2.20}$$

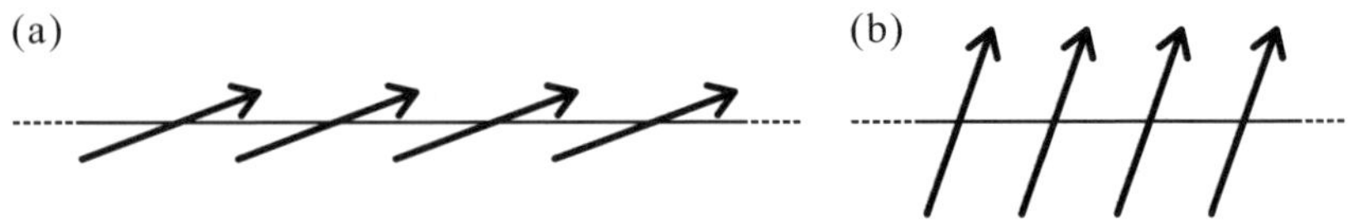

Fig. 2.3. Models for J-aggregate (a) and H-aggregate (b).

respectively.

Here, there is only one absorption band, *i.e.*,

$$\hbar\omega = \Delta E_1 + \sum_{(n_x \ \ n_y)\neq(0 \ \ 0)} J_{n_x,n_y}\ , \tag{2.21}$$

and the polarization is also parallel to the transition dipole of the dye molecule. For the systems where the unit cell contains more-than-one dye molecules, see Refs. 24 and 25.

2.4. *Optical Properties of J-aggregate*

The well-known optical characteristics of J-aggregate are summarized as follows:

(1) The optical absorption band (J-band) is remarkably red-shifted from that of the monomer band.
(2) The band shape is remarkably sharper compared to that of the monomer band.
(3) The fluorescence intensity is remarkably lager compared to that of the monomer.
(4) Zero or very small Stoke's shift is observed.

All the characteristics are well explained by the models presented in this section. The remarkable red-shift is explained well by Eqs. (2.17) and (2.18), when a head-to-tail arrangement [Fig. 2.3(a)] is realized by dye

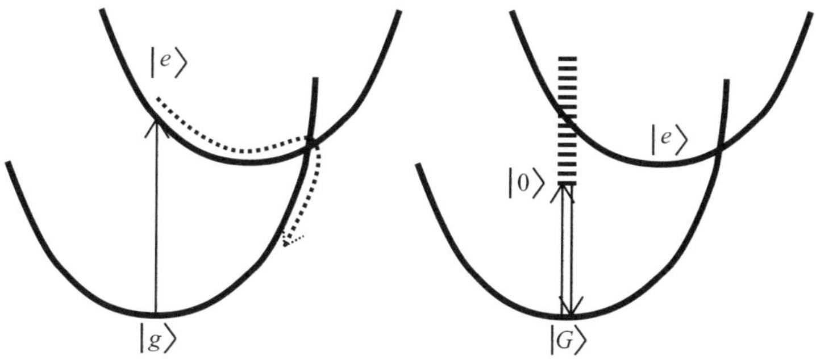

Fig. 2.4. Schematic picture of excitons in a monomer (a) and a J-aggregate (b).

molecules with relatively large transition dipole moments. Equation (2.16) tells us that photoexcited states of all the molecules have a common amplitude and that the exciton is delocalized. This leads to the "motional narrowing" effect that smears out small fluctuation in the monomer excitation energy ΔE_1 caused by inhomogeneity of the molecular environments.

Because atoms are much heavier than electrons and cannot move so fast, in general, the atomic geometry of the ground state is maintained during the photoexcitation (Frank-Condon principle). Equation (2.16) is based on this fact and all the molecules keep their original atomic geometry in each $|e_{\mathrm{m}}>$. However, the electron distribution of the excited state differs from that of the ground state. In the monomer case, therefore, the excited molecule changes its atomic geometry to minimize the elastic energy under the new electron distribution before emitting fluorescence light, thus loses a portion of energy transferred from the photon absorbed. This relaxed geometry is maintained in the photoemission process, adding a piece of elastic energy to the ground state level. Consequently, the wavelength of the emission becomes longer than that of the absorption (Stoke's shift). In the J-aggregate case, on the contrary, if the energy of the state $|0>$, which is given by Eq. (2.17), is lower than that of the monomeric excited state with the relaxed geometry, the elastic relaxation is suppressed. Thus, the photoemission occurs only from the state $|0>$ (Fig. 2.4), *i.e.*, there is no Stoke's shift. In other words, the exciton moves quickly before the elastic relaxation occurs. Since the possible internal conversion and intersystem crossing are suppressed in such a case, the intensity of fluorescence increases.

2.5. *Size and Edge Effects*

Light-harvesting antenna found in phototrophic bacteria is a circular aggregate of a chlorophyl derivative. In such a case there is no edges and the chromophores are equivalent as in the case of aggregates of infinite size. The Hamiltonian to be diagonalized in this case is given as

$$\mathbf{H}_0 + \mathbf{H}_{agg} = \begin{bmatrix} \Delta E_1 & J_1 & J_2 & J_3 & \cdots & J_2 & J_1 \\ J_1 & \Delta E_1 & J_1 & J_2 & \cdots & J_3 & J_2 \\ \vdots & \vdots & \vdots & \vdots & \vdots & \vdots & \vdots \\ J_2 & J_3 & J_4 & \cdots & J_1 & \Delta E_1 & J_1 \\ J_1 & J_2 & J_3 & \cdots & J_2 & J_1 & \Delta E_1 \end{bmatrix}. \tag{2.14'}$$

The finite size makes k discrete, but owing to the periodic condition Eq. (2.16) is not intrinsically changed.

Ends and edges in linear and planar aggregates give rise to other effect. The difference is clear when writing down the Hamiltonian in this case. For example in the 1D case it becomes

$$\mathbf{H}_0 + \mathbf{H}_{agg} = \begin{bmatrix} \Delta E_1 & J_1 & J_2 & J_3 & \cdots & J_{N-2} & J_{N-1} \\ J_1 & \Delta E_1 & J_1 & J_2 & \cdots & J_{N-3} & J_{N-2} \\ \vdots & \vdots & \vdots & \vdots & \vdots & \vdots & \vdots \\ J_{N-2} & J_{N-3} & J_{N-4} & \cdots & J_1 & \Delta E_1 & J_1 \\ J_{N-1} & J_{N-2} & J_{N-3} & \cdots & J_2 & J_1 & \Delta E_1 \end{bmatrix}. \tag{2.14''}$$

In the J-agggregate case, an exciton at its center is stabilized by $2\sum_{n=1}^{(N-1)/2} J_n$ (for odd N), but that at the edge is stabilized by $\sum_{n=1}^{N-1} J_n$. The effect of the former is apparently larger than that of the latter and as a a consequence the distribution of the exciton in the aggregate is no longer homogeneous. Diagonalization of Eq. (2.14") is difficult, but if we neglect J_n for $n \geq 2$, we obtain

$$|k\rangle = \sqrt{\frac{2}{N+1}} \sum_{m=1}^{N} \sin km |e_m\rangle \quad \left(k = \frac{j\pi}{N+1}\right), \tag{2.16'}$$

where $1 \leq j \leq N$ is an integer. Furthermore, not only the smallest k state $|\pi/(N+1)\rangle$, other states with odd j becomes optically allowed, causing an absorption band broadening.

3. *Basic Characterization of J-Aggregates of Amphiphilic Derivatives of Merocyanine Dye*

J-aggregates of the amphiphilic merocyanine dyes (DO, DS, DSe, and 6MeDS in Fig. 1.2) are usually formed in Langmuir films of the dye–

matrix mixtures prepared upon aqueous solutions of divalent cations. In the mixed systems, matrices such as arachidic acid and chelation of the carboxylic group born by the dye with cations such as Mg^{2+} stabilize the J-aggregates. However, both of matrices and metal chelation disturb basic characterization of the J-aggregates. Hence, basic characterization of their J-aggregates has been tried by generating those with as pure material as possible and as little metal cations as possible.

3.1. *J-aggregate Formation in Pure Langmuir Films of 6MeDS*

Among the amphiphilic derivatives of merocyanine dyes displayed in Fig. 1.2, 6MeDS has the strongest tendency to form J-aggregates, allowing us to prepare its J-aggregates without any matrices and metal cations. In this subsection, studies upon J-aggregates formed in metal-free pure L films of 6MeDS will be reviewed.

3.1.1. *Electrostatic Interaction for J-Aggregate Formation*

As pointed out in Sec. 1.1, since merocyanine dyes bear a large static dipole moment ($\boldsymbol{\mu}^{\mathrm{st}}$) along the long axis of chromophore. One of our interests is the contribution of the electrostatic interaction between large $\boldsymbol{\mu}^{\mathrm{st}}$'s to the J-aggregate formation.

For a 2D aggregate with an infinite size (Fig. 3.1, inset), the electrostatic energy per molecule E_{st} can be calculated based on the extended dipole model as,

$$\frac{E_{\mathrm{st}}}{|\boldsymbol{\mu}^{\mathrm{st}}|^2/(\varepsilon' L^3)} = \left(\frac{L}{l_{\mathrm{st}}}\right)^2 \sum_{(i,j)\neq(0,0)}^{(\pm\infty,\pm\infty)} \left[\frac{2L}{r_0(i,j)} - \frac{L}{r_+(i,j)} - \frac{L}{r_-(i,j)}\right], \quad (3.1)$$

where i and j index the row (head-to-tail lines) and column (side-by-side stacks), respectively. Here, ε' is the dielectric constant of the medium and the periods along the axes parallel and perpendicular to the head-to-tail lines are denoted by L and t. Eq. (2.21) sums up the static dipole moment–static dipole moment interaction between molecules m and n by fixing the molecule m at the (0,0) site (the center of the aggregate) and placing the molecule n at the $(i,j) \neq (0,0)$ site.

According to the extended dipole model, $\boldsymbol{\mu}^{\mathrm{st}}$ is treated as a pair of positive and negative charges ($\pm q_{\mathrm{st}}$) with the separation of l_{st}, and the distance between $+q_{\mathrm{st}}$'s on the molecules at (0,0) and (i,j) sites, $r_0(i,j)$, is given by,

$$r_0(i,j)/L = \sqrt{[i + j(s/L)]^2 + [j(t/L)]^2}\,, \tag{3.2}$$

where s is the slippage between adjacent rows. The distance between $-q_{\mathrm{st}}$'s is $r_0(i,j)$, too, and the distances between $\pm q_{\mathrm{st}}$ and $\mp q_{\mathrm{st}}$, $r_\pm(i,j)$, are given by,

$$r_\pm(i,j)/L = \sqrt{[i + j(s/L) \pm (l_{\mathrm{st}}/L)]^2 + [j(t/L)]^2}\,. \tag{3.3}$$

Numerical calculations assuming the t/L value to be 0.35/1.63 (the ratio between the thickness and length of the chromophore of 6MeDS) show that E_{st} is negative for a certain range of s/L as shown in Fig. 9. Since the transition dipole moment ($\boldsymbol{\mu}^{\mathrm{tr}}$) is nearly parallel to the long axis of the chromophore as well, and since the second term of the right-hand side of Eq. (2.21) can be calculated by applying Eq. (3.1) to $\boldsymbol{\mu}^{\mathrm{tr}}$'s, it can be concluded that E_{st} in a J-aggregate of 6MeDS is negative.

However, the minimum of its E_{st} calculated by slightly modified Eq. (3.1) is appreciably smaller than that for the parallel arrangement (Fig. 3.1), assuming an antiparallel arrangement with a certain l_{st} ($< L/2$).[25] The minimum of E_{st} for an antiparallel arrangement is always located at $s = 0$ and an H-aggregate should be formed if E_{st} solely determines the molecular arrangement. The difference between the electrostatic energies of J- and H-aggregates of 6MeDS is so large that π-π interaction may not be able to drive the J-aggregate formation, although it drives the J-aggregate formation of pseudoisocyanine and other "normal" J-aggregate-forming dyes. Thus, a basic and essential question arises: "why 6MeDS form J-aggregates?" The driving force for the J-aggregate formation will be discussed in following subsections.

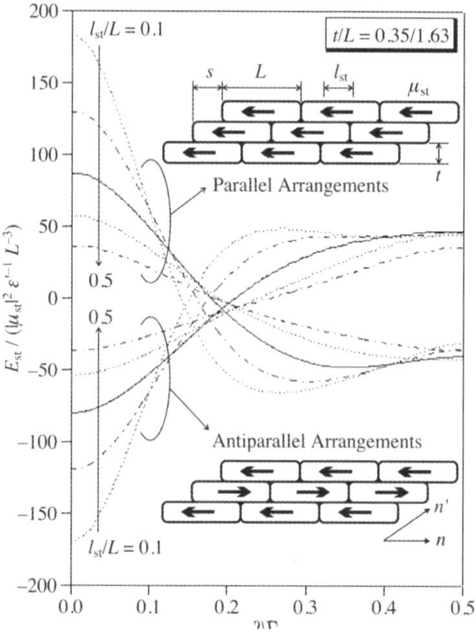

Fig. 3.1. Electrostatic energy due to interactions between mst's in a large aggregate with parallel and antiparallel arrangements ~insets! as a function of s/L with l_{st} $/L50.1$, 0.2, 0.3, 0.4, and 0.5. (Reprinted with permission from Ikegami[25]. © 2004, American Institute of Physics.)

3.1.2. *Spectroscopic Characterization*

Because both matrices and metallic cations disturb basic characterization of the J-aggregates of the amphiphilic merocyanine dyes, it is advantageous to form Langmuir films of the dyes without using any matrices upon pure water. L films of 6MeDS were prepared without using matrices on pure water and, fortunately, it exhibits a distinct J-band at ν_J = 16640 cm^{-1} (Fig. 3.2),[25] while only slight traces of J-bands were detected for any pure films (films without matrices) based on other merocyanine derivatives on pure water.

The intensity of the J-band, however, decreases with time and two absorption bands at ν_x= 19700 cm^{-1} and ν_y = 18400 cm^{-1} grow gradually. The absorption spectrum changes with an isosbestic point at 17900 cm^{-1}, showing that the growing two maxima have the same origin, which

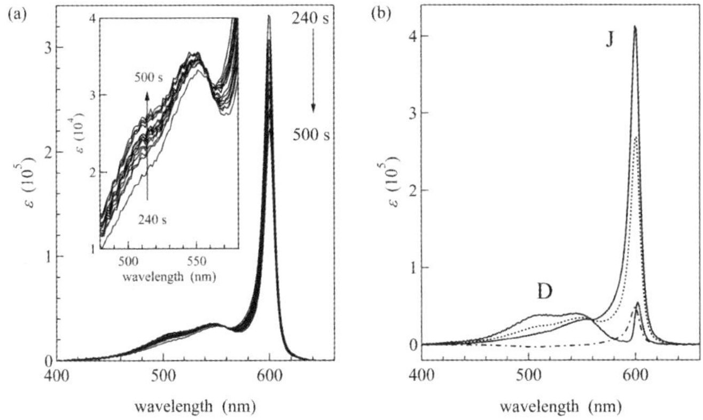

Fig. 3.2. (a) Time-dependent molar absorption coefficient (ε) spectrum observed for an L film of 6MeDS prepared upon a pure-water subphase. The measurements were started just after the surface pressure had reached to 5 mN/m (240 s after spreading the material) and 14 spectra were recorded with 20 s intervals under the surface pressure being kept at that value. Inset: expansion around the isosbestic point. (b) Dotted and chain lines are the averaged spectrum and optimum difference spectrum deduced by the ODS method, respectively. They are used to estimate the pure spectra of the two components (J and D) that are shown by solid lines. (The peak around 600 nm in the D spectrum is an artifact caused by the rapid and large variation of the refractive index.)

should be a herringbone-type aggregate, abbreviated as "D-aggregate" hereafter, because isolated monomers and dimers are not likely to be the major composition in the condensed pure films.

The 6MeDS films were transferred onto CaF_2 plates and infrared (IR) absorption spectra were measured. Although the metal-free J-aggregates in pure films tend to decompose during the transfer process, a certain amount of the J-aggregates can be transferred by the LS technique[14] and characterized.[25] The visible absorption spectra of the built-up films indicate that they are of mixtures of the J- and D-aggregates with different mixing ratios.

The IR spectra of the samples are different from one another and spectrum separation was absolutely necessary for further discussion. A spectrum separation method based on the least squares regression, named as "Optimized Difference Spectrum (ODS)" method[26] was developed for this purpose. The visible and IR spectrum of each sample were

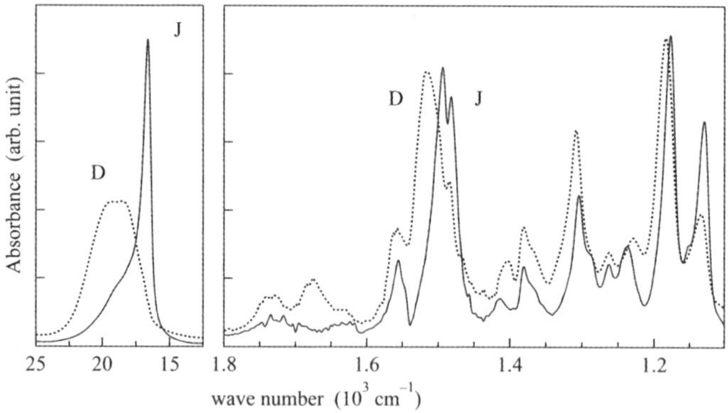

Fig. 3.3. Separated UV and IR spectra of the J- and D-aggregates extracted from those observed for metal-free LS films of 6MeDS using the ODS method.

connected into one "generalized spectrum", and the ODS method was applied to the generalized spectra.

The separated UV spectra obtained for the Langmuir film upon water (Fig. 3.2 (b)) and those for the LB film (Fig. 3.3) are reasonable, except for a spurious due to reflection, as pure spectra of J- or D-aggregates, demonstrating that the spectrum separation is successful. Thus the separated IR spectra (Fig. 3.3) can be considered reliable. To extract what modification of the electronic state is correlated with the J- and D-aggregation, they were compared with the spectra observed for 6MeDS in $CHCl_3$ and with the ab-initio vibrational calculation performed for an analogue of 6MeDS (Fig. 3.4), in which R' is replaced by a hydrogen, in $CHCl_3$.[27] The major IR absorption bands seen in these experimental and theoretical solution spectra are listed in Table 3.1. The vibrational calculation was carried out on the Gaussian 98 program[28,29] with the B3LYP model and 6-31+G† basis set by taking into account the solvent effect with the Onsager model.[30] As for the geometrical isomerism, the type indicated in Fig. 3.4 was assumed in the calculation because Fujimoto *et al.*[31] had suggested that this type is predominant in the DS case and that their suggestion was confirmed by the *ab-initio* calculation for DS (with substitution of the octadecyl group to an ethyl one).[32] The legitimacy of the calculation was proved by that it explains the ^{13}C-substitution effect on the IR absorption of DS.[32] The IR spectra in the

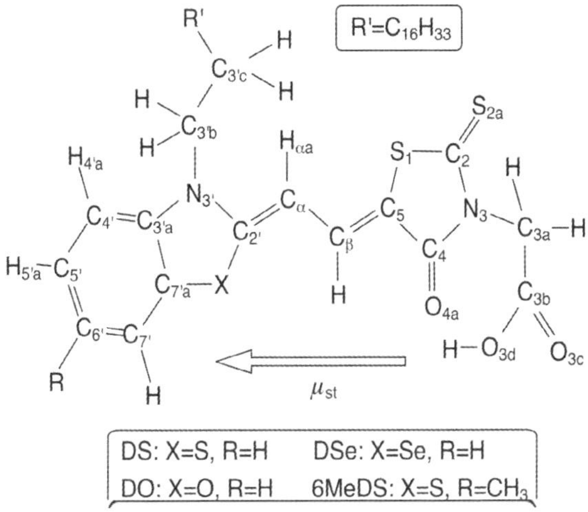

Fig. 3.4. Model molecule targeted instead of 6MeDS in the density-functional calculation. The octadecyl group in 6MeDS is substituted by a butyl group to reduce the calculation time. Similar model molecules were targeted in the calculations instead of DSe, DS and DO. (Reprinted with permission from Ikegami[25]. © 2004, American Institute of Physics.)

1800–1600 cm^{-1} region observed for the metal-free J- and D-aggregates of 6MeDS are drastically different from each other (Fig. 3.3). On one hand, absorption bands due to carboxylic and keto C=O groups are distinct for the D-aggregate. The positions of these bands (Table 3.1) reflect that the carboxylic O_{3c} (Fig. 3.4) is hydrogen bonded, but keto O_{4a} is not, respectively, suggesting that the formation of a dimer connected by intermolecular –COOH::::HOOC– hydrogen bonds [Fig. 3.5(a)] in the D-aggregate. On the other hand, the keto band in the J-aggregate spectrum is markedly broadened and shifted, suggesting that O_{4a} is hydrogen bonded to the carboxylic group either in the same molecule or in one of the adjacent molecules.

Bands assigned to the stretching modes of $C_5{=}C_\beta$ and $C_\alpha{=}C_{2'}$ in 6MeDS forming the J-aggregates show significant red shifts in comparison with those observed in chloroform solutions (Table 3.1). Another marked red shift is seen for the band assigned to the $C_2{=}S_{2a}$ stretching mode, suggesting a decrease and an increase in the electron density in the ethylidene and rhodanine groups, respectively, upon the

Table 1. IR bands[a] observed for 6MeDS and calculated for butyl-substituted 6MeDS.

	6MeDS			DO		
	Sol.[b]	D^c	J^c	Sol.[b]	D^d	J^d
$\nu(C_{3b}{=}O_{3c})$	1743	1734	(1730)	1742		
$\nu(C_4{=}O_{4a})$	1677	1678	(1630)	1682	1680	(1660)
$\nu_1(C{=}C)$	1564	1566 1556	1556			
$\nu_2(C{=}C)$	1516	1519 1514	1494	1546	1542 1530	1519
$\nu(C{=}S)$	1184	1185	1178	1186	1195 1189	1175
$\nu_{as}(CH_2)$	2927	2922	2919	2927	2920	2916
FWHM	22	21	19	22	17	13
$\nu_s(CH_2)$	2855	2852	2852	2855	2851	2851
FWHM	14	15	12	14	12	10

[a]Wave numbers are written in the unit of cm^{-1}.
[b]In chloroform solutions.
[c]In metal-free LS films.
[d]In Mg^{2+}-containing LS films.
(Reprinted with permission from Ikegami[25]. © 2004, American Institute of Physics.)

J-aggregate formation.[25] An enhancement of the intramolecular charge transfer is suggested upon the J-aggregate formation.

For the D-aggregate, the shifts of the vibrational bands are not significant. Instead, the C=C stretching modes, which are observed as singlet bands for the J-aggregate and solution, are observed as doublets. The full width at half maximum (FWHM) of the $C_2{=}S_{2a}$ band observed for the D-aggregate (23 cm^{-1}), which is much larger than that for the J-aggregate (17 cm^{-1}), may be due to a split of the band reflecting that the unit structure of the D-aggregate includes at least two molecules.

3.1.3. Intermolecular Hydrogen Bond as a Driving Force of J-aggregate Formation

Let us draw the structural models of the metal-free J- and D-aggregates of 6MeDS.

As for the D-aggregate of 6MeDS, the formation of intermolecular –COOH::::HOOC– hydrogen bonds is suggested from the split in the electronic absorption band and that of the C=C stretching modes. Therefore, it is concluded that the structural unit is a dimer and the chromophores are not parallel to each other in it.

On the contrary, the sharpness of the J-band and the singlet line shape of the C=C stretching modes imply the equivalence of the chromophores in the J-aggregate. The enhancement of the intramolecular charge transfer, *i.e.*, the enhancement of $\boldsymbol{\mu}^{st}$, upon the J-aggregate formation reflects that the molecular arrangement of the J-aggregate is favorable for large $\boldsymbol{\mu}^{st}$ because of negative E_{st}. The out-of-plane anisotropy of the visible and IR spectra suggest that the chromophore in the J-aggregate preferentially lies in the film plane.

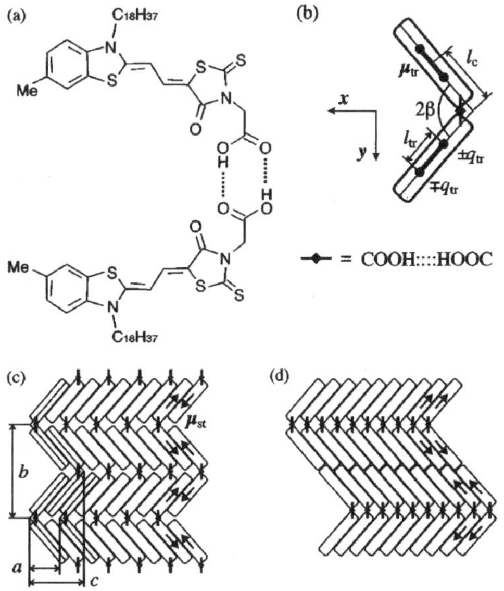

Fig. 3.5. (a) Two 6MeDS chromophores connected by hydrogen bonds. (b) Model of the dimer structure. The *x*-*y* plane of the introduced coordinate system contains both μ_{tr}'s. The intersecting line of the *x*-*y* plane and the plane of symmetry is selected as *x*. (c) and (d) Models of the metal-free *D* aggregate of 6MeDS based on the dimer modeled in (b). The lattice constants of (c) are supposed to be $a=t/\sin\beta$, $b=4l_c\sin\beta$, and $d=2l_c\cos\beta+a/2$. (Reprinted with permission from Ikegami[25]. © 2004, American Institute of Physics.)

As is well known, the wave numbers and widths of the antisymmetric and symmetric stretching modes of the CH_2 group [$\nu_{as}(CH_2)$ and $\nu_s(CH_2)$, respectively] provide information about the alkyl chain packing: the lower wave number and the narrower width correspond to the tighter packing.[33,34] From the values listed in Table 3.1, it is concluded that the alkyl chains are more tightly packed in the J-aggregate than in the D-aggregate. The SHG (Second Harmonics Generation) experiments upon DS performed by Kajikawa *et al.*[35] and Kato *et al.*[36] indicate that J-aggregated DS is highly SHG active, while non-J-aggregated DS is SHG inactive. The results indicate that J-aggregate domains keep mesoscopic large static dipoles, while D-aggregate domains do not.

Considering the formation of the intermolecular hydrogen bonds [Fig. 3.5 (a)], the unit structure of the D-aggregate can be illustrated as Fig. 3.5 (b), when restricting the direction of the chromophore parallel to the film plane. In the pure L films under a finite surface pressure, the dimers should be arranged in a certain manner to form mesoscopic domains. Two simplest molecular arrangements that are qualitatively consistent with the experimental results are illustrated as Figs. 3.5 (c) and 3.5(d).

For the J-aggregate, the brickstone model of Nakahara *et al.*[37], *i.e.*, a parallel arrangement (Fig. 3.1, inset), is consistent with the experimental results. Figure 3.6 (a) shows the structure of the chromophore plane of 6MeDS drawn based on the ab-initio optimization, which shows a high planarity of the chromophore plane being favorable for the brickstone arrangement. (In this drawing, hydrogen atoms bonded to $C_{6'a}$ and the carboxylic group are omitted because they are not in this plane and are not important for a moment. Since $C_{3'c}$ is important in the following discussion, it is displayed in this drawing, although it is not in the plane.)

According to the optimized model, L in the J-aggregate formed at an air-water interface is estimated to be 1.63 nm. Supposing that too close contact between the sulfur atoms in a chromophore and those in the adjacent chromophores are avoided, t is estimated at the sum of the van der Waals radii of carbon and sulfur atoms, *i.e.*, 0.35 nm. We postulate that the discrepancy between the tL value of 0.57 nm^2 and the area per 6MeDS observed for its Langmuir films on pure water under a surface pressure of 5 mN/m, A = 0.62 nm^2, is partly due to the vacancies between

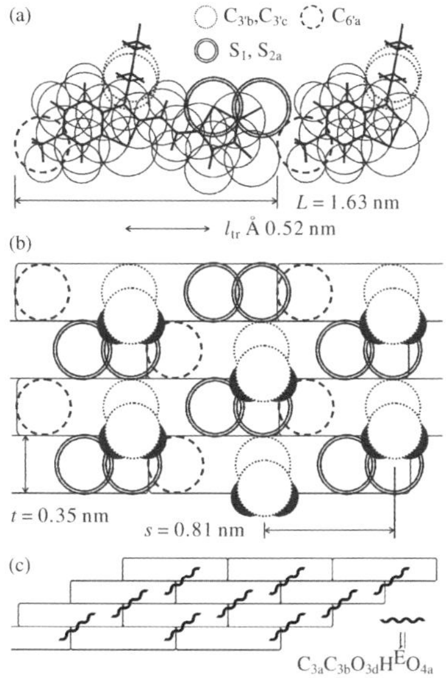

Fig. 3.6. (a) Chromophore plane drawn by superimposing circles with the corresponding Van der Waals radii on the atoms in the structure optimized for 6MeDS and its possible head-to-tail contact (side view). (b) A structural model (top view) of the metal-free J-aggregate of 6MeDS formed at the air-water interface. Hydrogen atoms bonded to $C_{3'c}$, which are not displayed in (a), are displayed in this drawing by filled crescents. (c) Possible intermolecular hydrogen bonds in the structural model. (Reprinted with permission from Ikegami[25]. © 2004, American Institute of Physics.)

the mesoscopic aggregates. Actually, A decreases down to 0.60 nm^2 with increasing the surface pressure to 25 mN/m.

The air-water interface is a highly anisotropic environment for amphiphilic molecules so that the hydrophilic part is immersed in the water and the hydrophobic part resides in air.[12-15] Accordingly, in the present Langmuir films, the second methylene group of the alkyl chain is located slightly above the border between the two chromophores [Fig. 3.6 (b)]. The hydrogen atoms in this methylene group in a molecule must avoid too close contact with $H_{4'a}$, $C_{3'b}$, $H_{\alpha a}$, S_1, and S_{2a} atoms in the adjacent molecule. In addition, too small s/L gives rise to positive E_{st} (Fig. 3.1). The plausible s value under these conditions is 0.81 nm ($s/L \approx 0.5$). The corresponding slip angle, $\alpha \equiv \arctan(t/s)$ =

23.4°, is compatible with that estimated for the J-aggregate of DS by the electron spin resonance study[38-40] and flow orientation study,[16,17,41], which are both in the range 25°–30°.

To examine the legitimacy of the proposed models of the molecular arrangements, numerical simulation of the excitation energies of the J- and D-aggregates (ν_J, ν_x, and ν_y) was performed. When the both aggregates are assumed to be large enough, ν_J and $\nu_{D\pm}$ can be calculated with Eq. (2.21). Namely, writing the second term in the right-hand side of this equation as ΔE_J and $\Delta E_{D\pm}$, ν_J and $\nu_{D\pm}$ can be calculated from $hc\nu_J = \Delta E + \Delta E_J$ and $hc\nu_{D\pm} = \Delta E + \Delta E_{D\pm}$, respectively. For two-dimensional J-aggregates, Eq. (3.1) gives ΔE_J, when $\boldsymbol{\mu}^{st}$ therein being replaced with $\boldsymbol{\mu}^{tr}$ and l_{st} in Eq. (3.3) with l_{tr}. In this case, the electric field oscillates rapidly and ε' can be approximated at a characteristic value for hydrocarbons, 2.5. A similar formula can be obtained for $\Delta E_{D\pm}$ by introducing geometrical parameters of the dimer structure, β and l_c [Fig. 3.5 (b)]. As well as L, t, and s, the parameters $|\boldsymbol{\mu}^{tr}|$ and l_{tr} are fixed in the simulation; the latter was estimated from Fig. 3.5 (a) and the former was estimated from the area of the visible absorption band observed for a $CHCl_3$ solution of 6MeDS.[42] Then, simulation of ν_J, ν_x, and ν_y was performed by adjusting l_c, β and $\Delta E/hc$.

In general, three values can be reproduced by adjusting three parameters, but whether the adjusted parameters are physically realistic or not is another story. Therefore, the parameters should be close to the roughly estimated values when the model used in the simulation is appropriate. Such a rough estimation of $\Delta E/hc$ is the excitation energy of 6MeDS in $CHCl_3$, 18800 cm^{-1}. The intensities of the maxima at ν_x and ν_y are approximately the same, giving the rough estimation of β at 45°. No realistic parameter sets have been found when either an isolated dimer or the model shown by Fig. 3.5 (d) was adopted. On the contrary, the parameters adjusted by adopting the model shown by Fig. 3.5 (c) are realistic (Table 3.2).

Based on the structural model proposed for the metal-free J-aggregate of 6MeDS, the nature of the hydrogen bond at the keto group can be elucidated by IR spectroscopy. The keto oxygen of a chromophore arranged in the proposed structural model can form a hydrogen bond with the carboxyl hydrogen of the adjacent molecule [Fig. 3.6 (c)]. On

Table 3.2. Calculated excitation energy of 6MeDS and DO aggregates.

	6MeDS[a]		DO[b]
	Fixed parameters		
L (nm)	1.63		1.59
t (nm)		0.35[c]	
s (nm)		0.81[c]	
l_{tr} (nm)		0.52[c]	
$\lvert\mu_{tr}\rvert$ (Debye)	10.3		11.1
	Adjusted parameters		
ν_0' (cm^{-1})	18 642		20 292
β (degree)	45.76		40.81
l_c (nm)	0.9291		0.9484
	Results		
ν_x (cm^{-1})	19 700		20 659
ν_y (cm^{-1})	18 400		19 339
ν_J (cm^{-1})	16 640		17 820

[a]In metal-free L films under 5 mN/m.
[b]In Mg^{2+}–containing L films under 5 mN/m.
[c]Common for both dyes.
(Reprinted with permission from Ikegami[25]. © 2004, American Institute of Physics.)

the other hand, to form an intramolecular hydrogen bond, the C_{3b} atom must protrude from the chromophore plane appreciably and thus the parallel arrangement of the dyes is distorted. That is why the former case is more probable.

In contrast to the calculation of the excitation energy, the electric field is static in the evaluation of Eq. (3.1) and ε' can be large in the L films on the water subphase. In this case, the difference between the minimum E_{st} of a parallel arrangement and that of the antiparallel one is not very large: it is estimated at 2×10^{-13} erg from Fig. 3.1 by putting $L = 1.63$ nm, $|\boldsymbol{\mu}^{st}| = 33$ Debye (obtained from the ab-initio calculation), $l_{st}/L = 0.3$ and $\varepsilon' = 40$ (half of water's value). The value is less than the formation energy of a usual hydrogen bond of 3–4×10^{-13} erg.[43] Furthermore, $|E_{st}|$ of a small aggregate is smaller than that evaluated from Eq. (3.1) because of edge and size effects, and thus, short range forces are more important in the early stage of the aggregation. Therefore, we can conclude that a carboxyl–keto intermolecular hydrogen bond compensates the electrostatic disadvantage of the parallel arrangement and expedites the J-aggregate formation in L films of 6MeDS. (To the knowledge of the authors, the presence of water or water vapor is indispensable for J-aggregation of the present merocyanine dyes.)

3.2 J-Aggregates of DO, DS and DSe in their Pure Langmuir Films

Although no trace of a J-band could be observed for pure L films of DO (Fig. 1.2) on Cd^{2+}– and Ca^{2+}–containing subphases, as well as for those on a pure-water subphase, it was found for DO spread on a Mg^{2+}–containing subphase (Fig. 3.7).[25] The observed ν_J = 17820 cm^{-1} is compatible with the discussion about the J-bands observed for LB films of 6MeDS–DO mutual mixtures.[44,45] However, formation of the J-aggregate is not uniformly seen in the DO films under a surface pressure lower than 5 mN/m. The development of J-aggregates depends

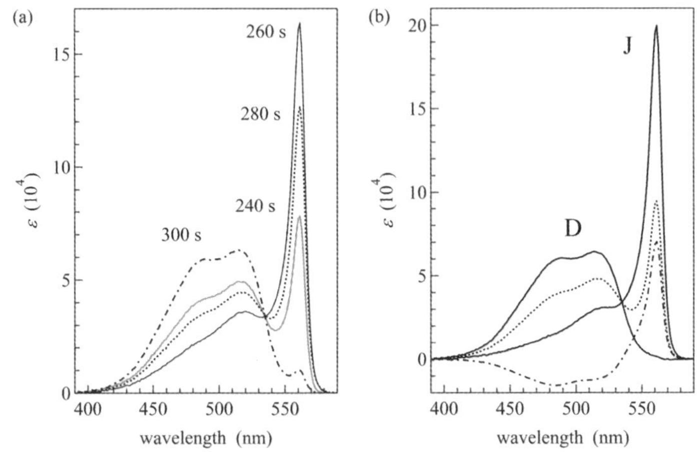

Fig. 3.7. (a) Irregularly time-dependent ε spectrum observed for an L film of DO under a surface pressure of 5 mN/m prepared upon a Mg^{2+}–containing subphase. Time is measured from the spreading of the material. (b) Dotted and chain lines are the averaged spectrum and optimum difference spectrum. They are used to estimate the pure spectra of the two components (J and D) that are shown by solid lines.

on domains and is competitive with the development of D-aggregates (ν_x = 20660 cm^{-1} and ν_y = 19340 cm^{-1}), which is shown by that the absorption spectrum varies largely and not monotonically with time due to slow flow of the films. An isosbestic point is clearly noted for this irregularly time-dependent spectrum (Fig. 3.7) reflecting that the present L films are of a two-component (the J- and D-aggregates) system, as in the case of the metal-free 6MeDS films. (It should be mentioned that the J-aggregate of DO with Mg^{2+} is stabilized by increasing the surface pressure up to 25 mN/m or by mixing a fatty acid.)

The J-aggregate of DO can be transferred onto CaF_2 substrates by the LS method. Visible and IR spectra observed for the obtained LS films are not pure spectra of the J-aggregate and then have been separated using the ODS method, as indicated in Fig. 3.8. The IR spectra of Mg^{2+}–containing films of DO are complicated due to metal chelation, but a discussion similar to that in the metal-free 6MeDS case can be made. Namely, the difference in the microscopic environment around O_{4a} in the J- and D-aggregates is evident: a relatively intense band is observed for the latter, while only a weak band with a shift is observed for the former. The difference in the 1650~1590 cm^{-1} region of the spectra observed for the J- and D-aggregates can be attributed to the difference in the microscopic environment around the (COO^-) group. These observations suggest that the keto group in the D-aggregate basically participates in

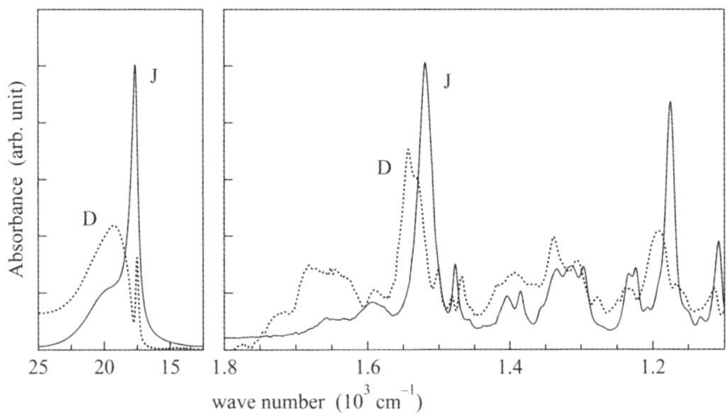

Fig. 3.8. Separated UV and IR spectra of the J- and D-aggregates extracted from those observed for Mg^{2+}–containing LS films of DO.

neither hydrogen bonding nor metal chelation, whereas that in the J-aggregate participates either of such bonds. Red shifts are clearly noted for some IR absorption bands of the J-aggregate, while those bands of the D-aggregate split.

Based on the experimental results and the ab-initio calculations done for a DO analogue (the octadecyl group is replaced by a butyl group),[25] a structural model similar to that in the 6MeDS J-aggregate can be proposed for the DO J-aggregate. The discussion concerning numerical simulation of ν_J, ν_x and ν_y for the Mg^{2+}–containing J- and D-aggregates of DO (Table 3.2) is also similar to that made for the metal-free 6MeDS case. The parameters t, s and l_{tr} are fixed to the values used in the 6MeDS case, while the $|\boldsymbol{\mu}^{tr}|$ and L values were estimated from an experimental solution spectra and ab-initio optimized structure of DO, respectively. The ν_J, ν_x and ν_y values are successfully simulated and the adjusted values of β and $\Delta E/hc$ are physically realistic, again.

Therefore, in the Mg^{2+}–containing DO system, it is concluded that bimolecular chelate structure with Mg^{2+} cation may play the same role as in the hydrogen bonds in the metal-free 6MeDS case, and expedites the formation of the J-aggregates. Such an effect of chelation must depend on the ionic radii of the metallic cations and the presence of Mg^{2+} in the subphase can drive the J-aggregate formation, while that of Cd^{2+} and Ca^{2+} cannot do it enough.

It should be mentioned that the shift of the keto C=O stretching mode upon J-aggregation can be more clearly observed by using the reflection-absorption spectroscopy (RAS). The RAS spectrum recorded for the J aggregate formed in pure LB films of DS made with Ca^{2+} exhibits the keto signal at 1659 cm^{-1}, while it is seen at 1681 cm^{-1} in the RAS spectra observed for the metal-free film of that dye.[46] (In addition, the RAS spectra detects that the keto bond in the J-aggregate is oriented along the normal axis of the LB-film plane, making the band due to it more and less intense in the RAS and usual transmission spectra, respectively.)

The addition of the counter cations Cd^{2+}, Ca^{2+} and Mg^{2+} to the Langmuir films of 6MeDS does not modify the J-band appreciably. Mixture of fatty acid to the dye system and the increase in the surface pressure (up to 25 mN/m) cause no clear changes in the spectral profiles for all the Langmuir films of 6MeDS suggesting that the molecular arrangement in the J-aggregate of 6MeDS is rigid, *i.e.*, it is not largely modified by a small change in the molecular environment. Such a character of 6MeDS and the strong tendency of 6MeDS to form J-aggregate must correlate.

A slight modification of the chromophore, however, drastically changes the dye's tendency to form J-aggregate. For instance, DSe and DS (Fig. 1.2) do not form J-aggregates on pure water without using matrices. The J-aggregates are obtained in their Langmuir films on aqueous solution of divalent cations such as Cd^{2+}, Ca^{2+} and Mg^{2+}, but the changes of the cations, subphase temperature, surface pressure and the mixture with inert surfactants (*e.g.*, as fatty acid) induce polymorphic behaviors of their J-aggregates.[26,47-55] Polymorphic behaviors of DS J-aggregates can be induced by secondary treatments with heat and basic vapor as well.[56,57] These results reflect that the molecular arrangements in the J-aggregates of DS and DSe are less stable. Furthermore, the Mg^{2+}–containing J-aggregate is the only one example of large J-aggregates found for DO. Thus, the tendency to form J-aggregates in their L films is as 6MeDS >> DSe > DS >> DO.

Looking at the head-to-tail contact of 6MeDS displayed in Fig. 3.6 (a) and similar drawings for DSe, DS and DO, it has been noted that the position of $C_{6'a}$ (or $H_{6'a}$) may be related to the stability of the J-aggregates. From the *ab-initio* calculations, the angles (ϕ) between the vectors of $\overrightarrow{H_{5'a}S_{2a}}$ and $\overrightarrow{H_{5'a}S_{6'a}}$ (or $\overrightarrow{H_{5'a}H_{6'a}}$) in dye molecules, 6MeDS, DSe, DS and DO (Fig. 3.4) were estimated as ϕ = 99.7°, 93.3°, 90.9° and 79.0°, respectively.[25] These ϕ values show a strong correlation with the merocyanine dyes' tendency to form J-aggregate, supporting the structural models proposed here.

3.3. *Nucleation of J-aggregate Found for DSe*

While only a trace of J-band is detectable for Langmuir films of DSe on pure water, a distinct J-band was found for Langmuir films of DSe–arachidic acid (C_{20}) mixtures prepared upon pure water even when the molar mixing ratio of C_{20} is much smaller than unity.[58] Furthermore, the mixing ratio-dependence of the J-aggregation rate provided a new insight into the formation mechanism and aggregation number of the

J-aggregates, which will be described in this subsection. In the followings, DSe–C_{20} mixtures with molar mixing ratio of 1 : x is abbreviated as $[DSe]_1[C_{20}]_x$.

3.3.1. Drastic Effect of Small Amount of Impurity

It was found that a distinct J-band, whose maximum is located around 16780 cm^{-1}, dominates the visible absorption spectrum of a $[DSe]_1[C_{20}]_1$ Langmuir film, immediately after the mixed solution is spread onto a pure-water subphase under air (Fig. 3.9 (a), x = 1).[58] The observation is contrastive to the case of pure DSe films (x = 0), whose visible absorption spectrum is dominated by a doublet band (maxima located around 19420 cm^{-1} and 18020 cm^{-1}) due to D-aggregates. The J-band formed in the mixed film is predominant under a relatively low surface pressure, but when the surface pressure is raised beyond 20 mN/m, it disappears in tens of seconds. In the course of the J-band's decay, a doublet band due to D-aggregates (maxima located around 19050 cm^{-1} and 17700 cm^{-1}) appears and grows. A clear isosbestic point at 17760 cm^{-1} can be recognized during this relatively rapid change in the spectrum, showing that the system consists of only two components: J- and D-aggregates.

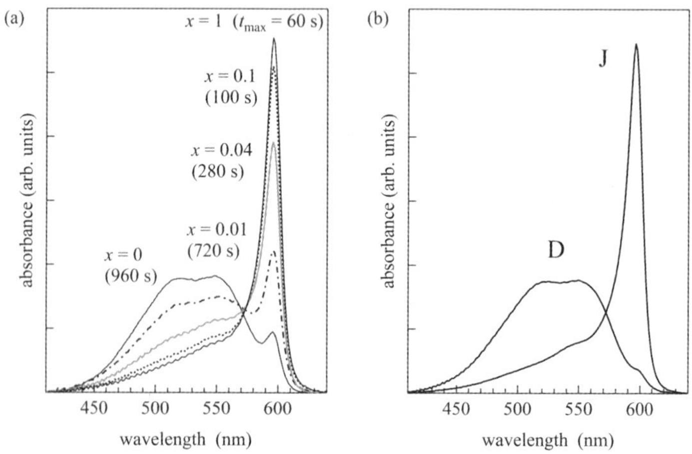

Fig. 3.9. (a) Representative UV spectra observed for $[DSe]_1[C_{20}]_x$ mixed Langmuir films observed at t_{max}. (b) Separated spectra corresponding to the J- and D-aggregates.

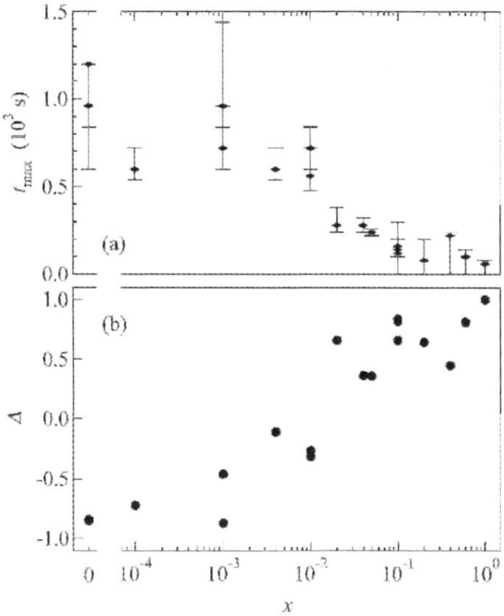

Fig. 3.10. (a) t_{max} as a function of x. A set of a rhombus and a bar represents a datum for one Langmuir film. (b) Predominance indicator, Δ, as a function of x. (Reprinted with permission from Ikegami[58]. © 2002, The Japan Society of Applied Physics.)

The predominance of the J-band shows that the most dye molecules are included in the J-aggregates in $[DSe]_1[C_{20}]_1$ films. On the other hand, the predominance of the doublet band shows that the most dye molecules are included in the D-aggregates in pure DSe films. Then, absorption spectra of Langmuir films of $[DSe]_1[C_{20}]_x$ with fractional molar mixing ratios of $x \leq 1$ were measured to investigate the crossover phenomenon between the two extremes.[58] The observed spectra are time-dependent and the time when the J-band becomes most distinct is denoted by t_{max}. t_{max} decreases monotonically with increasing x [Fig. 3.10 (a)]. The spectra observed at t_{max} for the mixed films with various mixing ratios are shown in Fig. 3.9.

Since $[DSe]_1[C_{20}]_x$ films are of two-component systems, information concerning the aggregation mechanism can be extracted by analyzing observed spectra by the ODS method. Eighteen mixed films were prepared and the ODS method was applied to the spectra observed at t_{max}. The obtained spectra of the J- and D- aggregates in pure L films are displayed in Fig. 3.9 (b). The ODS method also outputs the normalization factor and the predominance indicator for each spectra. The spectra displayed in Fig. 3.9 (a) are normalized by the corresponding normalization factors and the presence of a virtual isosbestic point guarantees the appropriateness of the analysis based on the ODS method. (No isosbestic point is noticed when the spectra are plotted without normalization.) The predominance indicator, denoted by Δ, has a value close to +1 when the component 1 (in this case, J-aggregates) is dominant in the two-component system, and has a value close to –1 when the component 2 (D-aggregates) is dominant. It is plotted as a function of x in Fig. 3.10(b) and the result that $\Delta \approx 0$ for $x \approx 10^{-2}$ implies that one C_{20} molecule can assist 10^2 DSe molecules in the J-aggregation. (This does *not deny* the possibility that one C_{20} molecule can assist *more than* 10^2 DSe molecules in the J-aggregation, especially when phase separation of C_{20} and DSe occurs.)

Development of the J-aggregates in the initial stage is rapid for $x \approx 1$ and gradual for $x << 1$. After t_{max}, the J-band diminishes gradually. Unlike the rapid spectral change seen for the $[DSe]_1[C_{20}]_1$ films under a high surface pressure, the diminishment of the J-band is not associated with development of any absorption bands, implying that the dye molecules are decomposed or isomerized. On the contrary, the doublet band survives for a long time (> 10^3 s) after the disappearance of the J-band. The phenomenon can be explained in the following scenario.[58] First, nuclei of J-aggregates are formed around C_{20} molecules. In other words, C_{20} assists nucleation of the J-aggregates. Second, they grow quasi one-dimensionally and stop growing when their endpoints reach other J-aggregates or isomerized DSe molecules. Consequently, non-J-aggregated DSe can remain even after all the J-aggregates stop growing, if the density of the J-aggregate nuclei is much smaller than that of the DSe molecule.

3.3.2. Estimation of Aggregation Number

Several reports attempted to determine the aggregation number N only from the absorption energy of the J-band and proposed relatively small values of N ($\leq$ 10). However, it should be emphasized that when physical quantities other than the absorption energy of the J-band are not known, determination of N is not straightforward because of the following reasons. (1) The number of unknowns is much larger than that of known quantities. Consequently, the estimation of N largely depends on the assumed values of other unknowns, such as geometrical parameters. (2) Accurate simulation of the shift of the J-band for a finite and non-periodic aggregate is not an easy task at all due to the edge effect as pointed in Sec. 2.5. The formula for an infinite system, *i.e.*, Eq. (2.21), may be used as an approximation, but the error introduced by such approximation is seldom evaluated.[59]

We treated the J-aggregate of 6MeDS as a large aggregate in Sec. 3.1. In other words, "N is large" was an assumption, but not a result of the numerical simulations. However, the fact that the simulation well reproduces the experimental results implies that the assumption is appropriate. The hypothesis that the J-aggregates of the present merocyanine dyes are large is also consistent with other observed phenomena, J-aggregate to J-aggregate transition in (pure or mixed) Langmuir films of DS and DSe prepared upon an aqueous solutions of

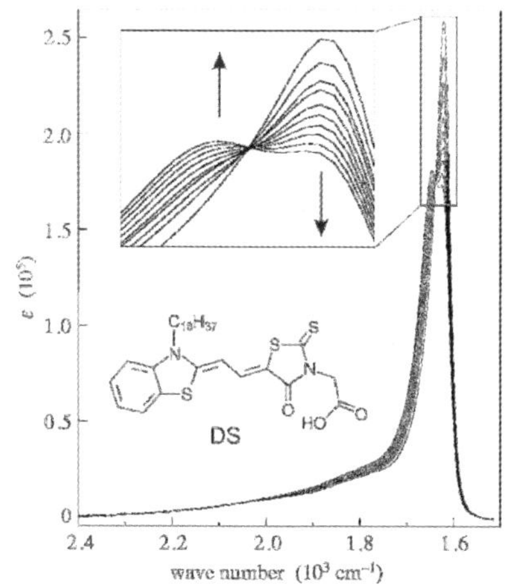

Fig. 3.11. Time evolution of ε spectrum observed for a DS–C_{20} 1:1 (molar ratio) mixed Langmuir film prepared upon an Mg^{2+}-containing subphase. The measurements were started just after the surface pressure had reached to 25 mN/m (180 s after spreading the material) and 10 spectra were recorded with 20 s intervals under the surface pressure being kept at that value. Inset: chemical structure of DS.

$MgCl_2$ (Fig. 3.11) and $CaCl_2$.[55] During these transitions, isosbestic points were observed in the in-situ visible absorption spectrum, showing that the major compositions of each film are two well-defined J-aggregates, whose excitation energies do not appreciably fluctuate. (If the excitation energies of the J-aggregates largely fluctuated, no isosbestic point would have been observed.) Hence, N of these J-aggregates should be large enough so that fluctuation of N does not appreciably affect the excitation energy. Since the fluctuation of N can be roughly estimated at $N^{1/2}$, $N^{1/2}/N << 1$, *i.e.*, $N \geq 10^2$ may be suggested.

A sophisticated approach for determining N is proposed by Hamanaka *et al.*[60] They performed nonlinear optical measurements upon the J-aggregate of 6MeDS and determined the excitation energy of the second exciton, as well as that of the first one. Since the difference in the excitation energies of the first and second excitons in a J-aggregate is determined by N and the bandwidth of the exciton band, their experimental results can provide indispensible information. Their estimation of N exceeds 10^{-3}. Unfortunately, however, they analyzed their data based on the theory for a periodic system.[61] If N is estimated appropriately based on the theory for a non-periodic system, it should be 25% smaller compared to their N value. The appropriate N value should be in the range of 10^{-2}-10^{-3}.

The discussion described in Sec. 3.3.1 provides an estimation of N based on totally different aspect: nucleation.[58] It implies that one J-aggregate can contain more than 10^2 DSe molecules because the number of J-aggregates in a film should be compatible with or smaller than that of C_{20} molecules for $x \approx 10^{-2}$. Furthermore, it is suggested that the J-aggregates in the films with $x \approx 1$ contain a considerable number of DSe, as well, because no significant x dependence of the position of the absorption maximum was detected.

Another totally different way for estimating N is from the characterization of the optical anisotropy of the LB films of merocyainine dyes forming J-aggregates. For instance, the mixed LB system of DS and C_{20} shows a distinct J-band at around 590 nm, which is remarkably red-shifted from the monomer band of DS located at 540 nm, when they are prepared on the subphase with Cd^{+2} cations and the J-band has an appreciable optical anisotropy. Figure 3.12 shows a typical

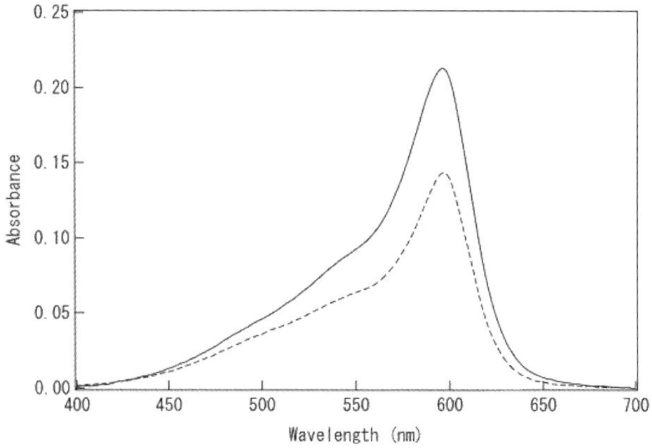

Fig. 3.12. The absorption spectra of DS-C_{20} mixed LB system (10 layers) with the molar mixing ratio [DS]:[C20]=1:2. The solid and dashed lines refer to the absorbance measured with the polarized incidents, whose electric vectors are parallel and perpendicular to the dipping and raising direction, respectively.

absorption spectra of the 1:1 mixture of the LB system of DS and C20 prepared on the subphase containing Cd^{2+}: the spectra $A_{//}$ and $A_{\perp}$ were measured using a normal incident with the electric vectors either parallel or perpendicular to the dipping and raising direction, respectively. The dichroic ratio R is larger than unity ($R \geq 1.3$, typically), where R is defined as $R = A_{//} / A_{\perp}$. The optical anisotropy can be quantitatively evaluated by the analytical model of flow orientation effect, *i.e.*, molecular aggregates elongated in shape tend to be aligned along the dipping and raising direction due to the torque created by the local velocity gradient of the film flow, which is inevitable in the conventional LB trough.[16-18] In this model, the flow orientation effect is expressed by six parameters as representing the LB vertical dipping process and the length of the rod-like aggregate is estimated to be ≈ 400 nm, which corresponds to a length of ~10^2 DS molecules aligned with a center-to-center distance of ~ 1 nm in accordance with the estimate from ESR line shape analysis. For a comprehensive review, please see Ref. 19.

4. *Reorganization of Merocyanine J-Aggregate by Hydrothermal Treatment*

Among various fluid-phase processes, hydrothermal treatment (HTT) based on the control of temperature, pressure, relative humidity and time, is well known as one of the powerful tools for modifying the physical and chemical properties of materials.[62] However, the application of HTT to the LB films had not come to our knowledge. In this subsection, we introduce that the HTT (annealing at 30-90° under relative humidity of 100%) process can reorganize the so far reported J-band of the mixed LB system of DS and C_{20} located at 590 nm.[63-75]

4.1. *HTT procedure*

HTT based on the following procedure is applied to the mixed LB system of DS and C_{20}. An aluminum tube (*ca.* 20 mm in diameter and 150 mm in length) with a screw top at one end shown in Fig. 4.1 was used. A small amount of water (~1 cm^3) was first introduced into the tube, and then, a spacer was set at the bottom to avoid direct contact between the water and the sample which was placed finally in the tube. The tube was immersed in a water bath kept at a constant temperature (T_H) for a given heating time (t_H) and then removed from the bath to be cooled down to room temperature.

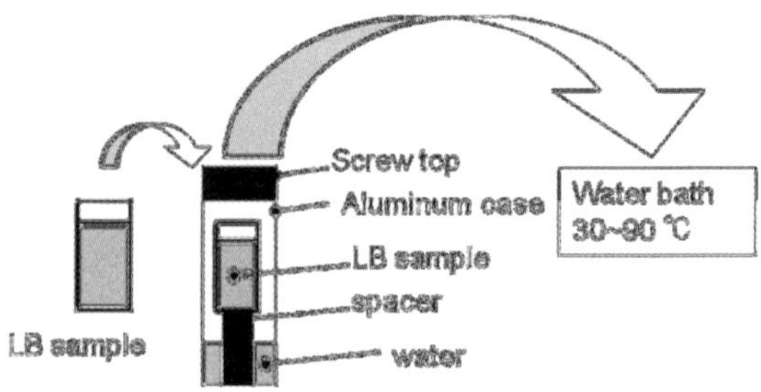

Fig. 4.1. Schematic representation of the hydrothermal treatment (HTT).

4.2. *Optical Properties*

Figure 4.2 shows typical absorption spectra $A_{\parallel}(\lambda)$ and $A_{\perp}(\lambda)$ of the sample after HTT, where $A_{\parallel}(\lambda)$ and $A_{\perp}(\lambda)$ refer to absorbance at wavelength of λ measured using linearly polarized incident light with the electric vectors parallel and perpendicular to the dipping direction in the film-transferring process, respectively. As shown in Fig. 4.2, the as-deposited spectra shown an in-plane anisotropy with a dichroic ratio $R= A_{\parallel}/A_{\perp} > 1$ (R=1.22 for this sample), each associated with a J-band centered around 590 nm, which is red-shifted by *ca.* 60 nm from the 530-nm band observed in DS solution in $CHCl_3$. By the HTT, the further red-shifted band located at 596 nm can be obtained reproducibly in the DS-C_{20} binary LB system.[67-75]

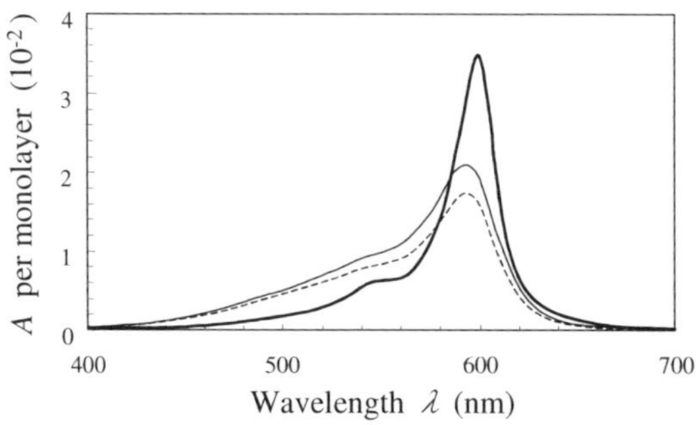

Fig. 4.2. Absorption spectra $A_{\parallel}(\lambda)$ and $A_{\perp}(\lambda)$ of the 10-layered mixed LB system of [DS]: [C_{20}]= 1 : 2 before and after HTT at 90 °C for 15 min. $A_{\parallel}(\lambda)$ and $A_{\perp}(\lambda)$ at the as-deposited state are shown by thin solid and thin dashed lines, respectively, while those after HTT are shown by the corresponding thick lines. (Reprinted with permission from Hasegawa *et al.*[70] © 2007, The Materials Research Society of Japan.)

4.3. *Morphology*

The photomicroscope images of as-deposited films are featureless within the resolution although they are associated with vague local inhomogeneities.[67] The brightness of the field under polarized light incident varies with a π-periodicity when the sample is rotated on the stage in accordance with the anisotropy in absorbance.

After HTT at T_H=30-90°C, there appear superstructures, round-shaped and as large as 100 μm or more in the diameter.[67] The occurrence of these round-shaped domains is found to depend on both T_H and t_H. Figures 23 (a)-(d) show the images from 10-layered LB films after 60-min HTT at T_H=40, 60, 80 and 90°C, respectively, where all samples belong to the same batch. Two different types of round-shaped domains are recognized: those surrounded by bright lines and those by dark lines, which will be hereafter referred to as the B- and the D-types, respectively.[70]

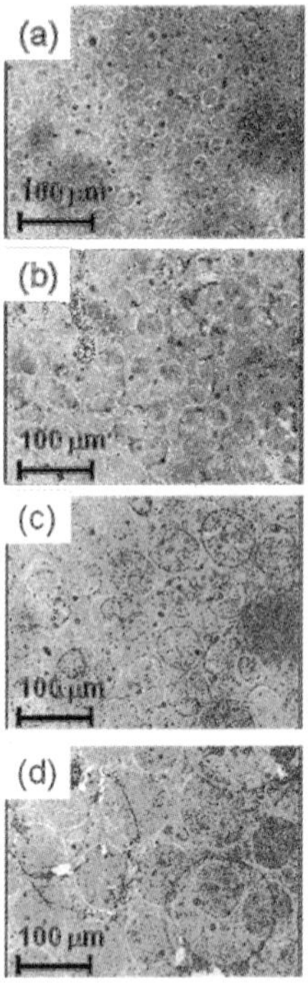

Fig. 4.3 Photomicrographs of 10-layered DS-C_{20} mixed LB films deposited on the C_{20}-hydrophobized substrates after HTT at T_H=40 °C (a), 60 °C (b), 80 °C (c) and 90 °C(d) for a fixed heating time of t_H=60 min. (Reprinted with permission from Hasegawa *et al.*[70] © 2007, The Materials Research Society of Japan.)

The image for the 40°C-HTT in Fig. 4.3 (a) represents an initial stage of the growing process of the round-shaped domains. The domains are of B-type and smaller in size than those seen in the later stages where the new phase of J-band is fully developed. It is noted that the remaining in-plane anisotropy is often observed in this stage as the anisotropic spectra with R>1 and the variation of field brightness with the π-periodicity under polarized light incident giving the maximum brightness when the dipping direction is parallel or anti-parallel to the polarization direction.

From T_H=60°C on, D-type domains appear in the field as seen in Figs. 4.3 (b)-(d). The B-type domains are occasionally stacked one on another in a random manner, while the D-type are located above the B-type. Stacks up to four-fold are so far recognized.[67,70] Both B-type and D-type domains tend to

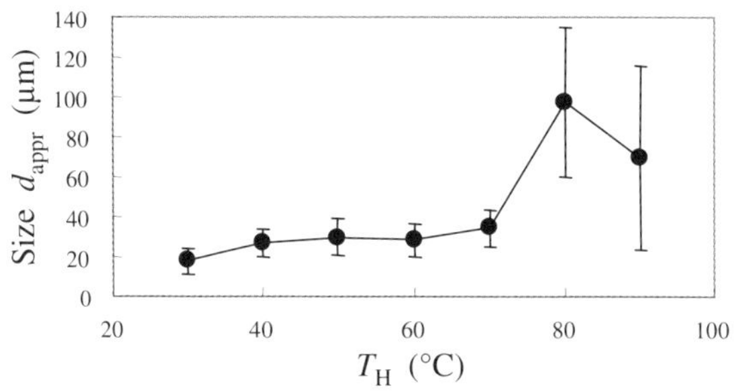

Fig. 4.4. Average size d_{appr} for 60-min-HTT plotted as a function of T_H. The standard deviations are given by bars. (Reprinted with permission from Hasegawa *et al.*[70] © 2007, The Materials Research Society of Japan.)

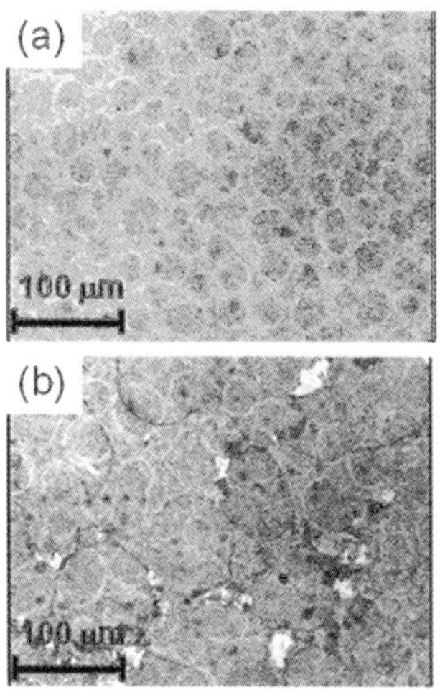

Fig. 4.5. Photomicrographs of 10-layered DS-C_{20} mixed LB films deposited on the C_{20}-hydrophobized substrates after HTT at a fixed heating temperature T_H=90°C. (a) and (b) refer to t_H=15 and 45 min, respectively (see also Fig. 35 (d) for t_H=60 min). (Reprinted with permission from Hasegawa *et al.*[70] © 2007, The Materials Research Society of Japan.)

grow in size with increasing T_H. This suggests that the B- and D-type domains are bound to the inner and the outer monolayers, respectively. Here, the difference in color between the B-type and D-type rims may originate from the different dielectric environments surrounding the domains: an inner monolayer is sandwiched between the adjacent monolayers while an outer monolayer is faced to air on one side.

The average value of the approximate domain diameter d_{appr} for t_H=60 min is given as a function of T_H in Fig. 4.4, where each point involves data from ~100 or more domains seen in randomly chosen fields. The d_{appr} value remains in a range of 20-30 μm below T_H=70°C, increases up to its maximum of 100 μm at T_H=80°C and then tends to decrease down to *ca.* 70 μm at T_H=90°C. The larger deviations of d_{appr} at T_H=80 and 90°C are due to the increased contribution from D-type domains whose sizes occasionally exceed 150 μm.

Figures 4.5 (a) and (b) show the images from 10-layered LB films at T_H=90°C for t_H=15 and 45 min, respectively. Here, Fig. 4.3 (d) is referred to as the case of t_H=60 min. The image for t_H=15 min is filled with B-type domains alone as is the case of T_H=40°C and t_H=60 min (Fig. 4.3 (a)). The new phase of J-band is, however, fully developed at T_H=90°C within 15 min as shown in Fig. 4.2. It is therefore indicated that the process of domain growth has a delay compared to the J-band reorganization process.

The image for t_H=45 min in Fig. 4.5 (a) (T_H=90°C) shares common features with Fig. 4.3 (d) (T_H=90°C) for t_H=60 min except for the larger domain sizes in the latter, showing that the domain growth is still under way.

We have prepared samples of DS-C_{20} mixed systems with reduced numbers of layers [4, 2 and single layer(s)] to investigate whether the round-shaped domains are confined to one single layer or not. Figures 5.2 (a)-(c) show the images of the 4-, 2- and 1-layered DS-C_{20} LB films after 80°C-HTT for 60 min, respectively. The 4-and the 2-layer cases are discriminated from the 10-layer case [Fig. 4.3 (c)] by their relative predominance of the D-type over the B-type, while the 1-layer case is characterized by the absence of the D-type. These observations lead to a picture that B-type and D-type domains represent those in the outermost monolayer and the inner ones, respectively, and that each is one monolayer thick or even less since it is confined to one single layer.[67,70]

4.4. *Kinetics*

The kinetics of the hydrothermally induced reorganization of J-aggregate in DS-C_{20} ternary system has been also studied extensively.[71-75] The as-deposited J-band is reorganized by HTT to form a new phase associated with a narrowing and a slight redshift of the peak. The change in the UV–visible absorption spectrum is found to involve two different stages: an increase in A(λ) at $\lambda < 540$ nm at the expense of the as-deposited J-component and a decrease in A(λ) for the shorter wavelengths compensated by the growth of the new phase of the J-band. By introducing a three-component model, each spectrum is deconvoluted into three bands, Bands I, II, and III, which are presumably assigned as the blue-shifted dimer, monomer and redshifted J-bands, respectively. The Band II fraction remains almost unchanged throughout the HTT process, indicating that the first and second stages of the spectral change can be interpreted as the dissociation of the as-deposited J-band into the dimers and as the reorganization of the J-band out of the dimers, respectively. The second-stage kinetics is found to obey a simple first-order law. The HTT induced superstructures have been quantitatively examined referring to the average size of the domains ($<d_{appr}>$) observed in the microscopic images. It has been suggested that the domain growth is a phenomenon occurring subsequent to the J-band reorganization process rather than being an aspect of the process, since a superlinear increase in $<d_{appr}>$ with heating time is observed with a long delay after the saturation of J-band growth.[75]

Following our series of reports on the effect of HTT to the DS-C_{20} binary LB system,[63-75] Hirano and coworkers also reported on effects of hydrothermal treatment in the liquid phase (HTTL) to the LB systems based on DS and DO dyes.[76,77] However, it should be noted that their liquid phase annealing process, where LB samples are directly immersed in heated water, is basically different from our HTT process because more complex kinetics, such as exchange of cations, should be involved in the former.

5. Control of J- and H-aggregate Formation in Langmuir-Blodgett Films Based on Amphiphilic Merocyanine Dyes

Considering the intermolecular π-π interaction between chromophores of merocyaine dyes, it is also expected that a formation of H-aggregate with a chromophore alignment of side-by-side type [Fig. 2.3(b)]. The H-aggregate, where the abbreviated "H" is from the term, "hypsochromic", is known as the counter extreme of the J-aggregate. The H-aggregate shows a blue-shifted band (generally is not as narrow as the J-band) with respect to the monomer absorption and it has a relatively large Stokes shift but with extremely low fluorescence yield, while the J-aggregate shows a red-shifted band with a small Stokes shift but with a high yield.

It should be stated here that the H-aggregate is intriguing not only from the viewpoint of basic research but also from that of device applications. For instance, the H-band absorption of an amphiphilic squarylium (SQ) dye at an LB film/metal interface can effectively contribute to the photocurrent in Schottky-type photovoltaic cells, while J-band cannot contribute to it effectively because the excitation in the J-aggregates tends to be quenched by coupling with surface plasmon polaritons.[78] Furthermore, for the SQ thin-film system of the H-aggregate, it has been found that a refractive index lower than glass is obtained at around the H-band and at the shorter-wavelength region. Such an effect cannot be observed for J-aggregates.[79]

So far, however, only a few reports had come to our knowledge until the mid 1990's as for the H-aggregate in L and LB films containing merocyanine dyes. In 1997, we reported that the addition of *n*-alkane molecules to the binary LB films of DS and arachidic acid [CH_3-$(CH_2)_{18}$-COOH, abbreviated as C_{20}] induces a formation of H-aggregate (Therefore, the LB films are of ternary systems.).[80] The *n*-alkane molecules, which are colorless in visible region, can somehow control the overall interaction between amphiphilic dyes and induce the H-aggregate formation. In this section, we introduce the novel way to obtain the H-aggregate in the dye LB system and discuss the structure and the formation mechanism.

5.1. *Formation of H-aggregate by Adding n-alkane as a Trigger*

Nakahara and coworkers observed a pronounced blue-shifted band in monolayers of the ternary systems of DS, C_{20} and *n*-hexadecane [CH_3-$(CH_2)_{14}$-CH_3, abbreviated as AL_{16}] However, the band is transient at the air/water interface, and no discussion was given as to its origin.[48]

Figure 5.1 (a) shows the typical absorption spectra of a 1:2 mixed LB film of the DS-C_{20} binary system.[81] The solid and the dashed lines represent the absorption spectra $A_{\parallel}(\lambda)$ and $A_{\perp}(\lambda)$, respectively. The absorption maximum is remarkably red-shifted from that of the DS monomer band (at 540 nm)[57] in the LB system. The J-band located at 590 nm is 0.20 eV red-shifted from the MS monomer peak and shows the in-plane anisotropy with the dichroic ratio $R > 1$, where R is defined as

$R=A_{\parallel}(\lambda)/A_{\perp}(\lambda)$. The main features of the absorption spectra coincide well with those of the J-band reported in the earlier works.[16-18,37-39,44,48,57,82]

Figures 5.1 (b)-(d) show the absorption spectra of the DS-C_{20}-octadecane (AL_{18}) ternary systems ($[DS]:[C_{20}]:[AL_{18}]=1:2:x$), where x=0.5, 1.0 and 1.5, respectively.[81] Each spectrum involves a prominent blue-shifted band at 505 nm and a red-shifted component at 590 nm. The 505-nm band is 0.15 eV blue-shifted from the DS monomer peak (540 nm). The spectral shape of the blue-shifted band is mirror-symmetric with that of the red-shifted one, J band. Then, we considered that the blue-shifted band is an H band due to the H-aggregate with a side-by-side alignment of DS dyes, as shown in Fig. 2.3(b).

It should be mentioned that the stable H-band formation was at first discovered for the DS-C_{20}- hexadecane (AL_{16}) ternary systems in 1997,[83,84] at around 510 nm. However, in this chapter we will focus on the DS-C_{20}-AL_{18} systems to discuss the mechanism of H-aggregation. This is because the energy shift of the H band from the monomer band is significantly greater in the DS-C_{20}-AL_{18} ternary systems compare to that in the DS-C_{20}-AL_{16} ones. Furthermore, the H band in the former systems is more stable than that in the latter ones, especially after transferred onto solid substrates as LB films. Thus more prominent characteristics are observed with AL_{18}.[81,85-89]

In Fig. 5.1 (d), the absorbance of the red-shifted component is more intense than those in Figs. 5.1 (b) and 5.1 (c). The red-shifted and the blue-shifted components show the in-plane anisotropy with the dichroic ratio $R>1$ and $R<1$ in the corresponding wavelengths, respectively. It is noted that the absorbance of the blue-shifted band at 505 nm of the freshly-prepared samples decreases by about 20% after keeping them in air ambient for 20 h, while the component around 550 nm tends to increase. On the other hand, the red-shifted band at 590 nm remains constant. The result indicates that the clusters of the 505-nm and the 590-nm components exist separately and that the possibility of Davydov splitting is discarded.

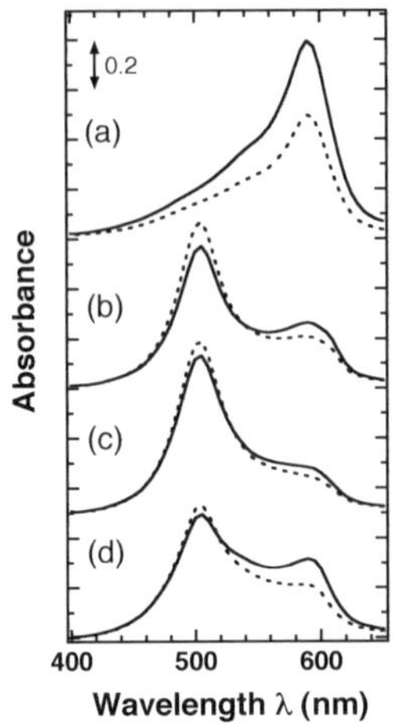

Fig. 5.1. The absorption spectra of the mixed LB films of the DS-C_{20}-AL_{18} ternary systems. The molar mixing ratio [DS]:[C_{20}]:[AL_{18}]=1:2:x, where (a) x=0.0, (b) x=0.5, (c) x=1.0 and (d) x=1.5. The solid and the dashed lines refer to the absorption spectra $A_{\parallel}(\lambda)$ and $A_{\perp}(\lambda)$ observed for a sample, both sides of which contain 20 monolayers deposited at 25 mN/m, respectively. (Reprinted with permission from Hirano *et al.*[81] © 2000, American Institute of Physics.)

As shown in Fig. 5.2, the AL_{18} content range of $2.0 \leqq x \leqq 4.0$, the blue-shifted band is broader in shape than that for $0.5 \leqq x \leqq 1.5$ (Fig. 5.1), and its absorption maximum is located at 520 nm, being 15 nm longer than that that for $0.5 \leqq x \leqq 1.5$. The peak heights of the blue-shifted and the red-shifted components are comparable with each other in Fig. 5.2 (a) and the red-shifted band then decreases gradually with increasing the AL_{18} content up to x=4.0, as shown in Figs. 5.2 (b)-(d). In Fig. 5.2 (a), the blue-shifted and the red-shifted components show the dichroic ratio R >1. R decreases and gradually approaches unity with increasing x up to 4.0 in the overall wavelength range. No further change is seen in the absorption spectra by adding the AL_{18} content more than x=4.0.

The observed spectra of the DS-C_{20}-AL_{18} ternary system shown in Figs. 5.1 and 5.2 are postulated to be superpositions of the blue-shifted and the red-shifted components. Then, each spectrum can be deconvoluted into two original components assuming that both transition moments in the blue-shifted and the red-shifted components lie in the film plane.[90]

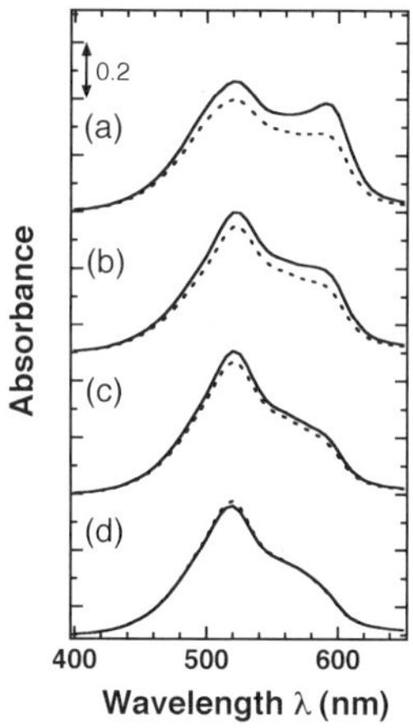

Fig. 5.2. The absorption spectra of the mixed LB films of the DS-C_{20}- AL_{18} ternary systems. The molar mixing ratio [DS]:[C_{20}]:[AL_{18}]=1:2:*x*, where *x*=(a) 2.0, (b) 2.5, (c) 3.0 and (d) 4.0. The solid and the dashed lines refer to the absorption spectra $A_{\parallel}(\lambda)$ and $A_{\perp}(\lambda)$ per 2×20 monolayers deposited at 25 mN/m, respectively. (Reprinted with permission from Hirano *et al.*[81] © 2000, American Institute of Physics.)

Figures 5.3 (a)-(c) show the deconvoluted spectra obtained for $0.5 \leqq x \leqq 1.5$. In each spectrum, the red-shifted band with R > 1 and the blue-shifted band with R < 1 are located at 590 nm and 505 nm, respectively. The red-shifted and the blue-shifted bands and their dichroic behavior can be interpreted as follows. There are two extremes of the dye aggregates recognized as to the orientation of the transition dipole moments and the spectral characteristics, as shown in Fig. 5.6. One is the J-aggregate with head-to-tail alignments of the transition dipole moments characterized by the red-shift from the monomer electronic transition.[91,92] The other is the H-aggregate characterized by side-by-side alignments of the transition dipole moments, showing a blue-shifted absorption from the monomer electronic transition.[93] It is noted that a sharp and narrow absorption is observed due to the motional narrowing effect in the electronic transition in the large aggregate.

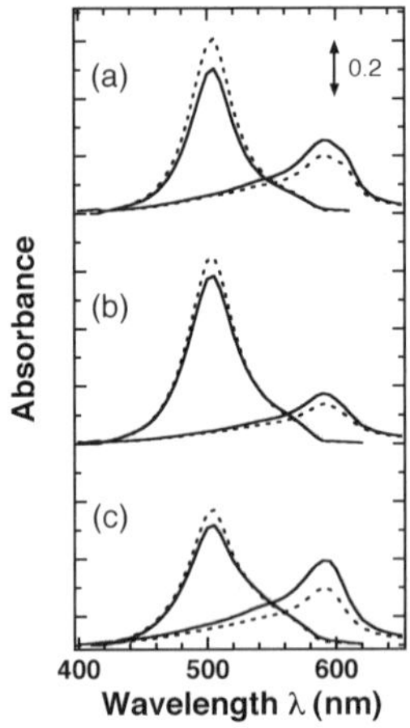

Fig. 5.3. The deconvoluted spectra of the mixed LB films of the MS-C_{20}-AL_{18} ternary systems. The molar mixing ratio [MS]:[C_{20}]:[AL_{18}]=1:2:x, where (a) x=0.5, (b) x=1.0 and (c) x=1.5. The solid and the dashed lines refer to the spectra with the electric vector parallel and perpendicular to the dipping direction, respectively. (Reprinted with permission from Hirano *et al.*[81] © 2000, American Institute of Physics.)

The results in Fig. 5.3 are in good agreement with predictions by Minari and coworkers (See Sec. 1.2), *i.e.*, the red-shifted J- and the blue-shifted H-bands should show R > 1 and R < 1, respectively, if both aggregates are elongated. Therefore, the 590-nm and the 505-nm bands in the range $0.5 \leqq x \leqq 1.5$ [Figs. 5.3 (b)-(d)] are assigned to the J-band and the H-band, respectively. If the 505-nm band were of relatively small aggregates such as the isolated dimers, the in-plane anisotropy with R < 1 at the 505-nm band and a sharp absorption would not be observed prominently because of the lack in the flow orientation effect in the film-transferring process.

Figures 5.4 (a)-(d) show the deconvoluted spectra obtained for $2.0 \leqq x \leqq 4.0$. In Fig. 5.4(a), the red-shifted band with R > 1 is located at around 590 nm, which is assigned to the J-band. This band shifts to shorter-wavelength, reaching 580 nm with increasing x as shown in Figs. 5.4(b)-(d). The peak shift of the J-band is associated with a decrease in the absorption with increasing x up to 4.0. On the other hand, the absorption of the blue-shifted band at 520 nm tends to remain constant for $x \geq 2.5$. In Figs. 5.4 (b)-(d), the 520-nm band is sharper than that in Fig. 5.4 (a),

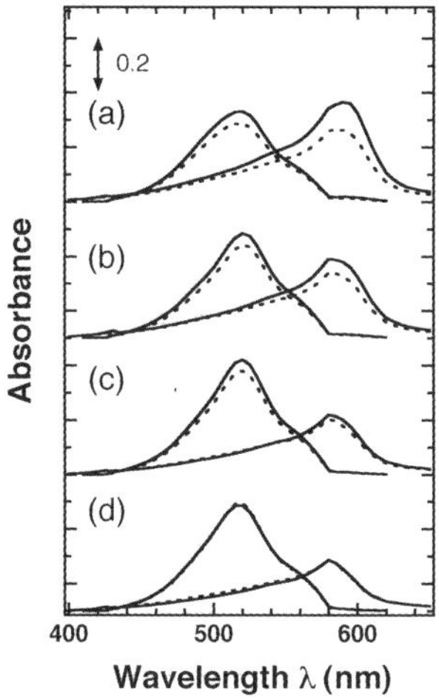

Fig. 5.4. The deconvoluted spectra of the mixed LB films of the DS-C_{20}-AL_{18} ternary systems. The molar mixing ratio [MS]:[C_{20}]:[AL_{18}]=1:2:*x*, where (a) *x*=2.0, (b) *x*=2.5, (c) *x*=3.0 and (d) *x*=4.0. The solid and the dashed lines refer to the spectra with the electric vector parallel and perpendicular to the dipping direction, respectively. (Reprinted with permission from Hirano *et al.*[81] © 2000, American Institute of Physics.)

with a slope at the shorter-wavelength being steeper. The 520-nm band show R > 1 in Figs. 5.4 (a)-(c), which contradicts the prediction from the flow orientation effect of the dye aggregates, *i.e.*, the blue-shifted band should show R < 1. As shown in Fig. 29, R(>1) of the 520-nm band decreases gradually, approaching unity with increasing the AL_{18} content up to *x*=4.0, in accordance with the R-behavior in the red-shifted band. This is possibly due to a correlation in the aggregate orientation between the red-shifted and the blue-shifted bands.

In order to interpret the optical isotropy in the overall UV-visible range, which includes both of the J- and H-bands, in case $x = 4.0$ [Fig. 5.4 (d)], there are alternative ways. One is that the two-dimensionally-developed aggregates, large in size but square- or disk-like in shape, are formed at the air-water interface. The other is that the small aggregates such as the isolated dimers, are formed there.

Contradictorily to the former of the aforementioned two interpretations, as shown in Fig. 5.4(d), in case *x*=4.0, the blue shift from the monomer band (540 nm→520 nm) is smaller than those in Figs. 5.4(a)-(c) (540 nm→505 nm). Furthermore, the spectral shape of the blue-shifted band at 520 nm [Fig. 5.4(d)] is appreciably broader than those in Figs. 5.4(a)-(c). Thus, it is suggested that the 520-nm band in Fig. 5.4(d) is due to the relatively small aggregates with side-by-side alignments rather than large, two-dimensionally-developed aggregates. The red-shifted band at 580 nm can be also ascribed to the small aggregates with the head-to-tail alignments referring to the in-plane isotropy, the broad spectral shape and the smaller energy shift than the J-band at 590 nm. In fact, it has been reported that the DS monomer band with the absorption maximum around 540 nm is isotropic (R=1) over the visible range in the LB system, and that the J-band with $R \simeq 1$ is formed in the analogous merocyanine dye with the substituted CH_3 group at the C-6 atom of the benzothiazole residue (6MeDS, Fig. 1.2) in the LB system.[94]

We hypothesize that the blue-shifted and the red-shifted bands are of side-by-side and head-to-tail oligomers for $2.0 \leqq x \leqq 3.0$, which represents the intermediary region between $0.5 \leqq x \leqq 1.5$, and $4.0 \leqq x \leqq 5.0$. There are several reports on the oligomeric character of the dye aggregates. Mizutani *et al.* have reported that the red-shifted band at 595 nm observed in an aqueous solution of DS, which is assigned to a tetramer.[95] Misawa and Kobayashi have proposed the hierarchical structure of the J-aggregates of cyanine dyes, where a macroaggregate is composed of mesoaggregates.[96,97]

5.2 Characterization using *the Extended Dipole Model*

We performed the numerical calculation using the extended dipole model[20,21,23] proposed by Kuhn and coworkers to estimate the structural parameters of DS dye molecules in the aggregate. In the model, as already touched upon in Sec. 2.1.2, the transition dipole moments of the dye molecules are simplified as classical dipoles of the length l_{tr} with the charges $-q$ and $+q$, as shown in Fig. 5.5 (a). The extended dipole length l_{tr} and the charge q are related to the transition moment μ_{tr} as

$$ql = \left|\mu^{\mathrm{tr}}\right| \tag{2.12}$$

The interaction integral J_{ij} between two dipoles is given as,

$$J_{ij} = \frac{q^2}{4\pi\varepsilon_0\varepsilon_r}\left(\frac{1}{r_{++}} + \frac{1}{r_{--}} - \frac{1}{r_{+-}} - \frac{1}{r_{-+}}\right) \tag{2.13'}$$

in SI unit, where ε_r is the relative dielectric constant of the system. We employ the ε_r=2.5 as the typical value for hydrocarbons. The transition dipole moment of DS is postulated to be parallel to the long axis of the dye chromophore. The excitation energy $\Delta E'$ of the absorption maximum for an N-mer aggregate is approximated by the sum of the excitation energy ΔE of the monomer peak and the interaction integrals J_{ij} as[98]

$$\Delta E' = \Delta E + J \approx \Delta E + \frac{1}{N}\sum_{i=1}^{N}\left(\sum_{j=1,2,\ldots,i-1,i+1,\ldots,N}^{N} J_{ij}\right). \tag{5.1}$$

The transition moment of μ^{tr}= 8.256×(3.33×10^{-30}) Cm given in Ref. 37 was used for DS in Eq. (2.12). In Fig. 5.5 (b), the area of the DS

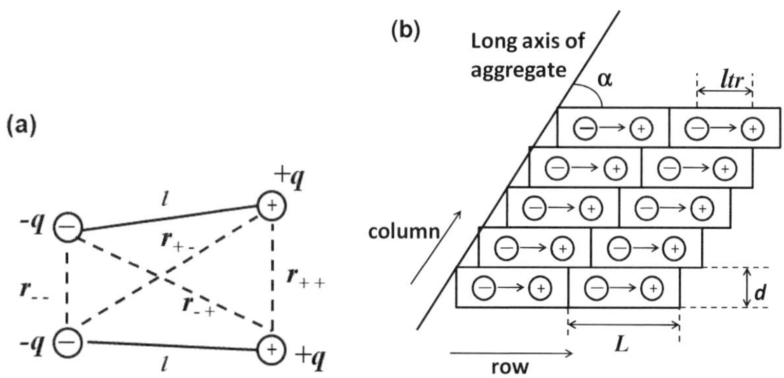

Fig. 5.5. (a) The configuration of the two extended dipole vectors with the length l_{tr} and the charge q for the calculation of J_{ij}. (b) Schematic representations of the two-dimensional arrangement of the dye molecules for the aggregate. α, d and L are the slip angle between the long axis of the aggregate and the dipole vectors, the separation between two adjacent dipole vectors and the length of the long axis of the dye chromophore, respectively.

chromophore ($= d \times L$) is identified with the occupied area per molecule ($=5.6\times10^{-19}$ m^2/molecule) of DS at 25 mN/m estimated from the surface pressure (π)-area (A) isotherm measurement.[100-102] Referring to one of earlier experimental results, the excitation energy ΔE of DS is assumed to be 3.69×10^{-19} J (540 nm).[57] We assume that $l_{tr} = 4.50\times10^{-10}$ m and q=6.11×10^{-20} C, which corresponds to 40% of the elementary charge. It should be noted here that energy shift to 590 nm for the J-band cannot be reasonably explained by the numerical calculation, if we assume the following parameters, $l_{tr} = 9.0\times10^{-10}$ m and q=3.05×10^{-20} C, which have been adopted by Nakahara and coworkers in their earlier work.[37]

Figure 5.6 shows the contour plot of the interaction J in the slip angle (α)-separation (d) plane with the aggregation number N=40 in one single row, where J for N=40 is found to be 95% or more than that for $N\rightarrow\infty$.

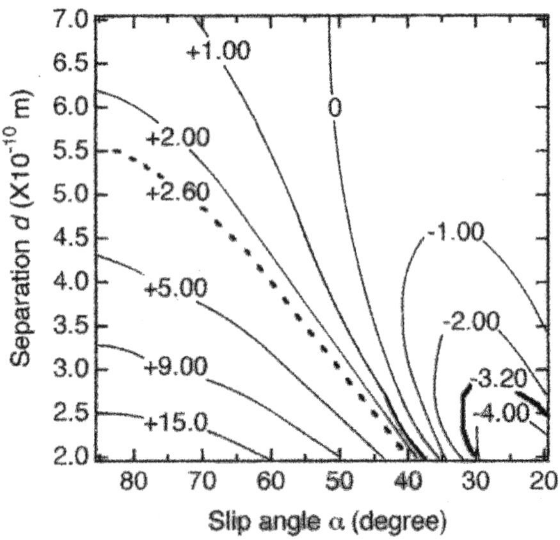

Fig. 5.6. Contour plot of the interaction J in the slip angle (□)-separation (d) plane with the aggregation number N=40 in one single row. The numeral added to each line denotes J of the order of 10^{-20} (J/molecule). The thick and the dashed curves refer to the energy shifts from the DS monomer peak, -3.20×10^{-20} J/molecule and $+2.60\times10^{-20}$ J/molecule, which correspond to the absorption bands of 590-nm and 505-nm, respectively. (Reprinted with permission from Hirano *et al.*[81] © 2000, American Institute of Physics.)

The numeral added to each line denotes J of the order of 10^{-20} J/mol. The thick and the dashed lines refer to the energy shifts -3.20×10^{-20} J/molecule and $+2.60\times10^{-20}$ J/molecule to the J-band at 590 nm and the H-band at 505 nm, respectively. The contour line for J=0 crosses the line, $d = 2.0\times10^{-10}$ m, at α=35°, and it approaches α=55° with increasing d, while the it remains constant at α=55° in our previous numerical calculation by the point dipole model.[85] At around $d = 3.0\times10^{-10}$ m, the minimum J is consistent with the energy shift of -3.20×10^{-20} *J/molecule* to the J-band peak at 590 nm. From the results, the values d and L are estimated to be $d = 3.0\times10^{-10}$ m and $L = 1.9\times10^{-9}$ m, respectively, allowing for the occupied area of DS=5.6×10^{-19} m^2/molecule.

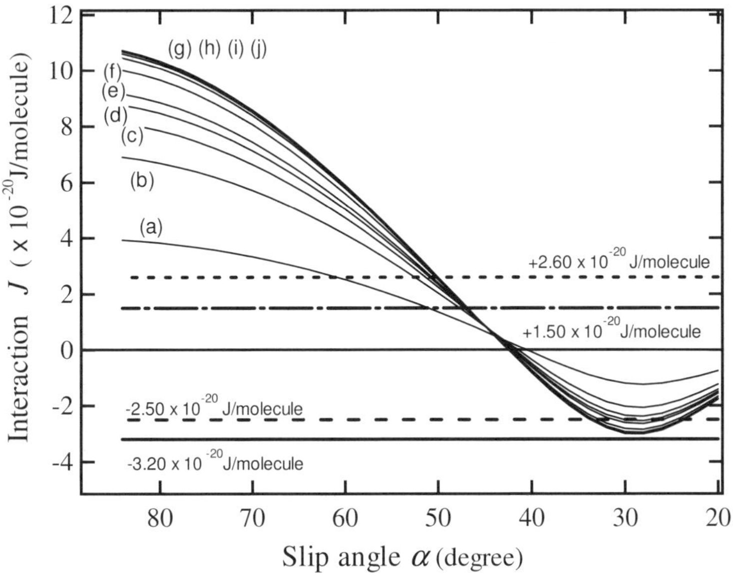

Fig. 5.7. Interaction J plotted against α. The curves denote the different values of the aggregation number N in one single row: N= (a) 2, (b) 4, (c) 6, (d) 8, (e) 10, (f) 20, (g) 40, (h) 60, (i) 80 and (j) 100. The solid, dashed, dashed-dotted and dotted lines denote the energy shifts from DS monomer band, -3.20×10^{-20}, -2.50×10^{-20}, $+1.50\times10^{-20}$ and $+2.60\times10^{-20}$ J/molecule, respectively, which correspond to the absorption bands at 590, 580, 520 and 505 nm, respectively.

Figure 5.7 shows J plotted against α, where d=3.0×10^{-10} m and L=1.9×10^{-9} m. The curves refer to the different values of the aggregation number N in one single row: N=2 (a), N=4 (b), N=6 (c), N=8 (d), N=10 (e), N=20 (f), N=40 (g), N=60 (h), N=80 (i) and N=100 (j). The energy shifts from the monomer peak, −3.20×10^{-20}, −2.50×10^{-20}, +1.50×10^{20} and +2.60×10^{-20} J/molecules, which correspond to those of 590, 580, 520 and 505 nm, respectively, are shown by the solid, dashed, dashed-dotted and dotted lines, respectively. Each J curve shows the sublinear dependence on N, converging to a constant value as $N\to\infty$. For $N \geqq 40$, J around α=30° is consistent with the energy shift (−3.20×10^{-20} J/molecule) to the J-band at 590 nm. Therefore, for the J-aggregate, the minimum aggregation number N_{min} and the slip angle α are estimated as N_{min}=40 and α=30°, respectively.

The estimated value of N_{min}=40 is compatible with N~10^2 deduced from an analysis of the line shape of electron spin resonance (ESR) measurement by Kuroda *et al.*[38,39] The estimate is also in agreement with N~10^2 deduced from the flow orientation model of Minari *et al.*[16,17]

The value of α=30° estimated for the J-aggregate is in good agreement with the value α=25-30° determined from the line shape of ESR by Kuroda *et al.*[38,39] From the analysis based on the flow orientation model, the slip angle is estimated as α=33.5 °, which is in fairly good agreement with our estimation. In one of the earliest works by Nakahara, the same slip angle α = 30° is also estimated,[x] However, it is obtained based on a misunderstanding in which a double summation of the interaction integral J_{ij} between two dipoles in the array of dye molecules simultaneously gives both the excitation energy shift and the energy gain brought about by the aggregation; then α makes their J_{ij} sum minimum with a given aggregation number was adopted. In addition, according to the calculation in Ref. 37, their J_{ij} sum diverges when $N\to\infty$, which is physically unreasonable.

On the other hand, for the H-aggregate, the slip angle α is estimated as α=50 °, assuming the same N_{min} value (=40), as shown by curves (g)-(j) in Fig. 5.7.

Let us also estimate the N and α values of DS for the red-shifted band at 580 nm and the blue-shifted band at 520 nm, shown in Fig. 5.4 (d). If one assumes the same α value of 30°, we obtain N=10 for the 580 nm band, as shown by the point of contact between the dashed line and curve (e), as shown in Fig. 5.7. On the other hand, as for the 520 nm band, α is estimated at 48°, assuming that N > 4, as shown in Fig. 32. The estimated α value (=48°) for the 520 nm almost coincides with that α=50° for the 505 nm band. We currently hypothesize that the emergence of the two kinds of blue-shifted bands (505 nm and 520 nm) is due to difference in aggregation number of DS rather than in the different configuration of DS transition dipole moments in the aggregate. Similar estimates are also obtained in case two and three rows are set in the calculation. It is noted that the possibility of the two-dimensionally-developed aggregate is not excluded in the red-shifted 580 nm band; for instance, J values for N=3×3 and 3×4 are -2.27×10^{-20} J/molecule and -2.57×10^{-20} J/molecule, respectively, assuming $\alpha=30^{\circ}$.

In Fig. 5.4(a)-(c), where $2.0 \leqq x \leqq 3.0$, which is identified with the intermediate region between $0.5 \leqq x \leqq 1.5$ and $4.0 \leqq x \leqq 5.0$., N of the blue-shifted band at 520 nm and that of the red-shifted band at 590 nm or 580 nm, are tentatively estimated to be in the range of tens or more

As stated in Sec. 1.1, the merocyanine dye has a large static dipole moment, and thus, for the evaluation of the energy shift upon the J-aggregate formation, the "monomer shift", which is due to the change of the electric dipole (μ^{st}) moment upon photoexcitation, should be also taken into account together with the interaction of the transition dipole (μ^{tr}) moments. As for present thin film system, however, the monomer shift is not taken into account because of the lack of exact structural parameters. The structural characterization of the present systems is not straightforward because they are of mixed ones. Kato and coworkers evaluated the energy shift upon the J-aggregate formation in a DS LB film (formed without matrix molecules) considering the monomer shift together with the μ^{tr} interaction, which will be described in Sec. 6.

5.3. *Where Are Octadecane Molecules?*

We have reported that the H-band is formed by adding AL_{18} molecules to the binary system of DS and C_{20} and that the H-band formation can be precisely controlled by the amount of AL_{18} keeping the molar mixing ratio of DS:C_{20}=1:2.[80,81,84-88,100] However, there arise questions, *i.e.*,

"Where are octadecane molecules (AL_{18}) ? and "How do they induce the H-aggregate formation?

Figure 5.8 shows the surface pressure (π)-area per molecule (A) isotherms of the DS-C_{20} binary and the DS-C_{20}-AL_{18} ternary systems with the C_{20} content of [DS]:[C_{20}]:[AL_{18}]=1:2:x at 17±1℃.[100,101] The lines (a), (b), (c), (d), (e) and (f) denote the π-A isotherms for x=0, 1.0, 2.0, 3.0, 4.0 and 5.0, respectively. The abscissa, area per molecule, refers to the average value of the area per molecule for the binary (x=0) or the ternary (x≠0) system at the air-water interface. Phase transitions are observed at 15 mN/m and 12 mN/m in the ternary systems for x=4.0 (e) and x=5.0 (f), respectively. For x=3.0 (d), a phase transition can be traced at around 35 mN/m.

Furthermore, in a ternary system, where molar mixing ratio of C_{20} was increased up to [DS]:[C_{20}]:[AL_{18}]=1:5:x, a similar phase transition was observed at 35, 35 and 25 mN/m for x=3.0, 4.0 and 5.0, respectively. We hypothesize that the phase transition is observed for $x \geqq 3.0$ (*i.e.*, [AL_{18}]/[MS]$\geqq$3.0) regardless of the C_{20} content used as the matrix. These results indicate that the microscopic structure of the monolayer of the ternary system changes depending on the surface pressure, and that the phase transition is governed by the molar mixing ratio between DS and AL_{18} in the ternary system.

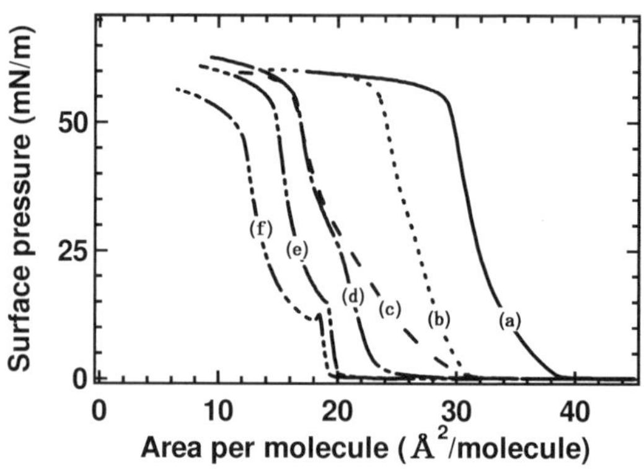

Fig. 5.8. Surface pressure (π)-area per molecule (A) isotherms of [DS]:[C_{20}]:[AL_{18}]=1:2:x on $CdCl_2$-containing water subphase of 17 ± 1 ℃; (a) x=0, (b) x=1.0, (c) x=2.0, (d) x=3.0, (e) x=4.0 and (f) x = 5.0 (Reproduced with permission from Hirano *et al.*[101] © 1999, The Japan Society of Applied Physics.)

From the π-A isotherms of the binary system (a), the occupied area of DS is estimated to be 56 Å^2/molecule, assuming that the occupied area of the cadmium arachidate (CdC_{20}) is 20 Å^2/molecule at 25 mN/m. The occupied area of DS estimated in each binary system is in agreement with those reported for pure DS[37,48] and the DS-C_{20} binary systems[57] in the earlier works, suggesting that the long axis of DS is nearly parallel to the air-water interface. Using these values, the occupied area per molecule of AL_{18} can be estimated for all the DS-C_{20}-AL_{18} ternary systems.

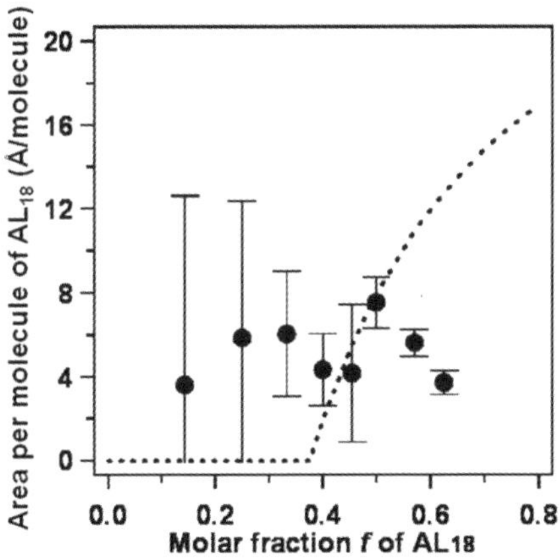

Fig. 5.9. Area per molecule at 25 mN/m plotted against AL_{18} molar fraction $f = x / (1+2+x)$. The dashed line represents a theoretical curve assuming that each DS molecule can accommodate two AL_{18} molecules at the "backside" (empty space). (Reproduced with permission from Hirano *et al.*[101] © 1999, The Japan Society of Applied Physics.)

Figure 5.9 shows the occupied area of AL_{18} at 25 mN/m plotted against the AL_{18} molar fraction $f = \{x /(1+2+x)\}$ in the DS-C_{20}-AL_{18} ternary systems with the molar mixing ratio of [DS]:[C_{20}]:[AL_{18}]=1:2:x. The average values of the occupied area are in the range of 2～7 Å^2/molecule, which is much smaller than the cross section of the hydrocarbon chain (20 Å^2/molecule), indicating that AL_{18} molecules do not significantly contribute to the area of the monolayer at the air-water interface.

According to earlier works,[4,7,17,37,48,56,57,102-105] each DS molecule in the binary system has an empty space which can roughly accommodate two straight-chain hydrocarbons. The present result suggests that AL_{18} molecules act as "fillers" to trigger off the change in the aggregate structure, where the fillers are defined as the additives that remain in the monolayer and fill the empty spaces of surface-active molecules[106,107]. In this case, the occupied area of AL_{18} should be zero if the molar fraction $f \leqq f_c$ (f_c=0.375), where the empty spaces in DS should be fully filled with AL_{18}. Above f_c, the occupied area of AL_{18} should increase monotonically and approaches to the asymptote line of 20 Å^2, as indicated by the dashed line in Fig. 5.9.[101]

The theoretical curve obtained from the simple filling model, shown by the dotted-line in Fig. 5.9, fits the occupied areas of AL_{18} well only in a narrow range of $2.0 \leqq x \leqq 3.0$ ($0.400 \leqq f \leqq 0.500$); the curve does not fit well outside the range. In ternary systems for x=0.5 (f=0.143), 1.0 (f=0.250) and 1.5 (f=0.333), the standard deviations of the occupied areas of AL_{18} are larger than those of other mixing ratios, indicating a complicated equilibrium of AL_{18} between inside and outside of the DS empty space: a certain amount of AL_{18} is outside the empty space of DS even for x=0.5 (f=0.143), 1.0 (f=0.250) and 1.5 (f=0.333). As seen in Fig. 5.9, the occupied areas of AL_{18} for x=4.0 (f=0.571) and 5.0 (f=0.625) are much smaller than the values estimated by the simple filling model. Similar results are also obtained in the ternary system with the C_{20} content of [DS]:[C_{20}]:[AL_{18}]=1:5:x. This may be related to the phase transition traced in the π-A isotherms, suggesting that one single DS molecule can accompany three or more AL_{18} molecules on average in the ternary system.

In order to characterize the structure of the DS-C_{20}-AL_{18} ternary LB systems precisely, we also utilized Fourier-transformed infrared (FT-IR) spectroscopy.[86,88,108,109] In the ternary DS-C_{20}-AL_{18} system, C_{20} and /or AL_{18} were substituted by deuterated arachidic acid (C_{20}-*d*) and/or deuterated *n*-octadecane (AL_{18}-*d*), respectively, and characterize the thermal mobility of the alky chain of each constituent by FT-IR spectroscopy. The CD_2 antisymmetric and symmetric bands are located at 2194 cm^{-1} and 2089 cm^{-1}, respectively, while The CH_2 antisymmetric and symmetric bands are located at 2920 cm^{-1} and 2850 cm^{-1}, respectively. Therefore, the substitution of C_{20} and/or AL_{18} to C_{20}-*d* and/or AL_{18}-*d* allows us to characterize the thermal mobility of the alkyl chain of each constituent separately.

First of all, the polarizing FT-IR transmission spectra of the mixed LB films with the molar mixing ratio of [DS]:[C_{20}-*d*]:[AL_{18}-*d*]=1:2:*x*, where *x*=0, 1.0, 2.0, 3.0, 4.0 and 5.0, were measured.[86,88,108,109] The full-widths of the half height (FWHH) of the CH_2 antisymmetric band (ν_{as}, at 2920 cm^{-1}) and symmetric one (ν_s, at 2850 cm^{-1}) are evaluated. Since the transfer ratio depends on the mixing ratios of the ternary system, the absolute values of absorbance are different from sample to sample. Thus, in order to avoid the effects of the transfer ratio, the absorption peak of each IR band is normalized to 1 and the normalized FWHH is used for the characterization. Both of the normalized FWHH gradually decrease with increasing the *x* and it becomes 40% smaller at *x*=2.0 compare to those at *x*=0, however, they do not show significant changes in the range of $2.0 \leqq x \leqq 5.0$.

The FWHH of a band is a useful indicator to evaluate the band sharpness, and a large value of the FWHH of the ν_{as} (or ν_s) stretching mode suggests a disordered state of a hydrocarbon chain, while a small value of the FWHH of the band suggests an ordered state of it.

As for the ternary LB system of [DS]:[C_{20}-*d*]:[AL_{18}-*d*]=1:2:*x*, in the range from x=0 up to 2, the both of the normalized FWHH values of ν_{as} and ν_s decrease with increasing the AL_{18}-*d* content *x*, indicating that the amount of Gauche bonds decreases in the hydrocarbon chains. We postulate that the thermal motion of the hydrocarbon chains of DS is hindered by the AL_{18} addition as the third component.[88,109] Furthermore, the tilt angles of hydrocarbon chains of DS and deuterated hydrocarbon chains of C_{20}-*d* were estimated based on inclination-angle-dependent FT-IR data, where projections of the transition dipole moments of CH_2 antisymmetric stretching band (ν_{as}, at 2920 cm^{-1}), CH_2 symmetric stretching one (ν_s, 2850 cm^{-1}), CD_2 antisymmetric stretching one (ν_{as}-*d*, 2194 cm^{-1}) and CD_2 symmetric stretching one (ν_s-*d*, 2089 cm^{-1}) to the plane perpendicular to the IR incident were considered.[88] In the analyses, the model proposed by P. Chollet, in which the refractive index, the reflection, and the optical rotation of the electromagnetic wave due to the molecular thin films are taken into account,[108,109] were utilized.

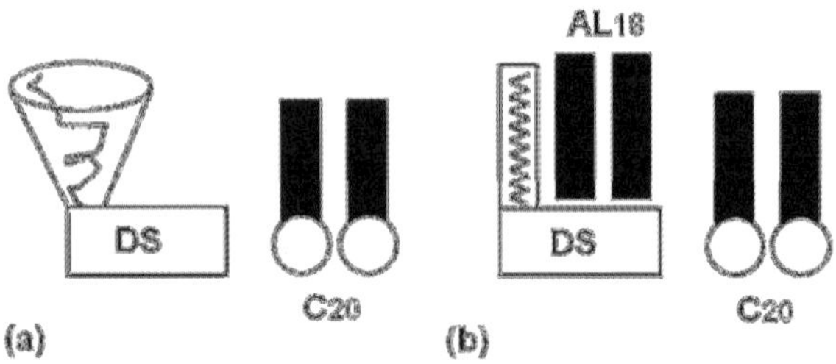

Fig. 35. Schematic representations of the orientation and order of the DS-C_{20}(or C_{20}-*d*) binary LB system (a) and DS-C_{20}(or C_{20}-*d*)-AL_{18}(or AL_{18}-*d*) ternary one (b).

The resultant average tilt angle, γ_{av}, of the hydrocarbon chain to the surface normal is γ_{av}=26° for the binary system of [DS]:[C_{20}-*d*]:[AL_{18}-*d*]=1:2:*x*, where *x*=0, while γ_{av}=17° for the ternary systems of [DS]:[C_{20}-*d*]:[AL_{18}-*d*]=1:2:*x*, where *x*=2, 3, 4, 5.[108,109] The estimation of the tilt angles coincides well with that the amount of Gauche bonds decreases in the hydrocarbon chains by the AL_{18} addition.[88,108,109]

The polarizing FT-IR transmission spectra of the mixed LB films with the molar mixing ratio of [DS]:[C_{20}-*d*]:[AL_{18}]=1:2:*x*, where *x*=0, 1.0, 2.0, 3.0, 4.0 and 5.0, were also measured. The FWHH of the CD_2 antisymmetric stretching band (ν_{as}-*d*, at 2194 cm^{-1}) and symmetric stretching one (ν_s-*d*, at 2089 cm^{-1}) were evaluated. Both of the normalized FWHH remains almost constant in the range from x=0 to x=5, indicating that the thermal motion of deutrated hydrocarbon chain of C_{20}-*d* retains constant irrespective of the AL_{18} content. Furthermore, the tilt angles of the deuterated hydrocarbon chains of C_{20}-*d* against the surface normal were estimated by the model proposed by P. Chollet, and the resultant tilt angle has a constant value of γ_{av}=11° regardless of the AL_{18} content.[109]

The schematic representations of the orientation and order of the DS-C_{20} binary LB system and DS-C_{20}-AL_{18} ternary one are shown in Figs. 5.10 (a) and 5.10 (b), respectively. In the binary system, the order of the hydrocarbon chains of DS is relatively lower compared to that of C_{20}. With adding AL_{18} to the binary system and increasing the content, the order the hydrocarbon chains of DS increases. The average tilt angle of the principal chains of DS to the surface normal is thus decreases with increasing the AL_{18} content. The AL_{18} molecules tend to "ride on the back of DS molecule" filling the empty space of 36 (= 56 – 20) Å^2. X-ray diffraction (XRD) analyses also revealed that the inter-layer spacing of the mixed LB systems tends to expand by increasing the AL_{18} contents,[101] which is in good agreement with the improvement of the hydrocarbon order by increasing the AL_{18} content.[88,108,109] Such controlling effects by adding *n*-alkane molecules have been also reported for cyanine dyes[110] and porphyrin molecules.[111] The addition of *n*-alkane molecules to amphiphilic molecules at the air/water interface has a potential to become a versatile technique to control the molecular organization.

6. *J-aggregate Structure of Merocyanine Dye LB Films Determined by Grazing Incidence X-ray Diffraction and the Evaluation of the Band Shift*

As stated in Sec. 1.1, since the merocyanine chromophore is a conjugated system sandwiched by electron withdrawing and repelling groups, it inevitably has a large static dipole moment along the long axis. Therefore, when the exciton band shifts to a lower energy in J-aggregate, the contribution from the static dipole moment (μ^{st}) should be taken into account as well as that from transition dipole moment (μ^{tr}). However, in order to discuss the contributions from the μ^{st} and μ^{tr} interactions quantitatively, structural parameters should be determined. Kato and coworkers utilized grazing incident x-ray diffraction (GIXD) method and analyzed the merocyaine (DS, Fig 1.2) J-aggregate monolayer floating on the water surface.[112,113]

The DS J-aggregate monolayer was prepared on the aqueous subphase of 0.5 mM $MgCl_2$ and 0.05 mM $NaHCO_3$, resulting in the exciton band at 2.006 eV (618 nm), which is red-shifted from the monomer band located at 2.366 eV (524 nm). The surface pressure was less than 5 mN/m and the temperature was kept at 17°C.

The structure of the two-dimensional (2D) system of DS was determined via following two stages:

(1) The in-plane diffraction pattern of the monolayer was obtained as a 2D powder diffraction and the 2D lattice parameters were determined [Fig. 6.1(a)].

(2) Molecular orientation parameters [angles (i), (ii), and (iii), in Fig. 37(b)] were optimized by simulating the diffraction intensities neglecting water molecules and ions in the subphase.[114] In the simulation, the DS molecule, whose geometry was optimized by the *ab initio* calculation using BLYP/6-31G**,[28] was placed at each lattice point with the dye plane set to be perpendicular to the lattice plane. The Debye-Waller factor is not taken into account.

Figure 6.1(c) shows the determined arrangement of the DS molecule in the unit cell. On the basis of the determined molecular arrangement, the

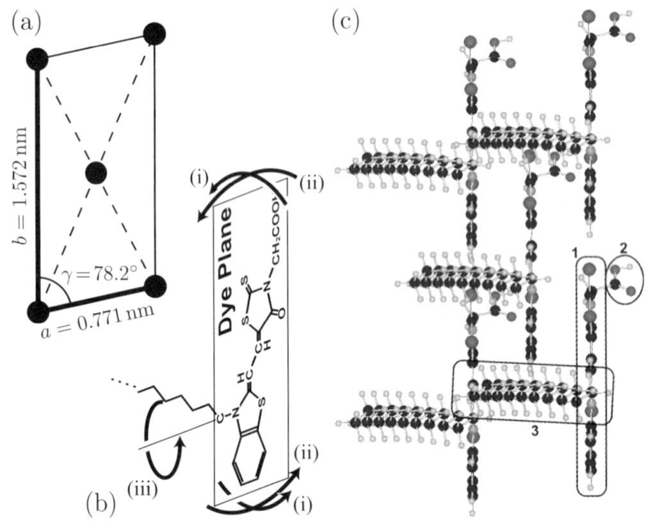

Fig. 6.1 (a) The 2D unit cell of the DS J aggregate. (b) Three parameters for simulating the diffraction intensities. (c) Determined arrangement of the DS molecules in the unit cell in (a). The dye group 1, the carboxylic acid group 2, and the alkyl chain 3 are highlighted for the molecule in the corner. (Reprinted with permission from Noritaka Kato, Kazuya Yuasa, Takeshi Araki, Ichiro Hirosawa, Masugu Sato, Naoshi Ikeda, Ken-ichi Iimura, and Yoshiaki Uesu, Phys. Rev. Lett., **94**, 136404(4pages), 2005 [Ref. 112]. © (2005) by the American Physical Society.)

energy shift upon aggregation has been discussed. By applying the extended dipole model, both of the μ^{st} and μ^{tr} contribution to the energy shift have been calculated. In the calculation, as for the μ^{tr} contribution, while electrons are treated as quantum-mechanical objects, the distribution of the positive charge is not, *i.e.*, Born-Oppenheimer approximation.

Consequently, the importance of μ^{st} interaction upon excitation is confirmed as well as the μ^{tr} interaction for the exciton band energy on the basis of the crystal structure determined by GIXD method: the μ^{st} interaction accounts for more than 50% of the observed energy shift in the DS J aggregate. As for the systems bearing a large μ^{st}, it has been proven that the μ^{st} interaction and the change upon photoexcitation should be taken into account together with the μ^{tr} interaction to discuss the band shift to lower energy from the monomer band. So far, however, evaluations on the exciton band shifts have been done only by μ^{tr} interaction for many 2D systems based on merocyanine dyes as in Sec. 5. It is expected that those 2D merocyanine systems will be characterized more precisely using structural parameters elucidated by GIXD method.

7. Concluding Remarks

Structure and optical properties of J-aggregates in the LB films based on amphiphilic derivatives of merocyanine dyes have been studied.

Section 1 briefly introduced the merocyaine dye family as an exotic constituent for forming J-aggregates. The J-aggregates based on merocyanine dyes are realized at the air/water interface and are transferred onto solid surfaces and accumulated by the technique called, Langmuir-Blodgett (LB) technique. A brief introduction was also given for the LB technique.

In Section 2, after giving a short review of the quantitative treatment of the optical properties of J-aggregates, analyses for the J-aggregates confined in the metal-free pure Langmuir film of 6MeDS were given, in Section 3. The importance of the intermolecular hydrogen bonds in the formation of J-aggregate based on the observed IR absorption bands together with the density-functional calculations. In addition, from the mixing effect of C_{20} in the DSe case, the aggregation number can be estimated at 10^2 or larger. We have pointed out that various interpretations of the structures and formation mechanisms of the J-aggregates should be amended by the more rigorous treatments presented here.

In Section 4, we have touched upon a noble reorganization process of J-aggregate called hydrothermal treatment (HTT), in which the LB samples are annealed at relative humidity of 100 %. By HTT in the temperature range of 30-90℃, the well-known J-band of the mixed LB system of DS and C_{20} located at 590 nm is found to be further red-shifted to 596 nm. The reorganized band at 596 nm is appreciably sharper compare to that before HTT and there appear round-shaped superstructures, which are as large as up to 100 μm or more in diameter.

In Section 5, we have also touched upon studies on control of J- and H-aggregate formation in LB films based on amphiphilic merocyanine dyes. It has been found that H-aggregate is formed when *n*-alkane molecules, such as hexadecane or octadecane, are added to the binary system of arachidic acid (C_{20}) and amphiphilic merocyanine dye (DS), where J-aggregate is solely formed when the film is formed under the typical condition with the presence of Cd^{2+} cations in the subphase. It has been further found that J- and H-aggregate can coexist in the ternary LB systems and the ratio can be controlled by changing the molar mixing ratio. It has been suggested that n-alkane molecules tend to "ride on the back of DS molecules" filling the empty space. The structural parameters, such as the aggregation number N of DS and the slip angle α between the long axis of the aggregate and the transition dipole moment, have been examined using the extended dipole model by the deconvolution of the UV/visible spectra. For fully-developed J-aggregates, estimated minimum aggregation number $N_{min.}$ and the slip angle α between the long axis of the aggregate and the transition dipole moment are $N_{min.}=40$ and $\alpha=30°$, respectively. On the other hand, for fully-developed H-aggregates, $N_{min.}=40$ and $\alpha=50°$, as seen for $x \leqq 1.5$. For both aggregate species, N shows the tendency to decrease with increasing x.

In Section 6, studies on determination of the structure of J-aggregate in merocyanine dye LB Films by grazing incidence X-ray diffraction (GIXD), which have been performed by Kato and coworkers, have been introduced. Consequently, the importance of static dipole (μ^{st}) interaction upon excitation is confirmed as well as the transition-dipole (μ^{tr}) interaction for the exciton band energy on the basis of the crystal structure determined by GIXD method: the μ^{st} interaction accounts for more than 50% of the observed energy shift in the DS J aggregate.

Acknowledgments

The present authors would like to thank Prof. M. Sugi, Prof. S. Kuroda, Dr. K. Saito, Dr. M. Lan, Dr. N. Minari, Prof. Y. Tabe, Dr. M. Saito, Prof. T. Ishii, Prof. Y. Hirano, and Dr. S. Morita for the fruitful collaboration. Thanks are due to Prof. N. Kato, Dr. S. Abe, Prof. H. Nakahara and Prof. K. Iriyama for their useful suggestions on the J-aggregate.

References

1. For example, N. F. Mott and R. W. Gurney, in Electronic Processes in Ionic Crystals, 2nd ed. (Clarendon Press, Oxford, 1948).
2. For example, S. Fujita, in Organic Chemistry of Photography (Springer, Heidelberg, 2004).
3. K. Ikegami and Y. Tatewaki, Trans. Mater. Res. Soc. Jpn. 33, 123 (2008).
4. M. Sugi and S. Iizima, Thin Solid Films 68, 199 (1980).
5. K. Iriyama, F. Mizutani, and A. Yoshiura, Chem. Lett. 1980, 1399 (1980).
6. M. Sugi, M. Saito, T. Fukui and S. Iizima, Thin Solid Films 99, 17 (1983).
7. M. Saito, M. Sugi, T. Fukui and S. Iizima, Thin Solid Films 100, 117 (1983).
8. M. Saito, M. Sugi and S. Iizima, Jpn. J. Appl. Phys. 24, 379 (1985).
9. K. Sakai, M. Saito, M. Sugi and S. Iizima, Jpn. J. Appl. Phys. 24, 865 (1985).
10. M. Sugi, K. Sakai, M. Saito, Y. Kawabata and S. Iizima, *Thin Solid Films* **132**, 69 (1985).
11. K. Saito, M. Yoneyama, M. Saito, K. Ikegami and M. Sugi, *Thin Solid Films* **160**, 133 (1988).
12. K. B. Blodgett, *J. Am. Chem. Soc.* **57**, 1007 (1936).
13. G. L. Gaines, Jr., *Insoluble Monolayers at Liquid-Gas Interfaces* (Interscience, New York, 1966).

14. A. Ulman, in *An Introduction to Ultrathin Organic Films* (Academic Press, Boston 1991).
15. K. Ikegami and M. Sugi, in *Bottom-up Nanofabrication: Supramolecules, Self-Assemblies, and Organized Films*, Eds. K. Ariga and H. S. Nalwa (American Scientific Publishers, New York, 2006)
16. N. Minari, K. Ikegami, S. Kuroda, K. Saito, M. Saito, and M. Sugi, *Solid State Commun.* **65**, 1259 (1988).
17. N. Minari, K. Ikegami, S. Kuroda, K. Saito, M. Saito, and M. Sugi, *J. Phys. Soc. Jpn.* **58**, 222 (1989).
18. Y. Tabe, K. Ikegami, S. Kuroda, K. Saito, M. Saito and M. Sugi, *J. Appl. Phys.* **70**, 1425 (1991).
19. For a comprehensive review, see, *e.g.*, M. Sugi, Y. Tabe and K. Ikegami, "Analytic Model of Flow Orientation in Langmuir-Blodgett Films", in *Organized Monolayers and Assemblies: Structure, Processes and Function* Ed. D. Möbius and R. Miller, (Elsevier, Amsterdam, 2002) p. 13.
20. H. Kuhn and C. Kuhn, in *J-Aggregates* Ed. T. Kobayashi (Word Scientific, Singapore , 1996), Chap. 1, p. 1.
21. P.O.J. Scherer, in *J-Aggregates* Ed. T. Kobayashi (World Scientific, Singapore, 1996), Chap. 4, p. 95.
22. A.S. Davydov, *Theory of Molecular Excitations* (Plenum Press, New York, 1971).
23. V. Czikklely, H.D. Förstering and H. Kuhn, *Chem. Phys. Lett.* **6**, 207 (1979).
24. S. Abe, *Chem. Phys.*, **264**, 355 (2001).
25. K. Ikegami, *J. Chem. Phys.*, **121**, 2337 (2004).
26. K. Ikegami, *Trans. Mater. Res. Soc. Jpn.* **30**, 147 (2005).
27. K. Ikegami and M. Lan, *Colloids Surf. A* **257-258**, 143 (2005).
28. M. J. Frisch, G. W. Trucks, H. B. Schlegel, G. E. Scuseria, M. A. Robb, J. R. Cheeseman, V. G. Zakrzewski, J. A. Montgomery, Jr., R. E. Stratmann, J. C. Burant, S. Dapprich, J. M. Millam, A. D. Daniels, K. N. Kudin, M. C. Strain, O. Farkas, J. Tomasi, V. Barone, M. Cossi, R. Cammi, B. Mennucci, C. Pomelli, C. Adamo, S. Clifford, J. Ochterski, G. A. Petersson, P. Y. Ayala, Q. Cui, K. Morokuma, D. K. Malick, A. D. Rabuck, K. Raghavachari, J. B. Foresman, J. Cioslowski, J. V. Ortiz, B. B. Stefanov, G. Liu, A. Liashenko, P. Piskorz, I. Komaromi, R. Gomperts, R. L. Martin, D. J. Fox, T. Keith, M. A. Al-Laham, C. Y. Peng, A. Nanayakkara, C. Gonzalez, M. Challacombe, P. M. W. Gill, B. Johnson, W. Chen, M. W. Wong, J. L. Andres, C. Gonzalez, M. Head-Gordon, E. S. Replogle and J. A. Pople, *Gaussian 98, Rev. A.1.*, Gaussian, Inc., Pittsburgh (1998).
29. http://unit.aist.go.jp/grid/QCgrid/.
30. J.B. Foresman and Æ. Frish, in *Exploring Chemistry with Electronic Structure Method,* 2nd Ed. (Gaussian, Pittsburgh, 1998).
31. Y. Fujimoto, Y. Ozaki, M. Takayanagi, M. Nakata, and K. Iriyama, *J. Chem. Soc. Faraday Trans.* **92**, 413 (1996).
32. K. Ikegami and S. Kuroda, *Chem. Phys.* **295**, 205 (2003).

33. H. Sapper, D. G. Cameron, and H. H. Mantsh, *Can. J. Chem.* **59**, 2543 (1981).
34. Y. Tian, *J. Phys. Chem.* **95**, 9985 (1991).
35. K. Kajikawa, H. Takezoe, and A. Fukuda, *Chem. Phys. Lett.* **205**, 225 (1993).
36. N. Kato, K. Saito, H. Aida and Y. Uesu, *Chem. Phys. Lett.* **312**, 115 (1999).
37. H. Nakahara, K. Fukuda, D. Möbius, and H. Kuhn, *J. Phys. Chem.* **90**, 6144 (1986).
38. S. Kuroda, K. Ikegami, M. Sugi, and S. Iizima, *Solid State Commun.* **58**, 493 (1986).
39. S. Kuroda, K. Ikegami, K. Saito, M. Saito, and M. Sugi, *J. Phys. Soc. Jpn.* **56**, 3319 (1987).
40. S. Kuroda, K. Ikegami, Y. Tabe, K. Saito, M. Saito, and M. Sugi, *Phys. Rev.* ***B*** **43**, 2531 (1991).
41. K. Ikegami, C. Mingotaud, and P. Delhaès, *Phys. Rev.* **E 56**, 1987 (1997).
42. $\int \varepsilon(\nu)\nu^{-1} d\nu$ (not $\nu^{-1}\int \varepsilon(\nu) d\nu$) was numerically obtained, where ε is the molar absorption coefficient.
43. K.P.C. Vollhardt and N.E. Schore, in *Organic Chemistry*, 2nd Ed. edited Jpn. (Japan UNI Agency, Tokyo, 1996).
44. K. Murata, H. Shin, K. Saito, and S. Kuroda, Thin Solid Films **327**, 446 (1998).
45. Murata *et al.* observed the J-bands in LB films of $[6MeDS]_{1-\gamma}[DO]_{\gamma}[FA]_2$ mutual mixtures with $\gamma \leq 0.8$ prepared upon a Cd^{2+}–containing subphase. They found a linear relationship between ν_J and γ: $\nu_J(\gamma) = 16530 + 710\ \gamma [cm^{-1}]$. In their system, some DO molecules are not included in the J-aggregates, while nearly all 6MeDS molecules are included in the J-aggregates. Therefore, the effective molar ratio of DO in the J-aggregates, γ', is smaller than γ: $\gamma' = \rho\gamma (0 < \rho < 1)$. Assuming that $\nu_J(\gamma' \to 1)$ converges to ν_J of pure DO films with Mg^{2+}, ρ is estimated at 0.55. This ρ value is compatible to the spectral line shapes of the mutually mixed films reported in Ref. 44.
46. K. Ikegami, T. Yoshiyama, K. Maejima, H. Tampo, S. Niki and H. Shibata, *Thin Solid Films* **518**, 462 (2009).
47. H. Nakahara and D. Möbius, *J. Colloid Surf. Sci.* **115**, 363 (1986).
48. T. Kawaguchi and K. Iwata, *Thin Solid Films* **165**, 323 (1988).
49. T. Kawaguchi and K. Iwata. *Thin Solid Films* **180**, 235 (1989).
50. T. Kawaguchi and K. Iwata. *Thin Solid Films* **191**, 173 (1990).
51. M. Yoneyama, T. Nagao, and T. Murayama, *Chem. Lett.* **1989**, 397 (1989).
52. A. Miyata, D. Heard, Y. Unuma, and Y. Higashigaki, *Bull. Chem. Soc. Jpn.* **66**, 999 (1993).
53. N. Kato, K. Saito, T. Serata, H. Aida, and Y. Uesu, *J. Chem. Phys.* **115**, 1473 (2001).
54. N. Kato, M. Yamamoto, K. Itoh, and Y. Uesu, *J. Phys. Chem.* **107**, 11917 (2003).
55. K. Ikegami, *Colloids and Surfaces A* **284-285**, 212 (2006).
56. T. Fukui, M. Saito, M. Sugi and S. Iizima, *Thin Solid Films* **109**, 247 (1983).
57. M. Sugi, M. Saito, T. Fukui and S. Iizima, *Thin Solid Films* **129**, 15 (1985).
58. K. Ikegami, *Jpn. J. Appl. Phys.* **41**, 5444 (2002).
59. An additional problem is inappropriate use of Kuhn's formula, which is sometimes seen in the literature: taking double summation for pairs of transition dipole moments

and confusion of one-electron transition dipole moment and molecular transition dipole moment. Both of them cause an unacceptable overestimation of the $|\Delta E_J|$ value for a given arrangement of μ^{tr}, and consequently bring about large underestimation of N.

60. Y. Hamanaka, H. Kurosawa, A. Nakamura, Y. Uchiyama, K. Marumoto, and S. Kuroda, *Chem. Phys. Lett.* **363**, 233 (2002).
61. In a periodic one-dimensional aggregate, for example, the difference between the energies of the second and first excited states can be written as $2J_{12}(\cos\frac{2}{N}\pi - \cos\frac{0}{N}\pi) \approx -4\pi J_{12}N^{-2}$ when only the nearest-neighbor interaction $J_{1,2}$ (< 0) is taken into account, while that in a non-periodic aggregate can be written as $2J_{12}(\cos\frac{2}{N+1}\pi - \cos\frac{1}{N+1}\pi) \approx -3\pi^2 J_{12}(N+1)^{-2}$ (See Ref. 21). The former is 4/3 times larger than the latter.
62. See, for example, *Proc. 7th Int'l Symp. Hydrothermal Reactions*, Jilin, P. R. China (2003).
63. J. Miyata, S. Morita, Y. F. Miura and M. Sugi, *Jpn. J. Appl. Phys.* **44**, 8110 (2005).
64. J. Miyata, S. Morita, Y. F. Miura and M. Sugi, *Colloids and Surfaces A* **284-285**, 509 (2006).
65. S. Mouri, J. Miyata, S. Morita, Y. F. Miura and M. Sugi, *Trans. Mater. Res. Soc. Jpn.* **31**, 573 (2006).
66. M. Sugi, H. Moshino, S. Hasegawa, S. Mouri and Y. F. Miura, *Trans. Mater. Res. Soc. Jpn.* **32**, 313 (2007).
67. S. Mouri, H. Moshino, S. Hasegawa, Y. F. Miura and M. Sugi, *Jpn. J. Appl. Phys.* **46**, 1650 (2007).
68. S. Mouri, S. Morita, Y. F. Miura and M. Sugi, *Jpn. J. Appl. Phys.* **45**, 7925 (2006).
69. H. Moshino, S. Hasegawa, S. Mouri, Y. F. Miura and M. Sugi, *Trans. Mater. Res. Soc. Jpn.* **32**, 305 (2007).
70. S. Hasegawa, H. Moshino, S. Mouri, Y. F. Miura and M. Sugi, *Trans. Mater. Res. Soc. Jpn.* **32**, 309 (2007).
71. Y. Koyano, Y. Sugano, H. Moshino, Y. F. Miura and M. Sugi, *Trans. Mater. Res. Soc. Jpn.* **33**, 95 (2008).
72. H. Moshino, Y. Koyano, Y. Sugano, Y. F. Miura and M. Sugi, *Trans. Mater. Res. Soc. Jpn.* **33**, 103 (2008).
73. Y. Sugano, Y. Koyano, H. Moshino, Y. F. Miura and M. Sugi, *Trans. Mater. Res. Soc. Jpn.* **33**, 107 (2008).
74. H. Moshino, S. Hasegawa, S. Mouri, Y. F. Miura and M. Sugi, Jpn. J. Appl. Phys. **47**, 1034 (2008).
75. H. Moshino, Y. Koyano, S. Mouri, Y. F. Miura and M. Sugi, *Jpn. J. Appl. Phys.* **48**, 051504 (2009).

76. Y. Hirano, A. Maio and Y. Ozaki, *Langmuir* **24**, 3317 (2008).
77. Y. Hirano, A. Yamazaki, A. Maio, Y. Kitahama and Y. Ozaki, *J. Phys. Chem. B.* **114**, 10782 (2010).
78. K. Saito, *J. Phys. Chem. B* **105**, 4235 (2001).
79. T. Wakamatsu, K. Watanabe and K. Saito, *Appl. Opt.* **44**, 906 (2005).
80. Y. Hirano, J. Kawata, Y. F. Miura, M. Sugi, and T. Ishii, Mol. Cryst. & Liq. Cryst. 294, 161 (1997).
81. Y. Hirano, T. M. Okada, Y. F. Miura, M. Sugi and T. Ishii, J. Appl. Phys. 88, 5194 (2000).
82. M. Sugi, T. Fukui, S. Iizima and K. Iriyama, Mol. Cryst. Liq. Cryst. 62, 165 (1980).
83. Y. Hirano, K. N. Kamata, J. Kawata, Y. F. Miura, M. Sugi and T. Ishii, Mol. Cryst. & Liq. Cryst. 322, 111 (1998).
84. Y. Hirano, J. Kawata, Y. F. Miura, M. Sugi and T. Ishii, Thin Solid Films 327-329, 345 (1998).
85. Y. Hirano, T. Okada, Y. F. Miura, M. Sugi and T. Ishii, Trans. Mater. Res. Soc. Jpn. 25, 417 (2000).
86. Y. Hirano, Y. S. Inadzuki, Y. F. Miura, M. Sugi and T. Ishii, Trans. Mater. Res. Soc. Jpn. 27, 521 (2002).
87. Y. Hirano, Y. F. Miura, M. Sugi and T. Ishii, Colloids and Surfaces A 198-200, 37 (2002).
88. Y. Hirano, S. Morita, Y. F. Miura and M. Sugi, *Thin Solid Films* **438-439**, 225 (2003).
89. S. Morita, Y. F. Miura, M. Sugi and Y. Hirano, *J. Appl. Phys.* **94**, 4368 (2003).
90. The deconvolution procedure of each absorption spectrum is described as follows. First, the deconvoluted spectra $A^{(1)}{}_{//}(\lambda)$ and $A^{(1)}{}_{\perp}(\lambda)$ of the blue-shifted bands are obtained by subtracting the greatest fraction of the J-band in Fig. 29(a) from the observed spectra of the DS-C_{20}-AL_{18} ternary system, satisfying $A^{(1)}{}_{//}(\lambda) \geqq 0$ and $A^{(1)}{}_{\perp}(\lambda) \geqq 0$. The treatment is carried out in the range 420-590 nm and the range 430-580 nm for $x \leqq 1.5$ and $x > 1.5$, respectively. Then, we obtain the deconvoluted spectra $A^{(2)}{}_{//}(\lambda)$ and $A^{(2)}{}_{\perp}(\lambda)$ of the red-shifted bands by subtracting the deconvoluted spectra $A^{(1)}{}_{//}(\lambda)$ and $A^{(1)}{}_{\perp}(\lambda)$ of the blue-shifted bands from the observed spectra $A_{//}(\lambda)$ and $A_{\perp}(\lambda)$ of the MS-C_{20}-AL_{18} ternary system. Further adjustment between the deconvoluted spectra $A^{(1)}(\lambda)$ and $A^{(2)}(\lambda)$ should be carried out, since the above procedure produces somewhat asymmetrical spectra $A^{(1)}{}_{//}(\lambda)$ and $A^{(1)}{}_{\perp}(\lambda)$. The spectra $A^{(2')}{}_{//}(\lambda)$ and $A^{(2')}{}_{\perp}(\lambda)$ of the red-shifted bands are the spectra $A^{(2)}{}_{//}(\lambda)$ and $A^{(2)}{}_{\perp}(\lambda)$ multiplied by a factor 0.95 in the range 410-620 nm, respectively. Finally, we obtain the spectra $A^{(1')}{}_{//}(\lambda)$ and $A^{(1')}{}_{\perp}(\lambda)$ of the blue-shifted bands by subtracting the spectra $A^{(2')}{}_{//}(\lambda)$ and $A^{(2')}{}_{\perp}(\lambda)$ of the red-shifted bands from the observed spectra $A_{//}(\lambda)$ and $A_{\perp}(\lambda)$ of the ternary system. Through the above procedure, we obtain two original bands which satisfy $A_{//}(\lambda) = \{ A^{(1')}{}_{//}(\lambda) + A^{(2')}{}_{//}(\lambda)\}$ and $A_{\perp}(\lambda) = \{ A^{(1')}{}_{\perp}(\lambda) + A^{(2')}{}_{\perp}(\lambda)\}$.
91. E. E. Jelly, *Nature* **138**, 1009 (1936).

92. G. Scheibe, *Angew. Chem.* **49**, 563 (1936).
93. See, for example, E. S. Emerson, M. A. Conlin, A. E. Rosennoff, K. S. Norland, H. Rodriguez, D. Chin and G. R. Bird, *J. Phys. Chem.* **71**, 2396 (1967).
94. T. Inoue, *Thin Solid Films* **132**, 21 (1985).
95. F. Mizutani, S. Iizima and T. Matsuda, *Bull. Chem. Soc. Jpn.* **55**, 1295 (1982).
96. K. Misawa and T. Kobayashi, *Nonlinear Opt.* **15**, 81 (1996).
97. K. Misawa and T. Kobayashi, in *J-Aggregates* Ed. T. Kobayashi (Word Scientific, Singapore, 1996), Chap. 2, p. 41.
98. The numerical calculation of the excitation energy, ΔE', for the dye aggregates has been carried out using Eq. (3) in Ref. 19 On the other hand, in Refs. 37 and 50, the calculation of ΔE' of the dye aggregates are expressed as follows:

$$\Delta E' \approx \Delta E \pm 2\sum J_{12} \ . \qquad \text{(R1)}$$

Here, let us write down the equation in its original form for calculating the excitation energy, *□E'*, in Refs. 20 and 23, which has been cited by the authors in Refs. 37 and 50, *i.e.*,

$$\Delta E' \approx \Delta E + 2\sum_{j \neq i} J_{ij} \ . \qquad \text{(R1')}$$

Although Eq. (R1) (in Refs. 37 and 50) is similar to Eq. (R1') (in Refs. 20 and 23) in its appearance, we should point out that the calculation done by authors in Refs. 37 and 50 made two major mistakes in common. First, the factor 2 in Eq. (R1), which is multiplied to the summation, should be removed because it is only necessary when the formulation starts from one-electron wavefunction, as in Refs. 20 and 23. In case a transition dipole moment of a dye molecule is estimated experimentally from an integral of an absorption spectrum of the solution using Eq. (11) as in Refs. 20 and 23, *the factor 2 is not necessary and it should be removed.* Second, in Refs. 37 and 50, Eq. (R1) has been interpreted as a calculation to estimate the energy shift by making a double summation of the interaction integral J_{ij} between two dipoles in the array of dye molecules as follows:

$$\Delta E' \approx \Delta E + 2\sum_{i=1}^{N} \Big(\sum_{j=1,2,\ldots,\, i-1, i+1, \ldots,\, N} J_{ij} \Big) \ . \qquad \text{(R1'')}$$

However, the original idea in Refs. 20 and 23 is that since each molecule *i* has the same environment for a dye aggregate consisting of sufficiently large number of molecule, *N*, the excitation energy, ΔE_{agg}, can be calculated by the sum of interaction of transition dipole moments between the *i*-th molecule and its' neighbors, *i.e.*,

$$\Delta E_{agg} \approx \Delta E + \sum_{j=1,2,\ldots,i-1,i+1,\ldots N} J_{ij} \quad . \qquad \text{(R2)}$$

in a planar aggregate with a small *N*, the average of the energy in the dye aggregate can be estimated by the following equation:

$$\Delta E_{agg} \approx \Delta E + \frac{1}{N}\sum_{i=1}^{N}\Big(\sum_{j=1,2,\ldots,i-1,i+1,\ldots,N} J_{ij}\Big) \quad . \qquad \text{(R3)}$$

We consider that Eqs. (R2) and (R3) are both appropriate and valid as order estimates when the dye aggregate is large enough. A weighted average should be calculated, instead of a simple average of $\frac{1}{N}\sum_{i=1}^{N}$, from the quantum-mechanical viewpoint because the exciton distribution is no longer homogenous in a finite system, where the edge effect should be also taken into account as pointed out in Sec. 2.5. It should be also pointed out that the excitation energy, *ΔE'*, calculated by Eq. (R1'') is significantly larger compared to those obtained by Eqs. (R2) and (R3) and diverges when $N \to \infty$, which is physically unreasonable, as pointed out in Ref. 99. Furthermore, in Ref. 81, the formula used to calculate the excitation energy in the dye aggregate is expressed as:

$$\Delta E' \approx \Delta E + 2 \times \frac{1}{2}\frac{1}{N}\sum_{i=1}^{N}\sum_{j=1}^{N} Jij \quad , \qquad \text{(R3')}$$

which should be rewritten as Eq. (R3) to avoid any confusion, although the numerals calculated from Eq. (R3') is appropriate and the same as those from Eq. (R3).

99. K. Saito, K. Ikegami, S. Kuroda, Y. Tabe and M. Sugi, *Jpn. J. Appl. Phys.* **30**, 1836 (1991).
100. Y. Hirano, K. N. Kamata, J. Kawata, Y. F. Miura, M. Sugi and T. Ishii, *Mol. Crys. & Liq. Crys.* **322**, 111 (1998).
101. Y. Hirano, K. Kamata, Y. Inadzuki, J. Kawata, Y. F. Miura, M. Sugi and T. Ishii, *Jpn. J. Appl. Phys.* **38**, 6024 (1999).
102. M. Sugi, T. Fukui, S. Iizima and K. Iriyama, *Mol. Cryst. Liq. Cryst.* **62**, 165 (1980).
103. S. Kuroda, M. Sugi and S. Iizima, *Thin Solid Films* **99**, 21 (1983).
104. S. Nishikawa, Y. Tokura, T. Koda and K. Iriyama, *Jpn. J. Appl. Phys.* **25**, L701 (1986).
105. K. Murata, S. Kuroda and K. Saito, *Thin Solid Films* **295**, 73 (1997).
106. H. Kuhn, *Thin Solid Films* **178**, 1 (1989).
107. T. Ishii, *Thin Solid Films* **178**, 47 (1989).
108. Y. Hirano, S. Morita, Y. F. Miura and M. Sugi, *Trans. Mater. Res. Soc. Jpn.* **28**, 55 (2003).

109. S. Morita, Y. F. Miura, M. Sugi and Y. Hirano, *J. Appl. Phys.* **94**, 4368 (2003).
110. H. Bücher, and H. Kuhn, *Chem. Phys. Lett.* **6**, 183 (1970).
111. R. Azumi, M. Matsumoto, Y. Kawabata, S. Kuroda, M. Sugi, L.G. King, and M. J. Crossley, *J. Am. Chem. Soc.* **114**, 12862 (1992).
112. N. Kato, K. Yuasa, T. Araki, I. Hirosawa, M. Sato, N. Ikeda, K. Iimura, and Y. Uesu, *Phys. Rev. Lett.* **94**, 136404 (2005).
113. N. Kato, I. Hirosawa, M. Sato, T. Koganezawa, and K. Yuasa, *Colloids and Surfaces A* **321**, 275 (2008).
114. In the optimization, the conformation is optimized when the molecule is isolated neglecting the presence of Mg^{2+} cations in the subphase. At the air/water interface, however, a chelation of carboxyl- and keto groups to Mg^{2+} is possible. Thus, as for the determination of the angle around the N-COOH bond, there still remains uncertainty.

Index